ETABS V23 Black Book

By
Gaurav Verma
(CADCAMCAE Works)

Edited by
Kristen

ISBN # 978-1-77459-222-9

DEDICATION

To teachers, who make it possible to disseminate knowledge
to enlighten the young and curious minds
of our future generations

To students, who are the future of the world

THANKS

To my friends and colleagues

To my family for their love and support

Table of Contents

Chapter 3 : Section Properties and Material

Chapter 5 : Creating Load Cases and Performing Analysis

Chapter 6 : Running Design Study

Chapter 7 : Detailing

Chapter 8 : Project

Chapter 9 : Editing and Advanced Structure Design

Preface

ETABS is the ultimate integrated software package for the structural analysis and design of buildings. Incorporating 50 years of continuous research and development, this latest ETABS offers unmatched 3D object based modeling and visualization tools, blazingly fast linear and nonlinear analytical power, sophisticated and comprehensive design capabilities for a wide-range of materials, and insightful graphic displays, reports, and schematic drawings that allow users to quickly and easily decipher and understand analysis and design results.

The **ETABS V23 Black Book**, the 6th edition of our book on ETABS, is written to help beginners learn the basics of ETABS structure modeling and analysis. The book is based on ETABS V23.2 software. The book follows a step by step methodology. This book explains the designing of structure, assigning various properties to structure, applying different load conditions, and performing analyses. The book covers almost all the information required by a learner to master basics of ETABS. Some of the salient features of this book are:

In-Depth explanation of concepts

Every new topic of this book starts with the explanation of the basic concepts. In this way, the user becomes capable of relating the things with real world.

Topics Covered

Every chapter starts with a list of topics being covered in that chapter. In this way, the user can easy find the topic of his/her interest easily.

Instruction through illustration

The instructions to perform any action are provided by maximum number of illustrations so that the user can perform the actions discussed in the book easily and effectively. There are about 600 illustrations that make the learning process effective.

Tutorial point of view

The book explains the concepts through the tutorial to make the understanding of users firm and long lasting. Each chapter of the book has tutorials that are real world projects.

Project

Projects and exercises are provided to students for practicing.

For Faculty

If you are a faculty member, then you can ask for video tutorials on any of the topic, exercise, tutorial, or concept. As faculty, you can register on our website to get electronic desk copies of our latest books. Faculty resources are available in the **Faculty Member** page of our website (**www.cadcamcaeworks.com**) once you login. Note that faculty registration approval is manual and it may take two days for approval before you can access the faculty website.

Formatting Conventions Used in the Text

All the key terms like name of button, tool, drop-down etc. are kept bold.

Free Resources

Link to the resources used in this book are provided to the users via email. To get the resources, mail us at ***cadcamcaeworks@gmail.com*** or ***info@cadcamcaeworks.com*** with your contact information. With your contact record with us, you will be provided latest updates and informations regarding various technologies. The format to write us e-mail for resources is as follows:

Subject of E-mail as ***Application for resources of .... Black Book***.

You can give your information below to get updates on the book.
Name:
Course pursuing/Profession:
Contact Address:
E-mail ID:

For Any query or suggestion

If you have any query or suggestion, please let us know by mailing us on ***cadcamcaeworks@gmail.com*** or ***info@cadcamcaeworks.com***. Your valuable constructive suggestions will be incorporated in our books and your name will be addressed in special thanks area of our books.

About Author

The author of this book, Gaurav Verma, has authored many books on CAD/CAM/CAE topics. He has authored books on **Autodesk Revit**, **Autodesk Fusion 360**, **Autodesk Fusion 360 PCB**, **Autodesk Inventor**, **SolidWorks**, **SolidWorks Simulation**, **SolidWorks Flow Simulation**, **SolidWorks CAM**, **SolidWorks Electrical**, **AutoCAD Electrical**, **Creo Parametric**, **Mastercam**, **MasterCAM for SolidWorks**, and many other CAD, CAM, and CAE topics. If you have query/doubt in any CAD/CAM/CAE package, then you can directly contact the author by writing at **cadcamcaeworks@gmail.com**.

Chapter 1

Starting with ETABS

Topics Covered

The major topics covered in this chapter are:

- ***Introduction***
- ***Downloading and Installing Educational ETABS***
- ***Starting ETABS***
- ***Creating a New Model***
- ***Steel Design and Concrete Design Codes***
- ***User Interface of ETABS***
- ***Opening a document***
- ***Closing documents***
- ***Basic Settings for ETABS***

INTRODUCTION

ETABS is a 3D Modeling and analysis software used to prepare and analyze the building models for real load conditions as per the building design codes. ETABS is capable of performing simple as well as largest & complex non-linear problems of multi-storey. Following are some of the tasks that can be performed easily with the help of ETABS:

- Multi-story commercial, government, and health care facilities
- Parking garages with circular and linear ramps
- Buildings with curved beams, walls, and floor edges
- Buildings with steel, concrete, composite, or joist floor framing
- Projects with multiple towers
- Complex shear walls and cores with arbitrary openings
- Performance based design utilizing nonlinear dynamic analyses
- Buildings based on multiple rectangular and/or cylindrical grid systems
- Flat and waffle slab concrete buildings
- Buildings subjected to any number of vertical and lateral load cases and combinations, including automated wind and seismic loads
- Multiple response spectrum load cases, with built-in input curves
- Automated transfer of vertical loads on floors to beams and walls
- Capacity check of beam-to-column and beam-to-beam steel connections
- P-Delta analysis with static or dynamic analysis
- Explicit panel-zone deformations
- Punching shear checks for concrete slabs
- Construction sequence loading analysis
- Multiple linear and nonlinear time history load cases in any direction
- Foundation/support settlement
- Large displacement analyses
- Nonlinear static pushover
- Buildings with base isolators and dampers
- Design optimization for steel and concrete frames
- Design of concrete slabs using mild reinforcement and post-tensioning
- Design capacity check of steel column base plates
- Floor modeling with rigid or semi-rigid diaphragms
- Automated vertical live load reductions

In this book, we are going to discuss most of these tasks and many others. Now, we will start with downloading and installing ETABS V23 evaluation version and then we will discuss the user interface of the software.

DOWNLOADING AND INSTALLING ETABS V23 EVALUATION

The procedure to download and install ETABS V23 (Evaluation) is given next.

- Open the link **https://www.csiamerica.com/products/etabs/trial** in your web browser. The CSI Software Download page will be displayed; refer to Figure-1.

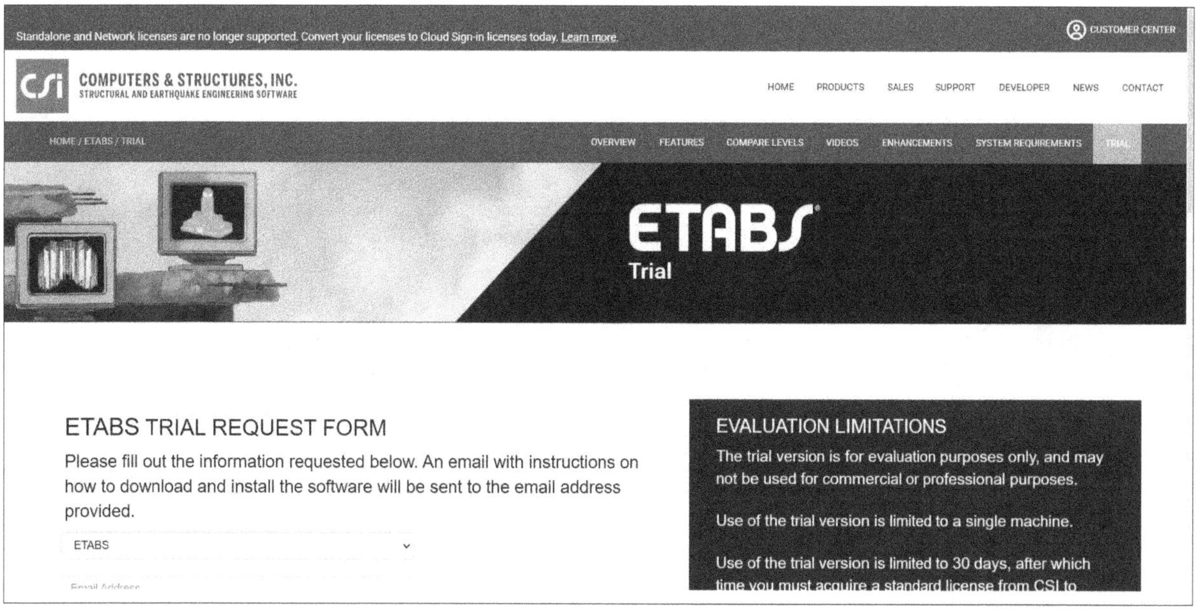

Figure-1. CSI Downloads page

- Enter your contact information in the form and click on the **SUBMIT** button. An e-mail will be sent to your provided e-mail id.
- Check the e-mail received from CSI and click on the **ETABS Trial** link in the e-mail. An executive file will start downloading. Once the file has downloaded, click on it to run and install the software.
- Once installation is complete, on first run you will be asked to sign in using your email address provided earlier for starting trial.
- Click on the Sign In button from first initialization page of application window and sign in to your account. An information box will be displayed asking you whether to start trial of software; refer to Figure-2. Click on the Yes button to start your 30 days software trial. Note that there are some restrictions on the usage of evaluation version of the software. If you want to work on live projects then you should buy the professional version of ETABS. For evaluation and basic learning, you can use the trial version of software.

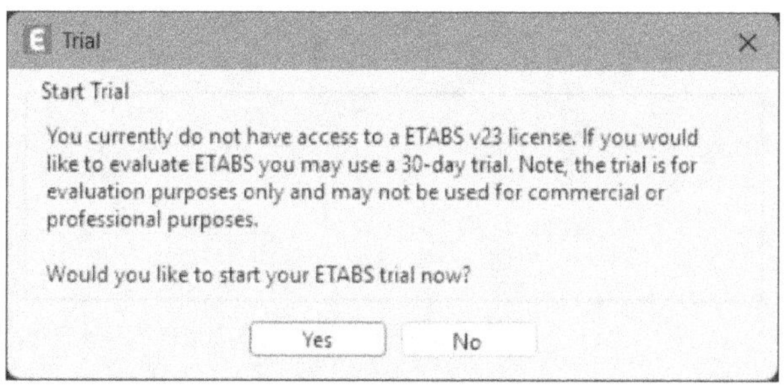

Figure-2. Trial information box

STARTING ETABS

- Click on the **Start** button in Windows 11 and type **ETABS**. List of the programs with the typed characters will be displayed; refer to Figure-3.

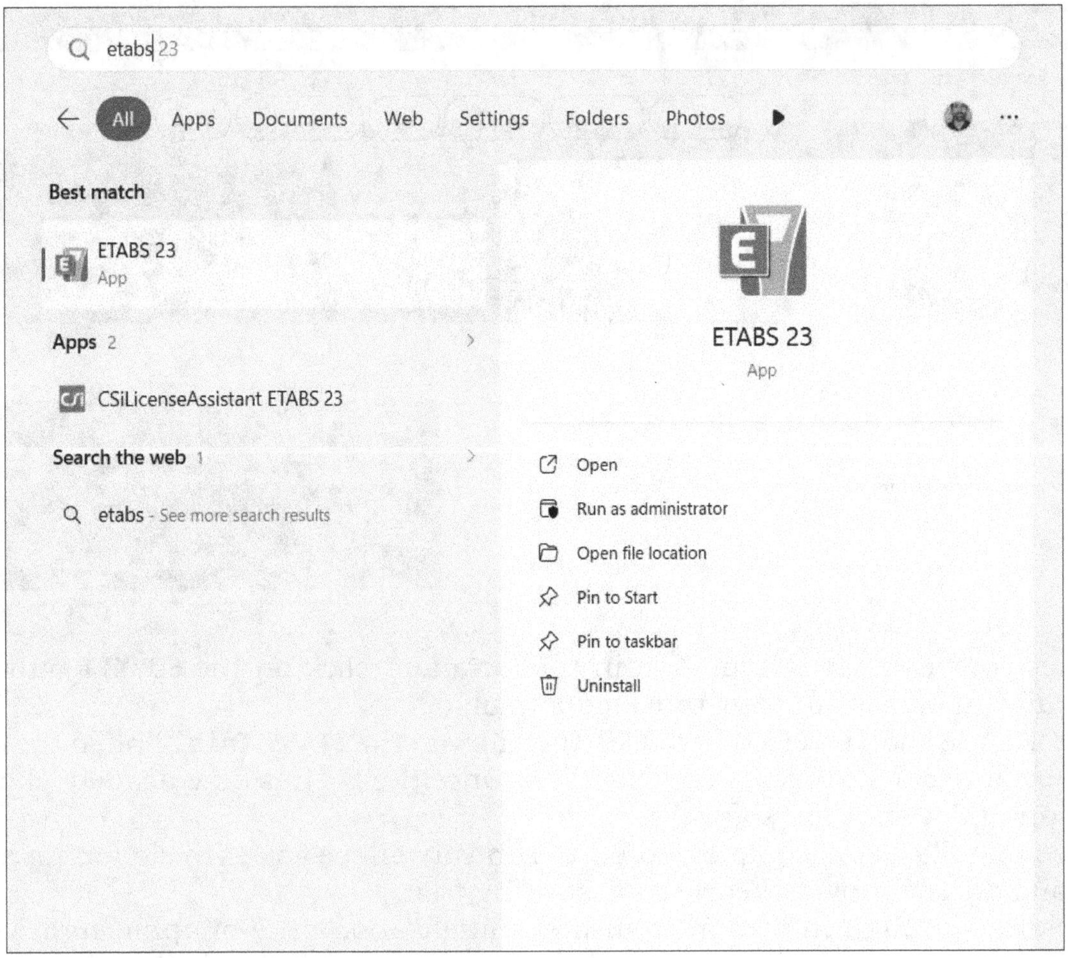

Figure-3. ETABS in Start menu

Click on the **ETABS 23** option from the menu. The user interface of ETABS will be displayed; refer to Figure-4.

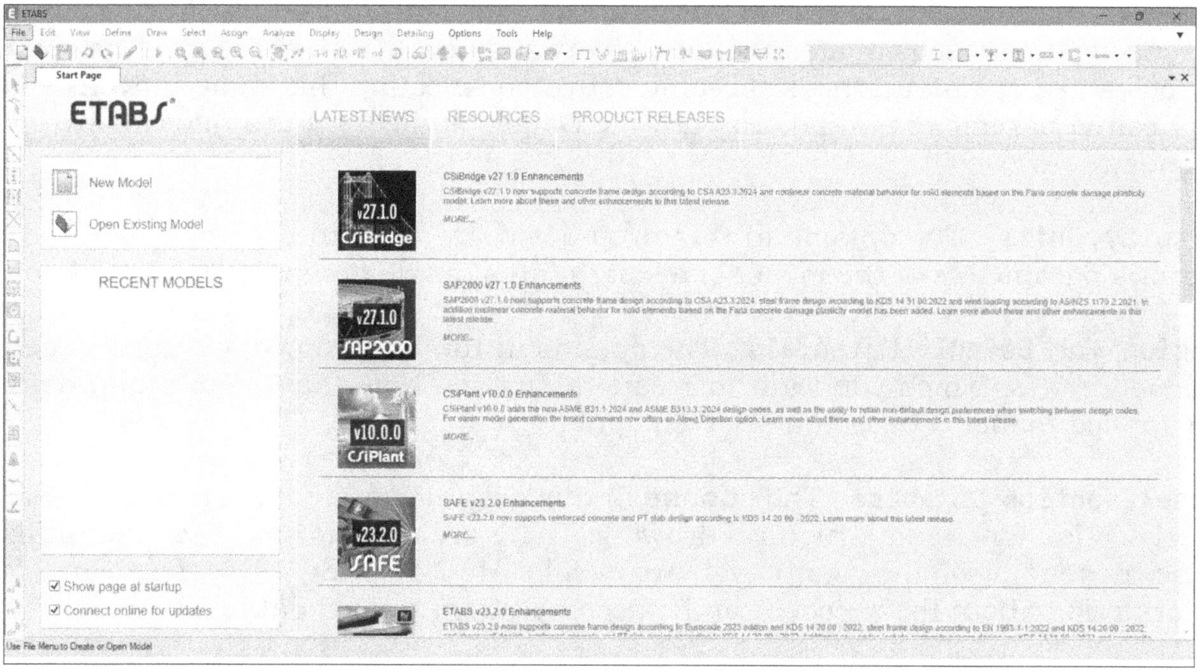

Figure-4. User interface of ETABS

STARTING A NEW MODEL

- Click on the **New Model** button from the Toolbar or click on the **New Model** option from the **File** menu or press **CTRL+N** from the keyboard. The **Model Initialization** dialog box will be displayed; refer to Figure-5.

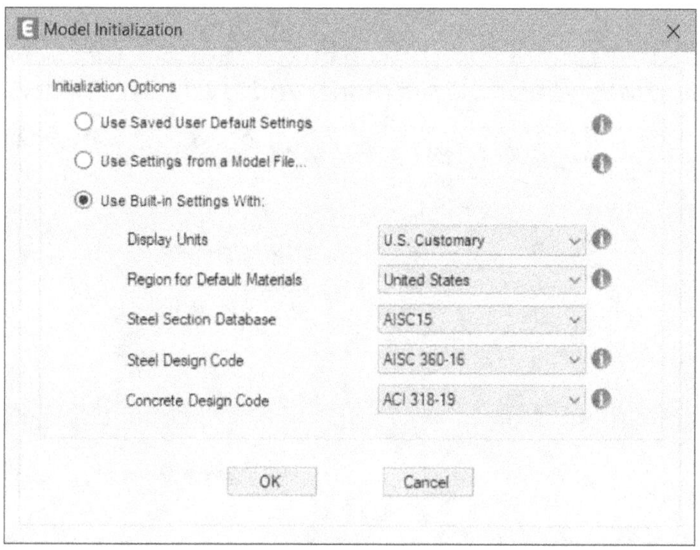

Figure-5. Model Initialization dialog box

- Select the **Use Saved User Default Settings** radio button if you want to use the default set options saved by you like units, preferences etc. Note that these options are set by using **Save User Default Settings** option in the **Options** menu. We will discuss more about these options later in the book.
- Select the **Use Settings from a Model File** radio button to use setting specified in earlier created model file of ETABS.

- Select the **Use Built-in Settings With** radio button if you want to manually specify the settings related to units, steel database, design codes etc. The options below the radio button will become active on selecting this radio button. Select desired parameters in the drop-downs. Details of different drop-downs are given next.

Display Units : The options in this drop-down are used to set units required for various parameters of the model like length, area, angle, temperature, and so on.

Region for Default Materials : The options in this drop-down are used to define the country for which you want to create the model. Note that default material will be applied to model based on selected option.

Steel Section Database : The options in this drop-down are used to select desired steel section for use in structural drawing. There are various steel section database codes depending on the country you are in. Like AISC 14, AISC 14M, and so on. More information about these codes can be accessed from related database.

Steel Design Code : The options in this drop-down are used to select the design code for steel as per the requirement.

Concrete Design Code : The options in this drop-down are used to set desired concrete design code. This code is used to define concrete frame design, shear wall design, and so on.

- Click on the **OK** button from the dialog box to start a new model. The **New Model Quick Templates** dialog box will be displayed; refer to Figure-6.

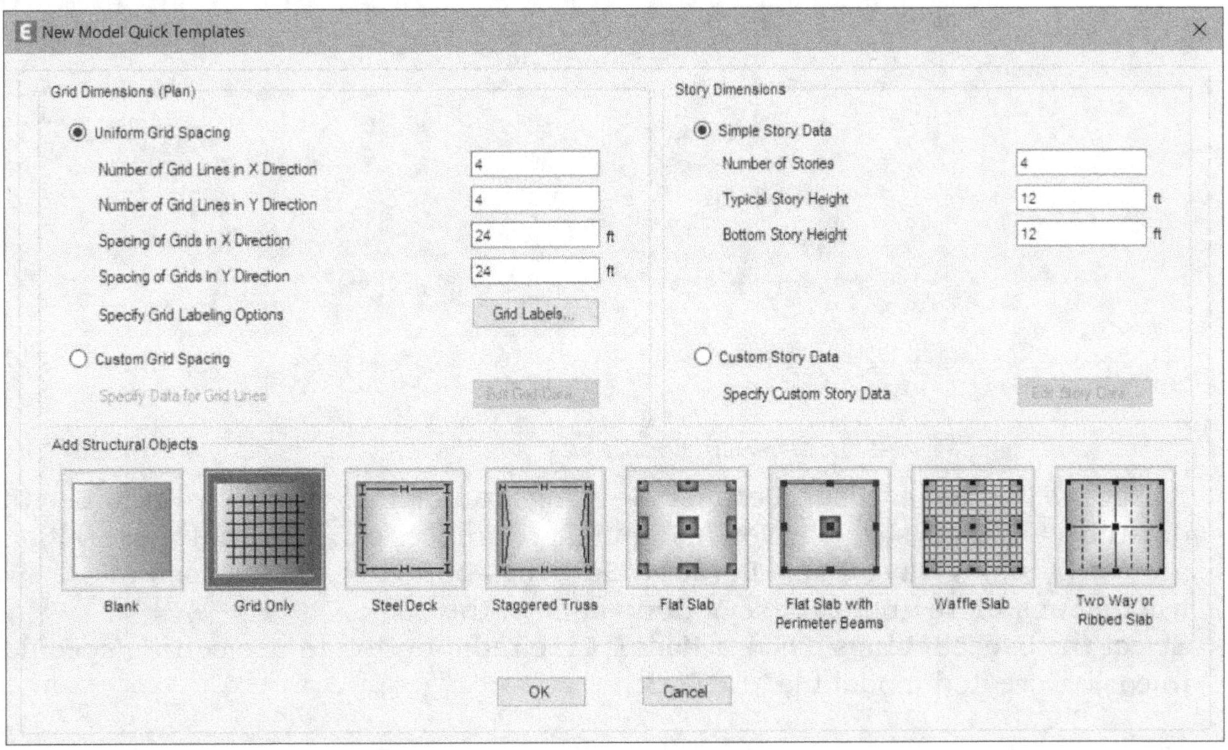

Figure-6. New Model Quick Templates dialog box

Grid Dimensions

- Select the **Uniform Grid Spacing** radio button if you want to create grids with uniform spacing vertically and horizontally, and specify desired parameters in the edit boxes below it. To set the labels of grid, click on the **Grid Labels** button. The **Grid Labeling Options** dialog box will be displayed; refer to Figure-7. Specify desired options in the dialog box and click on the **OK** button.

Figure-7. Grid Labeling Options dialog box

- If you want to create custom grid spacing then click on the **Custom Grid Spacing** radio button. The **Edit Grid Data** button will become active. On clicking this button, the **Grid System Data** dialog box will be displayed; refer to Figure-8. Set desired values for grid in the tables. To edit the values, double-click on them in the table.
- Click on the **OK** button to apply changes. The **New Model Quick Templates** dialog box will be displayed again.

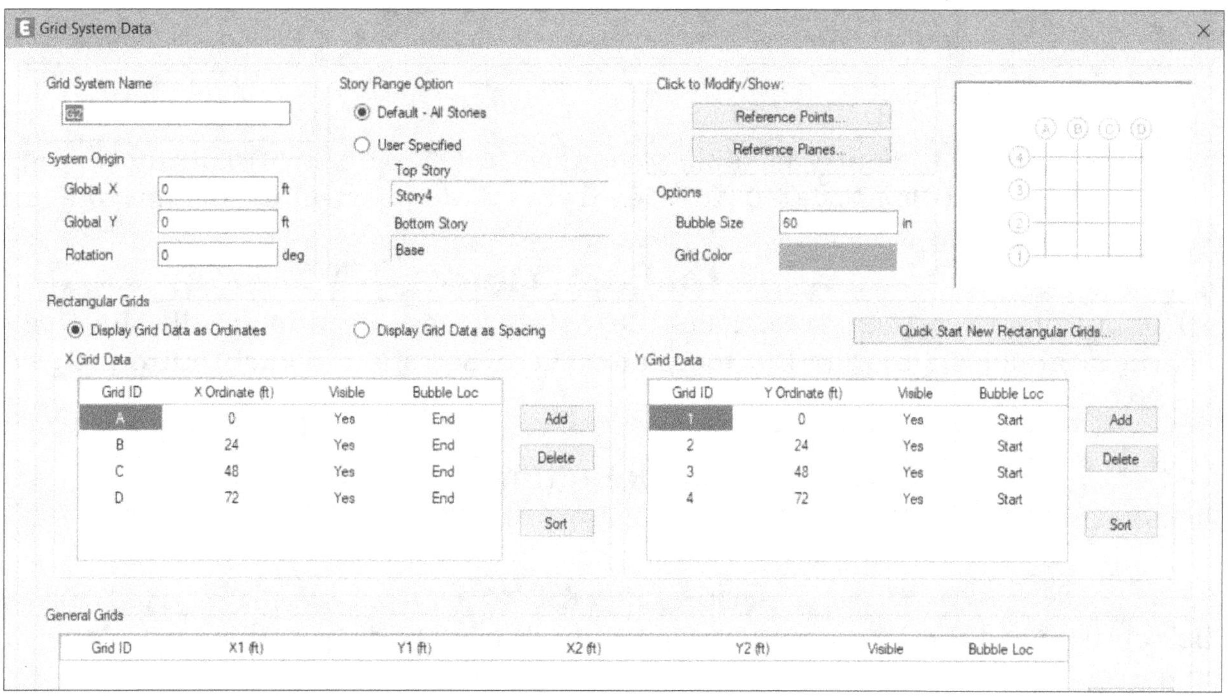

Figure-8. Grid System Data dialog box

Story Dimensions

- In the **Story Dimensions** area of the dialog box, select the **Simple Story Data** radio button to specify the story parameters like number of stories, typical story height, and bottom story height. For example, if you want a 4 story building with bottom height 15 ft and all other story heights 12 ft then specify **4** in **Number of Stories**, **12** in **Typical Story Height**, and **15** in **Bottom Story Height** edit boxes.
- Select the **Custom Story Data** radio button if you want to set the parameter of each story manually. On selecting this radio button, the **Edit Story Data** button will become active. On clicking this button, the **Story Data** dialog box will be displayed; refer to Figure-9.

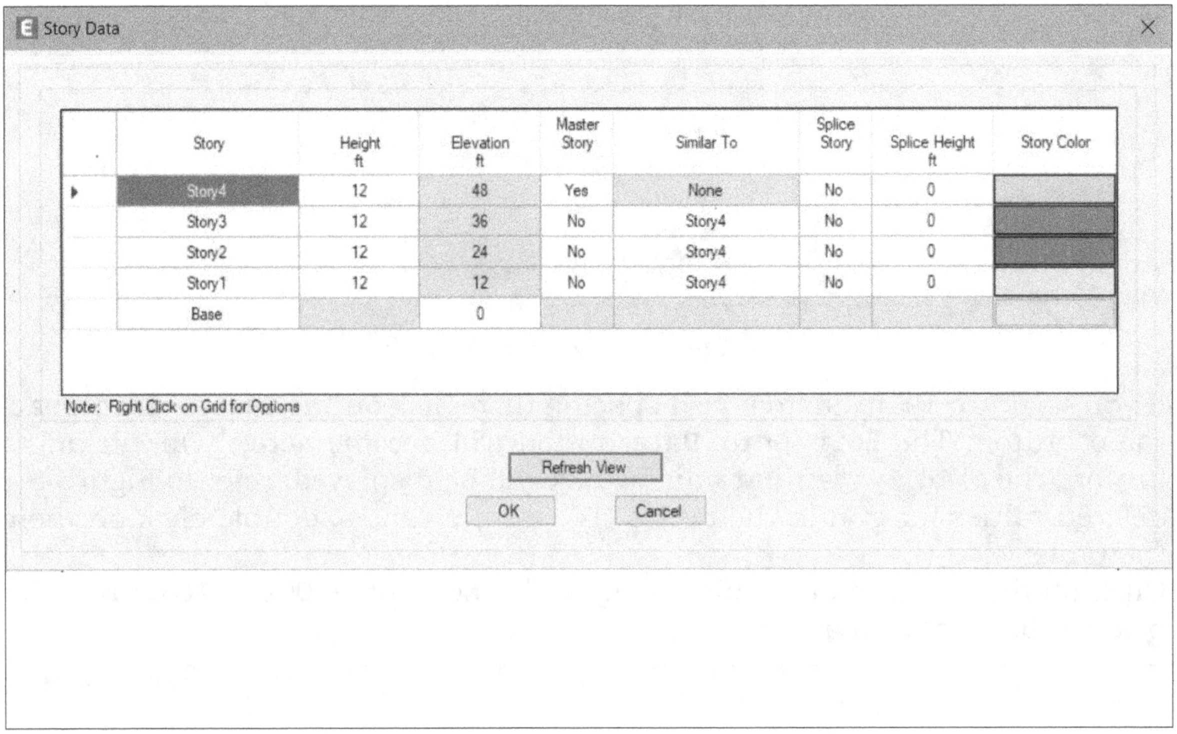

Figure-9. Story Data dialog box

- Double-click on the entry in table to change it. Click on the **OK** button to apply the changes. The **New Model Quick Templates** dialog box will be displayed again.

Add Structural Objects

- The buttons in the **Add Structural Objects** area are used to specify the type of structural objects required in template. The description of each button is given next.

Blank Button

Click on the **Blank** button if you want to start a blank model.

Grids Only Button

Click on the **Grids Only** button if you want to display grid lines along with labels in the model.

Steel Deck Button

Click on the **Steel Deck** button if you want to create floors with steel deck. On clicking this button, the **Structural Geometry and Properties for Steel Deck** dialog box will be displayed; refer to Figure-10. In the edit boxes of **Overhangs** area, specify the length up to which the steel desk should overhang from the building columns.

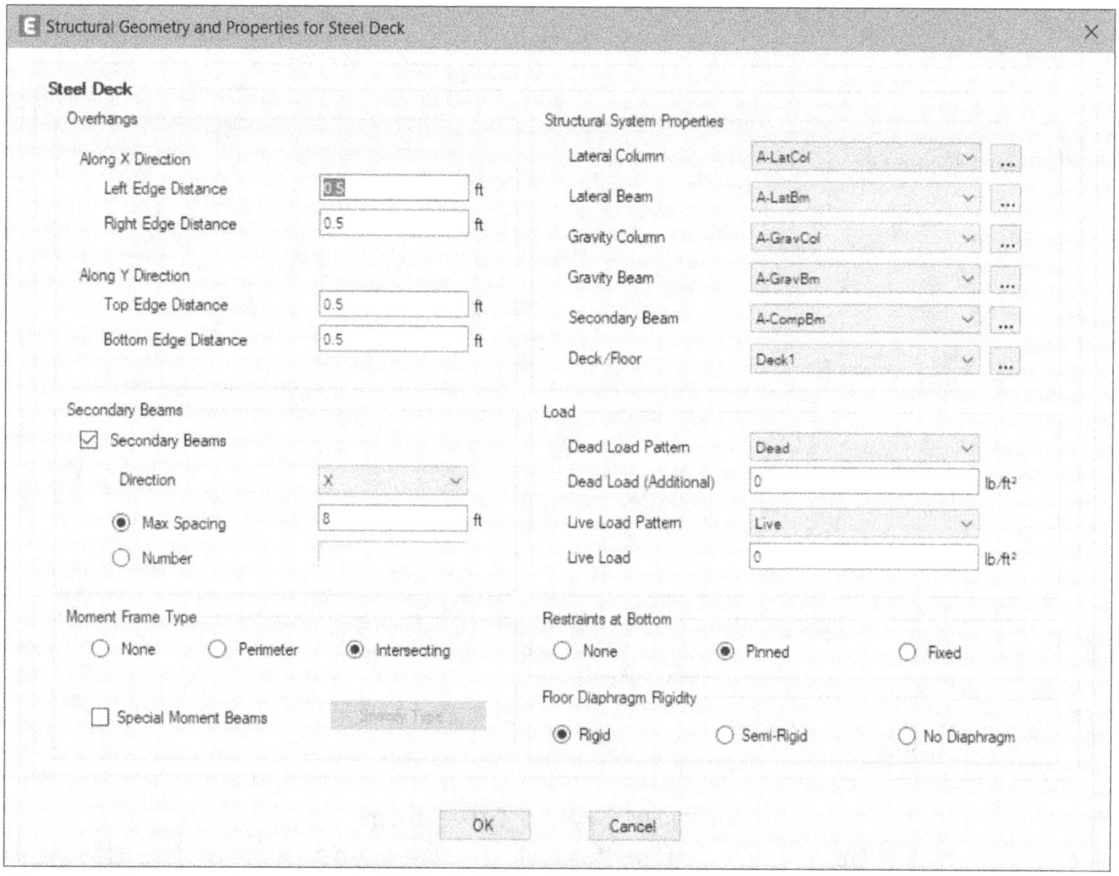

Figure-10. Structural Geometry and Properties for Steel Deck dialog box

Secondary Beams area

In the **Secondary Beams** area of the dialog box, select the **Secondary Beams** check box to add secondary beams in the model. On selecting this check box, the options below it will become active. Select desired direction along which the secondary beams are to be created from the **Direction** drop-down. Select the **Max spacing** radio button and specify the maximum spacing to be given between two secondary beams. Select the **Number** radio button and specify the number of secondary beams to be created along the selected direction. Various parameters of steel deck geometry are shown in Figure-11.

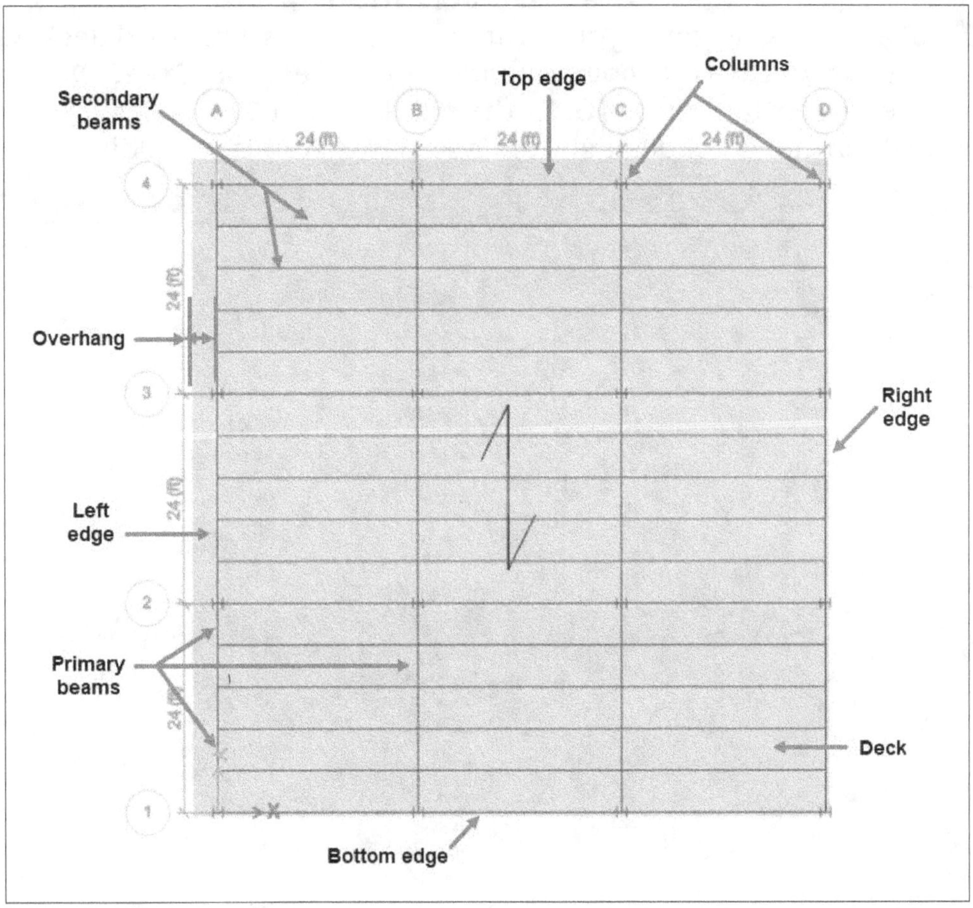

Figure-11. Steel deck parameters

Moment Frame Type area

Select desired radio button from the **Moment Frame Type** area of the dialog box to create moment resisting frame. To learn more about moment resisting frames, scan the QR code given next. If you select the **None** radio button then moment is released at beam-column joints of the model and all the connections are pinned. If you select the **Perimeter** radio button then beam-column connections at the perimeter are moment resisting and all the other connections are pinned. If you select the **Intersecting** radio button then all the column-beam connections are moment resisting. Select the **Special Moment Beams** check box if you want to assign special form to the beam at connections for analysis purpose. After selecting this check box, click on the **Specify Type** button. The **Frame Assignment - Moment Frame Beam Connection Type** dialog box will be displayed; refer to Figure-12.

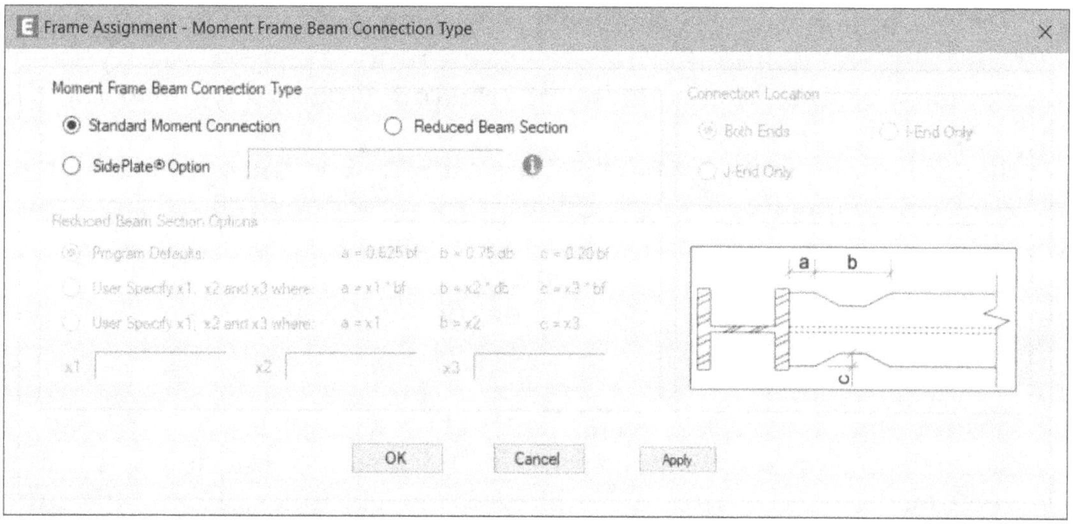

Figure-12. Frame Assignment-Moment Frame Beam Connection Type dialog box

Select desired connection type from the dialog box and specify the parameters as required. By default, the **Standard Moment Connection** radio button is selected and hence a standard full width beam connection is created. Select the **Reduced Beam Section** radio button to use special type of connection with reduced width of section. Set desired values of a, b, and c in the **Reduced Beam Section Options** area. Select the **SidePlate Option** radio button to use special type of connections provided by SidePlate Systems Inc. This type of connection consists of a series of plates welded to the beam end and the column to strengthen the beam-column connection and force hinging to occur in the beam away from the connection. After selecting this radio button, select desired option from the respective drop-down. Click on the **OK** button to apply changes. You will learn more about frame assignment later in the book.

Structural System Properties area

- Select desired option from the **Lateral Column** drop-down. The option selected in this drop-down will define frame section property of columns for beam-to-column connections that are fully moment resisting.
- Select desired option from the **Lateral Beam** drop-down. The option selected in this drop-down will define frame section property of beams for beam-to-column connections that are fully moment resisting. Wind load and other side loads act on lateral beam and column structures. Scan the code below to learn more about lateral systems.

- Select desired option from the **Gravity Column** drop-down. The option selected in this drop-down will define frame section property of columns for beam-to-column connections that are pinned and do not transfer any connection load to other frame members.
- Select desired option from the **Gravity Beam** drop-down. The option selected in this drop-down will define frame section property of beams for beam-to-column connections that are pinned and do not transfer any connection load to other frame members.

- Select desired option from the **Secondary Beam** drop-down. The option selected in this drop-down will define the frame section property of secondary frame. Note that secondary beams are also moment resisting type and transfer their loads to main beams/columns.
- Select desired option from the **Deck/Floor** drop-down. The option selected in this drop-down will define the frame section property for deck/floor.

Load area

The options in the **Load** area are used to define live/dead loads of the steel desk. These options are discussed next.

- Select desired option from the **Dead Load Pattern** drop-down. The **Dead** option is by default selected option in this drop-down. The dead load is used to represent the load caused by static structures like beams, columns, floors, and other fixed items. Specify desired load value in the **Dead Load (Additional)** edit box apart from load of structural members. The load of beams, columns, floors etc. is automatically included in Dead load.
- Select desired option from the **Live Load Pattern** drop-down. Live load is used to represent load caused by human movement, furniture, and other moving items. Specify desired value of load in the **Live Load** edit box below the drop-down.

Restraints at Bottom

The options in this area are used to specify the type of restraints at the bottom of columns of the deck. There are three radio buttons in this area; **None**, **Pinned**, and **Fixed**. Select the **None** radio button if you want no restraints to be applied at the bottom of the columns. Select the **Pinned** radio button if you want to specify pinned restraints at the bottom of the columns. Select the **Fixed** radio button if you want the columns to be fixed.

Floor Diaphragm Rigidity

The options in this area are used to specify the rigidity level of floor diaphragm. There are three radio buttons in this area. Select the **Rigid** radio button if you want to assume a fully rigid floor diaphragm for analysis purpose. Similarly, you can select **Semi-rigid** radio button for in-plane stiffness of floor. Select the **No Diaphragm** radio button to remove diaphragm rigidity of the floor.

After defining the parameters, click on the **OK** button from the dialog box. The **New Model Quick Templates** dialog box will be displayed again.

Similarly, you can use the other buttons in the **Add Structural Objects** area of the dialog box. Most of the parameters for other structure types are same so, we will now discuss the options different from earlier discussed.

Staggered Truss Button

Click on **Staggered Truss** button from the **Add Structural Objects** area in the **New Model Quick Templates** dialog box to create staggered truss structure; refer to Figure-13. On clicking this button, the **Structural Geometry and Properties for Staggered Truss** dialog box will be displayed; refer to Figure-14.

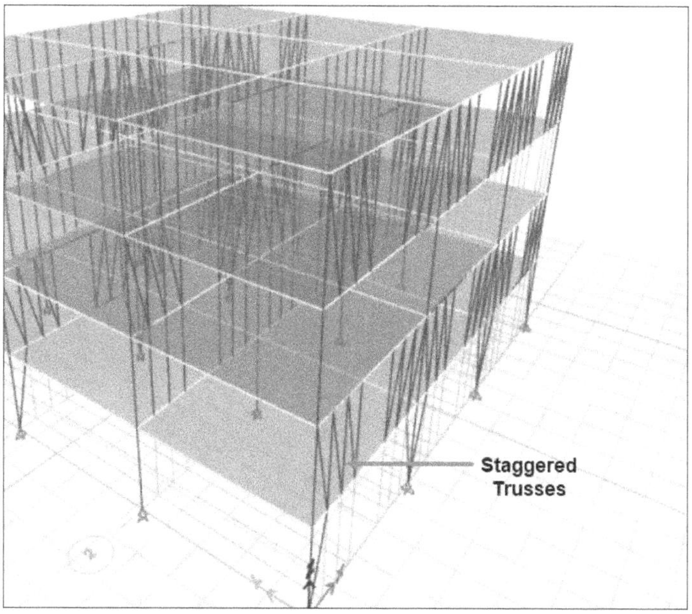

Figure-13. Staggered truss structure

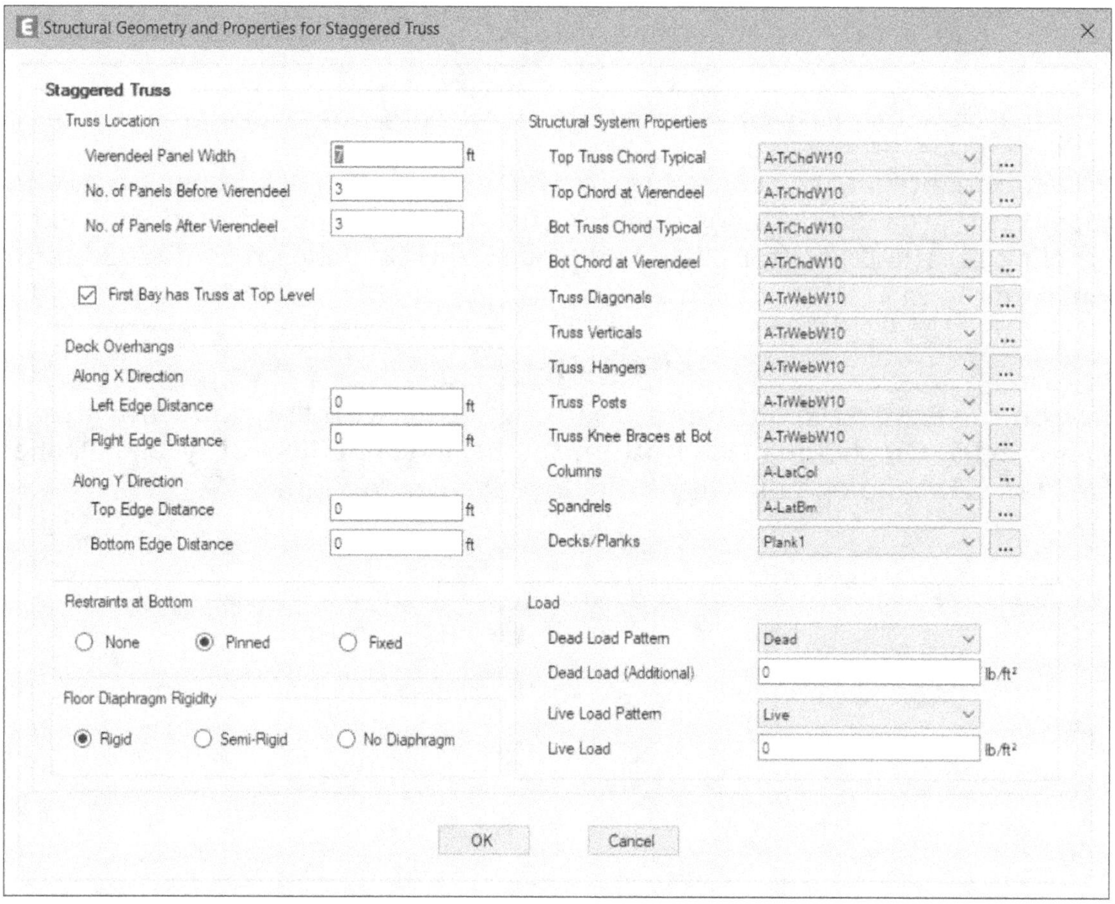

Figure-14. Structural Geometry and Properties for Staggered Truss dialog box

- Specify the width of Vierendeel panel in the **Vierendeel Panel Width** edit box. The Vierendeel truss is a structure where the members are not triangulated but form rectangular openings, and is a frame with fixed joints that are capable of transferring and resisting bending moments. The panel width defines width of this rectangular opening.

- Specify the number of truss panels before and after Vierendeel panel in respective edit boxes of **Truss Location** area.
- Select the **First Bay has Truss at Top Level** check box to create Vierendeel truss in the first bay of top level otherwise only columns will be used in first bay; refer to Figure-15. Scan the QR code next to check an analysis report on structure with Vierendeel trusses.

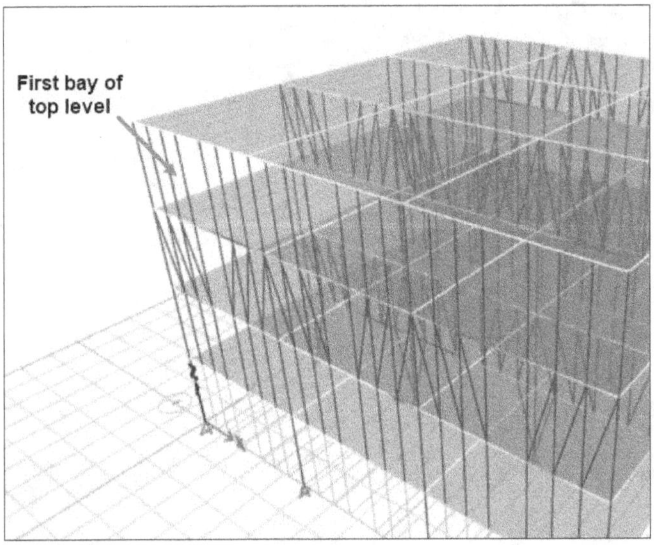

Figure-15. First bay of top level in structure

- Set desired structural elements in the drop-downs of **Structural System Properties** area in the dialog box. Set the other parameters as discussed earlier and click on the **OK** button. The **New Model Quick Templates** dialog box will be displayed again.

Flat Slab Button

- Click on the **Flat Slab** button to create structure with flat slabs and strips; refer to Figure-16. On clicking this button, the **Structural Geometry and Properties for Flat Slab** dialog box will be displayed; refer to Figure-17.

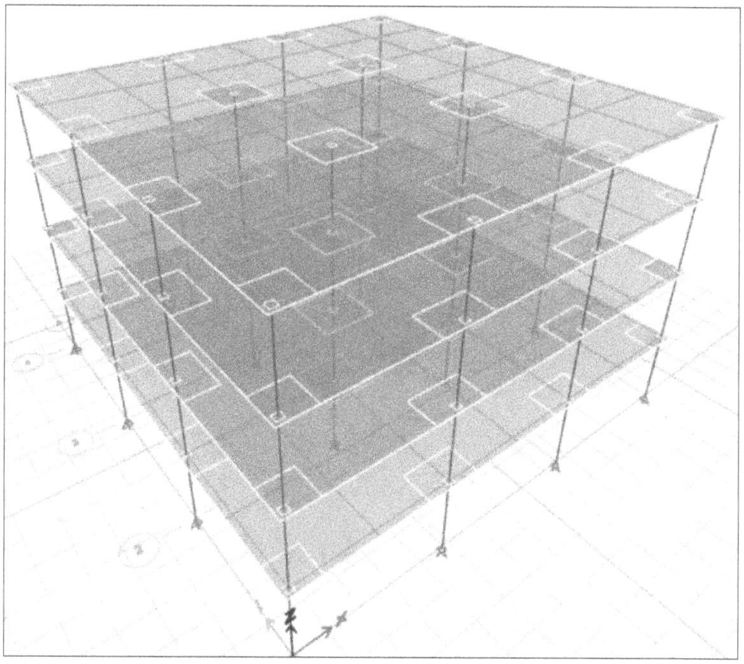

Figure-16. Flat slab structure

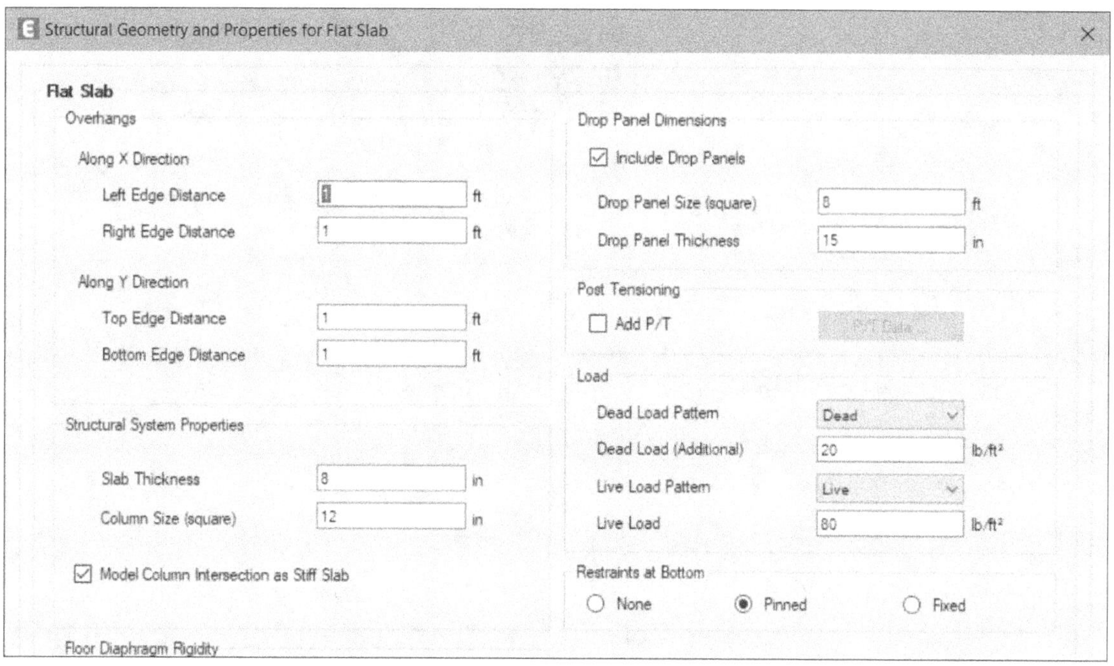

Figure-17. Structural Geometry and Properties for Flat Slab dialog box

- Select the **Include Drop Panels** check box to include drop-panels in structure for avoiding punching shear in flat slab; refer to Figure-18. After selecting the check box, specify side length of square drop panel and thickness in respective edit boxes of the **Drop Panel Dimensions** area.

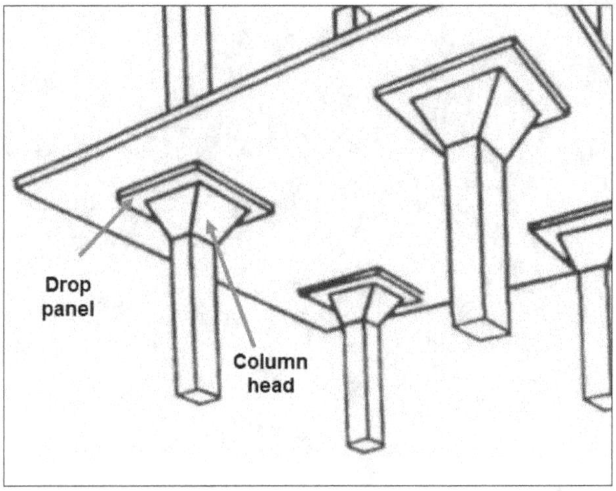

Figure-18. Drop panel

- Select the **Add P/T** check box to use post tensioning slabs (Check the next QR code about post tensioning slab). After selecting check box, click on the **P/T Data** button to specify construction data for post tensioning of slab. The **Quick Tendon Layout** dialog box will be displayed; refer to Figure-19.

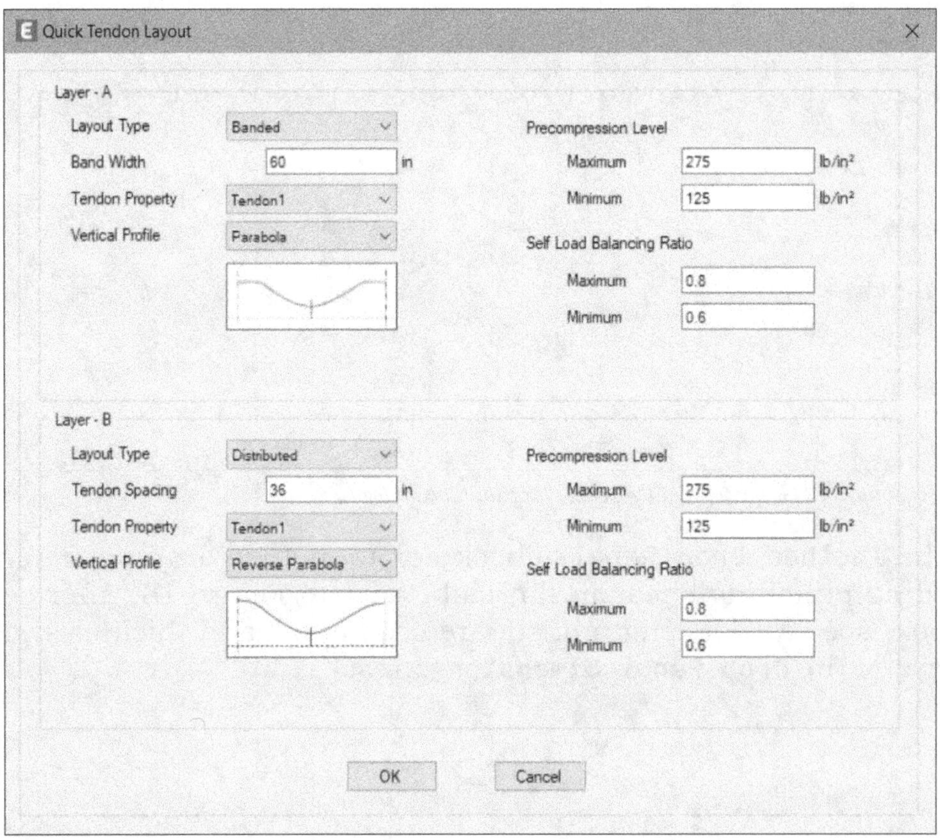

Figure-19. Quick Tendon Layout dialog box

- Specify the parameters as desired and click on the **OK** button to define post tensioning data. For post tensioning terminology, scan the QR given next.
- Specify the other parameters as discussed earlier and click on the **OK** button. The **New Model Quick Templates** dialog box will be displayed again.

Flat Slab with Perimeter Beams Button

- Click on the **Flat Slab with Perimeter Beams** button to create flat slab structure with perimeter beams as shown in Figure-20. Note that drop panels are not created for perimeter columns in this case. On selecting this button, the **Structural Geometry and Properties for Flat Slab with Perimeter Beams** dialog box will be displayed. The options in this dialog box are same as discussed earlier.

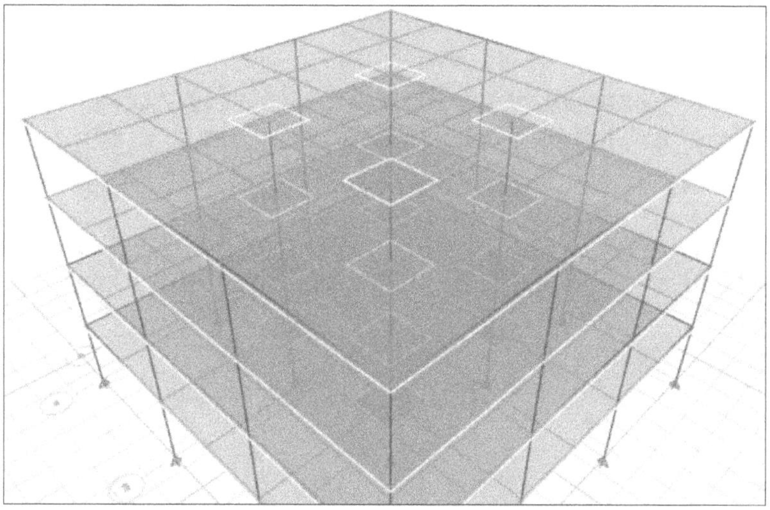

Figure-20. Flat slab structure. with perimeter beams

Waffle Slab Button

Click on the **Waffle Slab** button to create flat slab structure with ribs underneath running in two directions (X and Y); refer to Figure-21. The options for this structure template are similar to options discussed earlier.

Figure-21. Waffle slab

Two Way or Ribbed Slab Button

Click on the **Two Way or Ribbed Slab** button to create structure similar to waffle slab but without drop panels. The options are same as discussed earlier.

After setting the template parameters, click on the **OK** button from the **New Model Quick Templates** dialog box. A new document will open in the ETABS; refer to Figure-22.

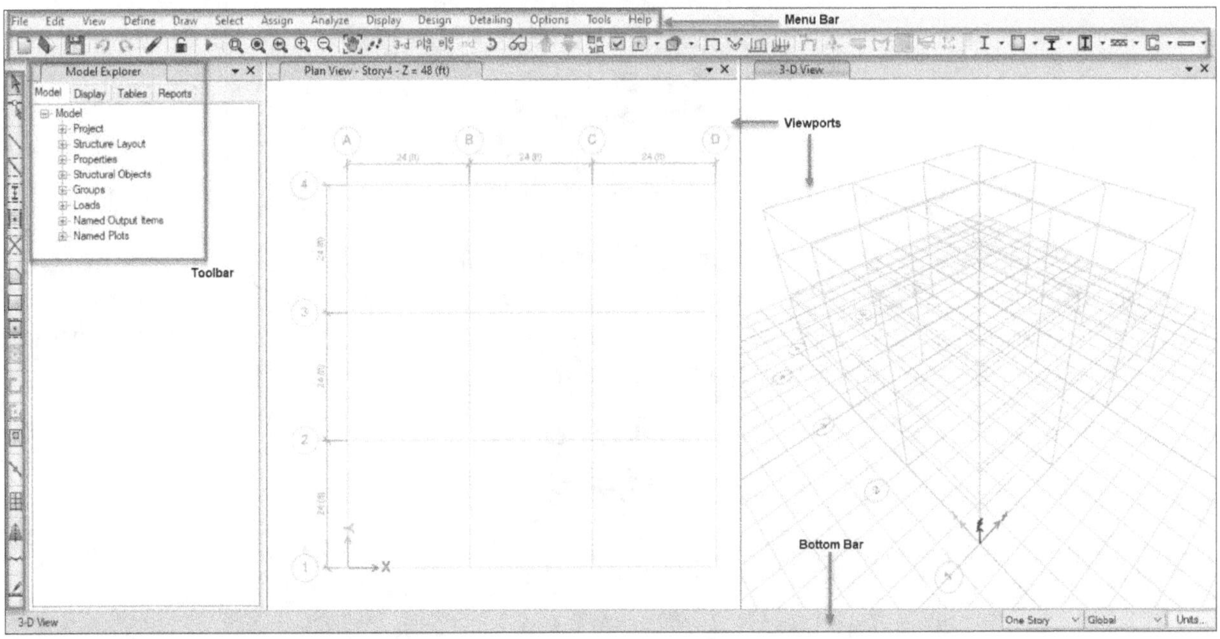

Figure-22. ETABS user interface

USER INTERFACE OF ETABS

As shown in the previous figure, ETABS has a very simple and optimized user interface providing you toolbars and menus for accessing various tools. Various components of user interface in ETABS are discussed next.

MENU BAR

Menu bar consists of all the tools required to perform an action in the software. Menu bar contains different menus named based on their collective functions which have tools of similar action; refer to Figure-23. Various tools and options of these menu will be discussed later in the book.

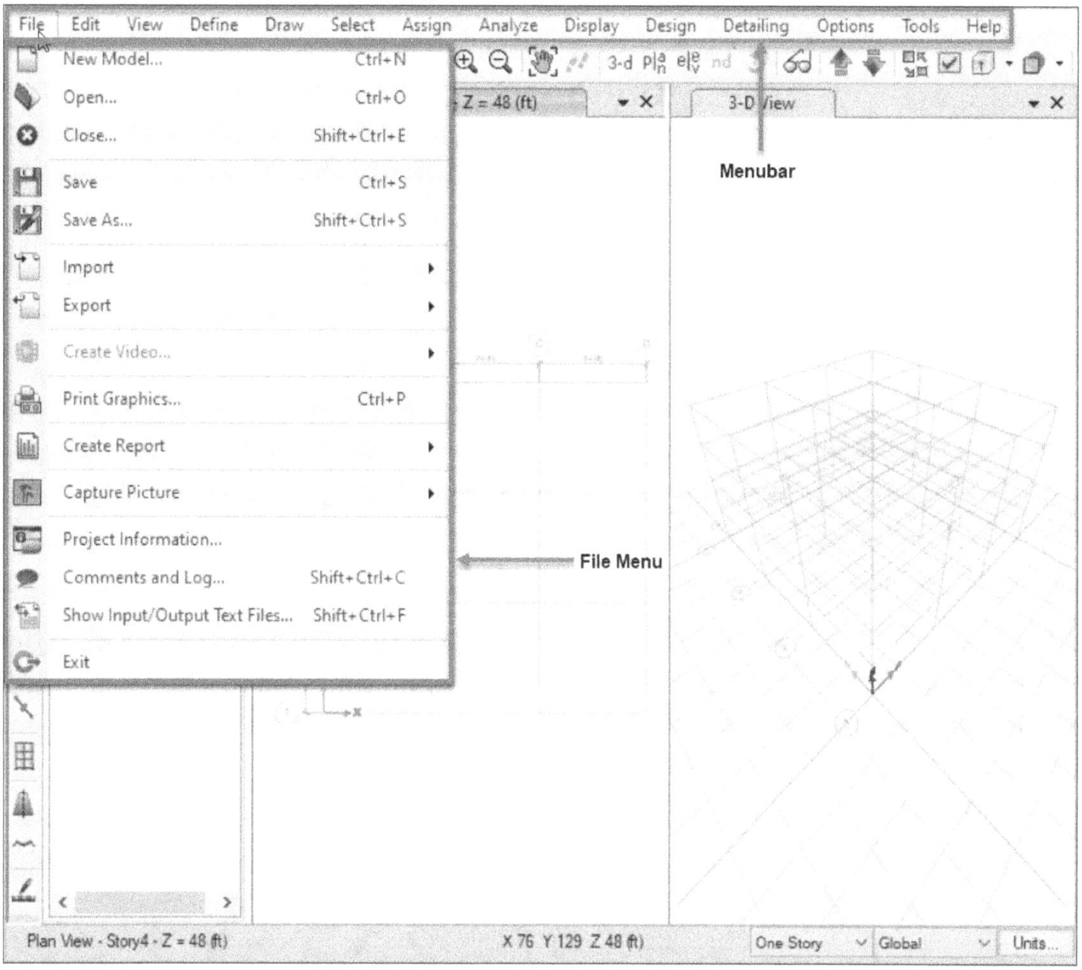

Figure-23. Menu bar

TOOLBAR

Toolbar consists of commonly used tools. Toolbar are available in left and right area of the viewports as well as below Menu bar. Tools in the toolbar will be discussed later in the book.

STATUS BAR

Status Bar is used to display information related to floor level and unit system used. Various interface elements available in the Status Bar are discussed next.

Current Working Plane

At the left-side in the bottom bar, you can check the current working plane on which any new geometry will be created. If you have selected 3D View in Viewports then 3D View will be displayed otherwise name of plan view will be displayed on which new geometries will be created. You will learn about changing the working plane later in this book.

Story drop-down

Select desired option from the **Story** drop-down to define which story of the building will be affected by your applied changes. Select the **One Story** option from the **Story** drop-down if you want the changes to be applied only on the selected story. Select the **All Stories** option from the drop-down if you want to select the elements on all stories of the building. Select the **Similar Stories** option from the drop-down if you want to select elements from the similar stories.

Grid System drop-down

Select desired grid system from the **Grid System** drop-down in the Status bar. You can edit the grid systems by using the **Edit Stories and Grid Systems** tool from the **Edit** menu.

Units

The **Units** button in the Status bar is used to change the unit system of current project. On clicking the **Units** button, a flyout will be displayed where you can select desired unit system; refer to Figure-24.

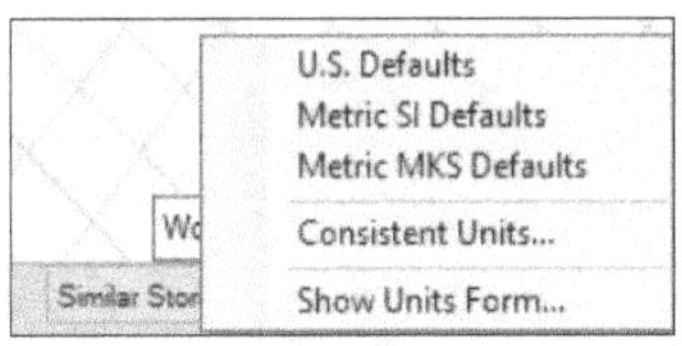

Figure-24. Units flyout

MODEL EXPLORER

Model Explorer is displayed at the left of the viewports. The options in the **Model Explorer** are used to check and modify the parameters related to model, display, tables, and reports. Various tabs of the **Model Explorer** are discussed next.

Model Tab

The options of the **Model** tab are used to check and edit various parameters of the model like project preferences, structure layout, loads, and so on; refer to Figure-25.

- Click on the **+** sign before a category in the **Model** tab of the **Model Explorer** to expand the list of parameters under the category.
- To edit the parameter, right-click on it from the **Model** tab and select the **Edit** option from the shortcut menu. Respective dialog box will be displayed to edit parameter.

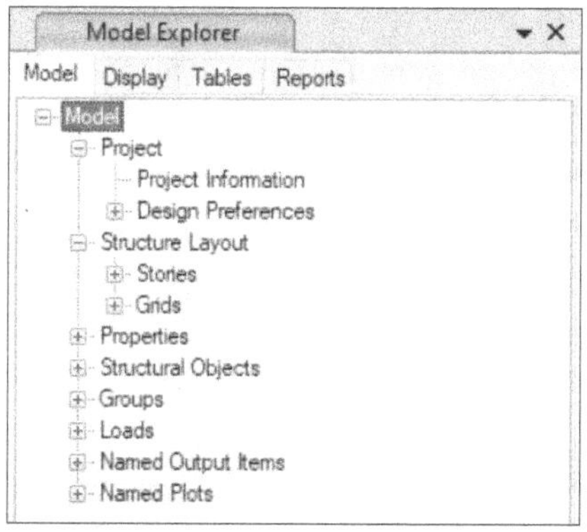

Figure-25. Model tab in Model Explorer

Detail about various categories displayed in the **Model** tab is given next.

Project Category

The project category in the **Model** tab is used to edit project information and design preferences. Expand the **Project** category and right-click on the **Project Information** option. A shortcut menu will be displayed. Click on the **Edit** option from the shortcut menu. The **Project Information** dialog box will be displayed; refer to Figure-26.

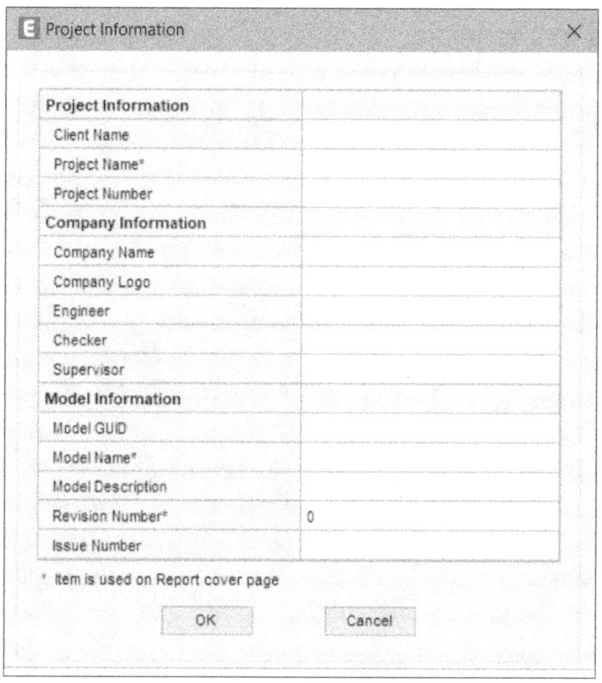

Figure-26. Project Information dialog box

Click in desired field to edit the option like click in the **Client Name** edit box and specify the name of your client. Note that information provided here will also be displayed in the final report. After specifying desired parameters, click on the **OK** button from the dialog box.

You can also display the information of project in table form. To do so, right-click on **Project Information** option in the **Project** category of the **Model** tab in the **Model Explorer** and select the **Show Table** option. The **Project Information** box will be displayed in the application window; refer to Figure-27.

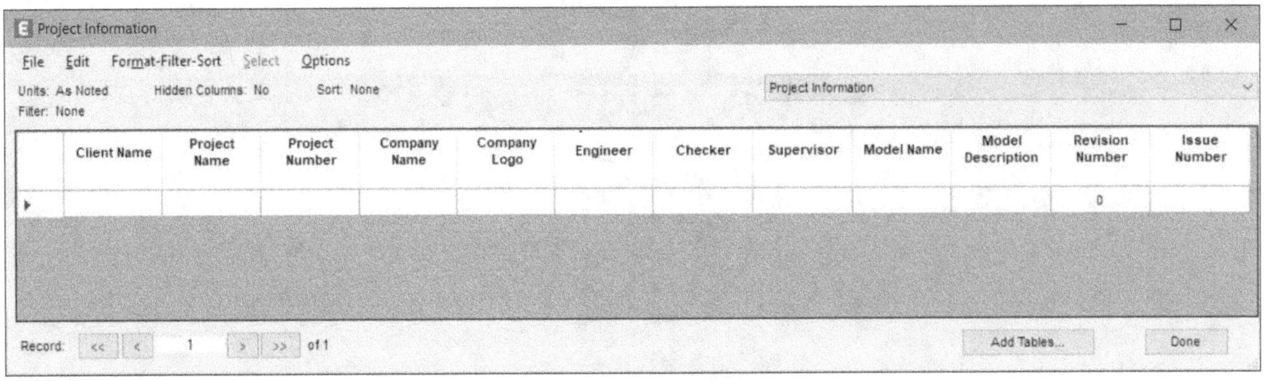

Figure-27. Project Information box

Here, you can check the project information provided by you. Click on the **Close** button (x) at the right corner of the box or click on **Done** button to close it.

Design Preferences

The options under this node are used to define parameters related to designing frames, beams, and shear walls. To edit any design parameter, expand the **Design Preferences** node and right-click on the design parameter which you want to edit. Like, right-click on the **Steel Frame Design** option and select the **Edit** option from the shortcut menu displayed. The related dialog box will be displayed; refer to Figure-28. To edit the parameters in the table, click in the field and specify desired value. Click on the **OK** button from the dialog box to apply the changes.

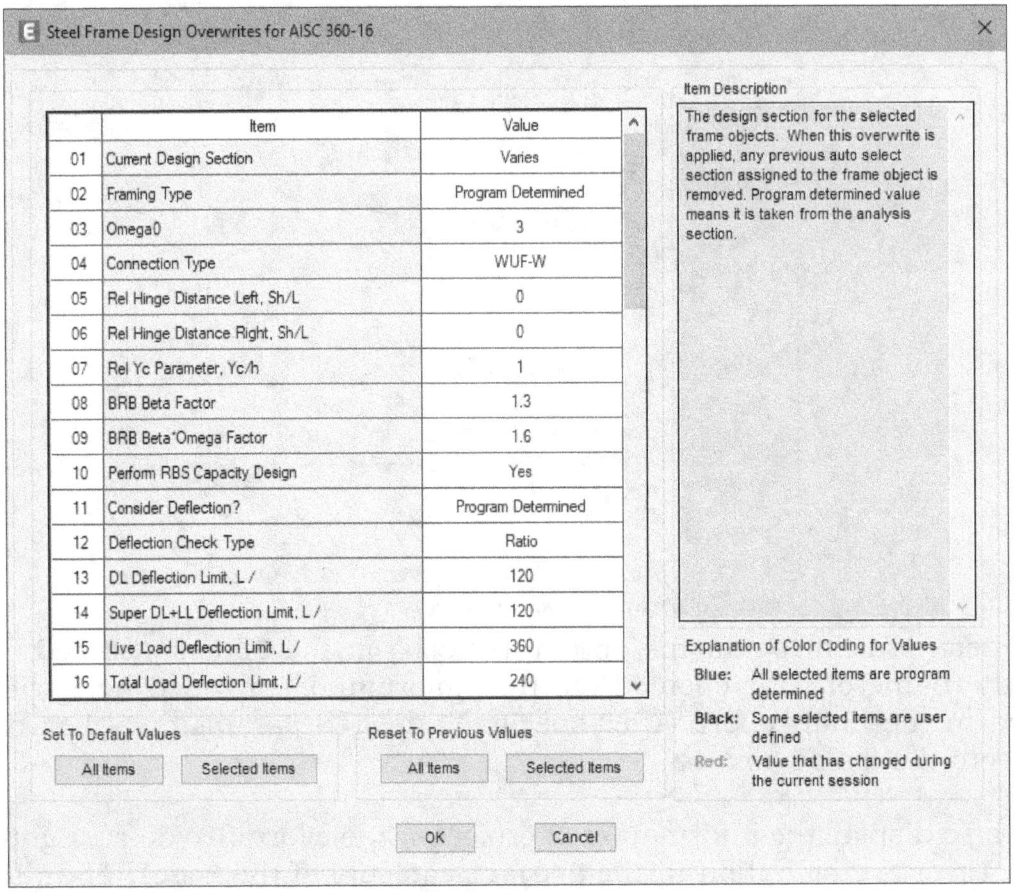

Figure-28. Steel Frame Design Preferences for AISC 360-10 dialog box

Note that the parameters in the above table are directly related to structural designing of building and analysis results so change these values very carefully. Details of each parameter can be found in your civil engineering books. We will discuss these parameters later in this book.

Structure Layout Category

Structure Layout consists of stories and grid system defined by you. The procedure to edit various parameters of structural layout is given next.

- Expand the nodes of **Structure Layout** to check various parameters defined for the structure layout. To edit the layout, right-click on **Structure Layout** node. The **Edit Story and Grid Systems** button will be displayed.
- Click on the **Edit Story and Grid Systems** option from shortcut menu. The **Edit Story and Grid System Data** dialog box will be displayed; refer to Figure-29.

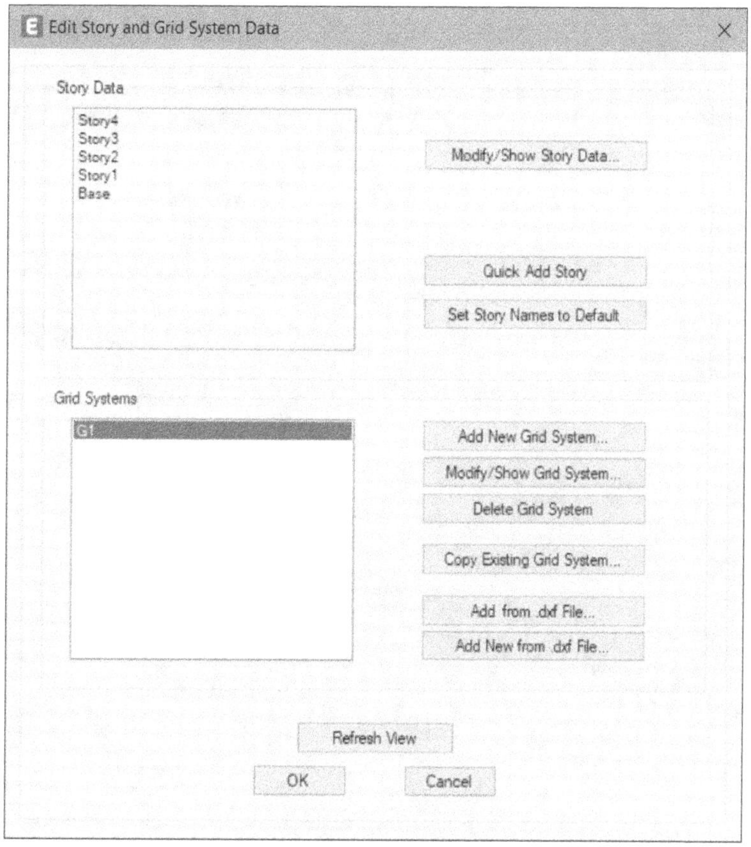

Figure-29. Edit Story and Grid System Data dialog box

- To edit story data, click on the **Modify/Show Story Data** button. The **Story Data** dialog box will be displayed; refer to Figure-30.
- Edit the story data as required and click on the **OK** button. The **Edit Story and Grid System Data** dialog box will be displayed again. You will learn more about these options later in the book.
- To modify existing grid, select the grid from the **Grid Systems** list box and click on the **Modify/Show Grid System** button. The **Grid System Data** dialog box will be displayed; refer to Figure-31.

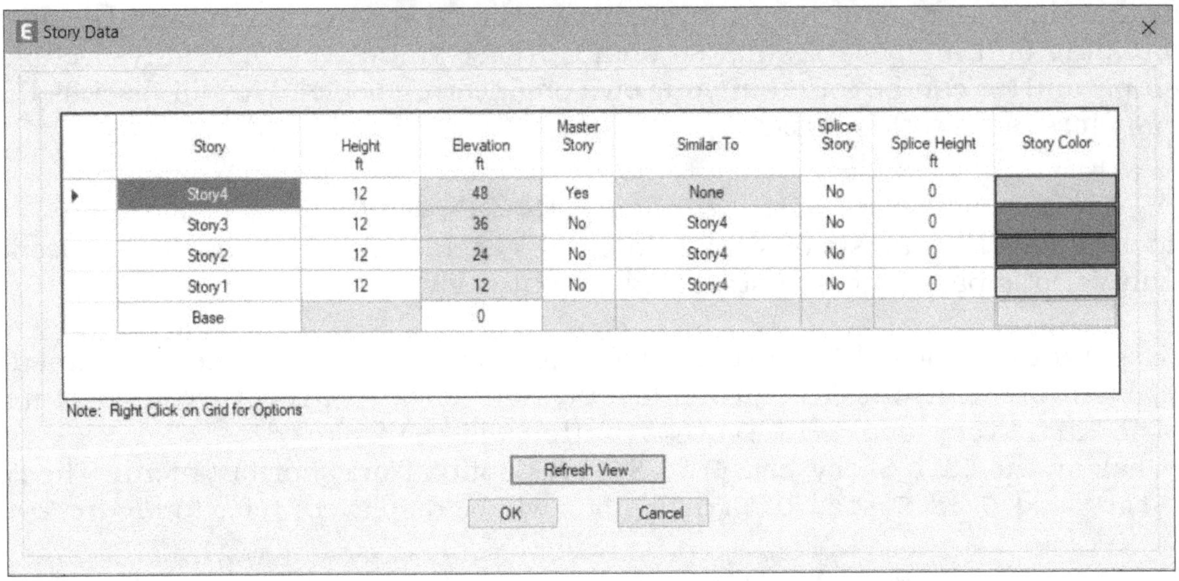

Figure-30. Story Data dialog box

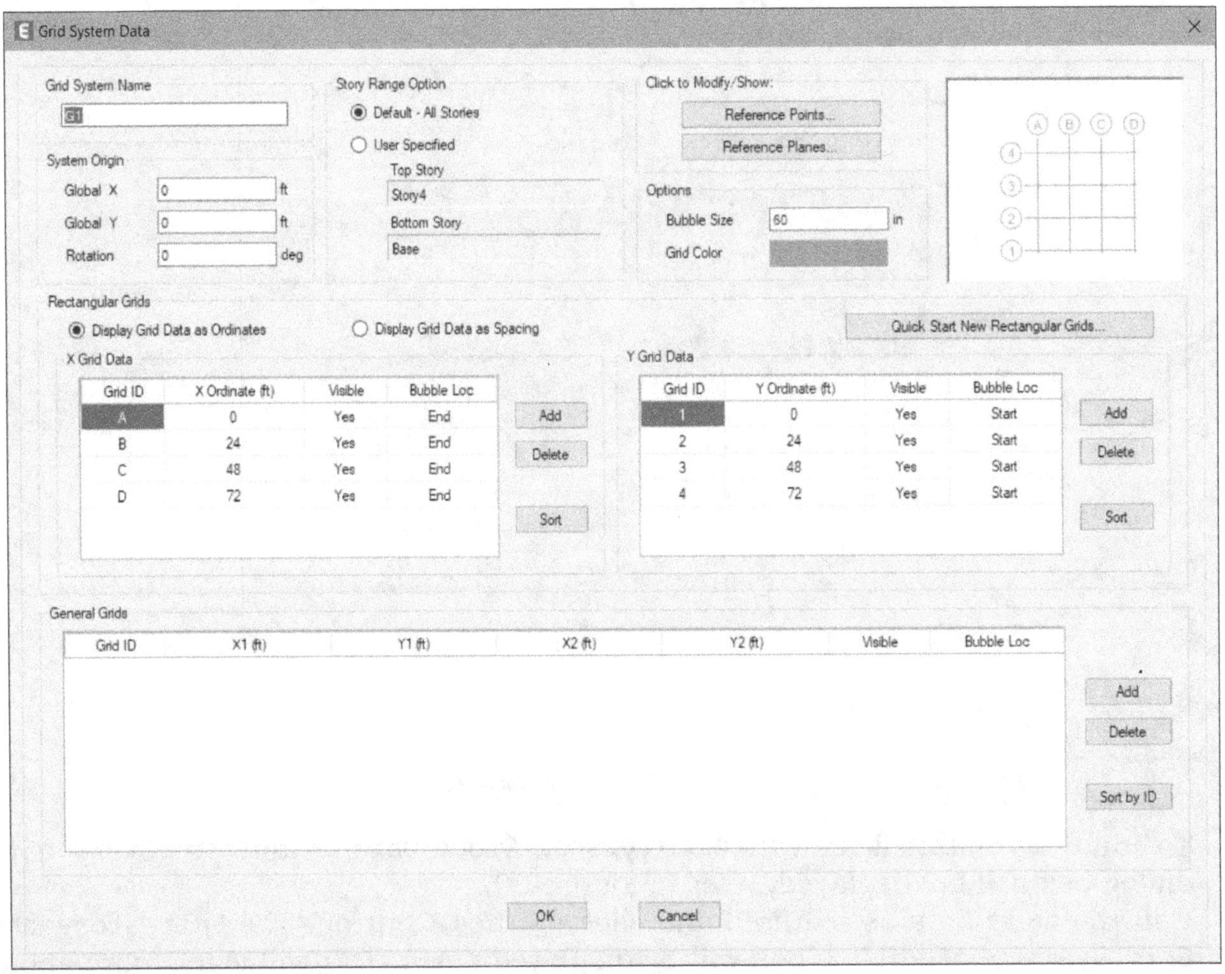

Figure-31. Grid System Data dialog box

- Set the parameters as required and click on the **OK** button. Note that you will learn more about this dialog box later in this book.
- Similarly, you can use the other options of this dialog box. Click on the **OK** button from the **Edit Story and Grid System Data** dialog box.

Similarly, you can edit the other parameters of **Model** tab in the **Model Explorer**.

Display Tab

- Click on the **Display** tab in the **Model Explorer** to check various views of the model in viewports; refer to Figure-32. The options in this tab are used to edit the parameters related to views.

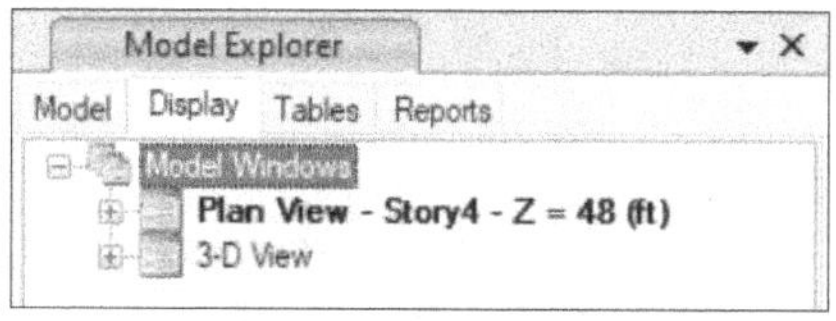

Figure-32. Display tab of Model Explorer

- By default, there are two views created in Model Windows (Viewports) viz. Plan View and 3-D View. Expand the view that you want to edit by clicking on the **+** button. Four categories of the view are displayed viz. View, Limits, Options, and Display; refer to Figure-33.
- Expand the **View** node and select desired radio button to change view to **3D**, **Plan**, or **Elevation**.

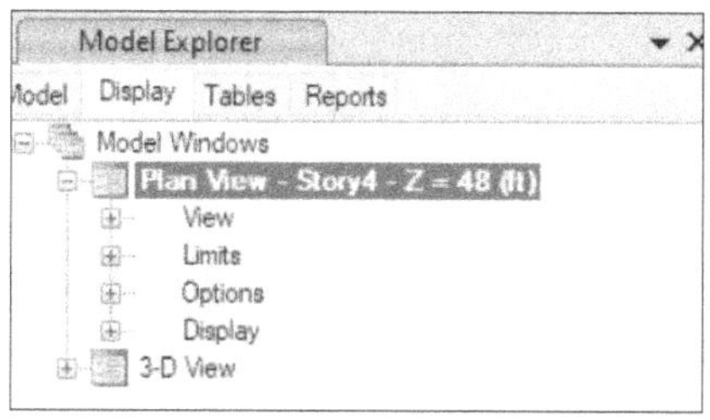

Figure-33. Options for Views in Display tab

- Expand the **Limits** node and click on the **Set Limits** option. The **Set Building View Limits** dialog box will be displayed; refer to Figure-34. Set desired minimum and maximum limits of the view and click on the **OK** button to apply changes. The portion of model within specified range will be displayed in graphics area.

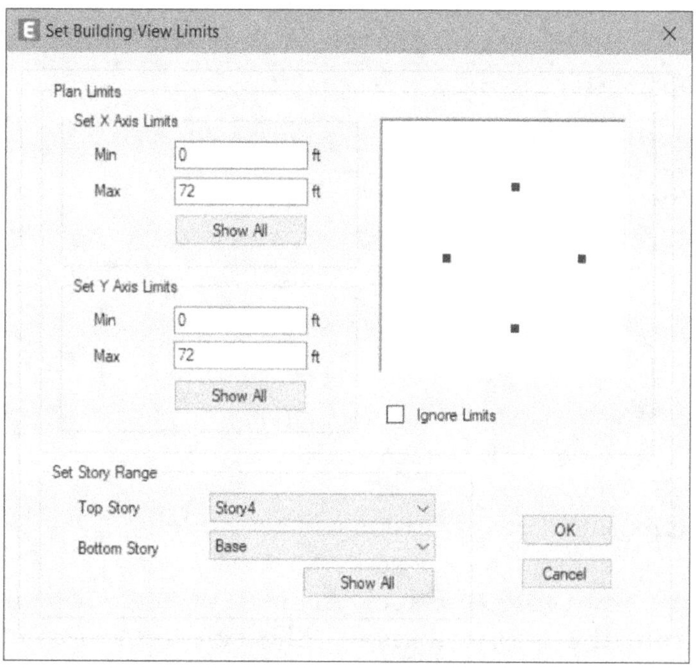

Figure-34. Set Building View Limits dialog box

Expand **Options** node in the **Display** tab of **Model Explorer**. The options to hide/show different elements of model are displayed. Expand desired category and select/de-select the check boxes to show/hide the elements of the model.

Expand the **Display** node and select desired radio button to display deformed and undeformed shape.

Tables Tab

The options in the **Tables** tab of the **Model Explorer** are used to show and export tables; refer to Figure-35.

- Expand desired category of **Tables** node and right-click on the table option that you want to be displayed; refer to Figure-36.

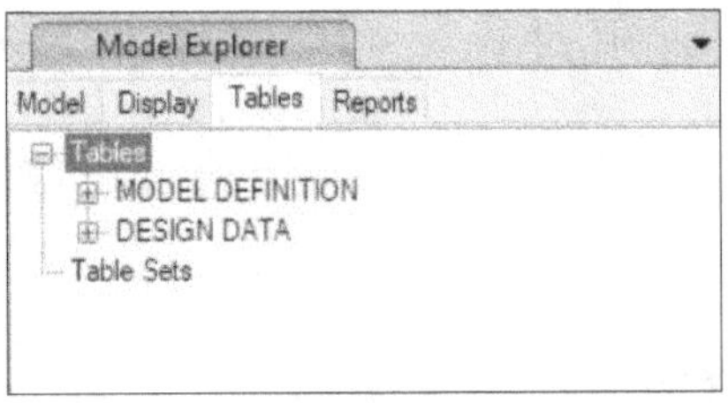

Figure-35. Tables tab of Model Explorer

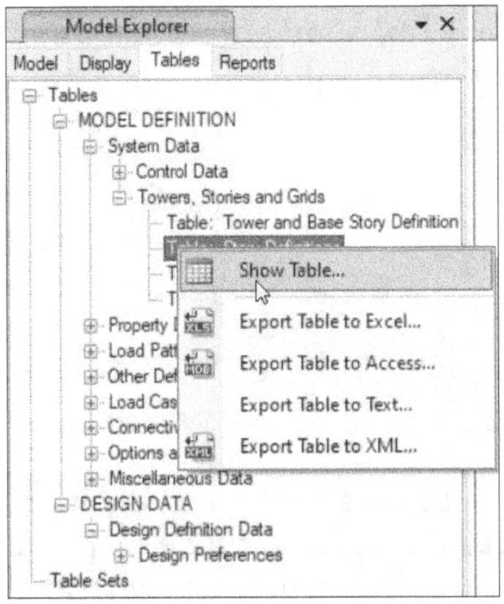

Figure-36. Shortcut menu for table

- Click on the **Show Table** button if you want to check the table of parameters for selected element in the **Model Explorer**. A box will be displayed containing the table; refer to Figure-37.

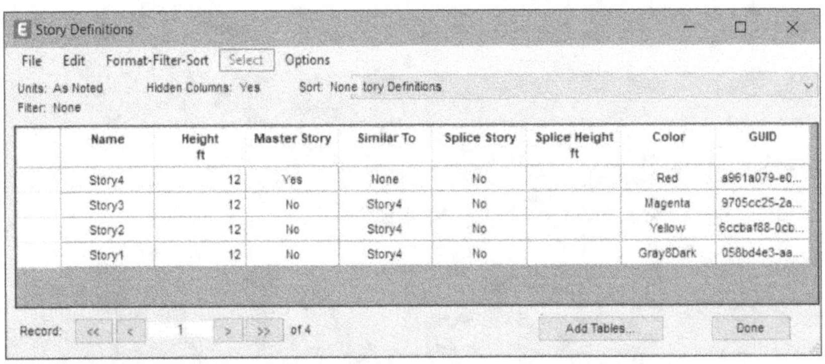

Figure-37. Story Definitions table box

Similarly, you can check the other tables of the model.

If you want to export the table to Excel, Access, or XML format then click on the respective button from the shortcut menu like select the **Export Table to Excel** option if you want to export the table in excel format.

Reports Tab

The **Reports** tab contains all the data of project compiled in report forms; refer to Figure-38. To generate the project report, right-click on **Project Report** category in **Reports** node of the **Reports** tab of **Model Explorer** and click on **Show/Refresh Report** option; refer to Figure-39. A report will be generated and displayed in the viewport; refer to Figure-40.

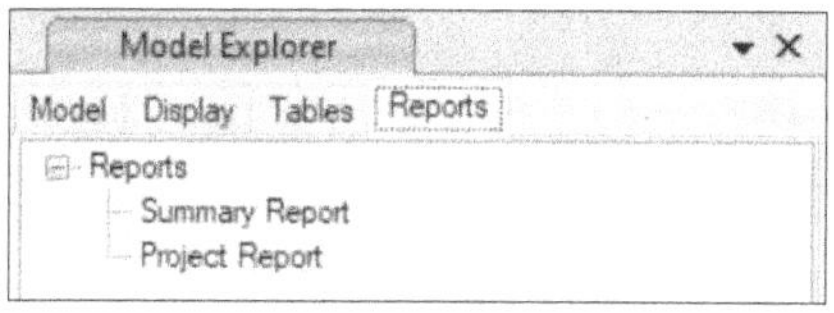

Figure-38. Reports tab of Model Explorer

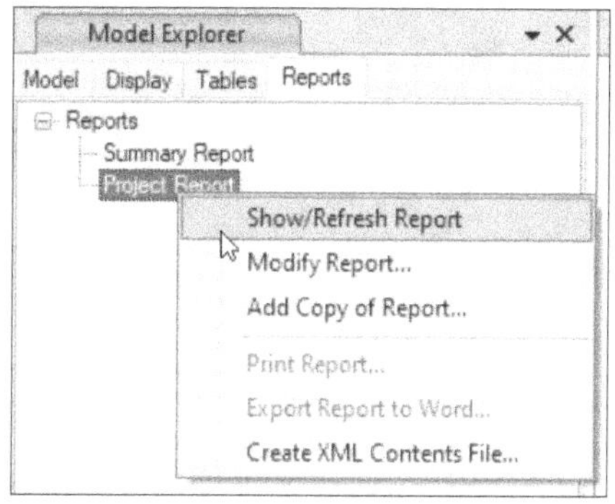

Figure-39. Project Report shortcut menu

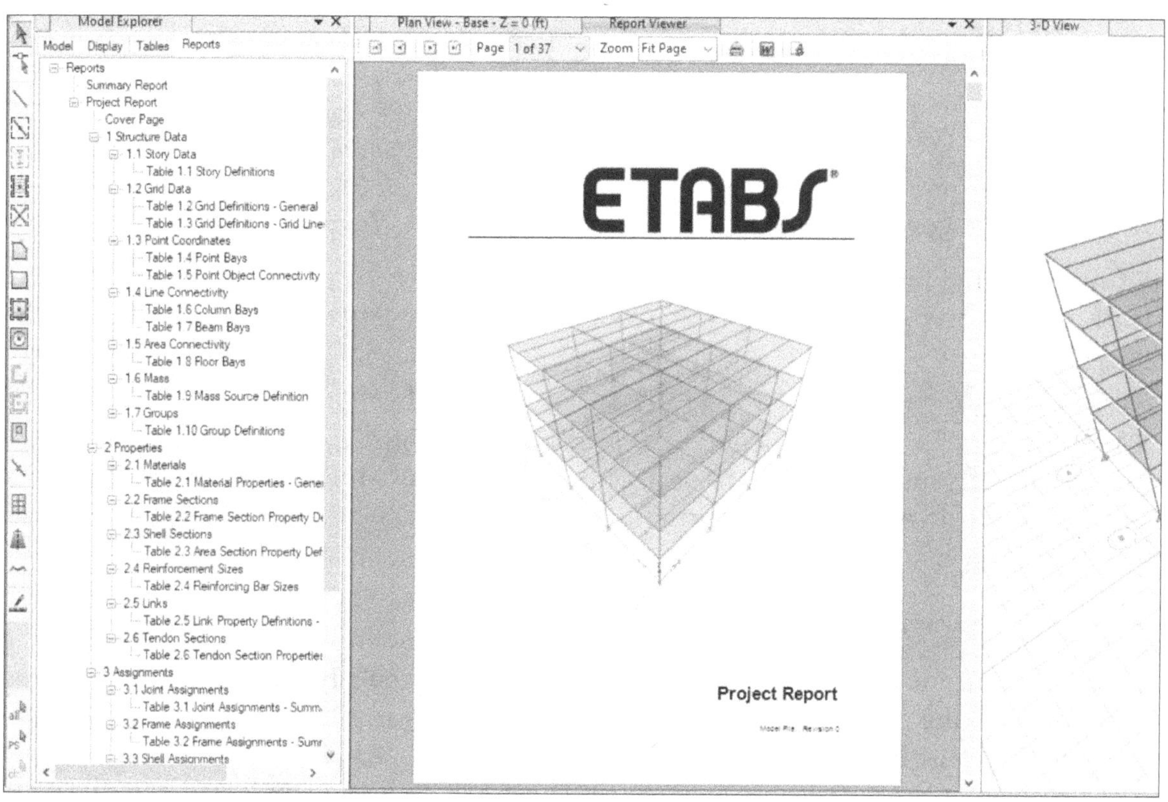

Figure-40. Project Report

FILE MENU

The tools in the **File** menu are used to manage file level operations like creating a new file, opening an existing ETABS file, importing or exporting files, and so on. Various functions of the **File** menu are discussed next.

Creating a New Model

The **New Model** tool in the **File** menu is used to create a new ETABS model. The procedure to start a new model has been discussed earlier.

Opening a Model File

The **Open** tool in the **File** menu is used to open existing models. You can also use the shortcut key **CTRL+O** to open a model file. The procedure is given next.

- Click on the **Open** tool from the **File** menu. The **Open Model File** dialog box will be displayed; refer to Figure-41.

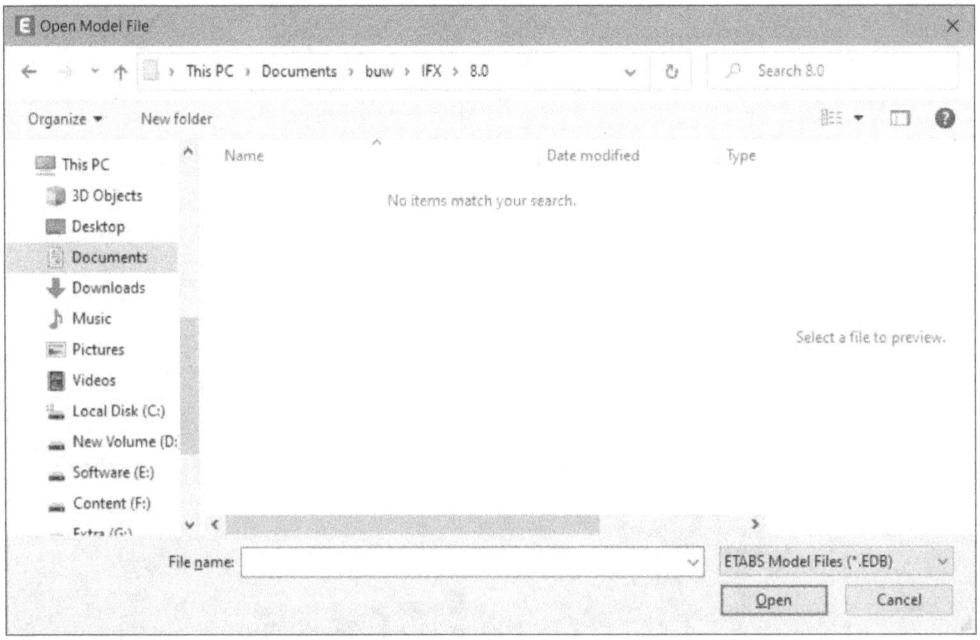

Figure-41. Open Model File dialog box

- Browse to the folder in which the file is available that you want to open and double-click to open it. Note that the extension for Etabs model file is ".EDB".

Closing a Model File

The **Close** tool in the **File** menu is used to close the model on which you are currently working. You can also use the shortcut key **SHIFT+CTRL+E** to close the model. If your file is not saved after making changes then you will be asked to save the changes via a dialog box.

Saving a File

You can save the file using **Save** and **Save As** tools in the **File** menu to save the model file. Using the **Save** tool, you can update the changes in the current file if you have performed any changes. Using the **Save As** tool, you can save the current file with different name or format. If you are using the **Save** tool for the first time then it will function similar to **Save As** tool. The procedure to save a file is given next.

- Click on the **Save** tool after making desired changes. The **Save Model File As** dialog box will be displayed (refer to Figure-42) if you are saving the file for the first time.
- Specify desired name and click on the **OK** button to save the file.
- If you have already saved the file and then made some changes, clicking on the **Save** button will save the file without showing any dialog box.
- Similarly, you can use the **Save As** tool in the **File** menu to save the file in different format or using the different name.

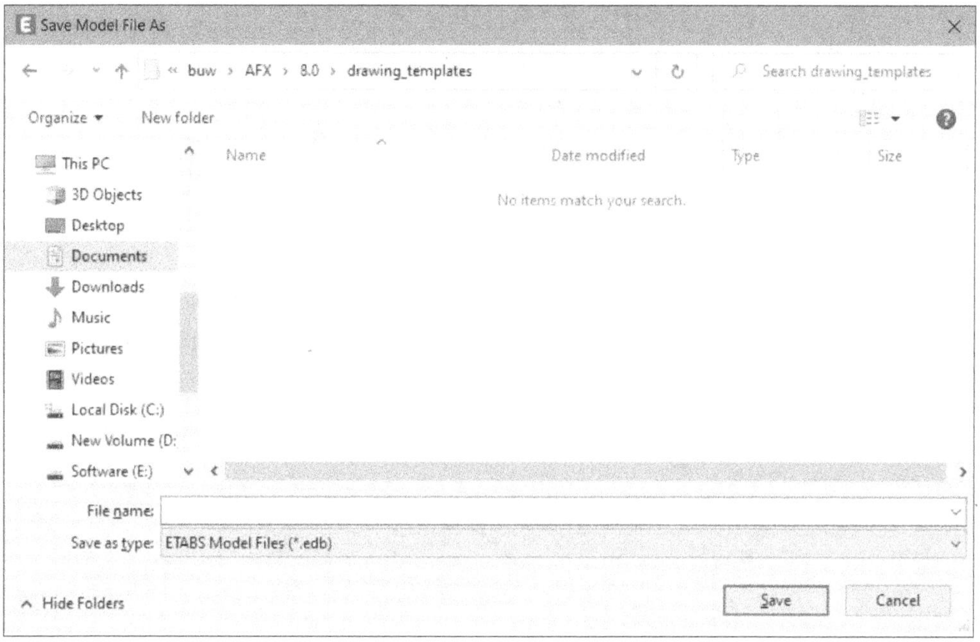

Figure-42. Save Model File As dialog box

Note that when you save the file then *.ebk file is also created as backup for the current file.

Importing Files/Data

The options to import files or data in the model are available in the **Import** cascading menu of the **File** menu; refer to Figure-43. The procedures to import different files are given next.

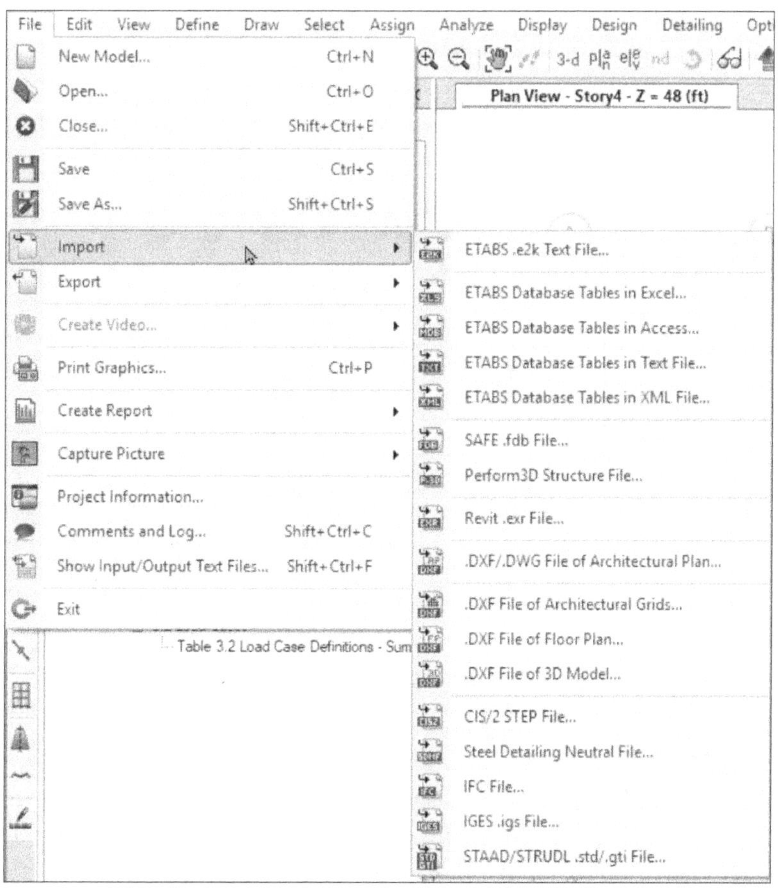

Figure-43. Import cascading menu

Importing ETABS .e2k Text File

There is one more file type associated with '.e2k' file format viz. "*.$et'. The .$et file is created every time you save a model in ETABS. This file contains the text commands for the model; refer to Figure-44. If you have worked with earlier versions of AutoCAD which worked on **Command Manager** only then you may find editing these files very interesting.

```
$ PROGRAM INFORMATION
  PROGRAM  "ETABS 2016"  VERSION "16.2.0"

$ CONTROLS
  UNITS  "LB"  "IN"  "F"
  TITLE2  "Beams"
  PREFERENCE  MERGETOL 0.1
  RLLF  METHOD "ASCE7-10"  USEDEFAULTMIN "YES"

$ TOWERS
  TOWER "T1"

$ STORIES - IN SEQUENCE FROM TOP - FOR EACH TOWER
  STORY "Story4"  HEIGHT 144 MASTERSTORY "Yes"  TOWER "T1"
  STORY "Story3"  HEIGHT 144 SIMILARTO "Story4"
  STORY "Story2"  HEIGHT 144 SIMILARTO "Story4"
  STORY "Story1"  HEIGHT 180 SIMILARTO "Story4"
  STORY "Base"  ELEV 0

$ GRIDS
  GRIDSYSTEM "G1"  TOWER "T1"  TYPE "CARTESIAN"  BUBBLESIZE 60
  GRID "G1"  LABEL "A"  DIR "X"  COORD 0 VISIBLE "Yes"  BUBBLELOC
"End"
  GRID "G1"  LABEL "B"  DIR "X"  COORD 288 VISIBLE "Yes"
BUBBLELOC "End"
  GRID "G1"  LABEL "C"  DIR "X"  COORD 576 VISIBLE "Yes"
BUBBLELOC "End"
  GRID "G1"  LABEL "D"  DIR "X"  COORD 864 VISIBLE "Yes"
BUBBLELOC "End"
  GRID "G1"  LABEL "1"  DIR "Y"  COORD 0 VISIBLE "Yes"  BUBBLELOC
"Start"
  GRID "G1"  LABEL "2"  DIR "Y"  COORD 288 VISIBLE "Yes"
```

Figure-44. Text commands in $et file

The procedure to import *.e2k or *.$et file is given next.

- Click on the **ETABS .e2k Text File** option from the **Import** cascading menu in the **File** menu. The **Open ETABS Text File** dialog box will be displayed; refer to Figure-45.

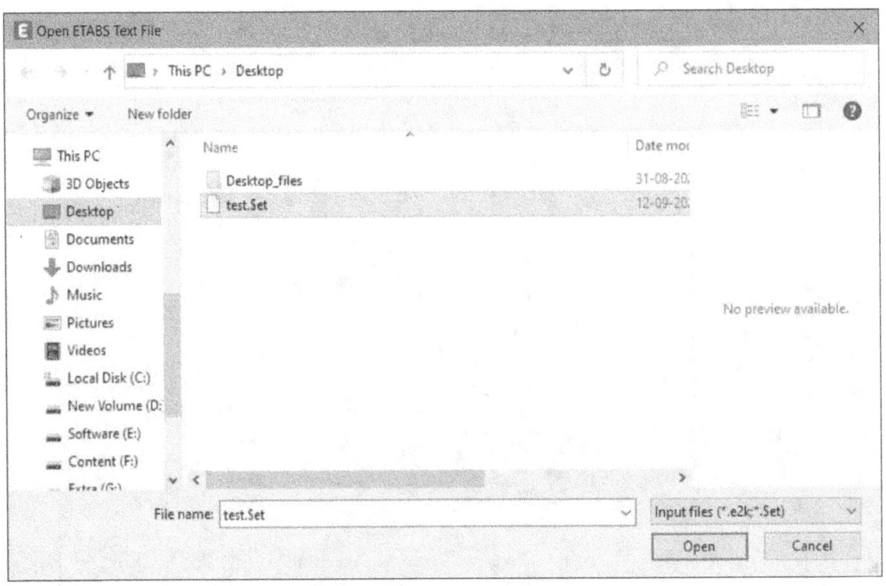

Figure-45. Open ETABS Text File dialog box

- Double-click on the ETABS text file that you want to import. The model will be generated based on your specified command text file.

Importing ETABS Database Tables

There are various tools available in the **Import** cascading menu of the **File** menu to import database tables for structure using Excel, Access, Text file, and XML file. The procedure to import ETABS database tab is same for different sources so we will discuss the procedure to use ETABS Database Tables in Excel tool and you can apply the procedure to others as well.

- Click on the **ETABS Database Tables in Excel** tool from the **Import** cascading menu of the **File** menu. The **Import Tabular Database** dialog box will be displayed; refer to Figure-46.

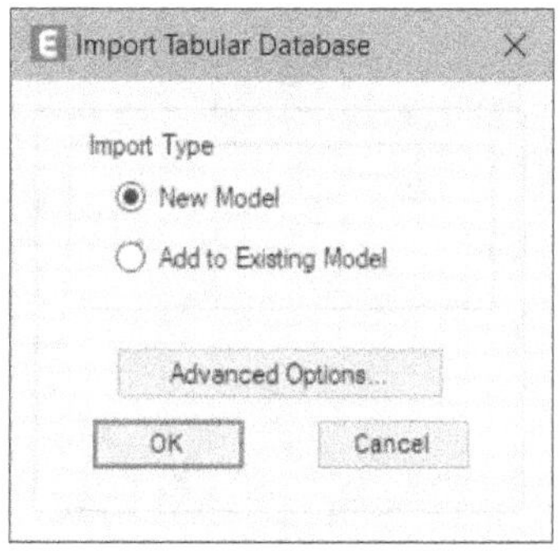

Figure-46. Import Tabular Database dialog box

- Select the **New Model** radio button to create a new model using database file. Select the **Add to Existing Model** radio button to modify current model and add the database to selected file.

Advanced Options

Click on the **Advanced Options** button from the **Import Tabular Database** dialog box to modify advanced options for importing database entries. The **Tabular Database Import Options** dialog box will be displayed; refer to Figure-47.

- In the **Message Control** tab, you can define what type of messages will be displayed when importing database entries. Select the **Display error messages** check box to display errors occurring when importing the file. Similarly, select the other check boxes in **Message Control** area of tab to display respective messages. In the **Import Control for Errors** area, you can specify whether importing will be aborted after specified number of errors or not. Similarly, you can specify abort condition for warnings in the **Import Control for Warnings** area.
- Click on the **Items with Same Name** tab in the dialog box to define how conflicting names of imported database will be resolved; refer to Figure-48. Select the **Overwrite previous item** radio button to overwrite duplicate name entries with last identified entry. Select the **Ignore subsequent items** radio button to keep the first encountered entry of duplicates in database and ignore the others.
- Similarly, set the parameters in **Joints and Structural Objects** area and **Other Items** area of the dialog box to manage duplicate objects.
- Click on the **Coordinate Transformation** tab to rotate or offset coordinate system of imported database.
- Using the **Reset Tab Defaults** button and **Reset All Defaults** buttons, you can reset parameters in respective tab or dialog box. After setting desired parameters, click on the **OK** button from the dialog box. The **Import ETABS Microsoft Excel Workbook** dialog box will be displayed; refer to Figure-49.

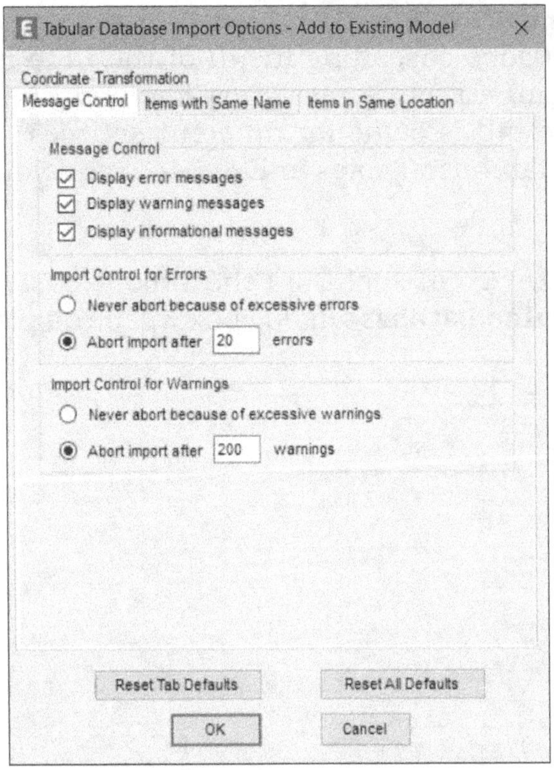

Figure-47. Tabular Database Import Options dialog box

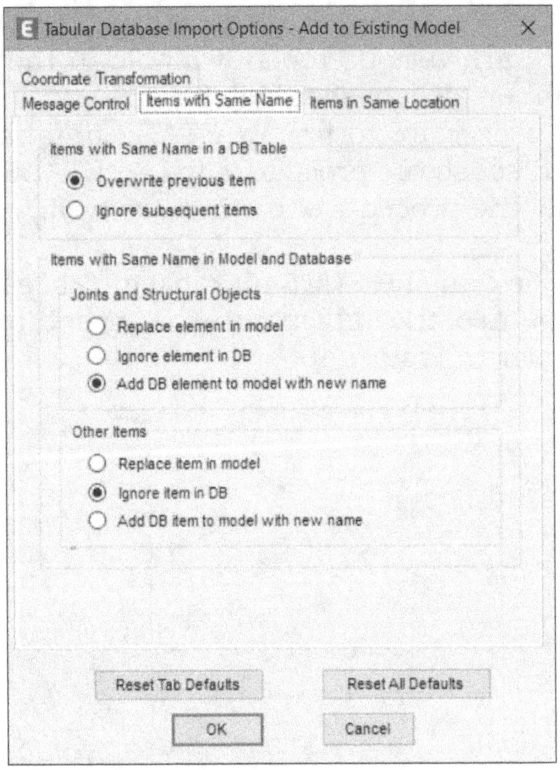

Figure-48. Items with Same Name tab

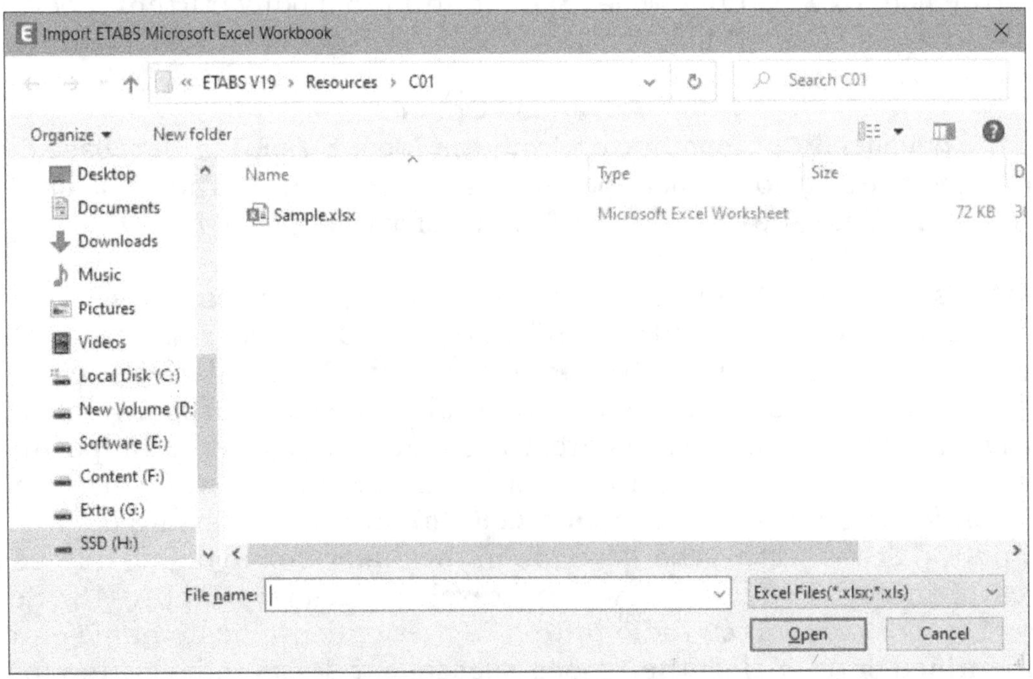

Figure-49. Import ETABS Microsoft Excel Workbook dialog box

- Select desired Excel sheet file and click on the **Open** button. If there are errors or warnings in importing which are more than specified threshold then a warning message will be displayed; refer to Figure-50. Click on the **Yes** button to abort importing or click on the **No** button to continue importing.

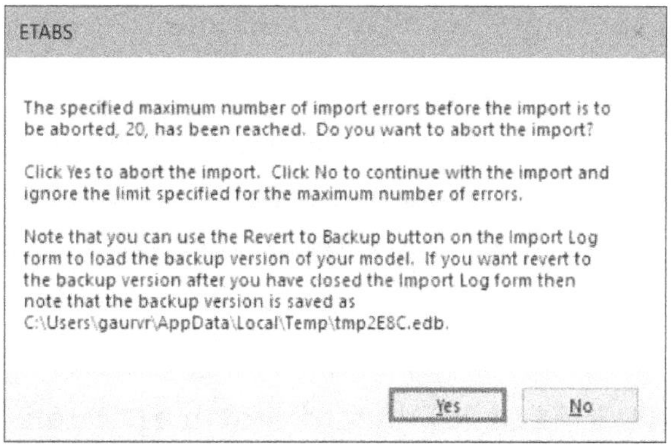

Figure-50. ETABS warning message

- Once the import is complete, a report will be displayed about importing. Click on the **Done** button to exit.

Importing Revit Structure Exchange File

Revit Structure is one of the most widely used structural designing software by Autodesk. You can import the exchange file generated in Revit Structure by following steps. Note that the exchange file must be created using the CSIxRevit application.

- Click on the **Revit Structure .exr File** option from the **Import** cascading menu in the **File** menu. The **Import ETABS/Revit Structure Exchange File** dialog box will be displayed; refer to Figure-51.

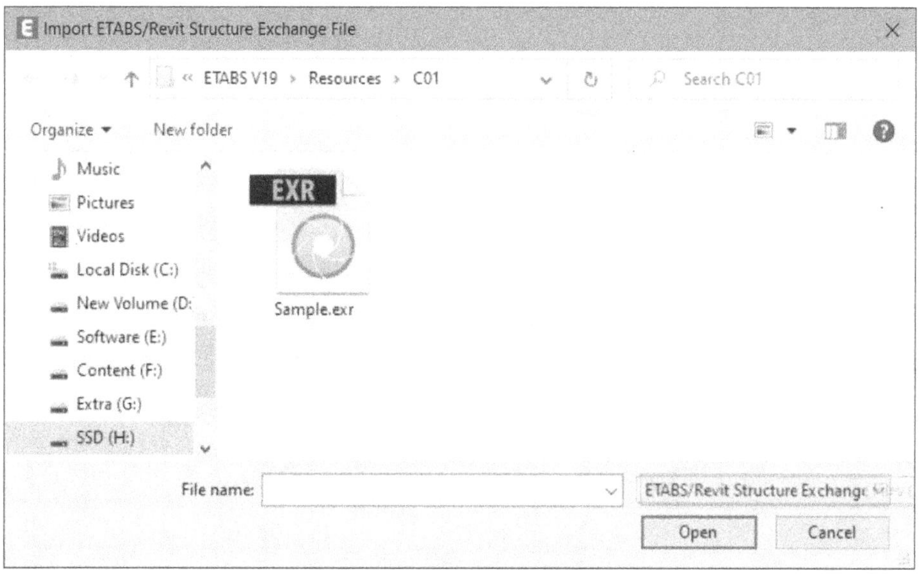

Figure-51. Import Etabs/Revit Structure Exchange File dialog box

- Double-click on desired *.EXR file to open and follow the steps as discussed earlier.

Importing DXF/DWG Files

There are four options in the **Import** cascading menu to import '.DXF' and '.DWG' files in ETABS:

- **.DXF/.DWG File of Architectural Plan**
- **.DXF File of Architectural Grid**
- **.DXF File of Floor Plan**
- **.DXF File of 3D Model**

Select the '**.DXF/.DWG File of Architectural Plan**' option from the **Import** cascading menu if you want to import a dxf or dwg file which contains architectural plan of a building. Select the '**.DXF File of Architectural Grid**' option if you want to import only architectural grid of the building. Select the '**.DXF File of Floor Plan**' option if you want to import the floor plans from other architectural designing software. Select the '**.DXF File of 3D Model**' option to import any 3D model for other CAD software. The procedure to import .dxf file of architectural plan is given next. You can use the same procedure with other options.

- Click on the **.DXF/.DWG File of Architectural Plan** option from the **Import** cascading menu in the **File** menu. The **Import DXF/DWG File** dialog box will be displayed; refer to Figure-52.

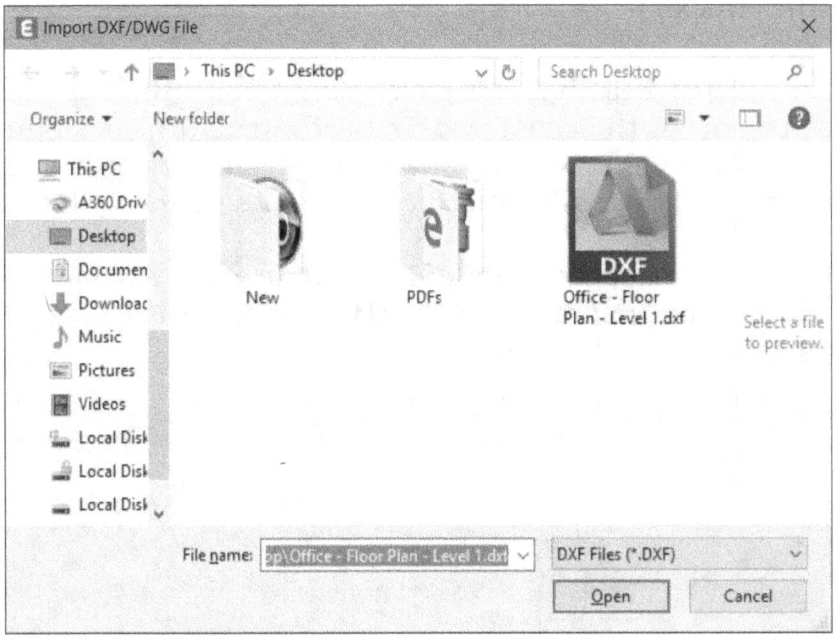

Figure-52. Import DXF DWG File dialog box

- Select the file that you want to import and click on the **Open** button. The **Architectural Plan Import** dialog box will be displayed; refer to Figure-53.

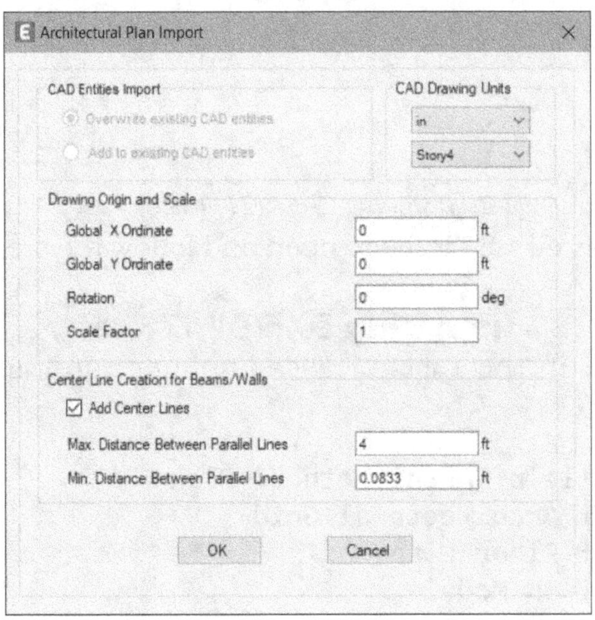

Figure-53. Architectural Plan Import dialog box

- Set desired parameters in the dialog box like units, level in ETABS, drawing origin, and so on.
- Click on the **OK** button. The drawing will be placed in the current ETABS model.

Similarly, you can use the other options in the **Import** cascading menu to import different type of files.

Exporting Files

Like importing files, there are various options to export the files. These options are available in the **Export** cascading menu of the **File** menu; refer to Figure-54.

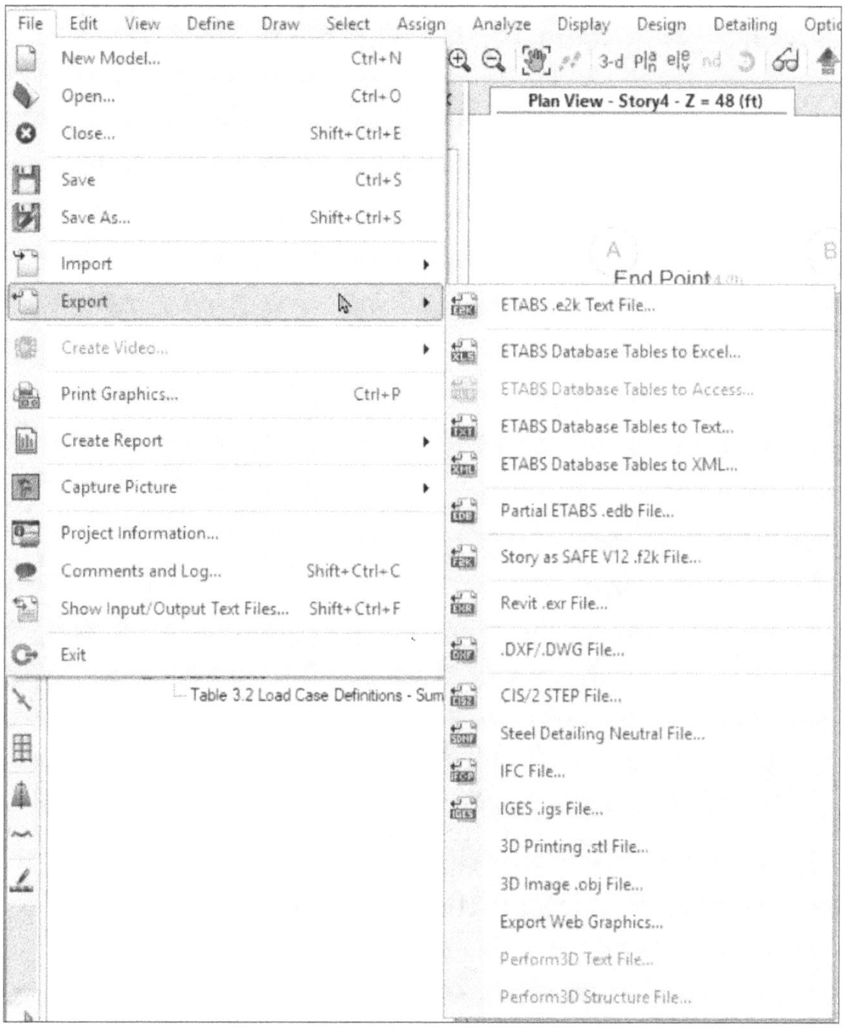

Figure-54. Export cascading menu

The procedure to export .exr file for Revit Structure is discussed next. You can apply the same procedure to other options.

- Click on the **Revit Structure .exr File** option from the **Export** cascading menu. The **Export to Revit EXR File** dialog box will be displayed; refer to Figure-55.

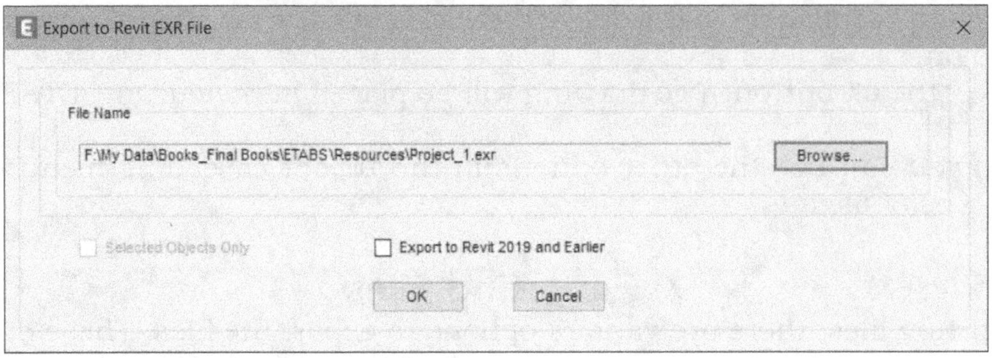

Figure-55. Export to Revit EXR File dialog box

- Click on the **Browse** button from the dialog box. The **Export ETABS/Revit Structure Exchange File** dialog box will be displayed; refer to Figure-56.

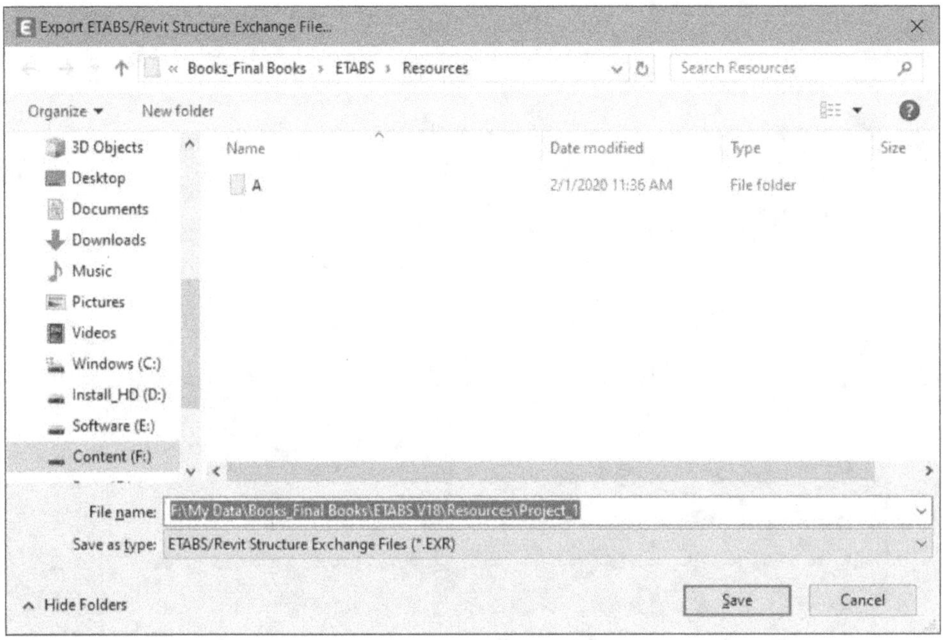

Figure-56. Export ETABS Revit Structure Exchange File dialog box

- Specify the name of file by which you want to save it in the **File name** edit box and click on the **Save** button. The file will be exported in the specified format.
- Select the **Export to Revit 2019 and Earlier** check box to save the file in new version of Revit software. Click on the **OK** button to save the file.

Upload to CSI Cloud

The **Upload to CSI Cloud** option in the **File** menu is used to upload the files on cloud. Note that you would need CSI Cloud Explorer for uploading the data.

Creating Video of Analysis

Once you have performed desired structural analyses, you can record a video of results to present in front of your customers. The tools to create videos are available in the **Create Video** cascading menu of the **File** menu; refer to Figure-57. Note that these options are active only after you have run the analysis on model. There are two options in the cascading menu: **Multi-step Animation** and **Cycle Animation**. Select the **Multi-step Animation** option if you have run a time history analysis and want to record the effect of loads at different time steps. Select the **Cycle Animation** option if you want to record the effect of loads on the model in cycle.

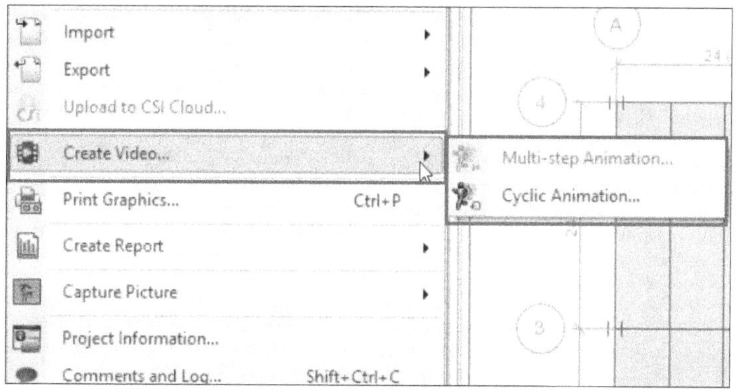

Figure-57. Create Video cascading menu

The procedure to create video of cyclic animation is given next. You can use the same procedure to create **Multi-step Animation**.

- After running the analysis, click on the **Cyclic Animation** option from the **Create Video** cascading menu. The **Video File** dialog box will be displayed; refer to Figure-58.
- Specify desired name of the file in the **File name** edit box and click on the **Save** button. The **Animation Video File Creation** dialog box will be displayed; refer to Figure-59.

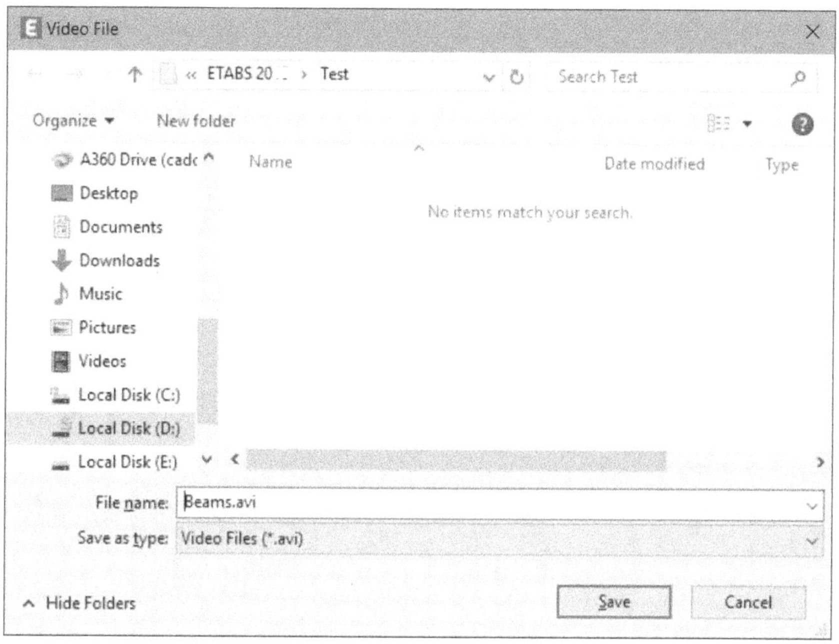

Figure-58. Video File dialog box

- Specify desired parameters in the dialog box and click on the **OK** button. The animation will run once and the video file will be saved at specified location.

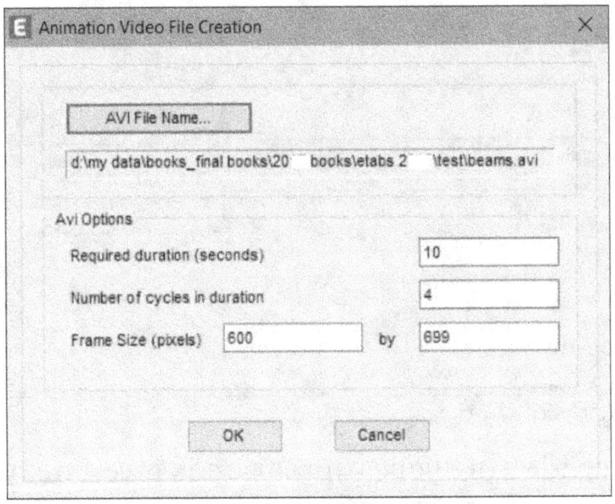

Figure-59. Animation Video File Creation dialog box

Printing Graphics

The **Print Graphics** tool in the **File** menu is used to print the object in selected viewport. The procedure to print graphics is given next.

- Click on the **Print Graphics** tool from the **File** menu or press **CTRL+P** from keyboard. The **Print Graphics** window will be displayed; refer to Figure-60.

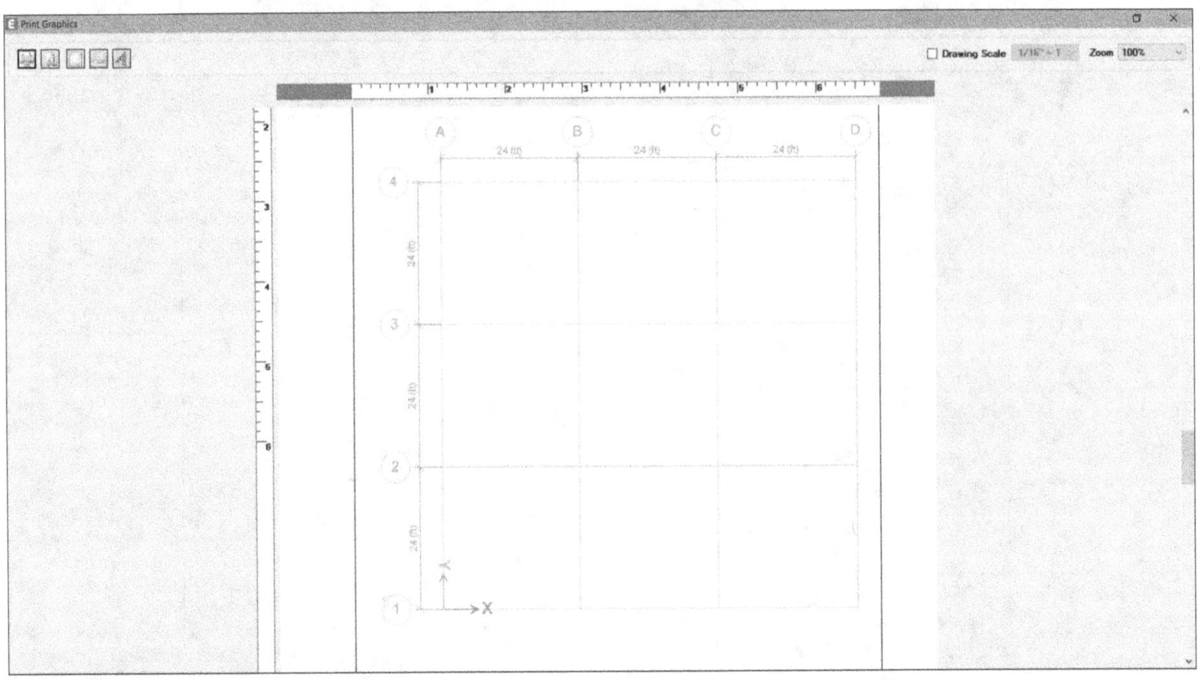

Figure-60. Print Graphics window

- There are various buttons in the top left corner of the window to perform printing and editing operations. Following are the operations that you can perform while in the **Print Graphics** window:

Adding Note to the Prints

- Click on the **Add Note** button from the **Print Graphics** window and click at desired location to place the text note. You will be asked to specify desired text and a toolbar will be displayed to specify font parameter.
- Specify desired parameter and type the text as desired.
- Click on the **Done** button to display text. You can drag this text at desired location.
- If you want to edit this text then double-click on it. The text will become edit-able and the toolbar will be displayed; refer to Figure-61. Modify the text as required and click on the **Done** button.

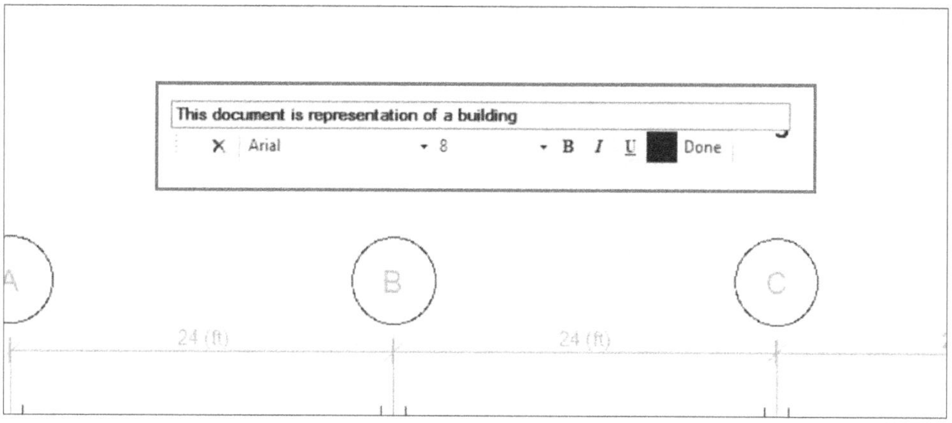

Figure-61. Editing text of note

Adding Image to Print

The **Add Image** button in the **Print Graphics** window is used to insert an image in the print. Follow the steps given next to add image.

- Click on the **Add Image** button from **Print Graphics** window and click at desired location. The **Open** dialog box will be displayed; refer to Figure-62.

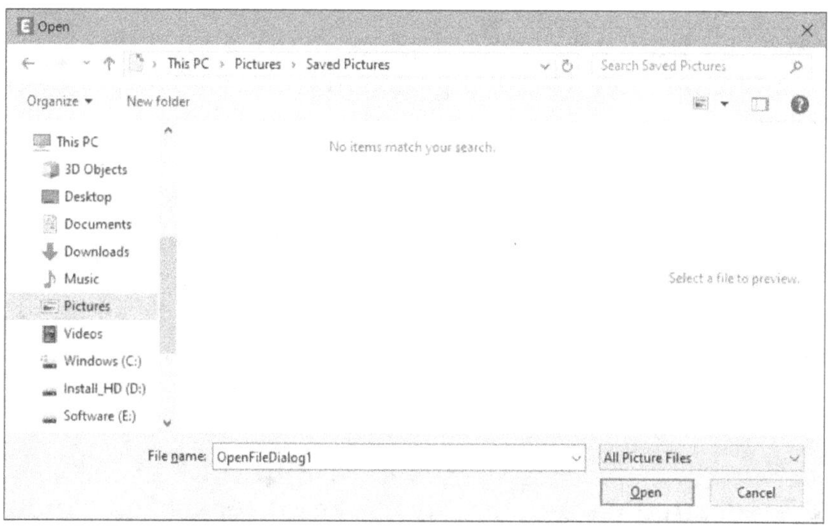

Figure-62. Open dialog box

- Select desired image file and click on the **Open** button. The image will be placed at specified location.
- Set desired parameters in the toolbar displayed; refer to Figure-63 and click on the **Done** button.

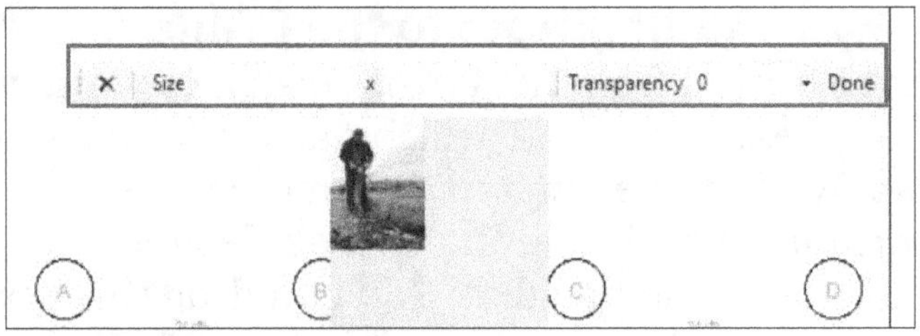

Figure-63. Toolbar displayed while placing image in print

Page Setup for Printing

- Click on the **Page Setup** button from the **Print Graphics** window. The **Page Setup** dialog box will be displayed; refer to Figure-64.

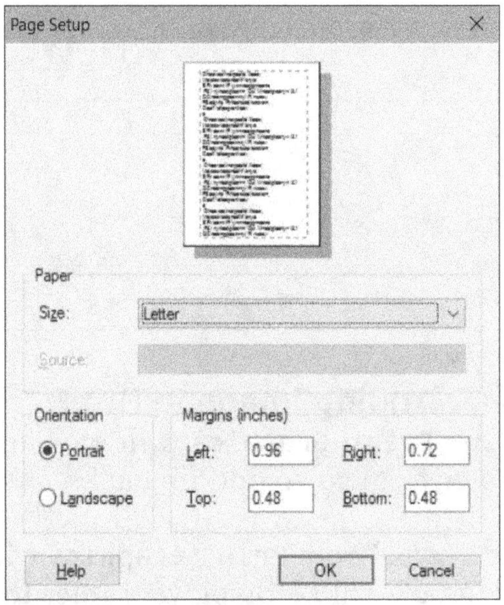

Figure-64. Page Setup dialog box

- Select the page size and specify other parameters as required.
- Click on the **OK** button after specifying the parameters. The page setup for printing will get changed.

Setting Printer

- Click on the **Printer Setup** button from the **Print Graphics** window. The **Print** dialog box will be displayed.
- Select the printer by which you want to print the current objects.
- Set the number of copies and other related parameters as required.
- Click on the **OK** button from the dialog box to apply changes.

Printing the Model

- After setting all desired parameters in the **Print Graphics** window, click on the **Print** button. The file will be printed by selected printer.

Generating Reports

The tools to generate reports are available in the **Create Report** cascading menu of the **File** menu; refer to Figure-65.

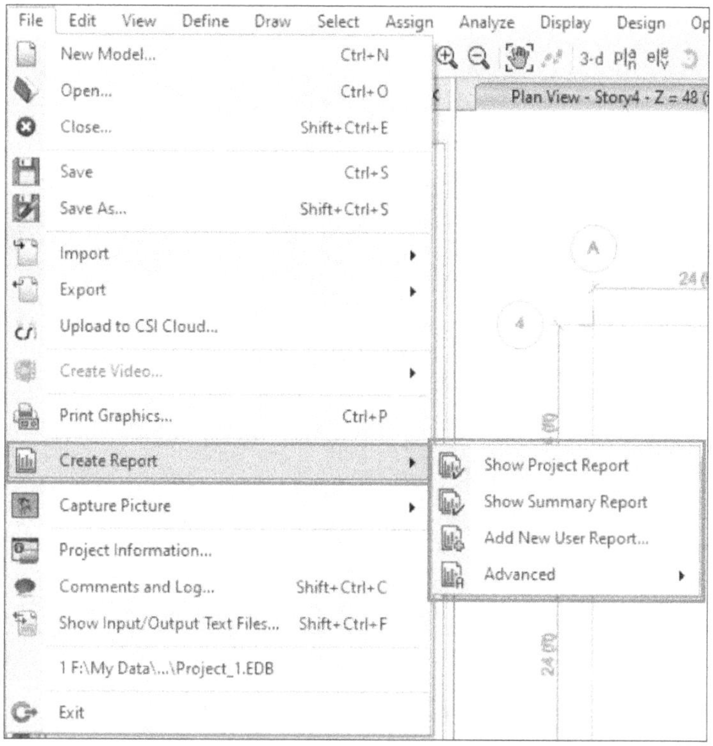

Figure-65. Create Report cascading menu

Various tools in this cascading menu are discussed next.

Show Project Report Tool

- Click on the **Show Project Report** tool from the **Create Report** cascading menu. The **Report Viewer** will be displayed with automatically generated report; refer to Figure-66.
- Click on the **Print** button in the **Report Viewer** if you want to take a printout of the report.
- Click on the **Export to Word Document** button if you want to export the report in MS Word format.
- Click on the **Regenerate Report** button if you want to update the report after making changes in the model.

Figure-66. Report Viewer

- To close the report, click on the 'x' button at the top right corner of the **Report Viewer**.

Similarly, you can use the **Show Summary Report** tool in the **Create Report** cascading menu.

Add New User Report Tool

The **Add New User Report** tool is used to generate the report based on user defined parameters. The procedure to use this tool is given next.

- Click on the **Add New User Report** tool from the **Create Report** cascading menu in the **File** menu. The **User Report** dialog box will be displayed; refer to Figure-67.
- Select the check boxes for all the parameters that you want to be displayed in the report. Note that there are 6 tabs for different categories in the dialog box.
- After setting the parameters, click on the **Create** button. The report will be generated.

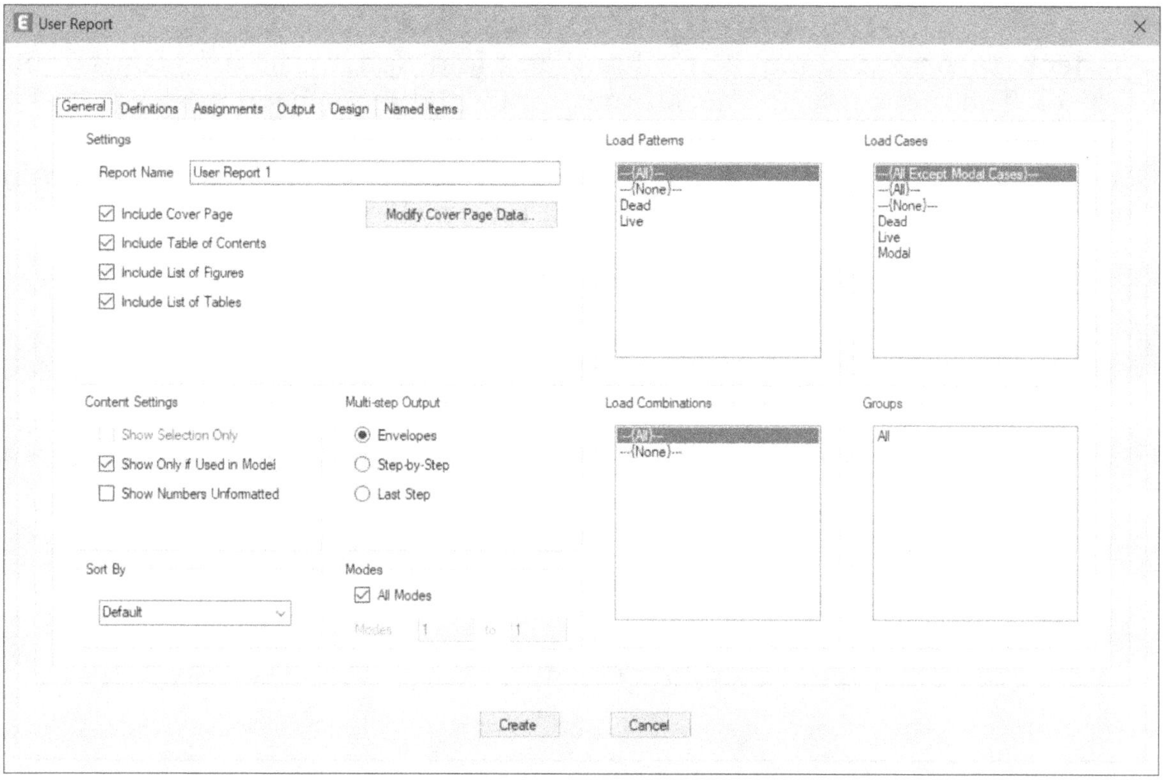

Figure-67. User Report dialog box

Capturing Picture

The tools to take screen shots are available in the **Capture Picture** cascading menu of the **File** menu; refer to Figure-68.

Select desired option from the cascading menu to take snapshot. Like, select the **Main Window** option from the cascading menu (or press **Shift+Ctrl+6**) to take a snapshot of the whole screen. Save the snapshot as image file at desired location as discussed earlier.

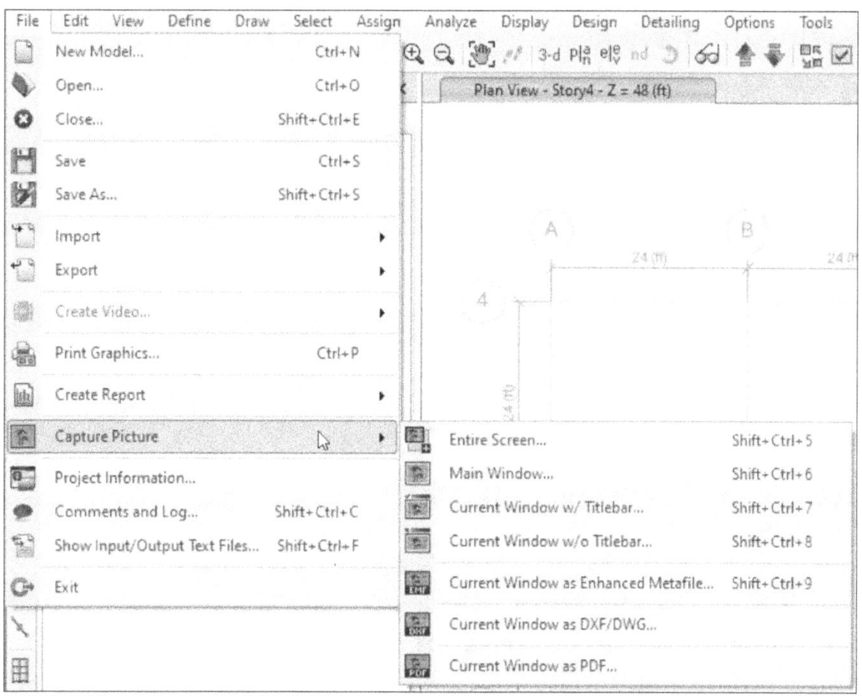

Figure-68. Capture Picture cascading menu

Checking the Project Information

The **Project Information** tool is used to check and update the information related to a project. The procedure to check project information is given next.

- Click on the **Project Information** tool from the **File** cascading menu. The **Project Information** dialog box will be displayed; refer to Figure-69.

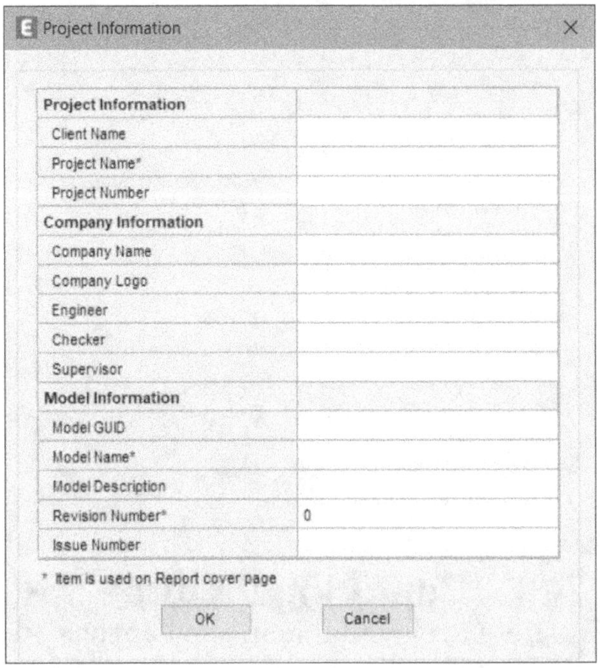

Figure-69. Project Information dialog box

- Specify the information as required in the dialog box.
- Click on the **OK** button to apply changes.

Using the **Comments and Log** tool of the **File** menu, you can check the log of operations generated by software related to file handling. This log is useful when there is any problem in loading the model.

Using the **Show Input/Output Text Files** tool, you can check the input and output files related to your model in text format.

PRACTICAL

Create a steel deck structure with following parameters:

Unit system: U.S. Customary
Region for Default Materials: Europe
Steel Section Database: Euro
Steel Design Code: Eurocode 3-2005
Concrete Design Code: Eurocode 2-2004
Floor size: 100 ft. x 100 ft.
Number of stories: 10
Story Heights: 15 ft.
Steel Deck Overhangs: 2 ft. on each sides.

Secondary Beams at distance of 5ft. along Y
Moment frame type for structures is Intersecting
Structure is pin restrained at bottom and Floors are rigid for load transfer.

Steps:

- Click on the **New Model** tool from the **File** menu. The **Model Initialization** dialog box will be displayed.
- Select the **Use Built-in Settings With** radio button to use specified database and code. Specify the parameters as shown in Figure-70.

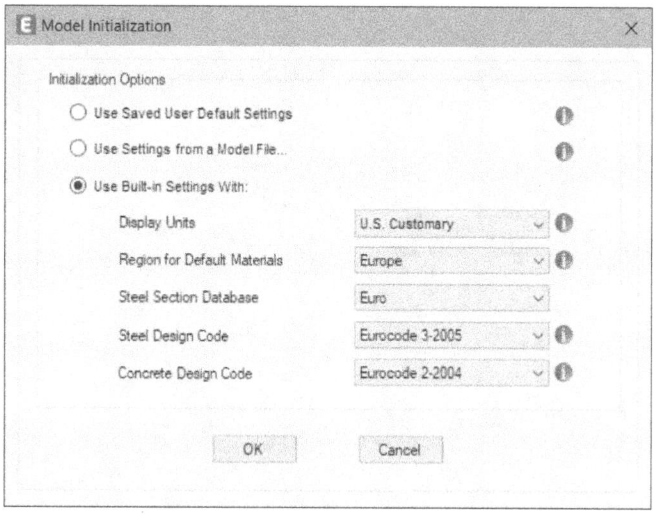

Figure-70. Specifying database and structural codes

- After specifying parameters, click on the **OK** button from the dialog box. The **New Model Quick Templates** dialog box will be displayed.
- Specify number of grid lines in X and Y directions as **5** in the **Number of Grid Lines in X Direction** and **Number of Grid Lines in Y Direction** edit boxes.
- Specify the spacing between grid lines along X and Y directions as 25 ft in the **Spacing of Grids in X Direction** and **Spacing of Grids in Y Direction** edit boxes.
- Specify total number of stories as **10** in the **Number of Stories** edit box.
- Specify the value of height as **15** ft in the **Typical Story Height** and **Bottom Story Height** edit boxes. The options specified in **New Model Quick Templates** dialog box should be as shown in Figure-71.

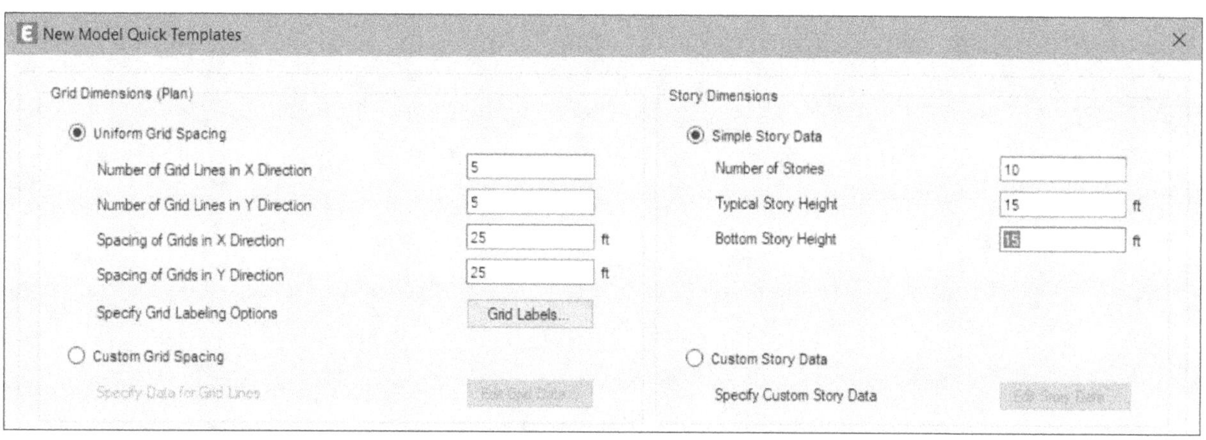

Figure-71. Parameters for template

- Click on the **Steel Deck** button from **Add Structural Objects** area at the bottom in the dialog box. The **Structural Geometry and Properties for Steel Deck** dialog box will be displayed.
- Specify the distance value as 2 ft in the **Left Edge Distance**, **Right Edge Distance**, **Top Edge Distance**, and **Bottom Edge Distance** edit boxes in the **Overhangs** area.
- Specify the maximum spacing between secondary beams as 5 ft. and direction of secondary beams as Y; refer to Figure-72.

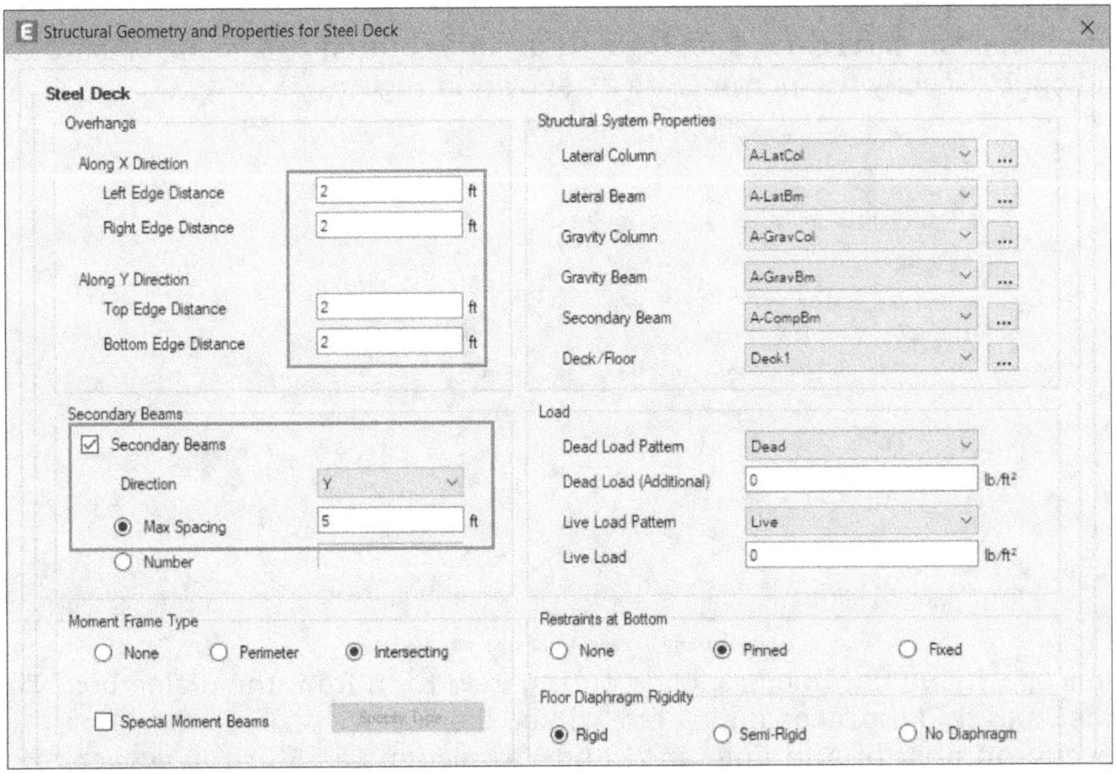

Figure-72. Defining steel deck geometry

- Select the **Intersecting** radio button from the **Moment Frame Type** area to allow load transfer between the elements at intersection.
- Select the **Pinned** radio button from the **Restraints at Bottom** area to constrain bottom of structure as pinned.
- Select **Rigid** radio button to make floor diaphragm rigid to allow load transfer.
- Click on the **OK** button from the dialog box to apply parameters. The **New Model Quick Templates** dialog box will be displayed again.
- Click on the **OK** button from the dialog box to create structure. The structure will be created as shown in Figure-73.

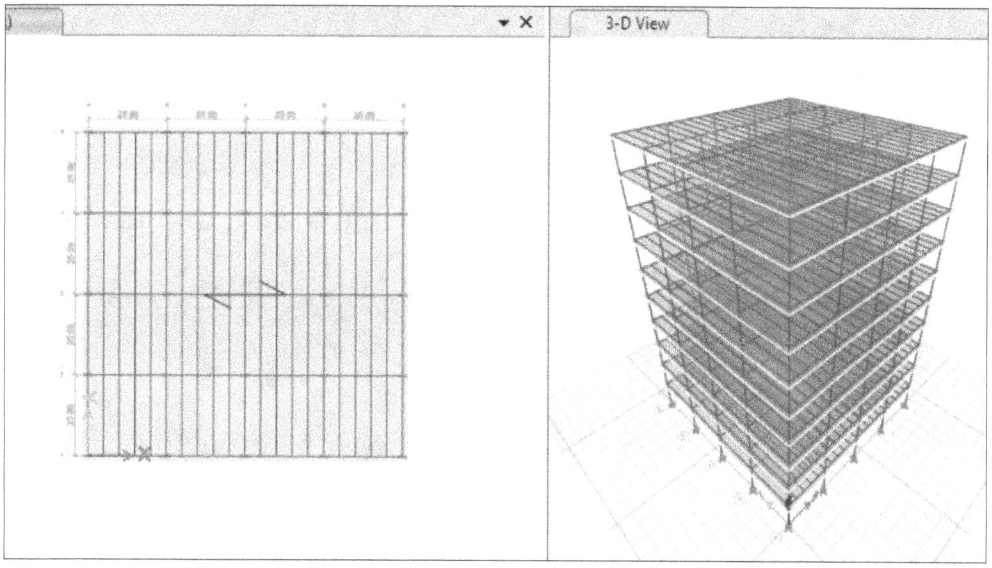

Figure-73. Practical structure

PRACTICE

Create Flat slab structure based on parameters specified next.
Units: Metric SI
Region for Default Materials: United States
Steel Section Database: AISC15M
Steel Design Code: AISC 360-10
Concrete Design Code: ACI 318-19
Floor Size: 75 mtr. x 100 mtr.
Number of stories: 8
Story height: 3.5m
Overhangs as 0.5m in X and Y directions for both sides
Slab thickness: 200 mm
Column Size: 300 mm
Drop Panel : size 4 sq. meter and thickness 300 mm
Apply post tensioning with banded tendons of size 1500 mm in parabola profile in Layer-A and Banded reverse parabola profile of same size in Layer-B.

SELF-ASSESSMENT

Q1. Which shortcut key is used to start a new model in ETABS?
A. CTRL+O
B. CTRL+N
C. CTRL+S
D. SHIFT+CTRL+E

Q2. Which dialog box appears after clicking the New Model option?
A. Story Data dialog box
B. Grid System Data dialog box
C. Model Initialization dialog box
D. Project Information dialog box

Q3. Which option is used to manually specify units, steel database, and design codes?
A. Use Saved User Default Settings
B. Use Built-in Settings With
C. Use Settings from a Model File
D. Uniform Grid Spacing

Q4. Which drop-down is used to set units like length, area, and temperature?
A. Display Units
B. Steel Design Code
C. Concrete Design Code
D. Grid System

Q5. Which radio button is selected to create grids with equal spacing?
A. Custom Grid Spacing
B. Uniform Grid Spacing
C. Simple Story Data
D. Grids Only

Q6. Which button becomes active after selecting Custom Grid Spacing?
A. Grid Labels
B. Edit Grid Data
C. Save As
D. Modify Grid

Q7. Which radio button is used to specify number of stories and story heights quickly?
A. Custom Story Data
B. Uniform Grid Spacing
C. Simple Story Data
D. Display Units

Q8. Which button is clicked to start a blank model?
A. Steel Deck
B. Grids Only
C. Blank
D. Flat Slab

Q9. Which button is used to create floors with steel deck?
A. Flat Slab
B. Steel Deck
C. Waffle Slab
D. Staggered Truss

Q10. Which check box is selected to add secondary beams in the model?
A. Include Drop Panels
B. Secondary Beams
C. Add P/T
D. Special Moment Beams

Q11. Which Moment Frame Type option creates moment resisting connections only at perimeter?
A. None
B. Perimeter
C. Intersecting
D. Fixed

Q12. Which radio button creates a standard full-width beam connection?
A. Reduced Beam Section
B. SidePlate Option
C. Standard Moment Connection
D. Intersecting

Q13. Which load pattern represents static loads from beams, columns, and floors?
A. Live Load
B. Wind Load
C. Dead Load
D. Fixed Load

Q14. Which diaphragm option assumes fully rigid floor behavior?
A. Semi-rigid
B. No Diaphragm
C. Rigid
D. Flexible

Q15. Which button creates structures with flat slabs and strips?
A. Steel Deck
B. Flat Slab
C. Waffle Slab
D. Grids Only

Q16. Which check box is selected to include drop panels in flat slab structures?
A. Add P/T
B. Include Drop Panels
C. Special Moment Beams
D. Secondary Beams

Q17. Which menu contains tools for creating and opening files?
A. Edit Menu
B. View Menu
C. File Menu
D. Display Menu

Q18. Which shortcut key is used to open an ETABS model file?
A. CTRL+N
B. CTRL+O
C. CTRL+S
D. SHIFT+CTRL+E

Q19. What is the extension of ETABS model files?
A. .DWG
B. .TXT
C. .EDB
D. .DOC

Q20. Which interface element displays information about floor level and units?
A. Toolbar
B. Menu Bar
C. Status Bar
D. Viewport

Q21. In which menu are the options to import files or data available?
A. Edit Menu
B. View Menu
C. File Menu
D. Analyze Menu

Q22. Which cascading menu in the File menu is used to import files?
A. Export
B. Import
C. Reports
D. Print Graphics

Q23. Which ETABS file format contains text commands for the model?
A. .EDB
B. .DWG
C. .e2k
D. .DOC

Q24. Which file is automatically created every time you save a model in ETABS?
A. .bak
B. *.$et
C. .txt
D. .xml

Q25. Which option in the Import menu is used to import ETABS text files?
A. Revit Structure .exr File
B. ETABS Database Tables
C. ETABS .e2k Text File
D. DXF File of Floor Plan

Q26. Which dialog box appears when importing an ETABS text file?
A. Open ETABS Text File dialog box
B. Import Tabular Database dialog box
C. Project Information dialog box
D. Video File dialog box

Q27. Which radio button is selected to create a new model using database files?
A. Add to Existing Model
B. New Model
C. Ignore Subsequent Items
D. Overwrite Previous Item

Q28. Which button is used to modify advanced import settings?
A. Export
B. Advanced Options
C. Reset Defaults
D. Show Table

Q29. Which tab in the Tabular Database Import Options dialog box controls import error messages?
A. Coordinate Transformation
B. Items with Same Name
C. Message Control
D. Structural Objects

Q30. Which radio button overwrites duplicate entries with the last identified entry?
A. Ignore subsequent items
B. Overwrite previous item
C. New Model
D. Add to Existing Model

Q31. Which tab is used to rotate or offset the coordinate system during import?
A. Coordinate Transformation
B. Message Control
C. Display
D. Reports

Q32. Which Autodesk software exchange file can be imported into ETABS?
A. AutoCAD
B. Inventor
C. Revit Structure
D. Maya

Q33. Which file extension is used for Revit Structure exchange files?
A. .EDB
B. .EXR
C. .DWG
D. .TXT

Q34. Which application must be used to create the Revit exchange file?
A. AutoCAD
B. CSIxRevit
C. ETABS Viewer
D. Navisworks

Q35. How many DXF/DWG import options are available in ETABS?
A. 2
B. 3
C. 4
D. 5

Q36. Which option is used to import only architectural grids?
A. .DXF File of 3D Model
B. .DXF File of Architectural Grid
C. .DXF File of Floor Plan
D. Architectural Plan Viewer

Q37. Which option is used to import a 3D CAD model?
A. .DXF File of 3D Model
B. .DXF File of Architectural Grid
C. ETABS Database Tables
D. Revit EXR File

Q38. Which dialog box appears after selecting a DXF/DWG architectural plan file?
A. Project Information dialog box
B. Video File dialog box
C. Architectural Plan Import dialog box
D. Story Data dialog box

Q39. In which cascading menu are export options available?
A. Import
B. Display
C. Export
D. Analyze

Q40. Which option exports files for Revit Structure?
A. ETABS Database Tables
B. Revit Structure .exr File
C. Print Graphics
D. Capture Picture

Q41. Which check box is selected to save files for older Revit versions?
A. Add to Existing Model
B. Export to Revit 2019 and Earlier
C. Overwrite Previous Item
D. Display Error Messages

Q42. Which option in the File menu uploads files to the cloud?
A. Create Report
B. Upload to CSI Cloud
C. Print Graphics
D. Capture Picture

Q43. Which cascading menu contains tools to create analysis videos?
A. Export
B. Create Video
C. Reports
D. Import

Q44. Which animation option records effects of loads at different time steps?
A. Cycle Animation
B. Multi-step Animation
C. Structural Animation
D. Frame Animation

Q45. Which animation option records load effects in cyclic form?
A. Multi-step Animation
B. Cycle Animation
C. Static Animation
D. Grid Animation

Q46. Which dialog box appears before saving an animation video file?
A. Project Information dialog box
B. Video File dialog box
C. Story Data dialog box
D. Print Graphics window

Q47. Which tool in the File menu is used to print objects in the selected viewport?
A. Print Graphics
B. Capture Picture
C. Create Report
D. Upload to CSI Cloud

Q48. Which shortcut key is used for Print Graphics?
A. CTRL+N
B. CTRL+O
C. CTRL+P
D. SHIFT+CTRL+6

Q49. Which button in the Print Graphics window is used to insert text notes?
A. Add Image
B. Add Note
C. Printer Setup
D. Page Setup

Q50. Which button in the Print Graphics window is used to insert images?
A. Add Image
B. Add Note
C. Print
D. Export

Q51. Which button is used to change page size and print settings?
A. Add Note
B. Printer Setup
C. Page Setup
D. Capture Picture

Q52. Which button is used to select the printer for printing?
A. Page Setup
B. Printer Setup
C. Add Image
D. Create Report

Q53. Which cascading menu contains report generation tools?
A. Export
B. Create Report
C. Display
D. Analyze

Q54. Which tool displays an automatically generated project report?
A. Show Project Report
B. Add New User Report
C. Print Graphics
D. Capture Picture

Q55. Which button in the Report Viewer exports reports to MS Word?
A. Print
B. Regenerate Report
C. Export to Word Document
D. Done

Q56. Which tool generates reports based on user-defined parameters?
A. Show Summary Report
B. Add New User Report
C. Project Information
D. Print Graphics

Q57. How many tabs are available in the User Report dialog box?
A. 4
B. 5
C. 6
D. 7

Q58. Which cascading menu contains screen shot tools?
A. Create Report
B. Capture Picture
C. Export
D. Import

Q59. Which shortcut key captures the Main Window snapshot?
A. CTRL+P
B. CTRL+N
C. SHIFT+CTRL+6
D. CTRL+O

Q60. Which tool is used to check and update project-related information?
A. Capture Picture
B. Project Information
C. Add Note
D. Create Video

FOR STUDENT NOTES

Chapter 2

Starting a Project in ETABS

Topics Covered

The major topics covered in this chapter are:

- ***Workflow in ETABS***
- ***Modifying Grid***
- ***Drawing Joint Objects***
- ***Drawing Grid Lines Manually***
- ***Snap Options***
- ***Drawing Beams, Columns, and Brace Objects***
- ***Drawing Floor and Wall Objects***
- ***Drawing Links***
- ***Drawing Dimension Lines***
- ***Drawing Reference Points***
- ***Setting Views in Viewports***
- ***Drawing Reference Planes***
- ***Drawing Developed Elevation Definition***
- ***Drawing Wall Stack***
- ***Auto Draw Cladding***

INTRODUCTION

In the previous chapter, you have learned about basic operations required for starting a model project in ETABS. In this chapter, you will learn to draw different types of components of building. The tools to draw different components are available in various menus of the ETABS interface. The workflow to create design in ETABS is given next.

WORKFLOW IN ETABS

Figure-1 below shows the common workflow used for structural designing by ETABS.

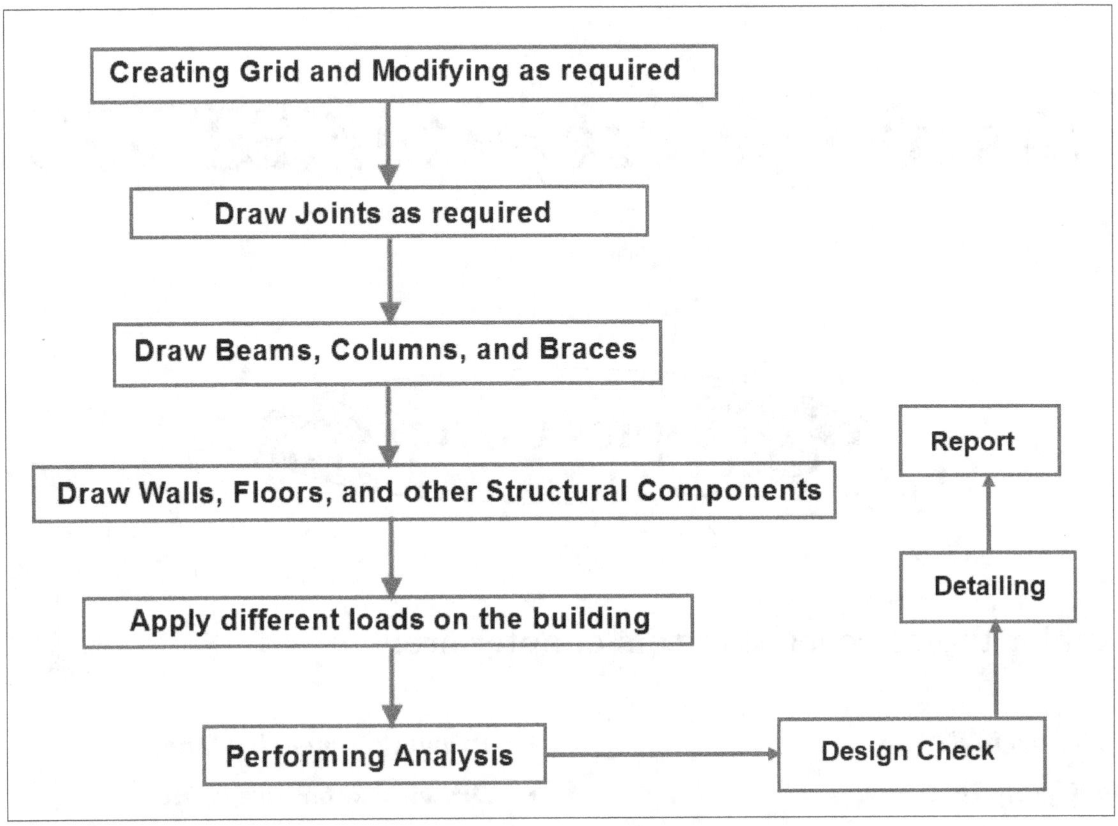

Figure-1. Workflow in ETABS

Now, we will discuss the tools as per the workflow. Start a new document with Grid lines.

MODIFYING GRID

The tools to edit/modify grid data are available in **Edit** menu. Various procedures to edit/modify data are given next.

Setting Grid Options

The options for grid setting are available in the **Grid Options** cascading menu of **Edit** menu which are **Glue Joints to Grids** and **Lock Onscreen Grid System Edit**. These options are discussed next.

- Click on the **Glue Joints to Grids** option from the **Grid Options** cascading menu to "glue" joints along grid lines so that if the grid line is moved then the joints move with it.

- Select the **Lock Onscreen Grid System Edit** option from the **Grid Options** cascading menu to enable or disable editing of grid points in the screen.

Editing Stories and Grid System

The **Edit Stories and Grid Systems** option in the **Edit** menu is used to edit the data related to grids and stories. On clicking this option, the **Edit Story and Grid System Data** dialog box will be displayed; refer to Figure-2.

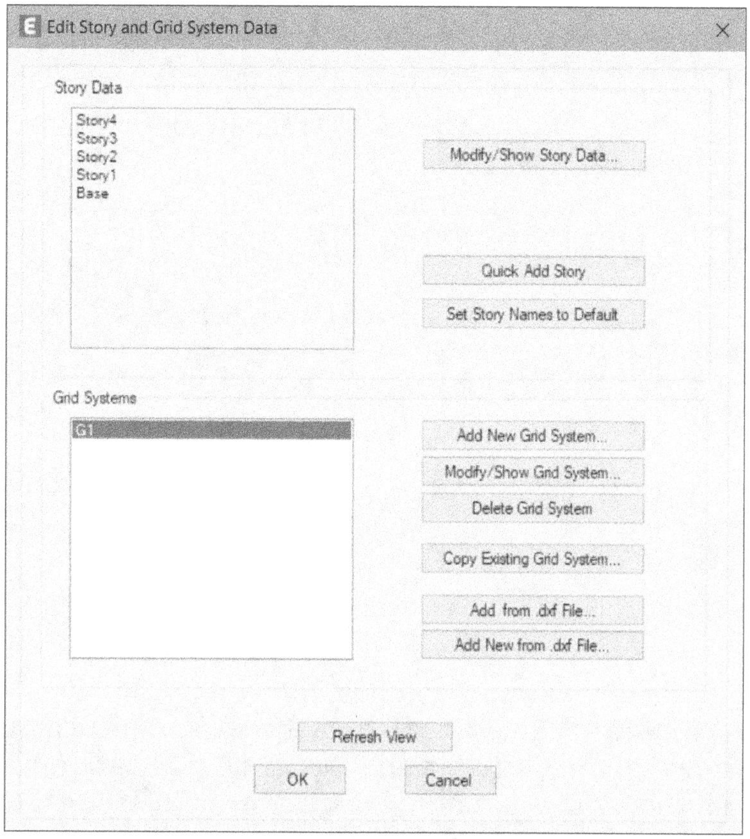

Figure-2. Edit Story and Grid System Data dialog box

The options of this dialog box have already been discussed in previous chapter.

DRAWING JOINT OBJECTS

The joints are used to locate beams and columns in the structure. It represents point of connection between structural members. The main benefit of using joints is the facilitation to provide loads, restraints, and additional masses for them. In this way, you can represent measured loads at specified locations of building in the analysis. In case of connected structure objects, the load is transferred between objects at joint locations. Joints are created automatically at the ends of tendons, links, and other line objects which will be discussed later. The procedure to create joint is given next.

- Click on the **Draw Joint Objects** tool from the **Draw** menu. You will be asked to specify the location of joint.
- Click at desired location in the structure to define joint location. The joint will be created; refer to Figure-3. Press **ESC** to exit the tool.

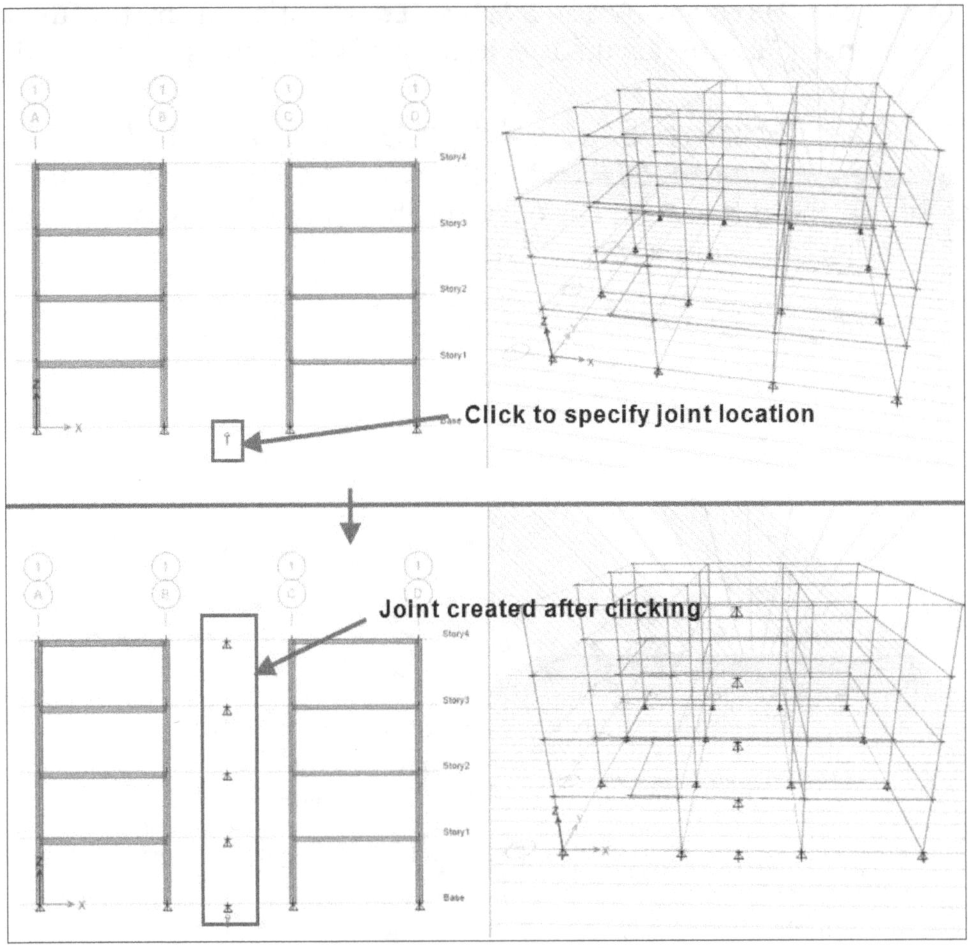

Figure-3. Joint created

By default, joint is created at the current story because **One Story** option is selected in the **Story** drop-down at the bottom right corner of application window. If you want to create joint on all the stories at same location, then select **All Stories** option. If you have different set of similar stories then you can also use **Similar Stories** to create joints only on the stories which are similar to current story. Similar stories are those which have same size and shape but are at different levels.
Sometimes, you might not see the result of created joints immediately in graphics area. In such cases, click on the **Refresh View** tool from **View** menu or press shortcut key **SHIFT+CTRL+V**.

Adding Grid Lines at Selected Joints

The **Add Grid Lines at Selected Joints** tool is used to add grid lines at the joints created in structure (discussed in previous topic). Grid lines can be used as reference for creating beams, columns, and other structure objects. The procedure to add grid lines at joints is given next.

- Select the joints using which you want to create grid lines and click on the **Add Grid Lines at Selected Joints** tool from the **Edit** menu. The **Add Grid Lines at Selected Joints** dialog box will be displayed; refer to Figure-4.

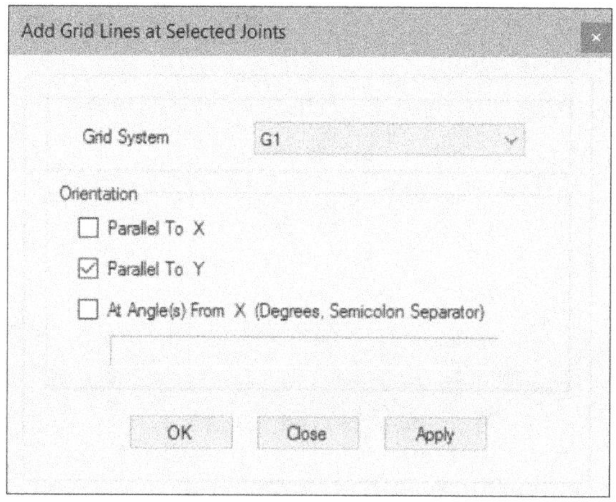

Figure-4. Add Grid Lines at Selected Joints dialog box

- Select the grid system in which you want to add the newly created grid from the **Grid System** drop-down. You can create new grid system by using the **Edit Stories and Grid Systems** tool discussed in previous chapter.
- Select the check box(es) to define direction of the grid lines, like select the **Parallel to Y** check box to create grid lines parallel to Y direction. Note that you can create grid lines in directions parallel to X direction, Y direction, and at an angle from X direction, simultaneously.
- Click on the **Apply** button to create current grid lines and keep the dialog box open. Click on the **OK** button from the dialog box to create the grid lines and exit the dialog box; refer to Figure-5.

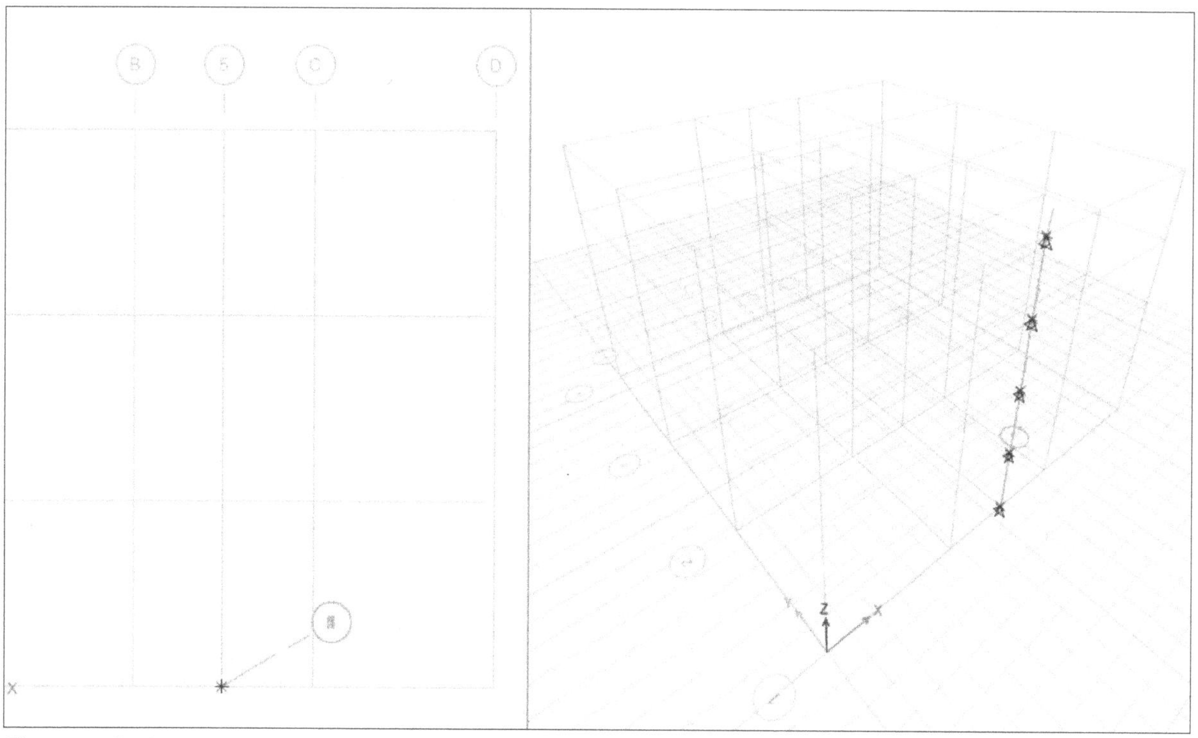

Figure-5. Grid lines created

DRAWING GRID LINES MANUALLY

In previous topics, we have discussed about creating grid lines automatically based on specified joints. In this section, we will discuss the procedure to manually create grid lines.

- Click on the **Draw Grids** tool from the **Draw** menu. You will be asked to select joints for creating grid lines.
- Click at desired joints one by one to create grid lines; refer to Figure-6.

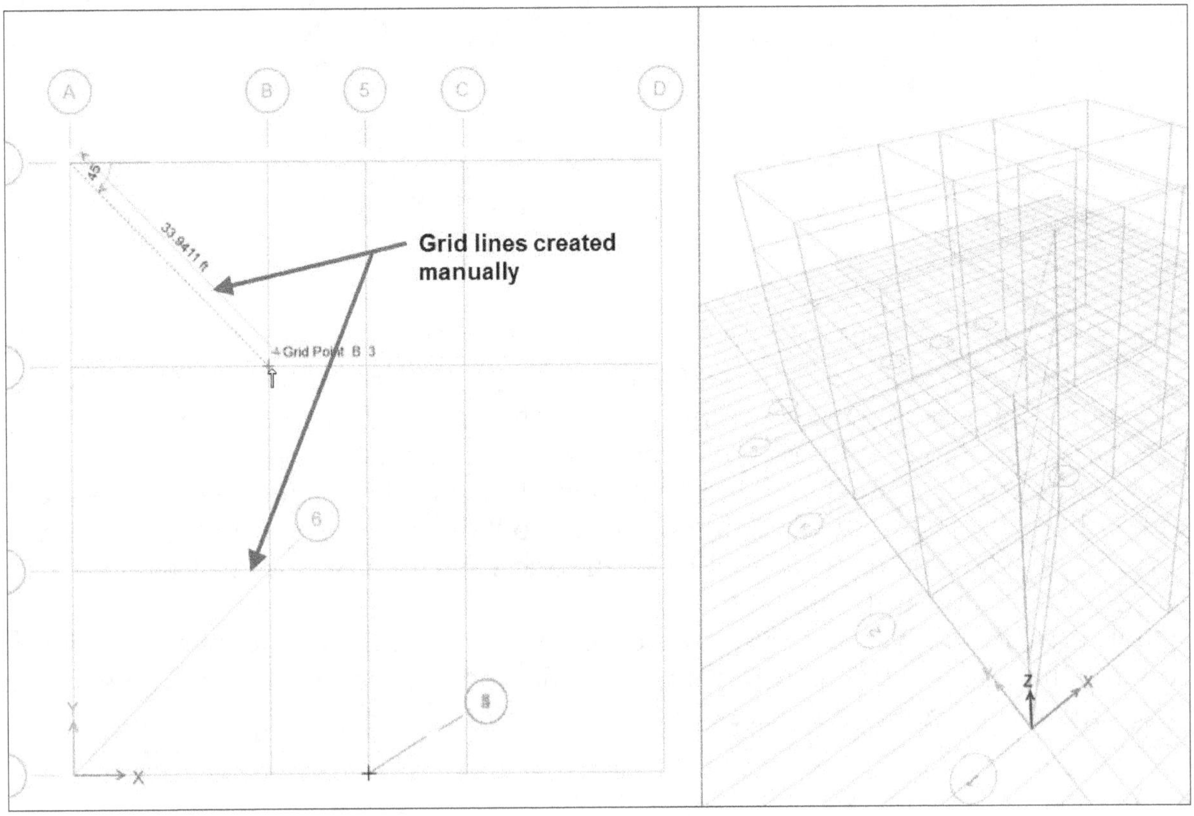

Figure-6. Grid lines created manually

SNAP OPTIONS

Snap options are used to restrict movement of cursor at some special points like mid points, end points, and so on. The cursor automatically sticks to the key points like magnet. Snapping allows easy selection of these key points when creating geometries. The procedure to modify which points to be snapped is given next.

Modifying Snap Options

- Click on the **Snap Options** tool from the **Draw** menu. The **Snap Options** dialog box will be displayed; refer to Figure-7.
- Select the check boxes for major points to which you want the cursor to snap. For example, select the **Line Ends and Midpoints** check box to snap cursor at the end points and midpoints of lines in graphics area. Select the **Arch. Layer** check box to snap automatically to key points of imported architectural objects.
- In the **Snap Increments** area, you can define length or angle at which cursor will automatically snap when creating line objects like beams, walls, and so on. This option is useful when creating objects of same lengths like creating walls and beams of same length.
- Specify desired value of gap between fine grid lines in the **Plan Fine Grid Spacing** edit box. Fine grid lines are imaginary lines between two consecutive grid lines.

- Specify desired value in the **Plan Nudge Value** edit box to specify distance by which nudged objects move using shortcut keys. To use nudge feature, select the objects to be moved, and press shortcut keys (**CTRL+UP** key, **CTRL+DOWN** key, **CTRL+RIGHT** key, and **CTRL+LEFT** key) to move the objects.
- Specify desired value in the **Screen Selection Tolerance** edit box to define distance around the object within which the object will get selected when clicked. The unit for this edit box is pixels.
- Specify desired value in the **Screen Snap To Tolerance** edit box to define distance in pixel around key points within which cursor will snap when clicked.
- Set the other parameters as desired and click on the **Apply** button.
- Click on the **Close** button to exit the dialog box.

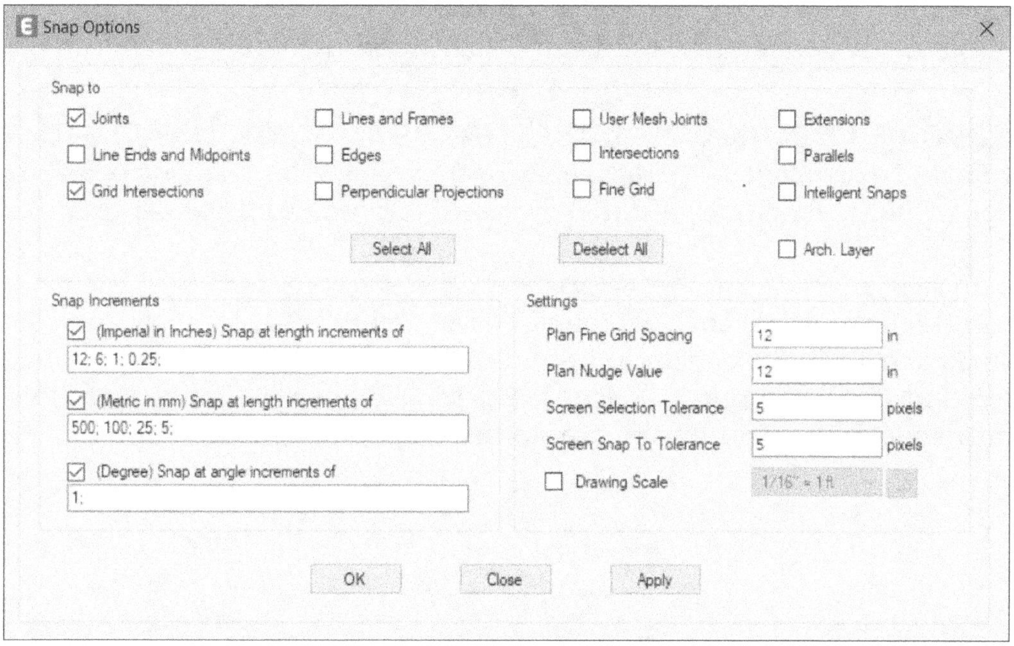

Figure-7. Snap Options dialog box

Drawing with Snap Only

The **Draw Using Snap Only** tool in the **Draw** menu is used to draw objects in ETABS using the snap points only. To enable or disable this function, click on the **Draw Using Snap Only** option from the **Draw** menu.

DRAWING BEAMS/COLUMNS/BRACE OBJECTS

The tools to draw beams, columns, and brace objects are available in the **Draw Beam/Column/Brace Objects** cascading menu; refer to Figure-8.

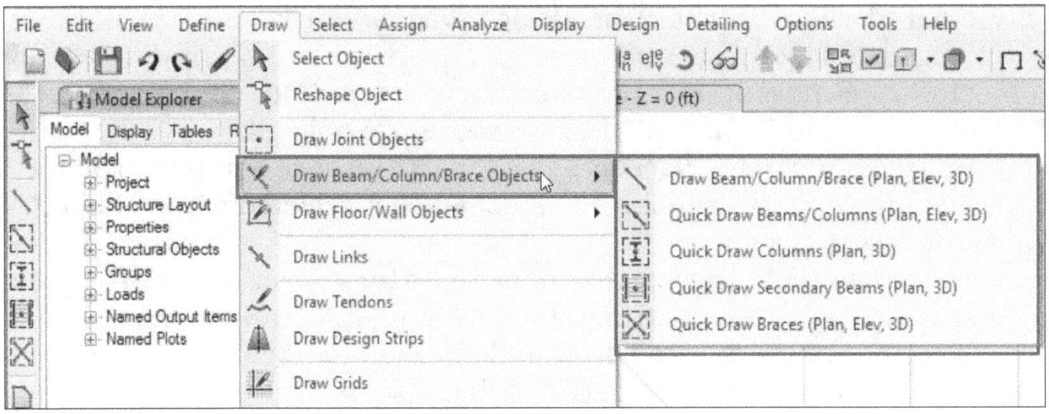

Figure-8. Draw Beam/Column/Brace Objects cascading menu

Various tools in this cascading menu are discussed next.

Drawing Beams/Columns/Braces in Plan, Elev, and 3D

The **Draw Beam/Column/Brace (Plan, Elev, 3D)** tool is used to draw beams, columns, and braces in plan, elevation, and 3D views. The procedure to use this tool is given next.

- Click on the **Draw Beam/Column/Brace (Plan, Elev, 3D)** tool from the **Draw Beam/Column/Brace Objects** cascading menu. The **Properties of Object** form will be displayed at the bottom in the **Model Explorer**; refer to Figure-9 and you will be asked to specify the grid line to create beams, columns, or braces.

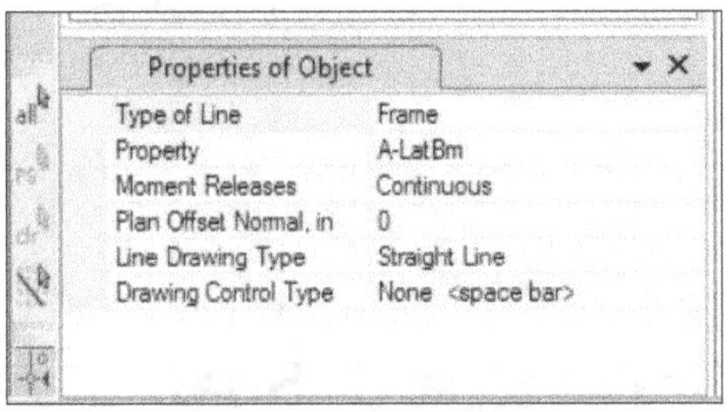

Figure-9. Properties of Object form

- Select the type of line from the **Type of Line** drop-down in the form.
- Click on the down button in the **Property** drop-down of the form and select desired section type from the list. The option selected in this drop-down will define the material and shape of beam/column/brace.
- Select desired moment release option from the **Moment Releases** drop-down. There are two options in this drop-down: **Continuous** and **Pinned**. Select the **Continuous** option if the beam is in continuous connection with the connecting beam. For analysis purpose, in this case loads in all 6 degrees of freedom are transferred to next element like a rigid body. Select the **Pinned** option if the beam/column/brace is pin joined with the connecting beam/column. It means rotational degree of freedom are locked for load transfer and only axial + shear loads will be transferred while no bending moment will be transferred.
- Select the offset from current plane as required if you have selected the **Straight Line** option from the **Line Drawing Type** drop-down.
- Select desired option from the **Line Drawing Type** drop-down. There are six options in the drop-down to draw different shapes: **Arc (3 Points)**, **Arc (Center & 2 Points)**, **Bezier**, **Multilinear**, **Spline**, and **Straight Line**. Based on your selection, you will be asked to specify desired points to create the shape.

- Specify the points in desired view to create beam/column/brace. Like, if you have selected the **Arc (3 Points)** option then specify the start point, end point, and middle point of arc to create the object; refer to Figure-10.

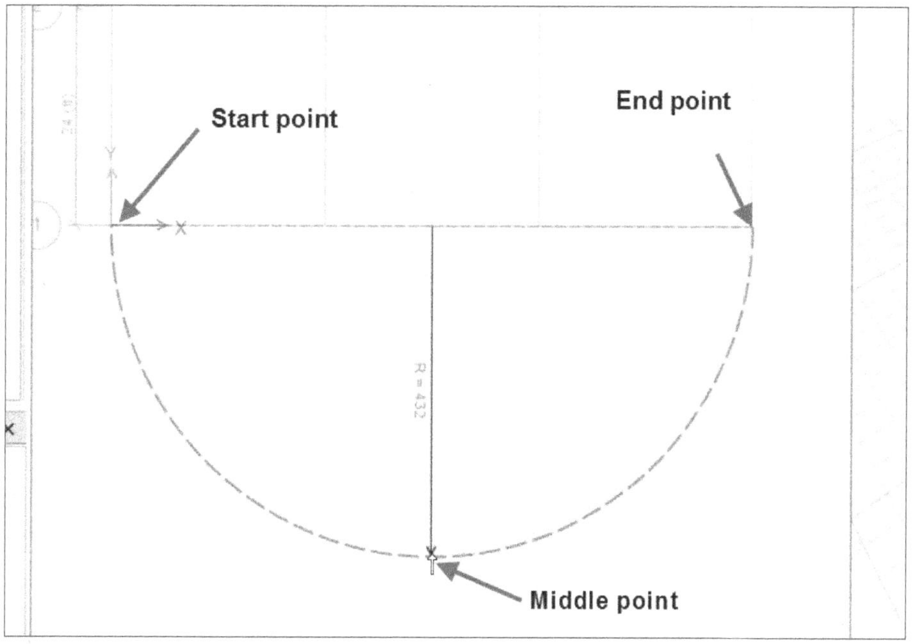

Figure-10. Creating 3 point arc beam

- Press **ESC** to exit the tool.

Note that by default, the object is created only on one story selected in Plan View. If you want the same object to be created on all stories then select the **All Stories** option from the **Story** drop-down at the bottom right corner of the software; refer to Figure-11 while creating the object.

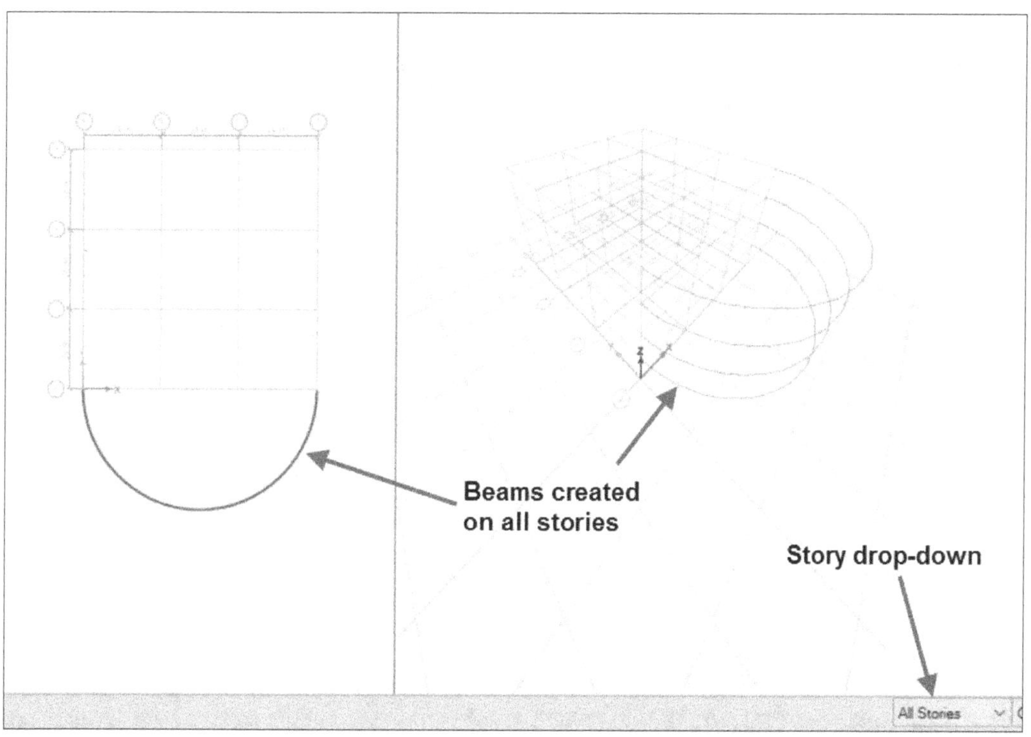

Figure-11. Beams created on all stories

Quick Draw Beams/Columns (Plan, Elev, 3D)

The **Quick Draw Beams/Columns (Plan, Elev, 3D)** tool is used to quickly draw beams/columns using grid lines and points of grids. The procedure to use this tool is given next.

- Click on the **Quick Draw Beams/Columns (Plan, Elev, 3D)** tool from the **Draw Beam/Column/Brace Objects** cascading menu in the **Draw** menu. The **Properties of Object** form will be displayed at the bottom in the **Model Explorer** as discussed earlier and you will be asked to select grids.
- Select the grids from desired viewport. The beams/columns will be created accordingly. The options in the **Properties of Object** form are same as discussed earlier.

Similarly, you can use the **Quick Draw Columns (Plan, 3D)** tool to create columns using Grids or grid points.

Quick Draw Secondary Beams (Plan, 3D)

The **Quick Draw Secondary Beams (Plan, 3D)** tool is used to draw secondary beams which transfer their loads on primary beams. The procedure to use this tool is given next.

- Click on the **Quick Draw Secondary Beams (Plan, 3D)** tool from the **Draw Beam/Column/Brace Objects** cascading menu in the **Draw** menu. You will be asked to specify the location of secondary beams and the **Properties of Object** form will be displayed at the bottom in the **Model Explorer**.
- Select desired option from the **Spacing** drop-down and enter respective value in the edit box below it. For example, if you have selected the **No. of Beams** option from the **Spacing** drop-down then specify the number of secondary beams required along the grid.
- Set desired orientation in the **Approx. Orientation** drop-down.
- Click at desired location in the grids to place the secondary beams; refer to Figure-12.

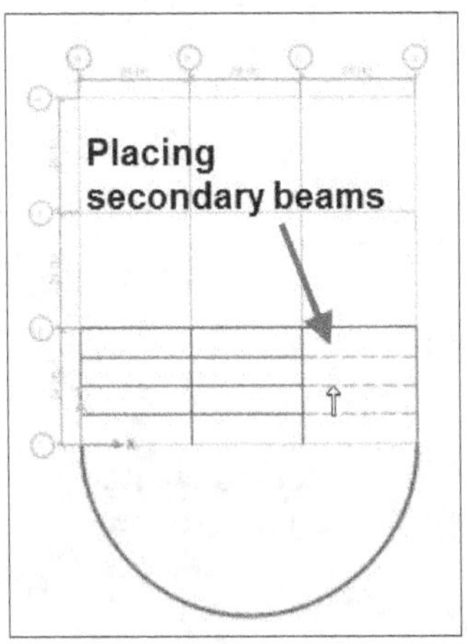

Figure-12. Placing secondary beams

Quick Draw Braces (Plan, Elev, 3D)

The **Quick Draw Braces (Plan, Elev, 3D)** tool is used to draw braces in the structure. Braces are used to increase the stability of structure in case of earth quakes or other dynamic loading cases. The procedure to draw braces in the structure is given next.

- Click on the **Quick Draw Braces (Plan, Elev, 3D)** tool from the **Draw Beam/Column/Brace Objects** cascading menu of **Draw** menu. You will be asked to specify the location of braces.
- Click in desired grid box to create braces; refer to Figure-13. You can modify the properties from the **Properties of Object** box as discussed earlier.

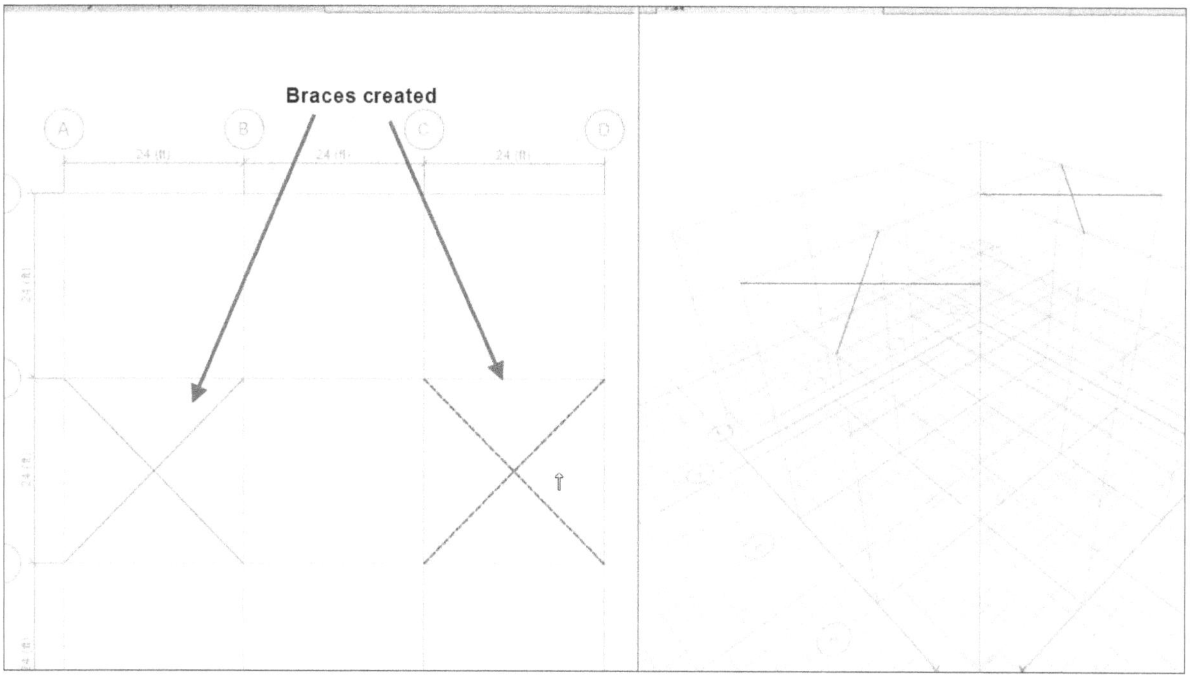

Figure-13. Braces created

DRAWING FLOOR AND WALL OBJECTS

The tools to draw floor and wall objects are available in the **Draw Floor/Wall Objects** cascading menu of the **Draw** menu; refer to Figure-14. Various tools of this cascading menu are discussed next.

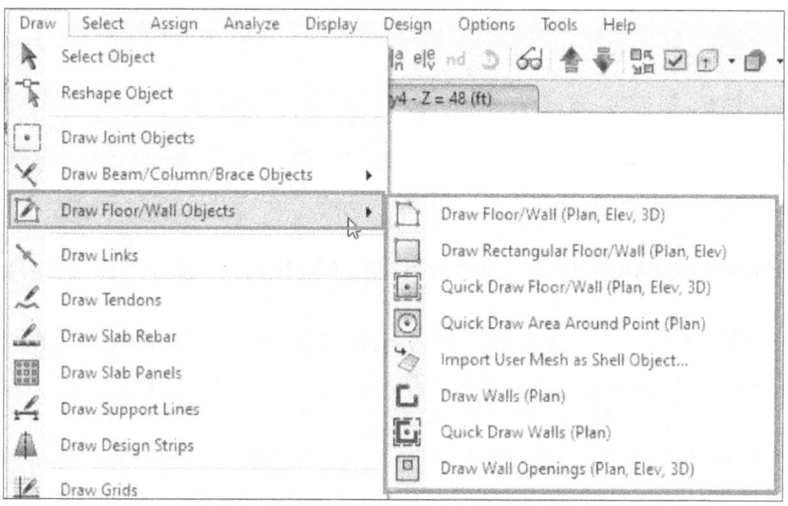

Figure-14. Draw Floor/Wall Objects cascading menu

Draw Floor/Wall (Plan, Elev, 3D) Tool

The **Draw Floor/Wall (Plan, Elev, 3D)** tool is used to draw floors and walls in plan, elevation, and 3D view by using the grid points. The procedure to use this tool is given next.

- Click on the **Draw Floor/Wall (Plan, Elev, 3D)** tool from the **Draw Floor/Wall Objects** cascading menu of the **Draw** menu. You will be asked to select the grid point and the **Properties of Object** form will be displayed.
- Select desired option from the **Property** drop-down. The options in this drop-down are Slab, Plank, Deck, and Opening.
- Select desired options from the **Edge Drawing Type** and **Drawing Control Type** drop-downs as discussed earlier.
- After specifying corner points of the floor/wall, press **ENTER** to create floor/wall; refer to Figure-15. Note that it is not necessary to specify 4 points, you can create a triangular floor or octagonal floor as well but you need to specify at least 3 non-coincident points.

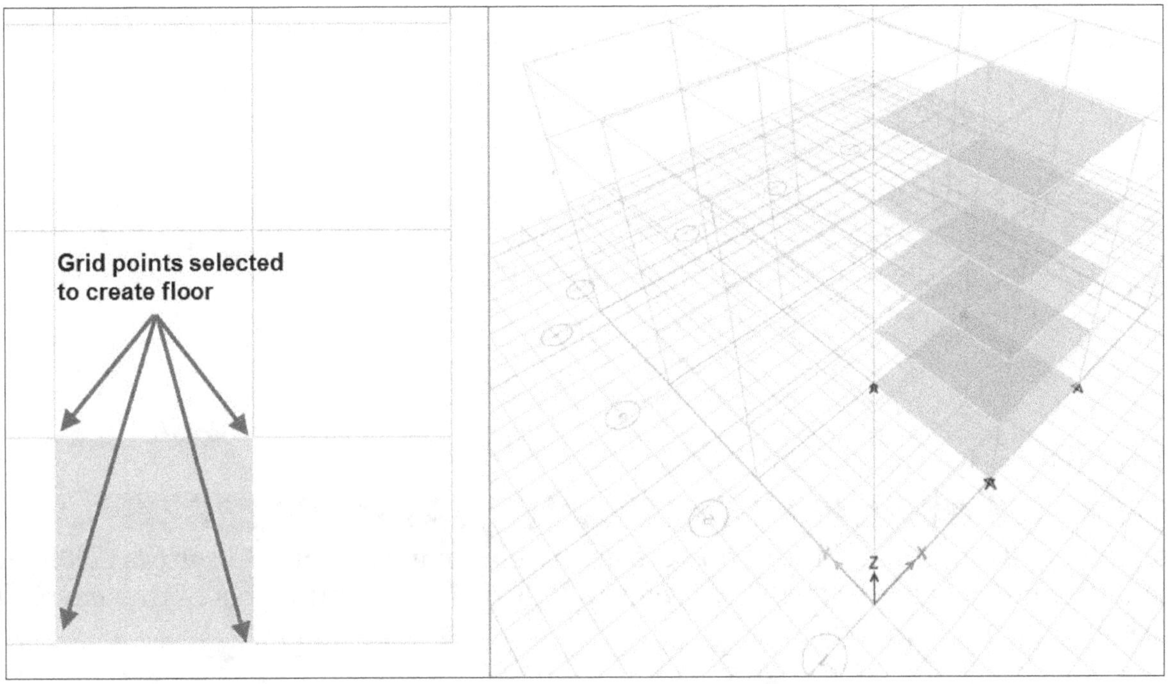

Figure-15. Grid points selected to create floor

Draw Rectangular Floor/Wall (Plan, Elev) Tool

The **Draw Rectangular Floor/Wall (Plan, Elev)** tool is used to draw rectangular floor and wall in plan and elevation views. The procedure to use this tool is given next.

- Click on the **Draw Rectangular Floor/Wall (Plan, Elev)** tool from the **Draw Floor/Wall Objects** cascading menu of the **Draw** menu. You will be asked to select the grid point and the **Properties of Object** form will be displayed.
- Click at desired grid point and drag the cursor while holding LMB to draw floor/wall; refer to Figure-16. If you want to specify dimensions of the floor/wall without dragging the mouse then enter desired value in **X Dimension** and **Y Dimension** edit boxes in the form.

Note that floors/walls can be created without selecting grid points from graphics area. You can create the objects by clicking anywhere in the graphics area.

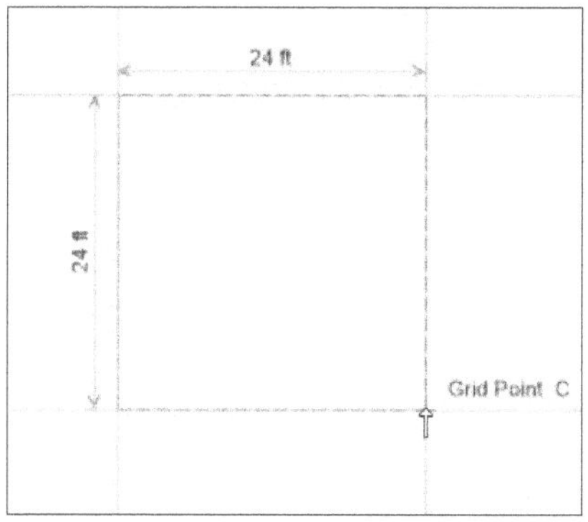

Figure-16. Creating rectangular floor/wall

Quick Draw Floor/Wall (Plan, Elev) Tool

The **Quick Draw Floor/Wall (Plan, Elev)** tool is used to quickly draw floors/walls in plan and elevation views using the grid blocks. The procedure to use this tool is given next.

- Click on the **Quick Draw Floor/Wall (Plan, Elev)** tool from the **Draw Floor/ Wall Objects** cascading menu of the **Draw** menu. You will be asked to select the grid block and the **Properties of Object** form will be displayed.
- Select desired option from the **Property** drop-down. You will learn about the options in this drop-down later in the book.
- Click at desired location in the grid block. The floor/wall will be created; refer to Figure-17.

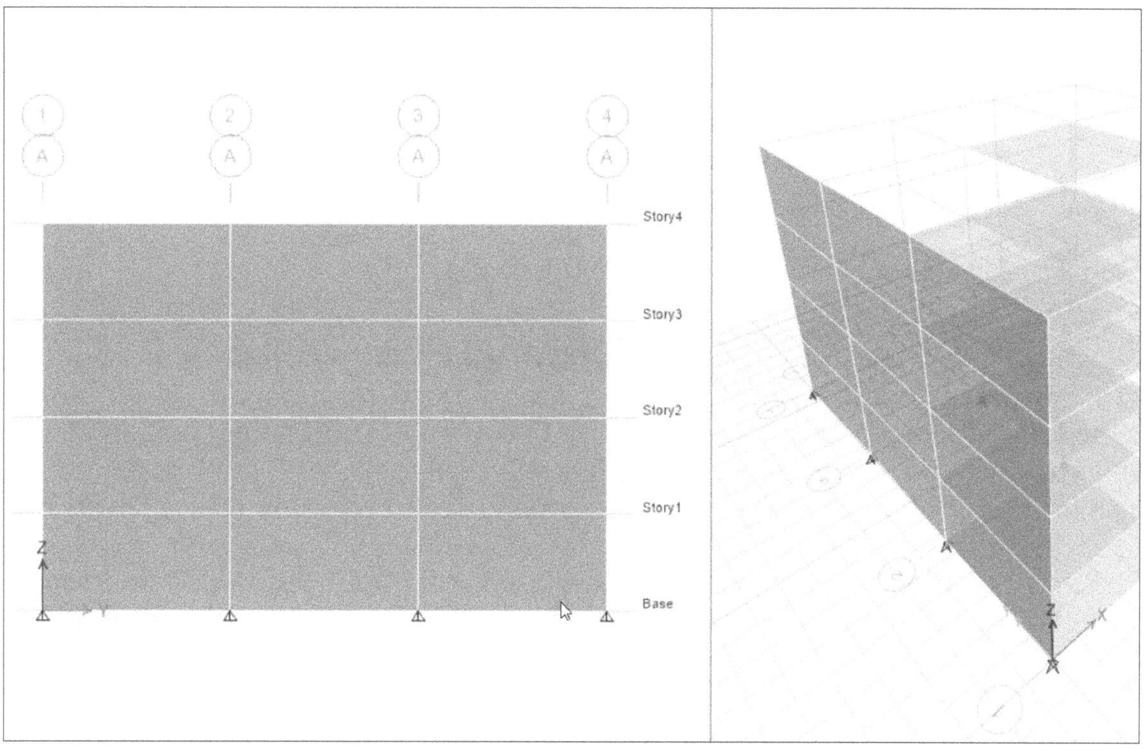

Figure-17. Walls created

- Press **ESC** to exit the tool.

Quick Draw Area Around Point (Plan) Tool

The **Quick Draw Area Around Point (Plan)** tool is used to create floor/slab of specified size around selected points in the plan view. The procedure to use this tool is given next.

- Click on the **Quick Draw Area Around Point (Plan)** tool from the **Draw Floor/ Wall Objects** cascading menu of the **Draw** menu. The **Properties of Object** form will be displayed; refer to Figure-18.

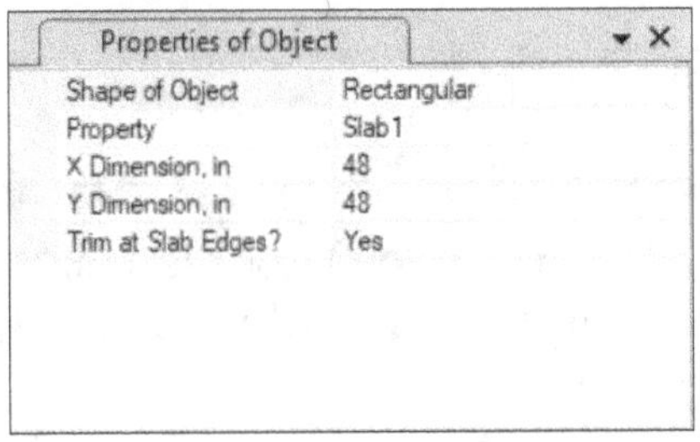

Figure-18. Properties of Object form

- Set desired size and parameters for the slab in the fields and click at desired location in the plan view to place the floor/slab; refer to Figure-19.

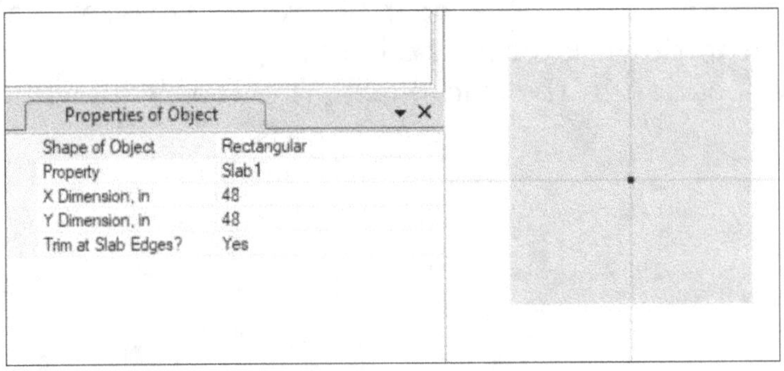

Figure-19. Slab placed around point

- After creating slabs, press **ESC** to exit the tool.

Importing User Mesh as Shell Object

The **Import User Mesh as Shell Object** tool is used to import mesh data of a dxf file. Shell objects are useful for simulating floor, wall, and bridge deck systems; 3D curved surfaces; and components within structural members, such as the web, and flanges of a W-Section in the model. The procedure to use this tool is given next.

- Click on the **Import User Mesh as Shell Object** tool from the **Draw Floor/Wall Objects** cascading menu of the **Draw** menu. The **Import DXF File** dialog box will be displayed; refer to Figure-20.
- Select desired DXF file and click on the **Open** button. The **Import User Mesh as Shell Object** dialog box will be displayed; refer to Figure-21.
- Set desired location, scale, and rotation parameters for mesh to be imported.
- Select desired option from the **Slab Property** drop-down to define the type of slab in which the mesh will be imported.
- After setting desired parameters, click on the **OK** button.

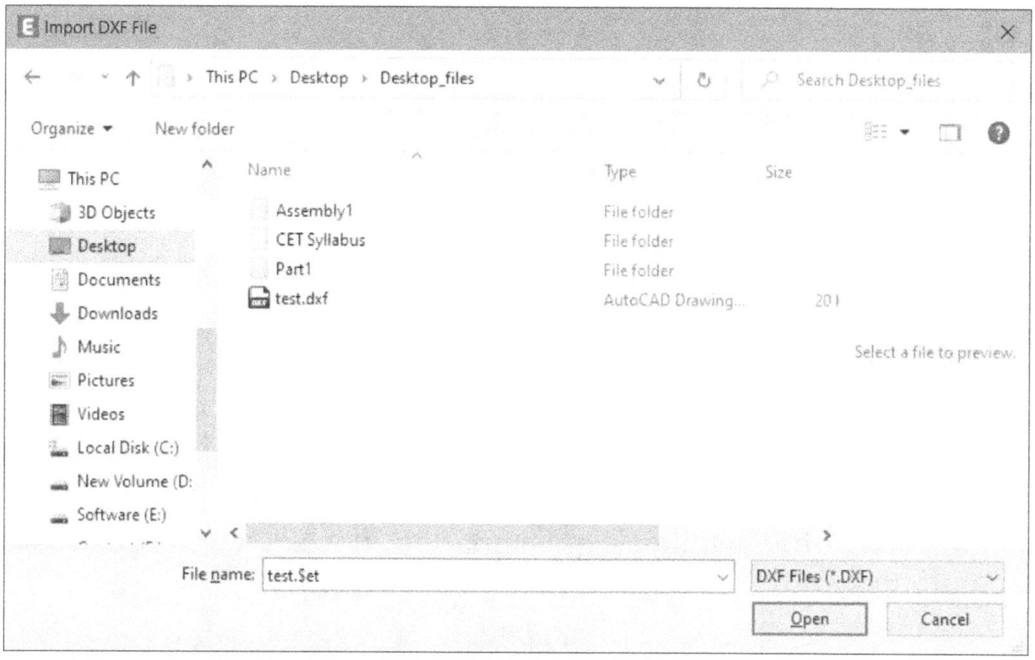

Figure-20. Import DXF File dialog box

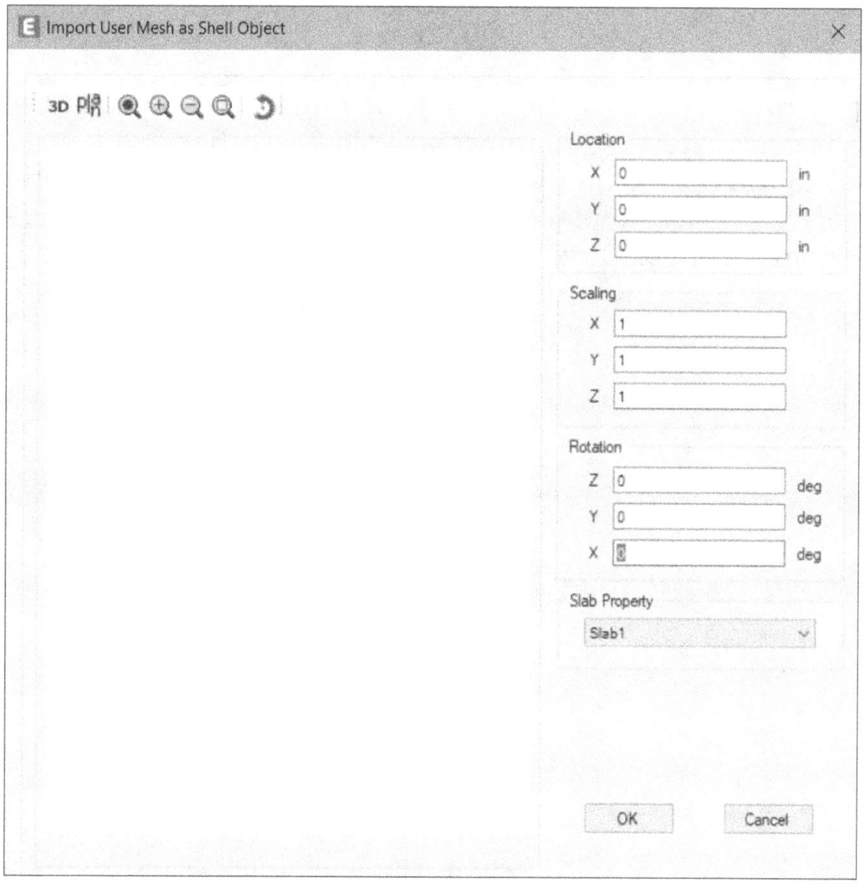

Figure-21. Import User Mesh as Shell Object dialog box

Draw Walls (Plan) Tool

The **Draw Walls (Plan)** tool is used to draw walls in plan view. The procedure to use this tool is given next.

- Click on the **Draw Walls (Plan)** tool from the **Draw Floor/Wall Objects** cascading menu of the **Draw** menu. You will be asked to specify start and end points of the wall. Also, the **Properties of Object** form will be displayed; refer to Figure-22.

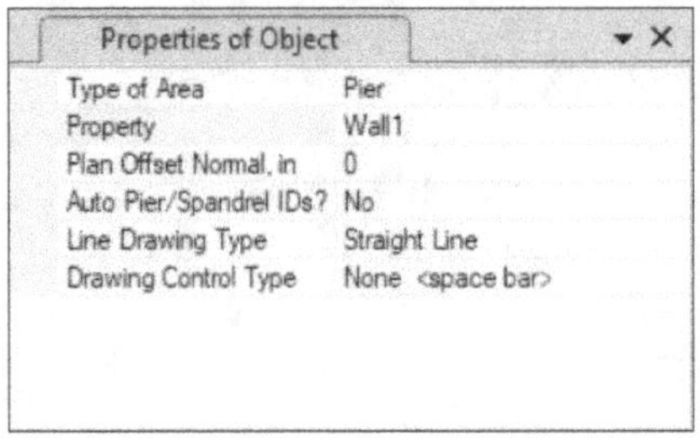

Figure-22. Properties of Object form

- Select desired option from the **Type of Area** drop-down. There are two options in this drop-down: **Pier** and **Spandrel**. Select the **Pier** option to create the most common type of permanent wall. Select the **Spandrel** option to create exterior curtain walls outside the beams of multistory buildings. Piers are vertical segments of walls that act like columns for analysis. Similarly, Spandrels are horizontal segments of walls that act like deep beams within walls. For more details on Pier and Spandrel, scan the QR code below.

- Set desired parameters in the fields of **Properties of Object** form and click on the model to specify starting point of the wall. The other end point of the wall will get attached to the cursor and you will be asked to specify the other point.
- Click at desired location to place the other end point of the wall. The wall will be created; refer to Figure-23 and you will be asked to specify the other end point.

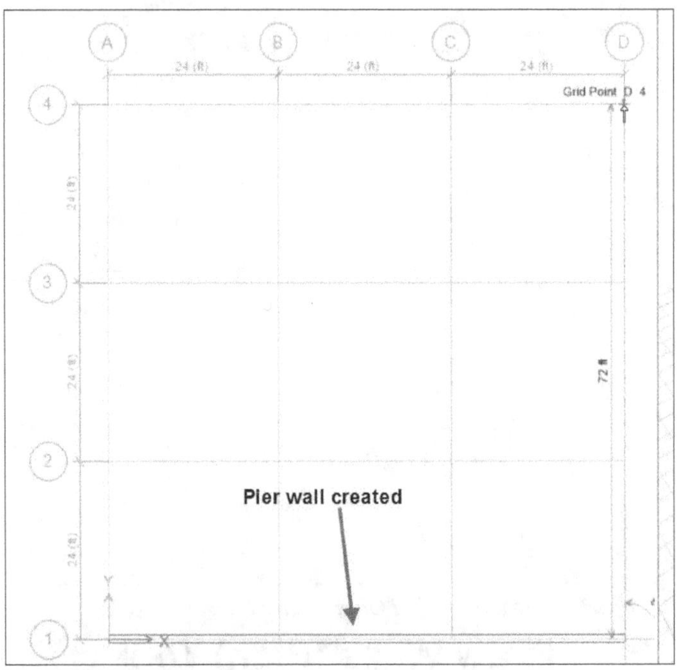

Figure-23. Creating wall in Plan view

- Click to specify the next end point or press **ESC** to exit.

Quick Draw Walls (Plan) Tool

The **Quick Draw Walls (Plan)** tool is used to quickly draw walls based on selected line elements in the graphics area. The procedure to use this tool is given next.

- Click on the **Quick Draw Walls (Plan)** tool from the **Draw Floor/Wall Objects** cascading menu of the **Draw** menu. The **Properties of Object** form will be displayed and you will be asked to select line elements for creating walls.
- Set desired parameters in the form and click in the plan view at desired line elements to create walls.

Creating Wall Opening

The **Draw Wall Openings (Plan, Elev, 3D)** tool is used to create opening in walls while in Plan, Elevation, or 3D view. A wall opening is generally created for architectural purpose. The procedure to use this tool is given next.

- Switch to 3-D view or Elevation view as they are easier for creating wall opening. To do so, click on the **Set Elevation View** tool from the **View** menu. The **Set Elevation View** dialog box will be displayed; refer to Figure-24.

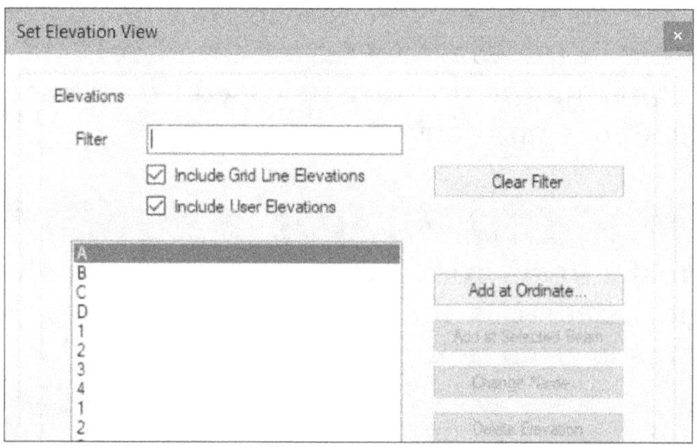

Figure-24. Set Elevation View dialog box

- Select the side based on grid for which you want to view elevation view and click on the **OK** button.
- Click on the **Draw Wall Openings (Plan, Elev, 3D)** tool from the **Draw Floor/Wall Objects** cascading menu in the **Draw** menu. The **Properties of Object** box will be displayed as shown in Figure-25.

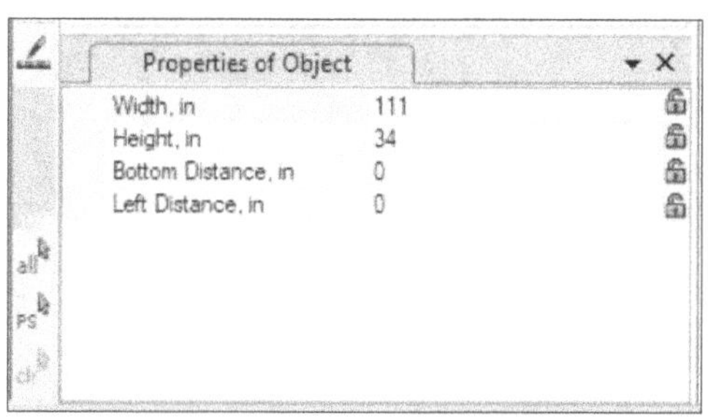

Figure-25. Properties for wall opening

- Specify desired values of Width and Height of opening in respective edit boxes of the **Properties of Object** box and click on the lock buttons to lock dimensions.
- Click at desired wall in which you want to create opening. Dimensions to place opening will be displayed; refer to Figure-26.

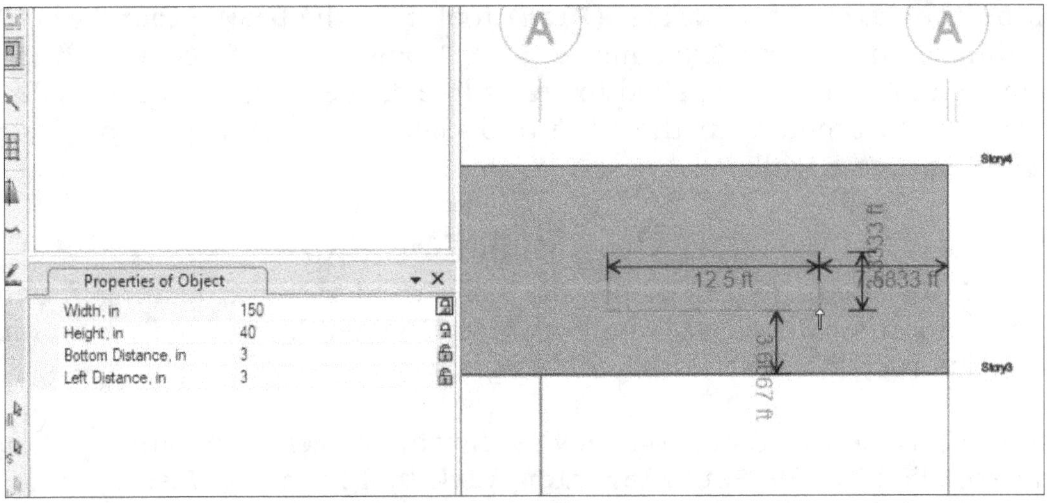

Figure-26. Dimensions for placing wall opening

- Click at desired location to place the opening or specify the distance values for locating wall opening in the **Bottom Distance** and **Left Distance** edit boxes (in inches). If you are placing the opening using edit boxes then make sure to lock them and then click once on the wall.

DRAWING LINKS

Links are used to transfer loads from one point to other. In physical world, you can assume links as rigid wires/rods that connect two objects for load transfer. The procedure to draw links is given next.

- Click on the **Draw Links** tool from the **Draw** menu. You will be asked to specify the points to be linked.
- Click at desired locations to connect them by link; refer to Figure-27.

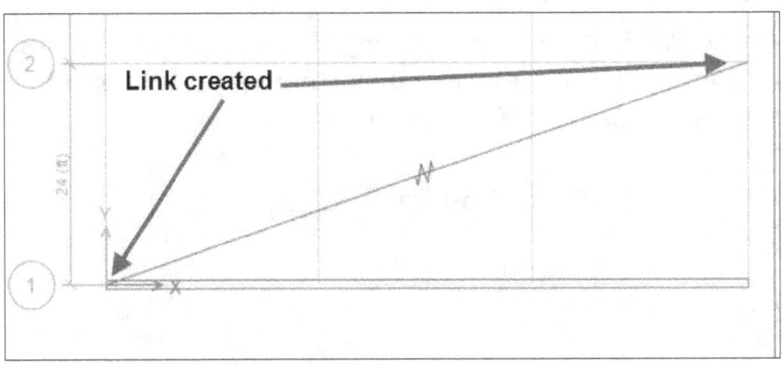

Figure-27. Link created

You will learn about assigning properties to these links later in the book.

DRAWING TENDONS

Tendons are used to simulate the effect of prestressing and post tensioning on the objects like beams, columns, frames etc. The procedure to draw tendons is given next.

- Click on the **Draw Tendons** tool from the **Draw** menu. You will be asked to specify start point for tendon and properties will be displayed as shown in Figure-28.

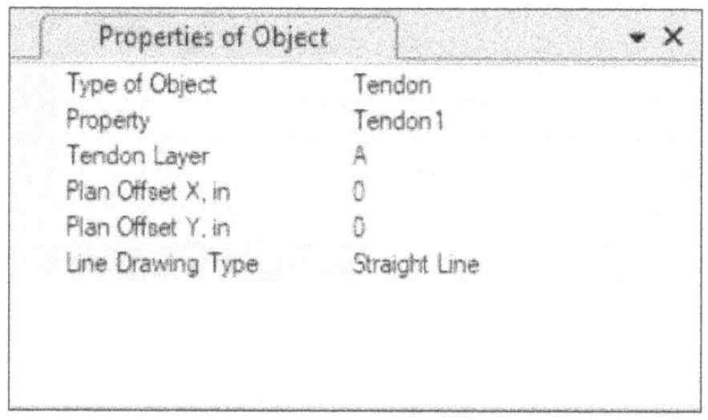

Figure-28. Properties for tendons

- You can create tendons in two shapes, using straight line or using arc. Select desired option form the **Line Drawing Type** drop-down in the **Properties of Object** box.
- Click at desired locations to draw tendons. Press **ESC** to exit the tool. You will learn about other options later in the book.

DRAWING SLAB REBAR

The **Draw Slab Rebar** tool is used to lay rebar in slabs for increasing strength of slab and keep concrete stable after cracks. The procedure to use this tool is given next.

- Click on the **Draw Slab Rebar** tool from the **Draw** menu. You will be asked to specify starting point for rebars and the properties will be displayed as shown in Figure-29.

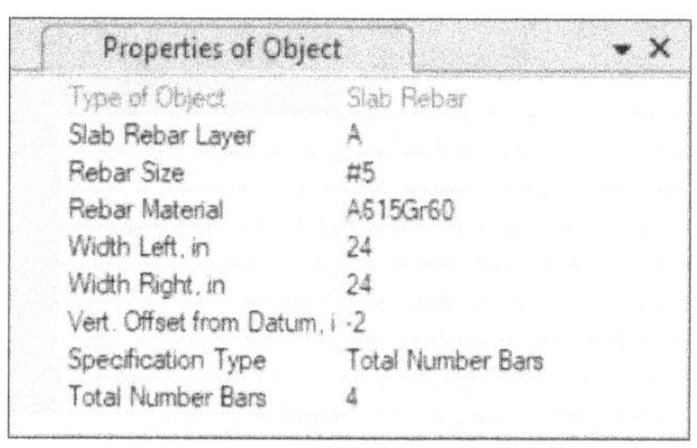

Figure-29. Properties for slab rebar

- Select desired layer in which you want to create rebar from the **Slab Rebar Layer** drop-down in the **Properties of Object** box.
- Select desired option from the **Rebar Size** drop-down to define size of rebar.
- Select the material for rebars from the **Rebar Material** drop-down in **Properties of Object** box.
- Similarly, set the number of rebars and other parameters in the **Properties of Object** box.
- Click at desired location on slab to create rebar. Press **ESC** to exit the tool.

DRAWING SLAB PANELS

The **Draw Slab Panels** tool is used to create slab panels by clicking at specified points. The procedure to use this tool is given next.

- Click on the **Draw Slab Panels** tool from the **Draw** menu. You will be asked to specify the corner points of slab panel.
- Click in the plan view at desired locations; refer to Figure-30 and press **ENTER** to create the slab panel.

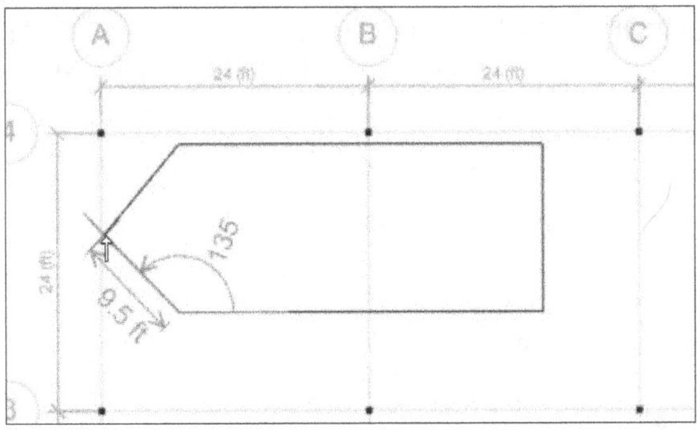

Figure-30. Drawing slab

DRAWING SUPPORT LINES

The **Draw Support Lines** tool is used to create supporting lines for later creating design strips, slabs, and other elements.

- Click on the **Draw Support Lines** tool from the **Draw** menu. You will be asked to select points for creating support lines.
- Click at desired locations to create multiple support lines; refer to Figure-31 and press **ENTER** to exit the tool.

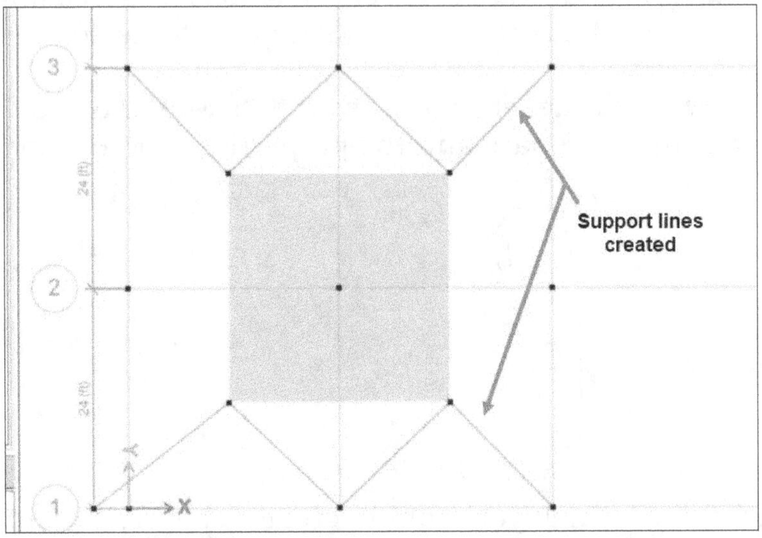

Figure-31. Support lines created

DRAWING DESIGN STRIPS

Design Strips are used to form regions in a slab where design cuts will be automatically generated and flexural reinforcing bars will be designed. If the moment varies along the length of design strip then the individual design cuts will show varying moment at each section. The results for the entire Design Strip is governed by the maximum moment demand at an individual design cut. Note that the same number and size of reinforcing bars are used for the entire Design Strip. The procedure to draw design strip is given next.

- Click on the **Draw Design Strips** tool from the **Draw** menu. You will be asked to specify the location points for strip and options related to design strip will be displayed in the **Properties of Object** form.
- Specify desired parameters in the **Properties of Object** form and click at desired points to create design strip; refer to Figure-32.

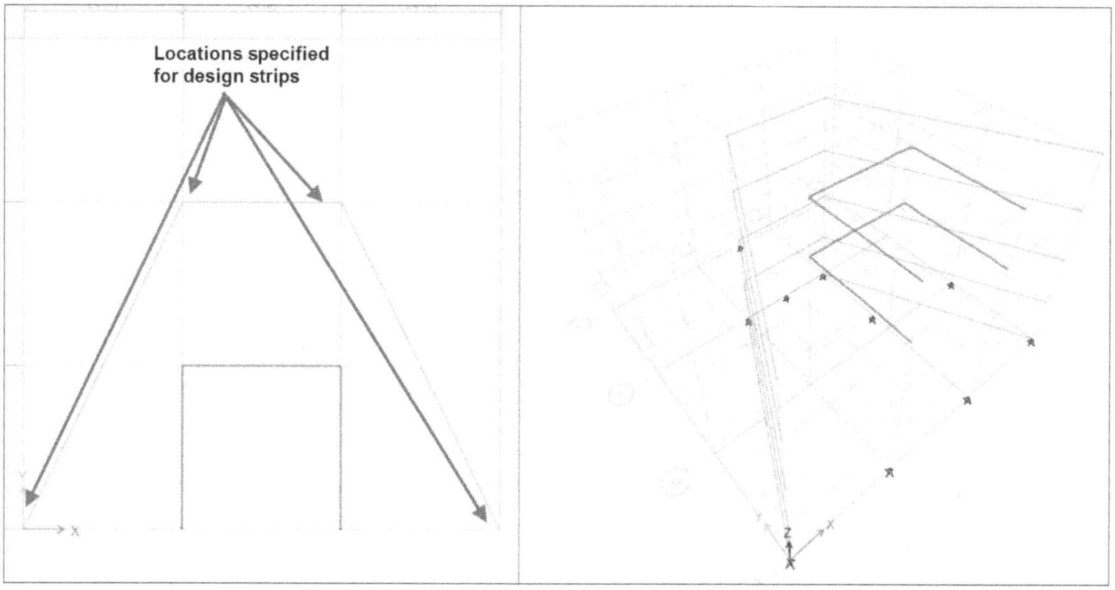

Figure-32. Design strips created

- Press **ESC** to exit the tool.

DRAWING DIMENSION LINES

Dimension lines are used to annotate distance between two points on the drawing or size parameters of any geometric object. The procedure to draw dimension lines is given next.

- Click on the **Draw Dimension Lines** tool from the **Draw** menu. You will be asked to specify the points to measured.
- Click at desired locations to specify the start and end points of the dimension line. A dashed line will get attached to the cursor; refer to Figure-33.

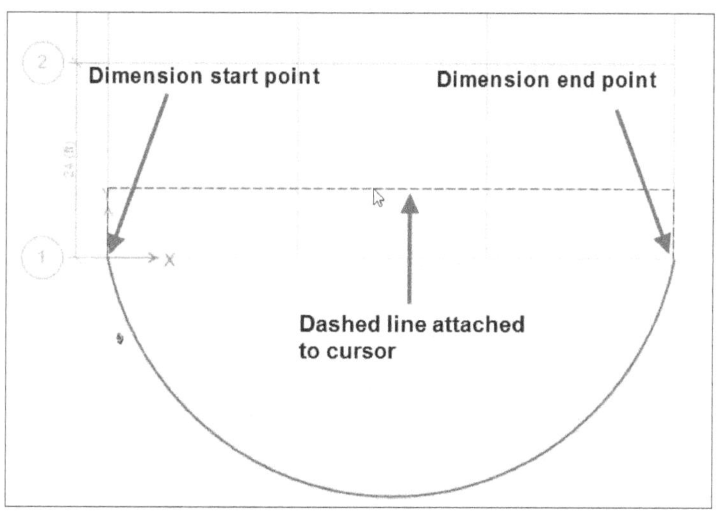

Figure-33. Dashed line for placing dimension

- Click at desired location to place the dimension.

DRAWING REFERENCE POINTS

The **Draw Reference Points** tool is used to draw reference points for creating other objects. These reference points can be used as snap points for creating various structural objects in the model. The procedure to use this tool is given next.

- Click on the **Draw Reference Points** tool from the **Draw** menu. You will be asked to specify the location of reference point and the **Properties of Object** form will be displayed.
- Specify desired offset values in the X, Y, and Z direction if you want the reference point to be created at specified distance in X, Y, and Z axis from selected point. Select desired grid system from the **Grid System** drop-down in the form.
- Click at desired locations in the viewport to create the reference points.

SETTING VIEWS IN VIEWPORTS

Using the tools in toolbar, you can set different views in the viewports as required. The procedures to set different views are given next.

Setting 3D View

- Click in desired viewport while no tool is active.
- Click on **Set Default 3D View** tool from the top toolbar or press **CTRL+SHIFT+F3** shortcut key or click on the **Set 3D View** tool from **View** menu to set the current viewport to 3D view.

Setting Plan View

- Click on the **Set Plan View** tool from the top toolbar or press **CTRL+SHIFT+F1** shortcut key or select **Set Plan View** tool from **View** menu to set the current viewport to plan view. The **Select Plan View** dialog box will be displayed; refer to Figure-34.

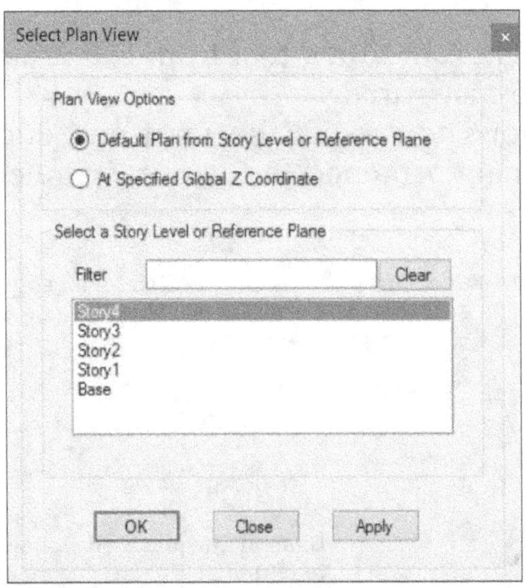

Figure-34. Select Plan View dialog box

- Select desired story from the list to set it in plan view as working plane.
- If you want to set a user defined height as working plane location then select the **At Specified Global Z Coordinate** radio button from the top area of the dialog box. The **Global Z Coordinate** edit box will be displayed below it.
- Specify desired height value in the edit box using the Global coordinate system.
- After setting desired options, click on the **OK** button.

Setting Elevation View

- Click on the **Set Elevation View** tool from the top toolbar or press **CTRL+SHIFT+F2** shortcut key or select the **Set Elevation View** tool from **View** menu after clicking in the viewport whose view is to be changed. The **Set Elevation View** dialog box will be displayed; refer to Figure-35.

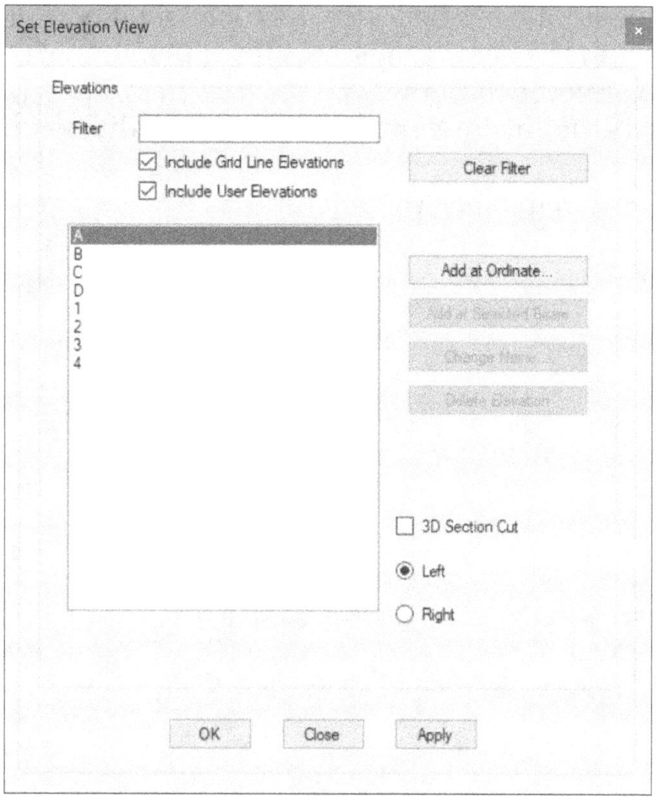

Figure-35. Set Elevation View dialog box

- Select desired option from the list.
- If you want to create elevation view at desired X or Y ordinate then click on the **Add at Ordinate** button from the dialog box. The **Add Elevation View At Ordinate** dialog box will be displayed; refer to Figure-36.

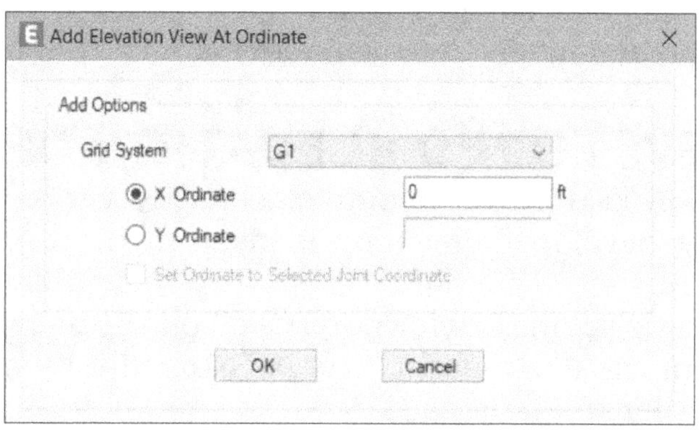

Figure-36. Add Elevation View At Ordinate dialog box

- Select desired radio button and specify the ordinate value. Click on the **OK** button to add the ordinate to current list of elevations.
- Click on the **OK** button from the **Set Elevation View** dialog box to set the current viewport to elevation view.

DRAWING REFERENCE PLANES

The reference planes are used to provide horizontal planes/lines that can be used to snap to when drawing objects in elevation views. The procedure to draw reference plane is given next.

- Set a viewport to desired elevation view as discussed earlier.
- Click on the **Draw Reference Planes** tool from **Draw** menu. You will be asked to specify location for reference plane.
- Set desired offset value in the **Vertical Offset Z** edit box of the **Properties of Object** form.
- Click at desired location to place the reference. The reference plane will be created; refer to Figure-37.

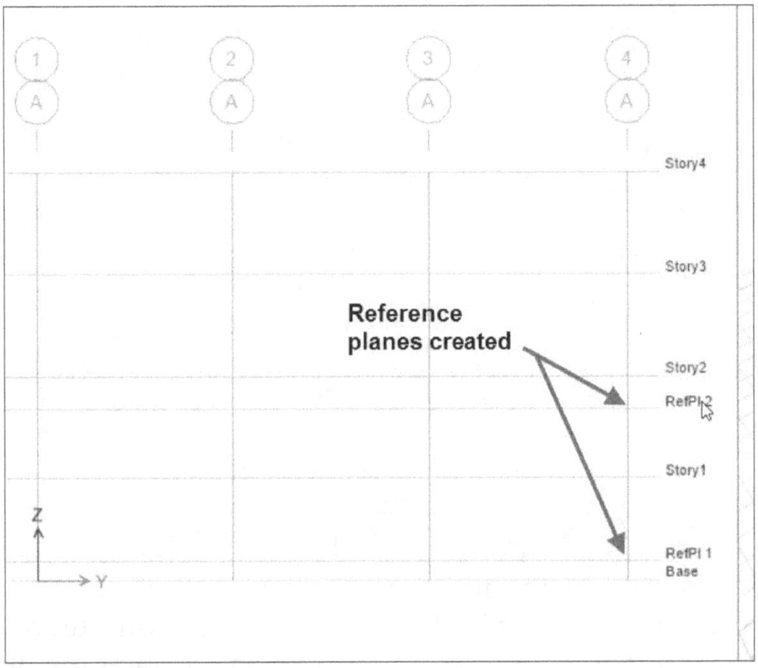

Figure-37. Reference plane created

The **Draw Section Cut** tool of **Draw** menu is used to create section view of the deformed model showing results of analysis. You will learn about this tool later in the book.

DRAW DEVELOPED ELEVATION DEFINITION

The developed elevation is used to display selected elevation in unfolded position. The procedure to draw developed elevation is given next.

- Make sure plan view is set in the viewport. Click on the **Draw Developed Elevation Definition** tool from the **Draw** menu. The **Developed Elevation Name** dialog box will be displayed; refer to Figure-38.

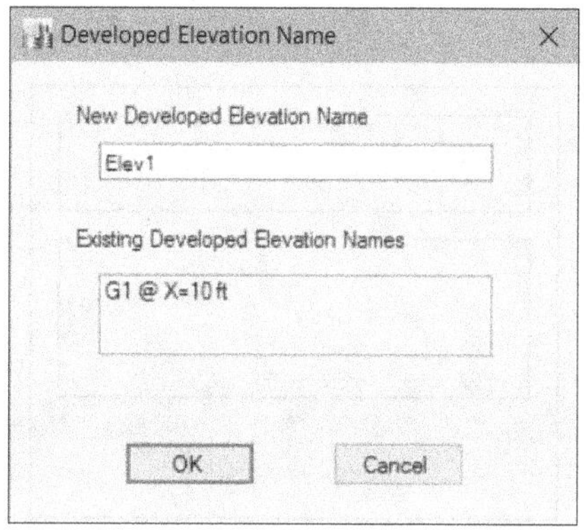

Figure-38. Developed Elevation Name dialog box

- Specify desired name in the edit box and click on the **OK** button. You will be asked to create a multiline over all the walls/beams/objects to be included in the developed elevation.
- Create a multiline as required; refer to Figure-39.

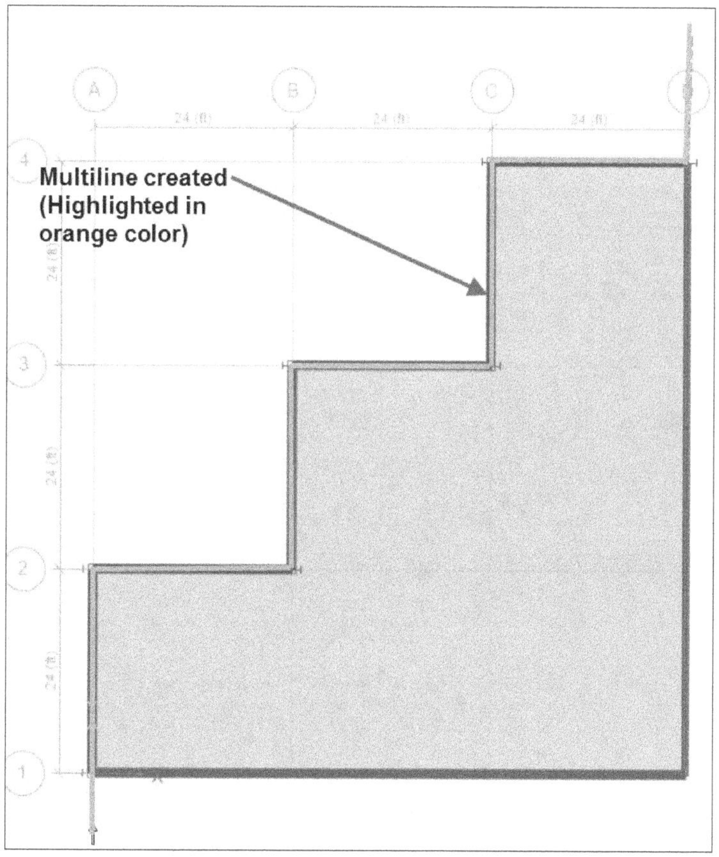

Figure-39. Multiline created for developed elevation

- Press **ENTER** after drawing the multiline for developed elevation. The developed elevation will be displayed in the viewport; refer to Figure-40.

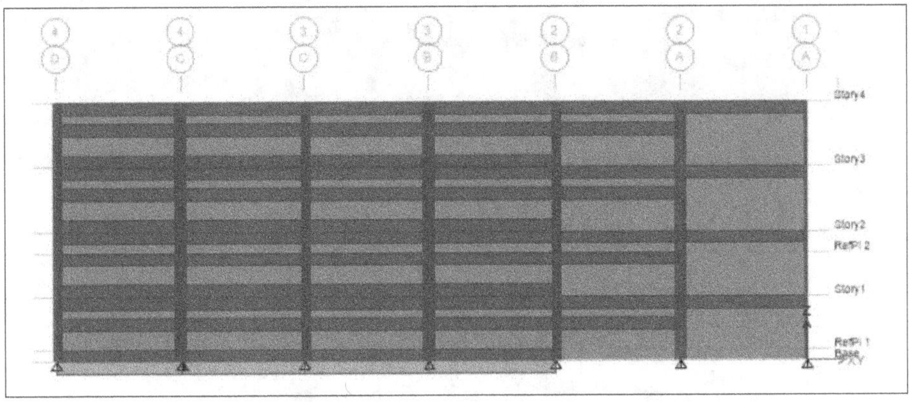

Figure-40. Developed elevation created

DRAWING WALL STACK

Wall stack is a complex combination of walls. Make sure plan view is active in the viewport as it makes easy to place wall stacks in the building. The procedure to draw wall stack in ETABS is given next.

- Set the viewport to plan view. Click on the **Draw Wall Stacks (Plan, Elev, 3D)** tool from the **Draw** menu. The **New Wall Stack** dialog box will be displayed; refer to Figure-41.

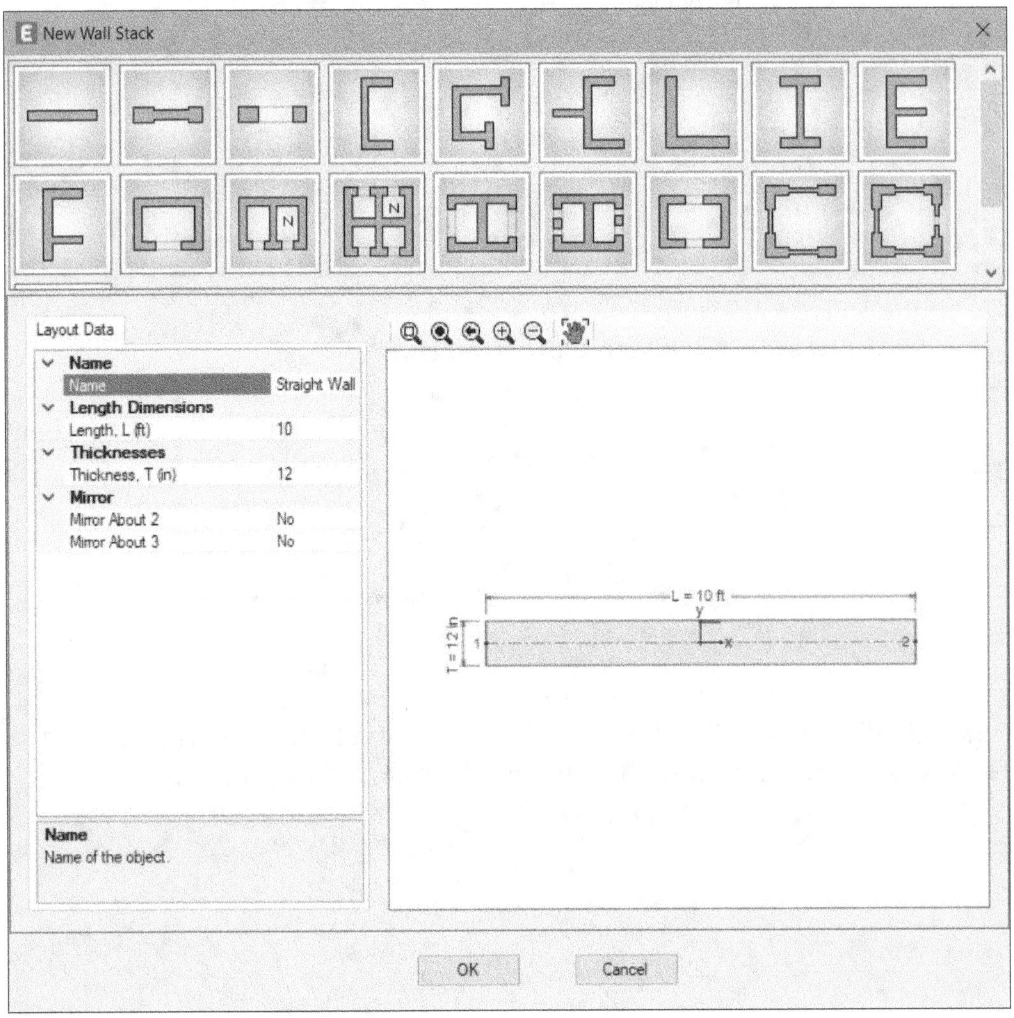

Figure-41. New Wall Stack dialog box

- Select desired button from the palette to select a shape of walls. The parameters for selected shape will be displayed along with preview in the dialog box.
- Modify the parameters in **Layout Data** tab of the dialog box and click on the **OK** button. The selected wall shape will get attached to cursor and you will be asked to specify the insertion point for the wall stack.
- Click at desired locations in the plan view to place wall stacks; refer to Figure-42.

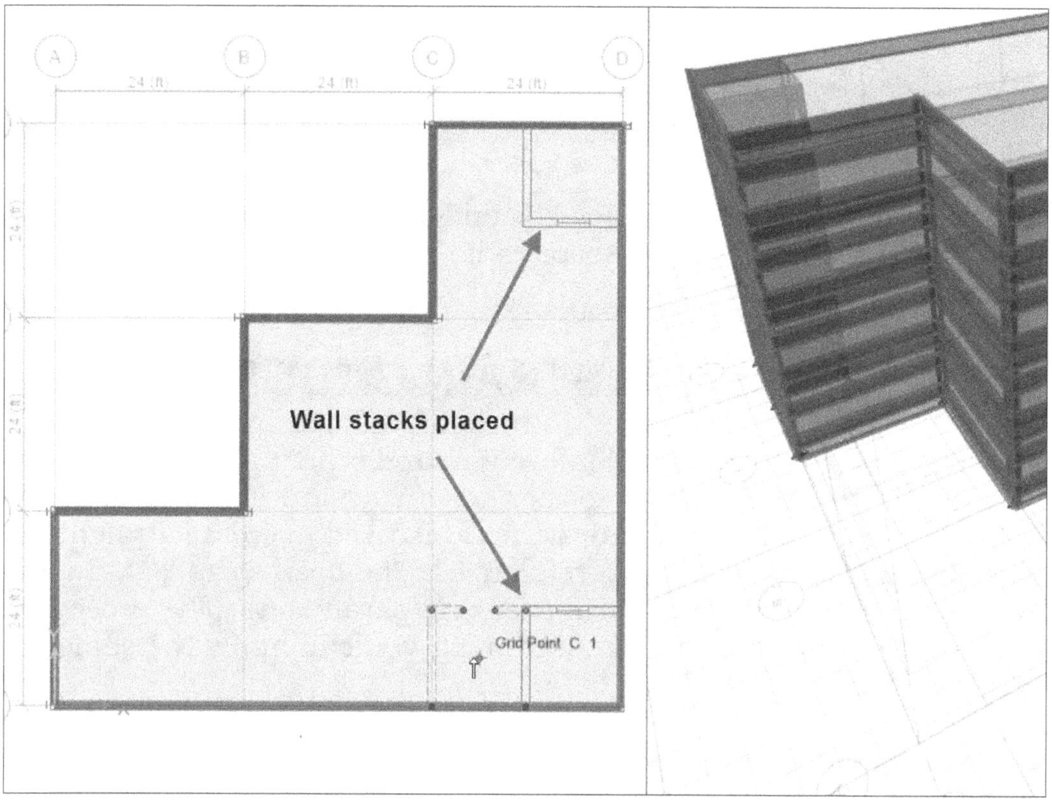

Figure-42. Drawing wall stacks

- Note that you can change the orientation of wall stack by specifying desired angle value in the **Angle, deg** edit box of the **Properties of Object** form. To define the limit of wall stacks vertically, select desired options in the **Top Story** and **Bottom Story** drop-downs.
- Press **ESC** to exit the tool.

AUTO DRAW CLADDING

The **Auto Draw Cladding** tool in the **Draw** menu is used to draw cladding based on specified parameters automatically. Cladding is used to provide thermal insulation and weather protection. The procedure to use the **Auto Draw Cladding** tool is given next.

- Click on the **Auto Draw Cladding** tool from the **Draw** menu. The **Cladding Options** dialog box will be displayed; refer to Figure-43.

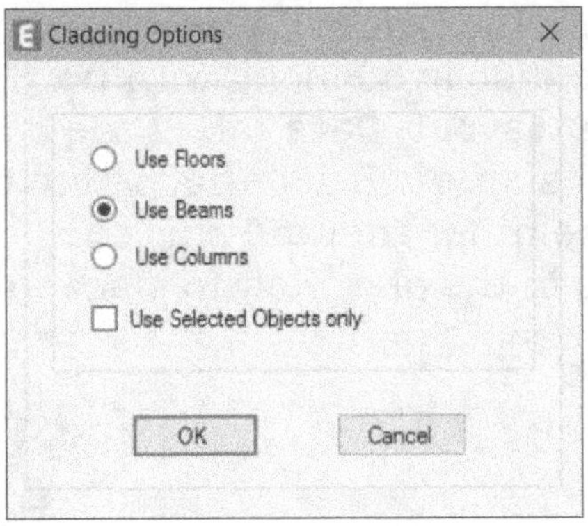

Figure-43. Cladding Options dialog box

- Select desired option from the dialog box and click on the **OK** button (We have used beams to create cladding so **Use Beams** radio button is selected). The cladding will be created automatically.

DRAWING STRAIN GAUGES

As the name suggests, gauge is used to measure something. In case of buildings, gauge converts force, pressure, tension, weight, and other parameters into electrical resistance which can be measured to monitor structural integrity of the building. The **Draw Line Strain Gauge** tool in **Draw** menu is used to create straight line gauge to measure parameters between two references. The **Draw Quad Strain Gauge** tool is used to create polygonal gauge for measuring parameters. The procedure to use **Draw Quad Strain Gauge** tool is discussed next, you can apply the same procedure to **Draw Line Strain Gauge** tool.

- Click on the **Draw Quad Strain Gauge** tool from the **Draw** menu. You will be asked to specify start point of gauge.
- Click at desired locations to create polygon; refer to Figure-44 and press **ENTER**. The strain gauge will be created; refer to Figure-45. Note that you need to specify at least four corners of polygon to create gauge. To know more about strain gauges, scan the QR given next.

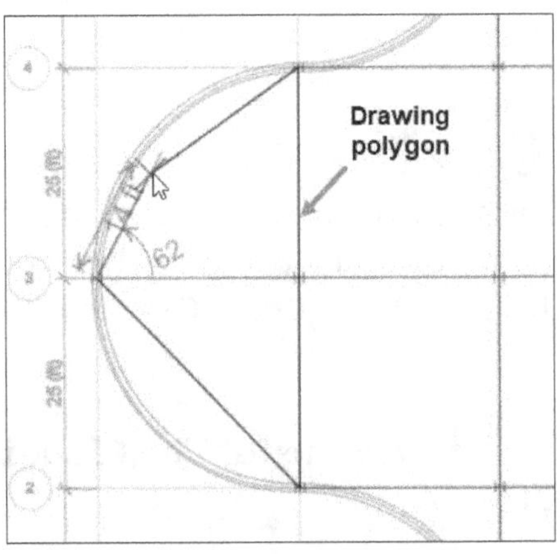

Figure-44. Polygon for strain gauge

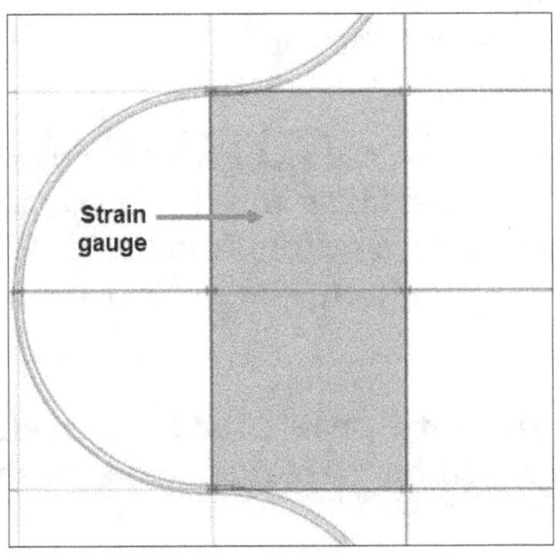

Figure-45. Quad strain gauge created

CREATING WORKING PLANES

The working planes are used to create reference planes for drawing other objects. The procedure to create working planes is given next.

- Click on the **Working Planes** tool from the **Draw** menu. The **Working Planes** dialog box will be displayed; refer to Figure-46.

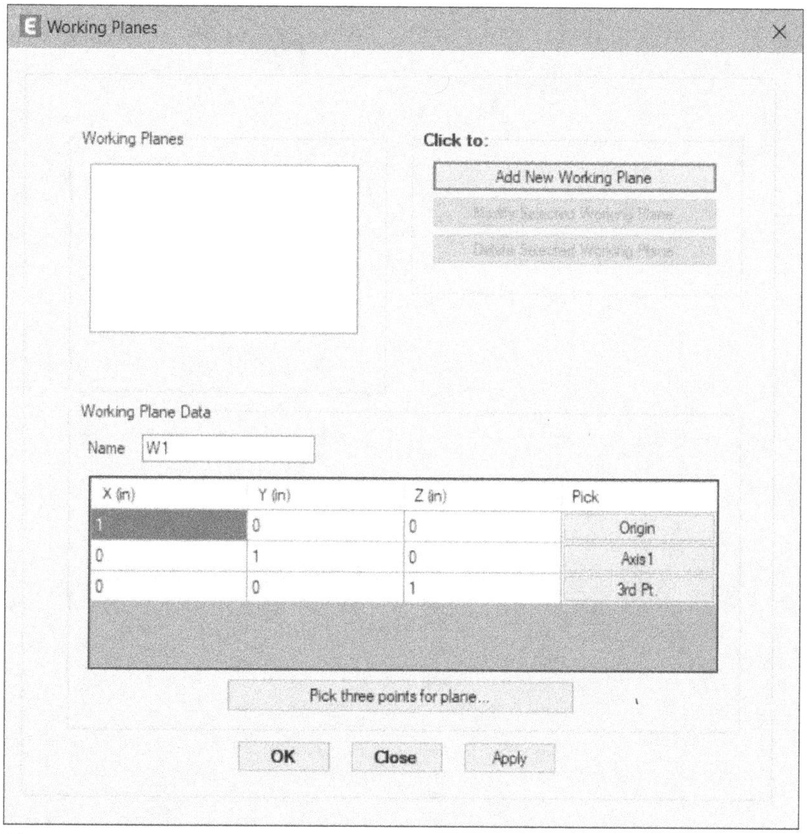

Figure-46. Working Planes dialog box

- Click on the **Origin** button from table to pick origin location for plane from graphics area or specify the coordinates in X, Y, and Z fields of table for origin.
- Similarly, set plane data for axis and third point.
- You can also use the **Pick three points for plane** button from the dialog box to create plane by selecting three points.
- After setting parameters, click on the **Add New Working Plane** button from the dialog box. The new plane will be added in the list. You can modify the new created plane by using the **Modify Selected Working Plane** button or delete the plane using **Delete Selected Working Plane** button.
- Click on the **OK** button from the dialog box to create the plane.

PRACTICAL

Create the structure as shown in Figure-47.

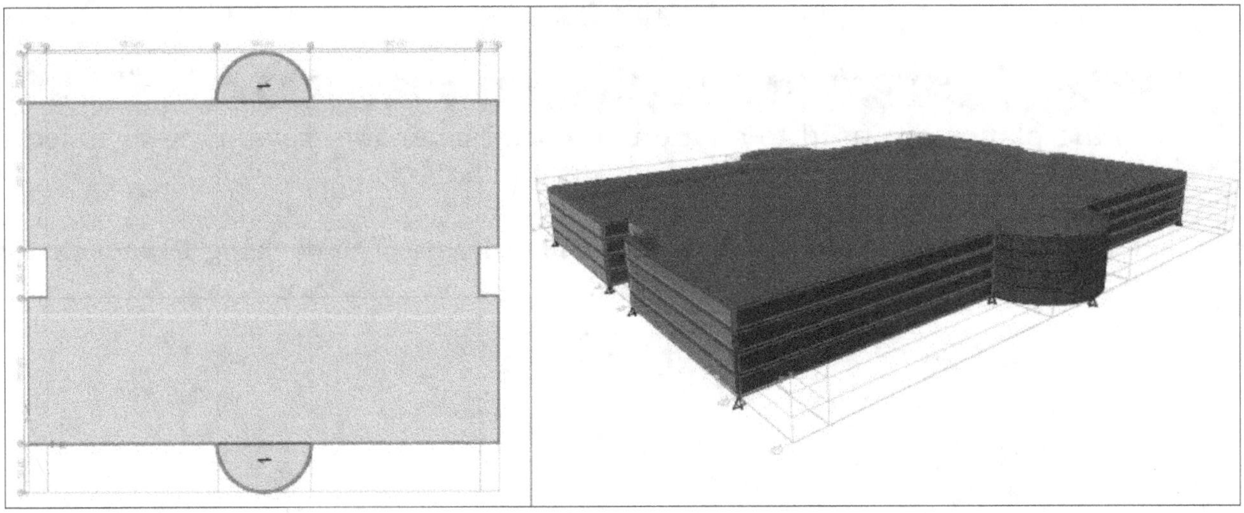

Figure-47. Model

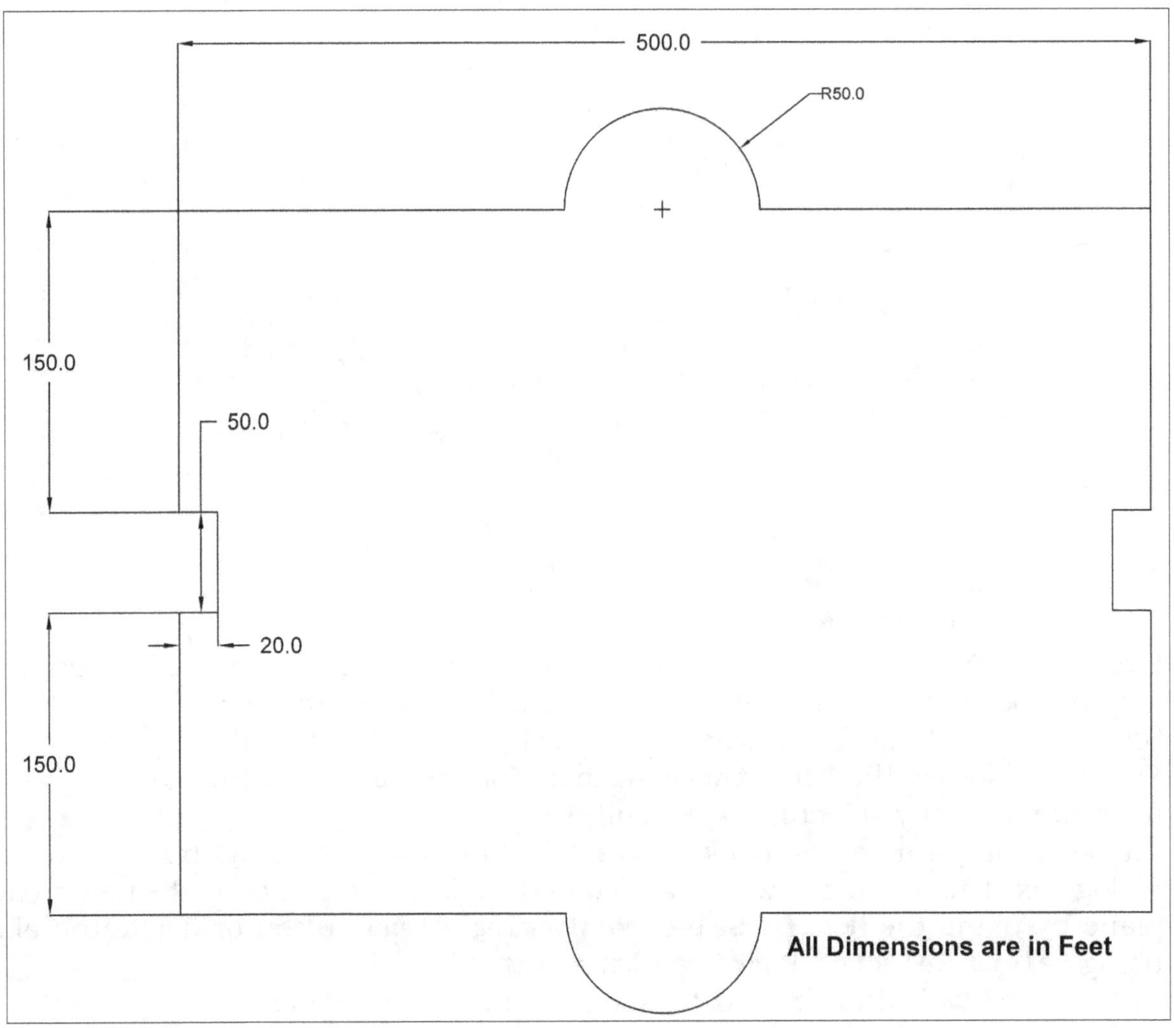

Figure-48. Dimensions for model

Creating Grid Data

- Start ETABS if not started yet.
- Click on the **New** button from the toolbar. The **Model Initialization** dialog box will be displayed.
- Select the **Use Built-in Settings With:** radio button and set the parameters as U.S. Customary, AISC14, AISC 360-10, and ACI 318-14 in the drop-downs. Click on the **OK** button. The **New Model Quick Templates** dialog box will be displayed.
- Select the **Custom Grid Spacing** radio button and click on the **Edit Grid Data** button. The **Grid System Data** dialog box will be displayed; refer to Figure-49.

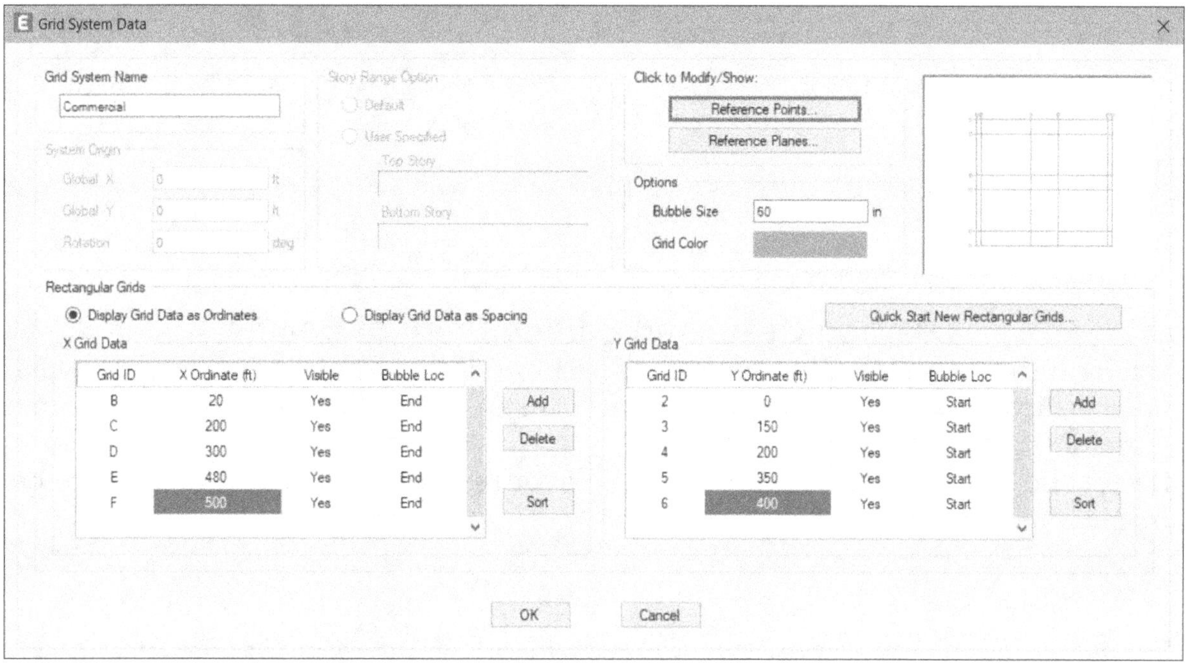

Figure-49. Grid System Data dialog box

- Set the name as **Commercial** in the **Grid System Name** edit box.
- Set the grid points in X Grid Data table as **0**, **20**, **200**, **300**, **480**, and **500**; refer to Figure-49.
- Set the grid points in Y Grid Data table as **-50**, **0**, **150**, **200**, **350**, and **400**; refer to Figure-49.
- After setting desired parameters, click on the **OK** button. The **New Model Quick Templates** dialog box will be displayed again.
- Click on the **OK** button from the dialog box. The grids will be displayed in the plan and 3D view; refer to Figure-50.

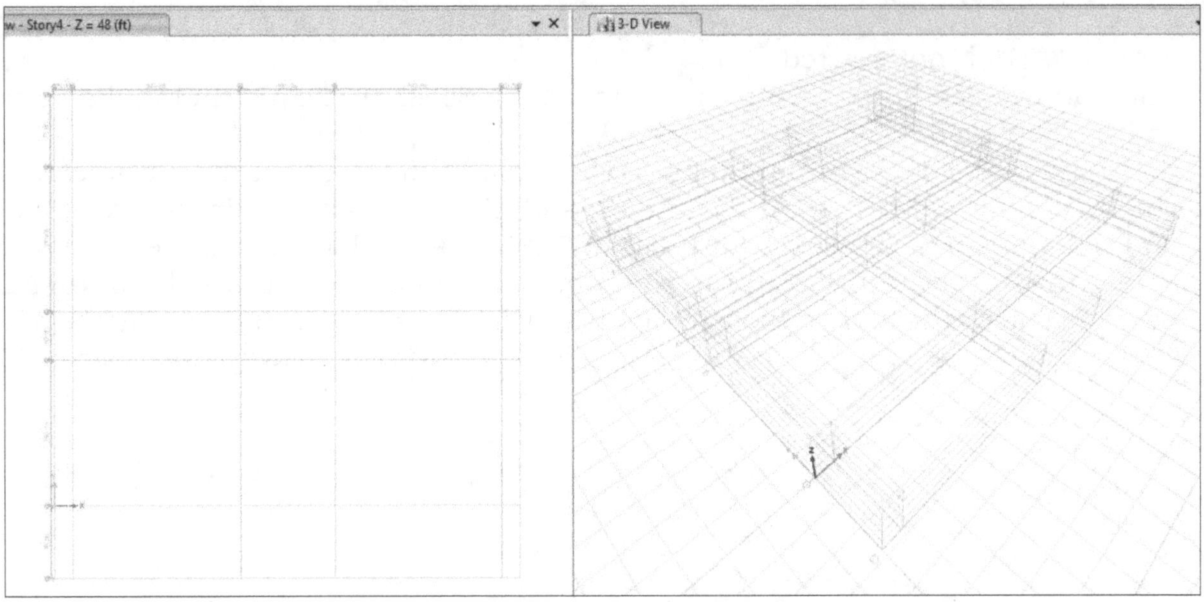

Figure-50. Grids displayed in viewport

Drawing Beam and Editing Point of Beams

- Click on the **Draw Beam/Column/Brace (Plan, Elev, 3D)** tool from the **Draw Beam/ Column/Brace Objects** cascading menu of **Draw** menu. You will be asked to specify starting point of the beam.
- Select the **All Stories** option from the **Stories** drop-down and create the beam frame as shown in Figure-51.

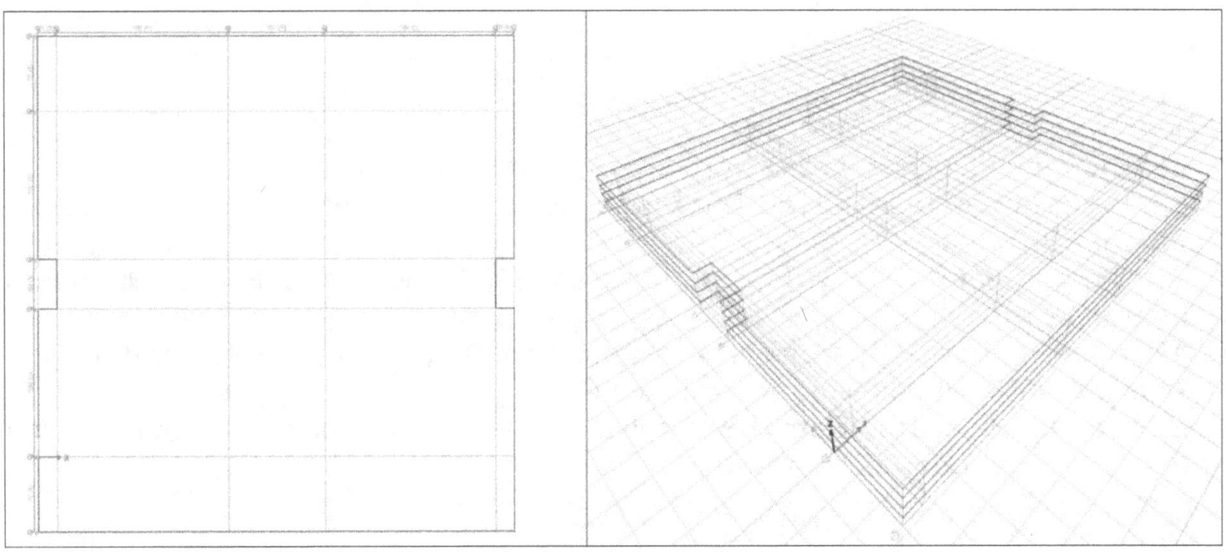

Figure-51. Creating beam structure

- Delete the outer beams and draw new beams as shown in Figure-52.

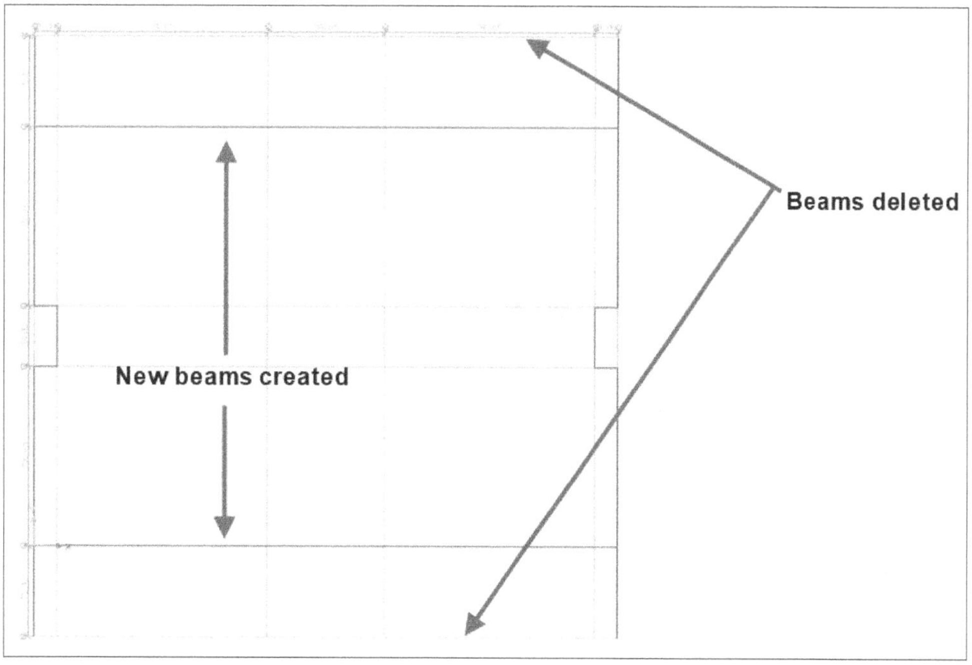

Figure-52. New beams created

- Now, click on the **Reshape Object** tool from the **Draw** menu. You will be asked to select objects to be reshaped. Select the beam as shown in Figure-53.

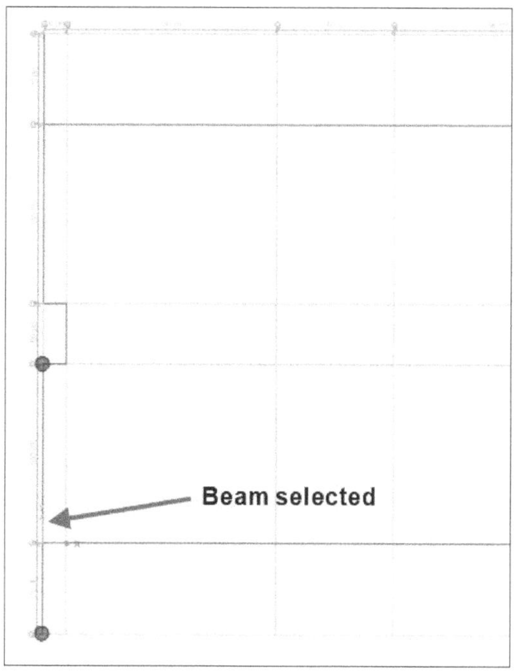

Figure-53. Beam selected

- Drag bottom end point of the beam selected and link it to the newly created beam; refer to Figure-54. Make sure you modify the beams in 3D view so that beam do not coincide at single point.

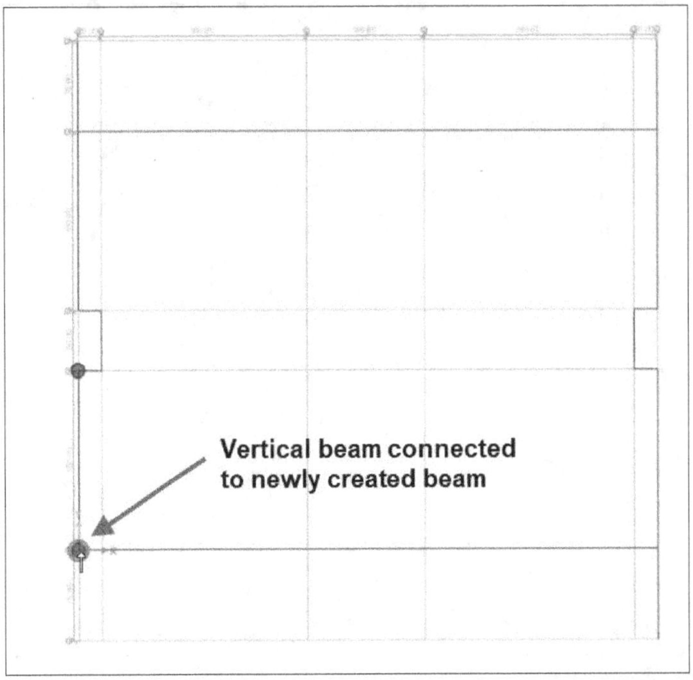

Figure-54. Dragging point to connect to beam

- Set the other extending beams in the same way; refer to Figure-55.

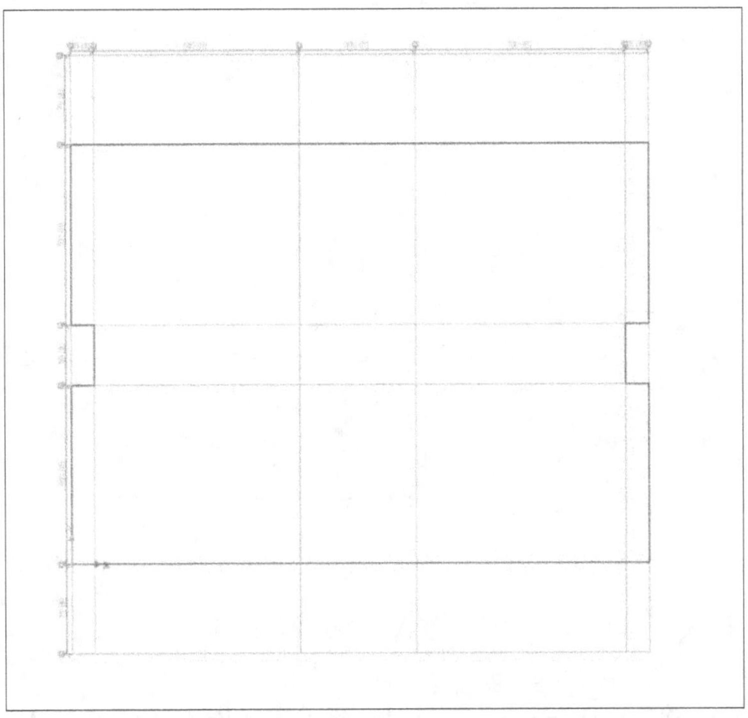

Figure-55. Reshaping beams

- Click on the **Draw Beam/Column/Brace (Plan, Elev, 3D)** tool from the **Draw Beam/Column/Brace Objects** cascading menu of **Draw** menu. You will be asked to specify points of beam.
- Select the **Arc (3 Points)** option from the **Line Drawing Type** drop-down of **Properties of Object** tab and create arc beams as shown in Figure-56.

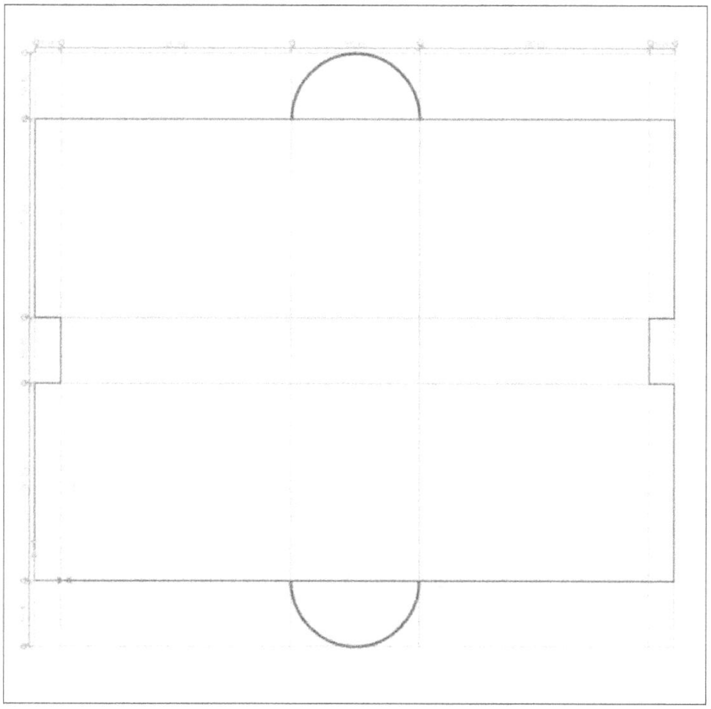

Figure-56. Arc beams created

Drawing Floors

- Click on the **Draw Floor/Wall (Plan, Elev, 3D)** tool from the **Draw Floor/Wall Objects** cascading menu. You will be asked to specify start point of the floor.
- Draw the floor as shown in Figure-57. Press **ESC** to exit the tool.

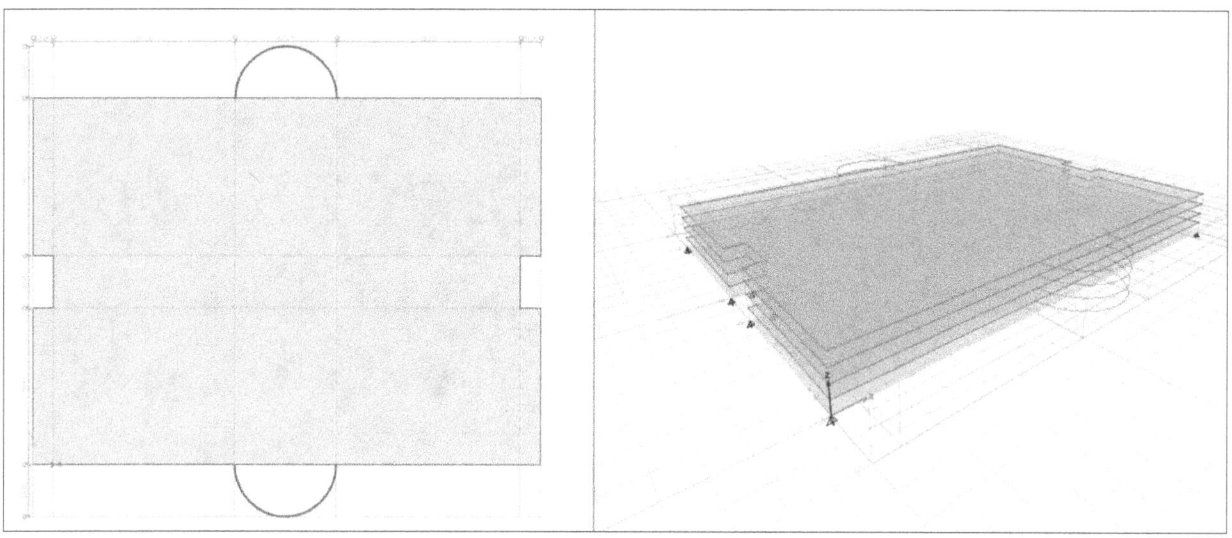

Figure-57. Creating floor slab

- Activate the tool again and select the **Deck 1** option from the **Property** drop-down in the **Properties of Object** tab.
- Select the **Arc (3 Points)** option from the **Edge Drawing Type** drop-down in **Properties of Object** tab and draw the deck in arc beams; refer to Figure-58.

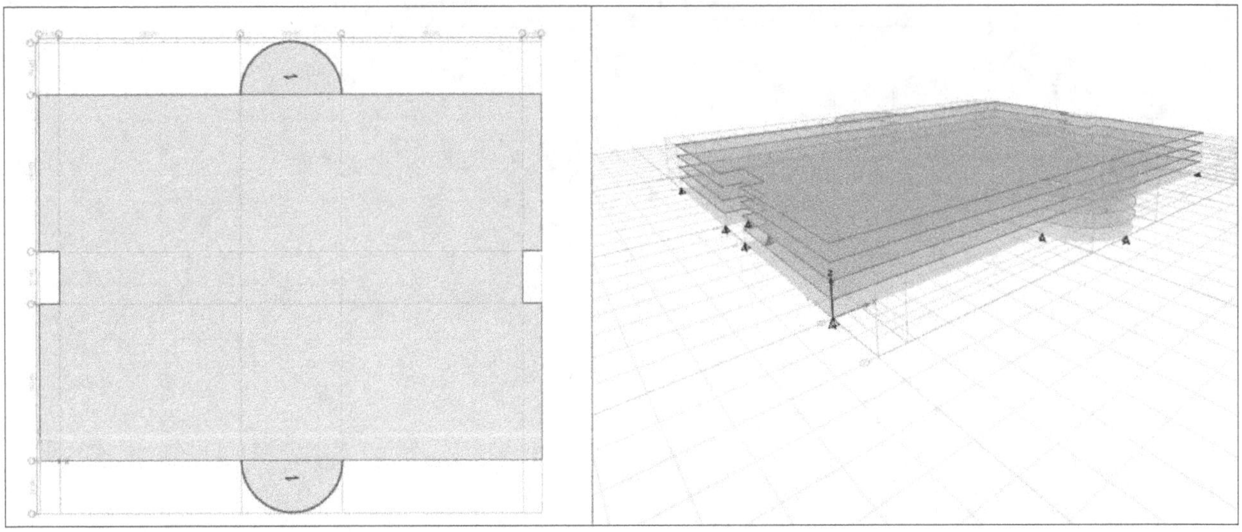

Figure-58. Structure after creating deck

Creating Columns

- Click on the **Quick Draw Columns (Plan, 3D)** tool from the **Draw Beam/Column/Brace Objects** cascading menu of the **Draw** menu. You will be asked to select points for creating columns.
- Click at the grid points encircled in Figure-59. The columns will be created; refer to Figure-60.

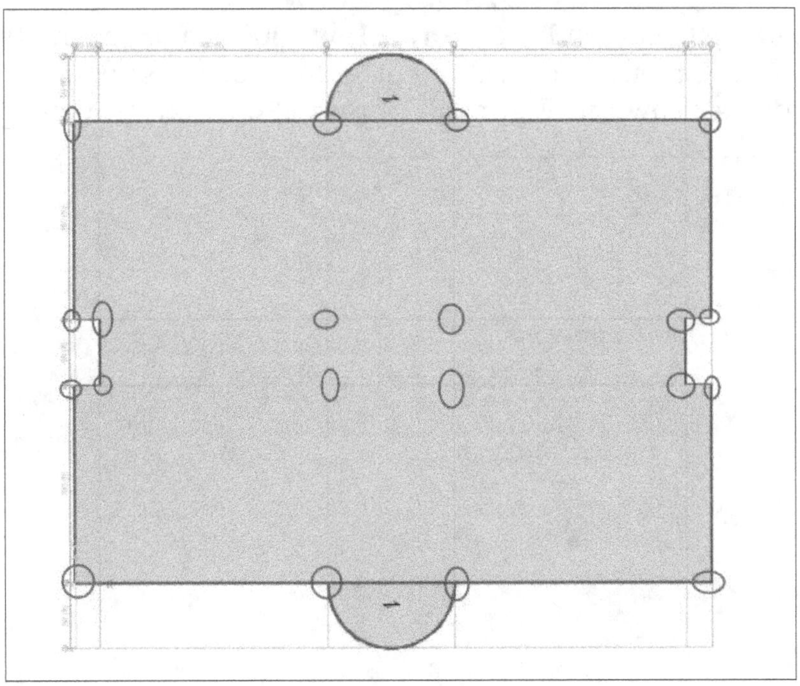

Figure-59. Points to be selected for column

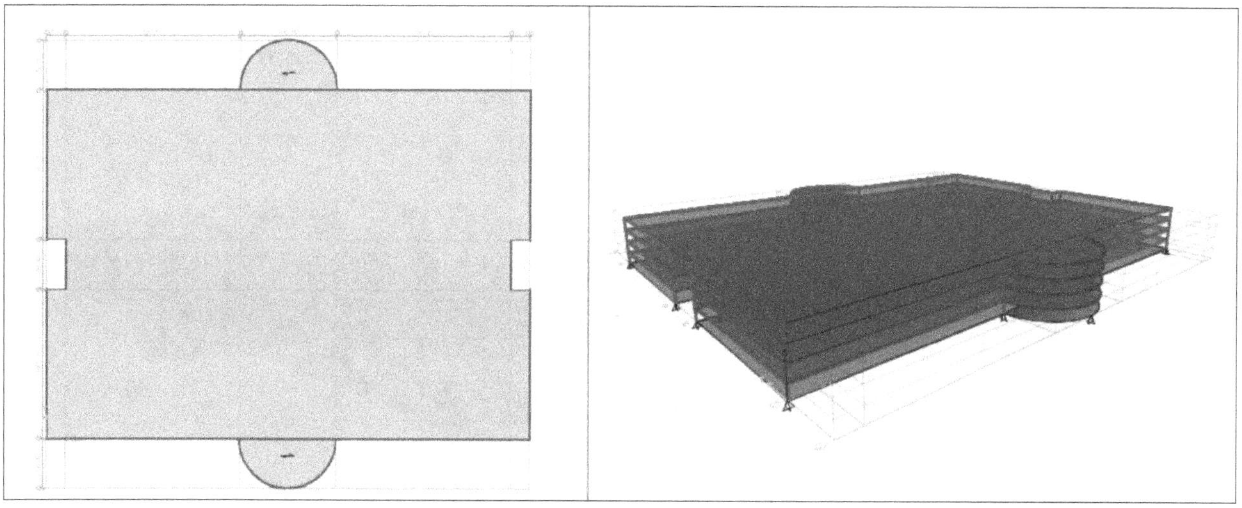

Figure-60. Columns created

Creating Walls

- Click on the **Draw Walls (Plan)** tool from the **Draw Floor/Wall Objects** cascading menu of the **Draw** menu. You will be asked to select start point for wall.
- Select the **Straight Line** option from the **Line Drawing Type** drop-down in **Properties of Object** tab and create the wall as shown in Figure-61.

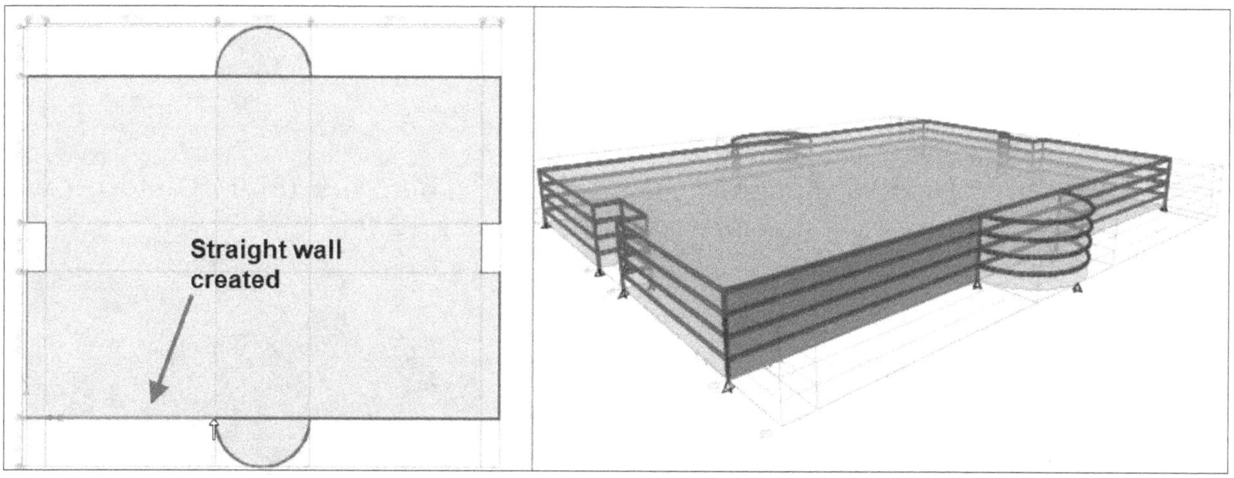

Figure-61. Straight wall created

- Select the **Arc (3 Points)** option from the **Line Drawing Type** drop-down in the **Properties of Object** tab and create semicircular wall as shown in Figure-62.

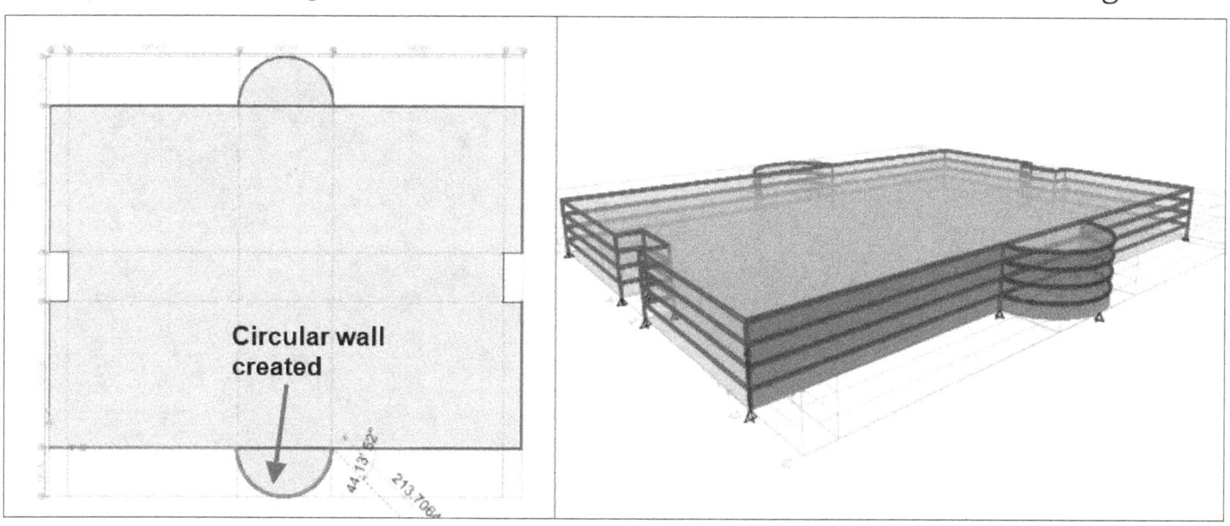

Figure-62. Circular wall created

- Select the option from the **Line Drawing Type** drop-down as required and create rest of the walls as shown in Figure-63. Press **ESC** to exit the tool.

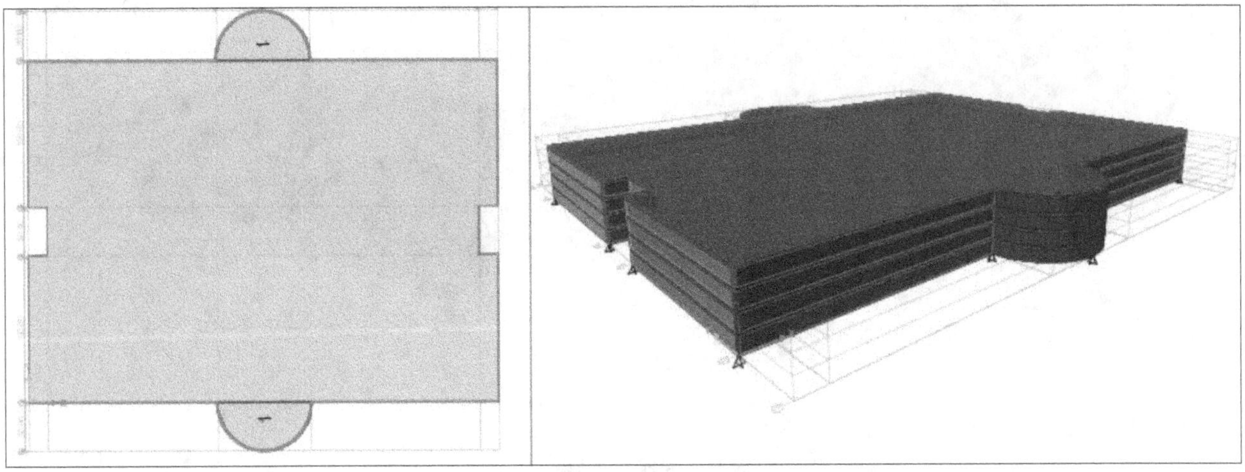

Figure-63. Creating walls

- Save the model and close the file.

PRACTICE 1

Create the structure of any high-rise building nearby you. Use common measurement instruments like measuring tape for dimensions.

PRACTICE 2

Create structure of model as shown in Figure-64. There are total 10 stories with bottom story height as 15 ft and rest story heights as 12 ft.

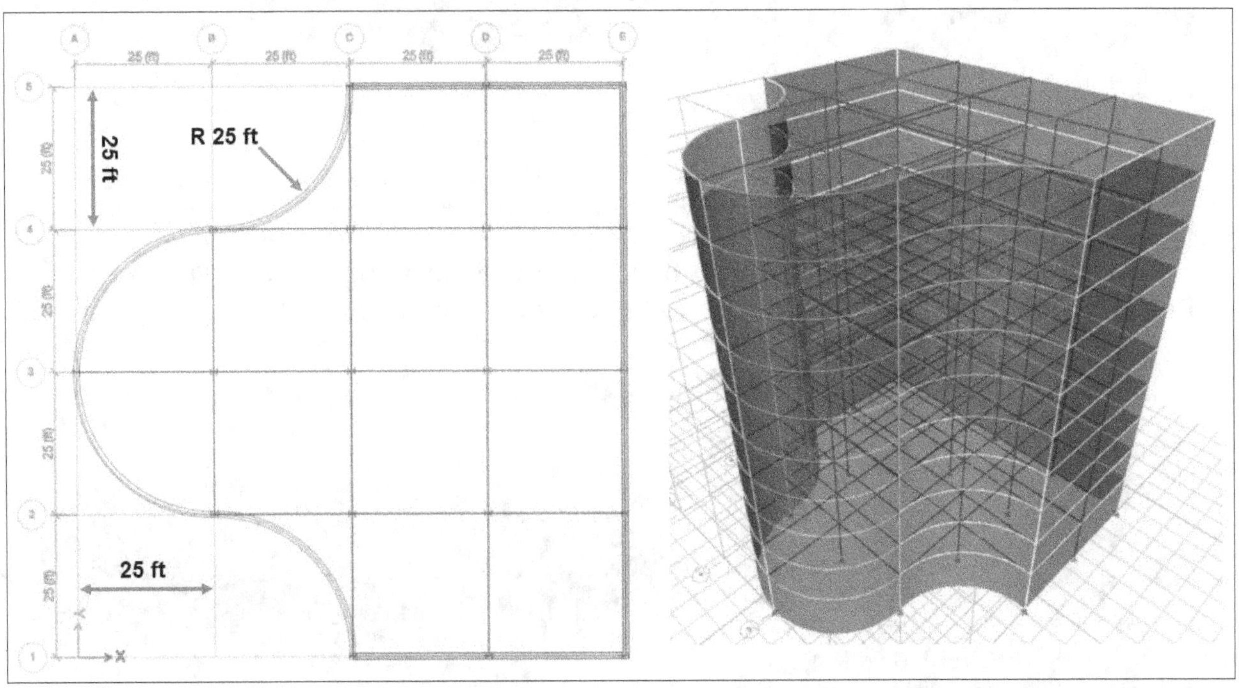

Figure-64. Practice 2

SELF-ASSESSMENT

Q1. In which menu are the tools to modify grid data available?
A. File Menu
B. Draw Menu
C. Edit Menu
D. Analyze Menu

Q2. Which option glues joints along grid lines so they move with the grid?
A. Lock Onscreen Grid System Edit
B. Glue Joints to Grids
C. Snap Options
D. Draw Grids

Q3. Which option enables or disables editing of grid points on screen?
A. Draw Using Snap Only
B. Refresh View
C. Lock Onscreen Grid System Edit
D. Draw Joint Objects

Q4. Which option in the Edit menu is used to edit stories and grid systems?
A. Draw Grids
B. Add Grid Lines at Selected Joints
C. Edit Stories and Grid Systems
D. Draw Walls

Q5. What do joints represent in a structure?
A. Slab openings
B. Grid labels
C. Connection points between structural members
D. Floor panels

Q6. Which tool is used to create joints manually?
A. Draw Joint Objects
B. Draw Grids
C. Draw Tendons
D. Draw Walls

Q7. Which Story option creates joints at the same location on all stories?
A. One Story
B. Similar Stories
C. All Stories
D. Current Story

Q8. Which shortcut key refreshes the graphics view?
A. CTRL+P
B. SHIFT+CTRL+V
C. CTRL+O
D. CTRL+N

Q9. Which tool is used to create grid lines at selected joints?
A. Draw Grids
B. Add Grid Lines at Selected Joints
C. Draw Walls
D. Snap Options

Q10. Which check box creates grid lines parallel to the Y direction?
A. Parallel to X
B. Parallel to Z
C. Parallel to Y
D. Parallel to Grid

Q11. Which button creates grid lines and keeps the dialog box open?
A. Close
B. Apply
C. Done
D. Cancel

Q12. Which tool is used to manually draw grid lines?
A. Draw Grids
B. Draw Links
C. Draw Walls
D. Draw Slab Panels

Q13. What is the purpose of Snap Options?
A. To export files
B. To print graphics
C. To restrict cursor movement to key points
D. To generate reports

Q14. Which check box snaps the cursor to line ends and midpoints?
A. Fine Grid
B. Arch. Layer
C. Line Ends and Midpoints
D. Nudge Value

Q15. Which edit box defines spacing between fine grid lines?
A. Plan Fine Grid Spacing
B. Screen Snap To Tolerance
C. Screen Selection Tolerance
D. Plan Nudge Value

Q16. Which shortcut keys are used for nudging objects?
A. ALT + Arrow Keys
B. CTRL + Arrow Keys
C. SHIFT + Arrow Keys
D. TAB + Arrow Keys

Q17. Which tool allows drawing only using snap points?
A. Draw Using Snap Only
B. Draw Grids
C. Snap Increment
D. Draw Walls

Q18. In which cascading menu are beam, column, and brace tools available?
A. Draw Floor/Wall Objects
B. Draw Beam/Column/Brace Objects
C. Grid Options
D. File Import

Q19. Which tool is used to draw beams, columns, and braces in 3D view?
A. Draw Floor/Wall
B. Draw Beam/Column/Brace (Plan, Elev, 3D)
C. Draw Links
D. Quick Draw Walls

Q20. Which drop-down defines the material and shape of beams or columns?
A. Story
B. Property
C. Spacing
D. Type of Area

Q21. Which Moment Releases option transfers loads in all six degrees of freedom?
A. Pinned
B. Continuous
C. Fixed
D. Semi-rigid

Q22. Which Moment Releases option transfers only axial and shear loads?
A. Continuous
B. Pinned
C. Rigid
D. Arc

Q23. Which line drawing type creates curved objects using three points?
A. Straight Line
B. Spline
C. Arc (3 Points)
D. Bezier

Q24. Which tool quickly draws beams and columns using grid lines?
A. Draw Joint Objects
B. Draw Walls
C. Quick Draw Beams/Columns (Plan, Elev, 3D)
D. Draw Tendons

Q25. Why are braces mainly used in structures?
A. To decorate walls
B. To reduce slab thickness
C. To increase stability during earthquakes
D. To create openings

Q26. Which tool is used to create supporting lines for design strips and slabs?
A. Draw Dimension Lines
B. Draw Support Lines
C. Draw Reference Points
D. Draw Design Strips

Q27. What is the main purpose of Design Strips in ETABS?
A. Creating grids
B. Generating reinforcement design regions in slabs
C. Importing DXF files
D. Adding wall openings

Q28. Which key is pressed to exit the Draw Design Strips tool?
A. ENTER
B. TAB
C. ESC
D. SHIFT

Q29. Dimension lines are mainly used to:
A. Apply loads
B. Measure and annotate distances
C. Create beams
D. Generate reports

Q30. Which tool is used to create snap reference points for other objects?
A. Draw Reference Points
B. Draw Support Lines
C. Draw Section Cut
D. Draw Design Strips

Q31. In the Draw Reference Points tool, offset values can be specified in which directions?
A. X only
B. X and Y only
C. X, Y, and Z
D. Z only

Q32. Which shortcut key is used to set the current viewport to Default 3D View?
A. CTRL+SHIFT+F1
B. CTRL+SHIFT+F2
C. CTRL+SHIFT+F3
D. SHIFT+F3

Q33. Which view is set using the Set Plan View tool?
A. Perspective View
B. Plan View
C. Elevation View
D. Section View

Q34. In the Select Plan View dialog box, which radio button is used to define a custom working plane height?
A. At Default Height
B. At Story Height
C. At Specified Global Z Coordinate
D. At Local Coordinate

Q35. Which shortcut key is used for Set Elevation View?
A. CTRL+SHIFT+F1
B. CTRL+SHIFT+F2
C. CTRL+SHIFT+F3
D. CTRL+ALT+F2

Q36. Which button in the Set Elevation View dialog box allows creation of elevation at a desired X or Y ordinate?
A. Add View
B. Add at Ordinate
C. Create Elevation
D. New Elevation

Q37. Reference planes are mainly used for:
A. Printing reports
B. Snapping objects in elevation views
C. Exporting files
D. Creating loads

Q38. Which edit box is used to specify vertical offset while drawing reference planes?
A. Horizontal Offset
B. Global Z
C. Vertical Offset Z
D. Elevation Distance

Q39. The Draw Developed Elevation Definition tool is used to:
A. Create slabs
B. Display selected elevations in unfolded position
C. Draw strain gauges
D. Create support lines

Q40. Which view should be active before using the Draw Developed Elevation Definition tool?
A. 3D View
B. Elevation View
C. Section View
D. Plan View

Q41. Wall stack is best placed while which view is active?
A. Perspective View
B. Plan View
C. Section View
D. Isometric View

Q42. Which dialog box appears when using Draw Wall Stacks tool?
A. Working Planes
B. Developed Elevation Name
C. New Wall Stack
D. Cladding Options

Q43. Which edit box is used to change wall stack orientation?
A. Rotation Value
B. Angle, deg
C. Direction Angle
D. Orientation Box

Q44. Auto Draw Cladding tool is mainly used to:
A. Create foundations
B. Automatically generate cladding
C. Import walls
D. Add dimensions

Q45. Cladding is generally used to provide:
A. Reinforcement only
B. Thermal insulation and weather protection
C. Water supply
D. Electrical connections

Q46. Which tool is used to create polygonal strain gauges?
A. Draw Line Strain Gauge
B. Draw Quad Strain Gauge
C. Draw Reference Plane
D. Draw Section Cut

Q47. Minimum how many corners are required to create a polygonal strain gauge?
A. 2
B. 3
C. 4
D. 5

Q48. Working planes are mainly used for:
A. Printing graphics
B. Creating reference planes for drawing
C. Generating reports
D. Importing models

Q49. Which button allows selection of three points to define a working plane?
A. Pick Plane Origin
B. Pick Axis Point
C. Pick three points for plane
D. Create Plane by Grid

Q50. Which button is used to add a newly created working plane into the list?
A. Add Working Grid
B. Add New Working Plane
C. Create New Plane
D. Insert Plane

For Student Notes

Chapter 3

Section Properties and Material

Topics Covered

The major topics covered in this chapter are:

- ***Introduction***
- ***Defining Material Properties***
- ***Defining Section Properties***
- ***Section Designer***
- ***Drawing Shapes***
- ***Defining Tendon Sections***
- ***Defining Slab Sections***
- ***Defining Deck Sections***
- ***Defining Wall Sections***
- ***Defining Reinforcing Bar Sizes***
- ***Defining Link/Support Properties***
- ***Defining Frame/Wall Hinge Properties***
- ***Defining Panel Zone***
- ***Defining Spring Properties***
- ***Diaphragm Properties***
- ***Pier Labels and Spandrel Labels***
- ***Creating Groups***

INTRODUCTION

In the previous chapter, you have learned to draw different objects in ETABS. But we have not discussed about the type of sections used in drawing these objects. Like, we have not discussed the section type being used to draw beam/column. We have also not discussed about the materials and their properties for objects. In this chapter, we will discuss all these topics.

DEFINING MATERIAL PROPERTIES

The **Material Properties** tool in **Define** menu is used to create and manage material properties for structure. New materials are being researched in construction industry everyday for making building stronger and lighter. To use these materials in analysis using ETABS, you might need to create new material properties. The procedure to create and manage material properties is given next.

- Click on the **Material Properties** tool from the **Define** menu. The **Define Materials** dialog box will be displayed; refer to Figure-1.

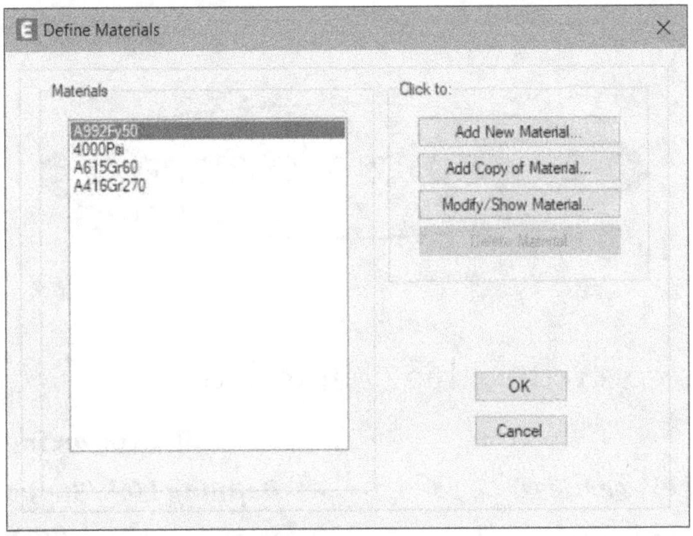

Figure-1. Define Materials dialog box

Adding New Material

- Click on the **Add New Material** button from the dialog box. The **Add New Material Property** dialog box will be displayed; refer to Figure-2.

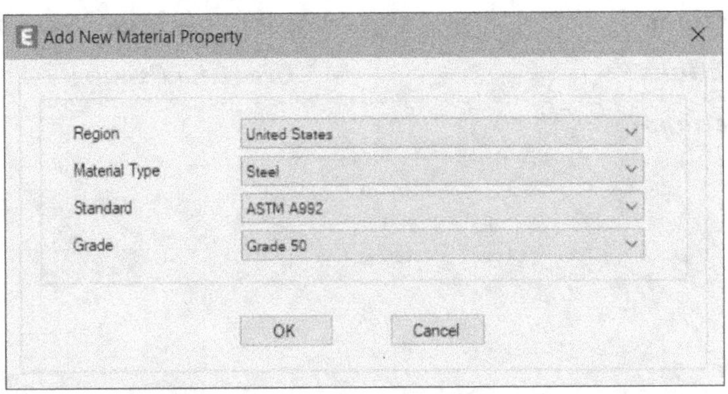

Figure-2. Add New Material Property dialog box

- Select desired option from the **Region** drop-down to define country/region for which you want to create the material. Note that standard and grade of material will change based on region selected for material.
- Select desired option from the **Material Type** drop-down. You can use standard materials available in the list or you can select the **Other** option to create a material of user defined type.
- After setting parameters, click on the **OK** button. The **Material Property Data** dialog box will be displayed; refer to Figure-3.

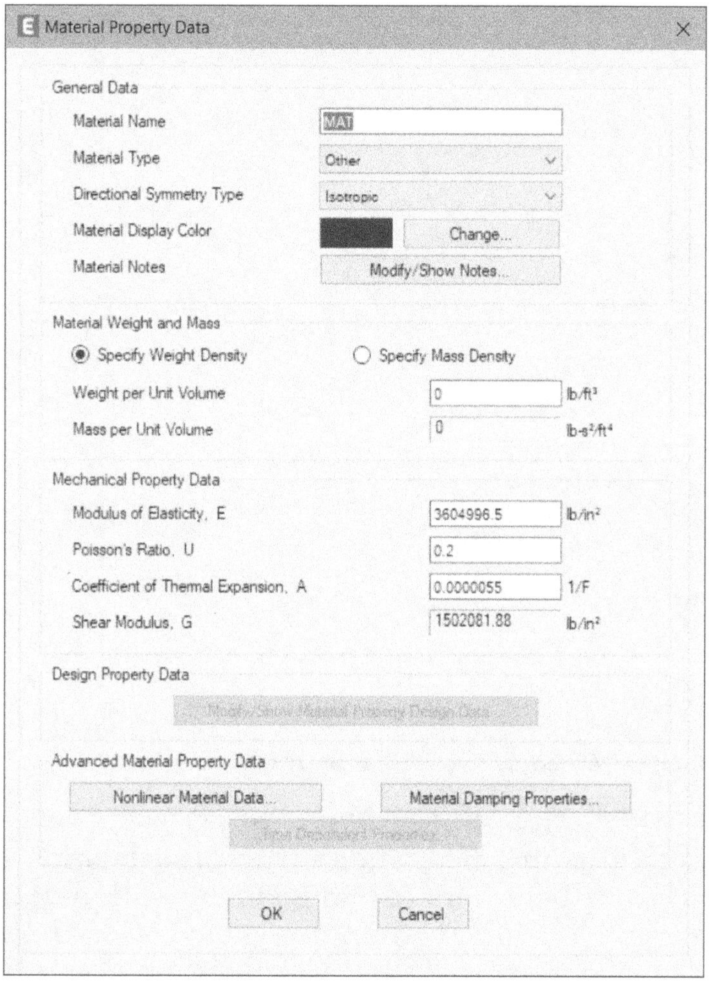

Figure-3. Material Property Data dialog box

- Click in the **Material Name** edit box and specify desired name.
- Select desired option from the **Directional Symmetry Type** drop-down to define whether material has same properties in all directions or not. Select the **Isotropic** option if material has same properties in all directions. Select the **Orthotropic** option if material has different properties in X, Y, and Z directions.
- Select the **Specify Weight Density** radio button if you want to define density in the form of weight per unit or select the **Specify Mass Density** radio button if you want to define mass per unit volume.
- If you have selected **Isotropic** option then **Mechanical Property Data** area will be displayed. Specify desired parameters like modulus of elasticity, Poisson's ratio, coefficient of thermal expansion, and shear modulus in respective fields.

- If the **Orthotropic** option is selected then **Modify/Show Mechanical Property Data** button will be displayed in **Mechanical Property Data** area. Click on the button to modify properties for each direction. The **Material Mechanical Property Data** dialog box will be displayed; refer to Figure-4. Specify desired values in the edit boxes of the dialog box and click on the **OK** button.

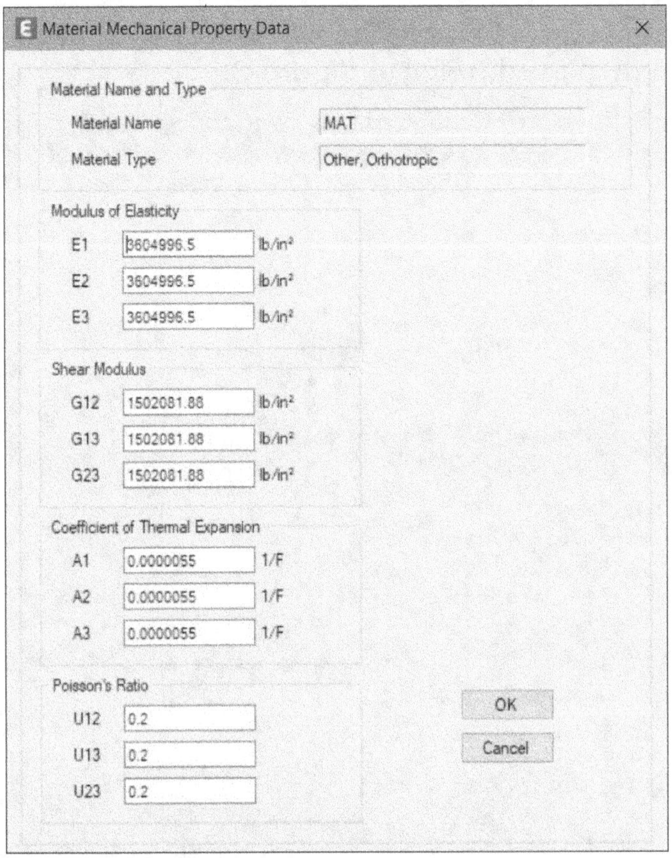

Figure-4. Material Mechanical Property Data dialog box

Nonlinear Material Data

Nonlinear material data is specified when structural deforms non-linearly under stress. If you need to study behavior of building material that does not deform as per linear stress-strain graph and after certain threshold material deforms permanently with lesser amount of load. Steel, aluminum, and other metals have nonlinear data associated with them. Note that this data is used only for non-linear analysis which is not generally required in common cases of structure analysis.

- Click on the **Nonlinear Material Data** button from the **Advanced Material Property Data** area in the dialog box. The **Nonlinear Material Data** dialog box will be displayed; refer to Figure-5.
- Set desired values in the **Tension** and **Compression** edit boxes of **Acceptance Criteria Strains** area. Immediate Occupancy (IO), Life Safety (LS), and Collapse Prevention (CP) levels in Nonlinear Analysis are set based on specified values.
- Select the **Ignore Tension Acceptance Criteria** check box if you do not want to include tension levels in analysis.

- The options in **Hysteresis Type** drop-down are used to define non-linearity of material based on history of its use. Some of the material retain shape after applying load for a long time like rubber, shape-memory alloys, and so on. You can define this behavior using Force-Displacement graphs. You can learn more about this non-linearity using next QR code.

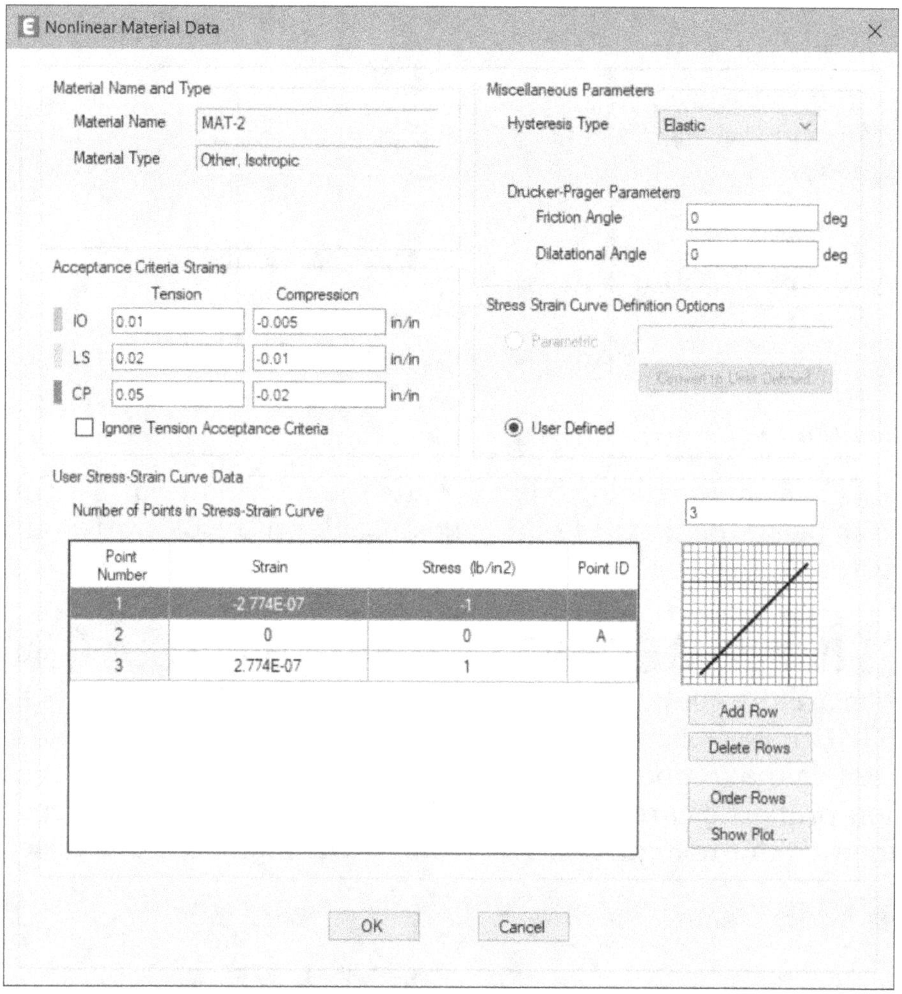

Figure-5. Nonlinear Material Data dialog box

- Specify desired parameters in the table of **Number of Points in Stress-Strain Curve** table based on data provided by your material supplier. Use the **Add Row** button to add new row in table. Use the **Delete Rows** button to delete selected rows. Similarly, you can use the other buttons in the dialog box.
- Click on the **OK** button from the dialog box to apply the data.

Material Damping Properties Data

- Click on the **Material Damping Properties** button from the **Advanced Material Property Data** area to define how material absorbs vibrations. The **Additional Material Damping** dialog box will be displayed; refer to Figure-6.
- There are two types of damping coefficients, Modal damping and Viscous Proportional damping. Specify the parameters based on material data provided by your supplied and click on the **OK** button. The **Material Property Data** dialog box will be displayed again. Check the QR code next to learn more about damping coefficients.

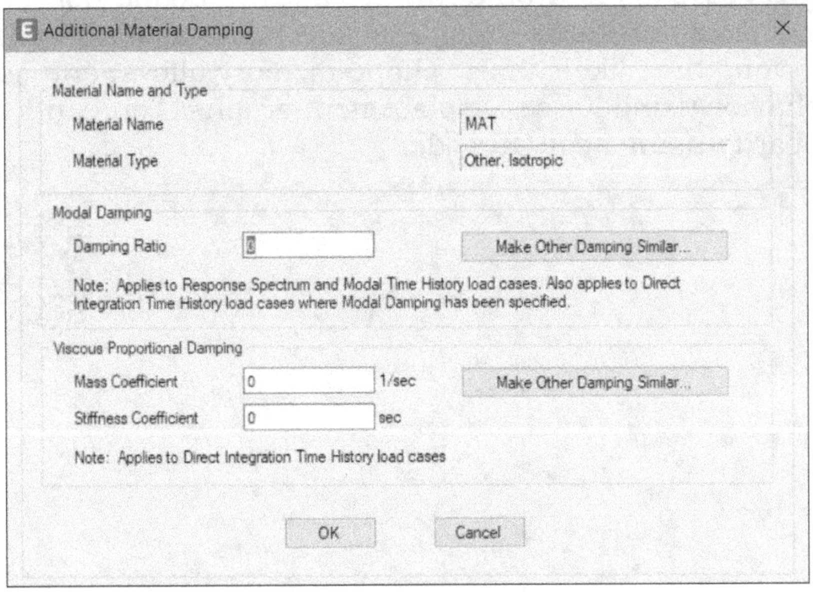

Figure-6. Additional Material Damping dialog box

- Click on the **OK** button from the dialog box to create the new material. The material will be added in the **Define Materials** dialog box. Click on the **OK** button from the dialog box to exit the dialog box.

DEFINING SECTION PROPERTIES

The section properties are the shape and material definitions of building blocks used to draw objects in ETABS. For examples, beams used in building have some section properties, slabs used to make floor have some section properties, and so on. The tools to define section properties are available in the **Section Properties** cascading menu of the **Define** menu; refer to Figure-7. The tools in this menu are discussed next.

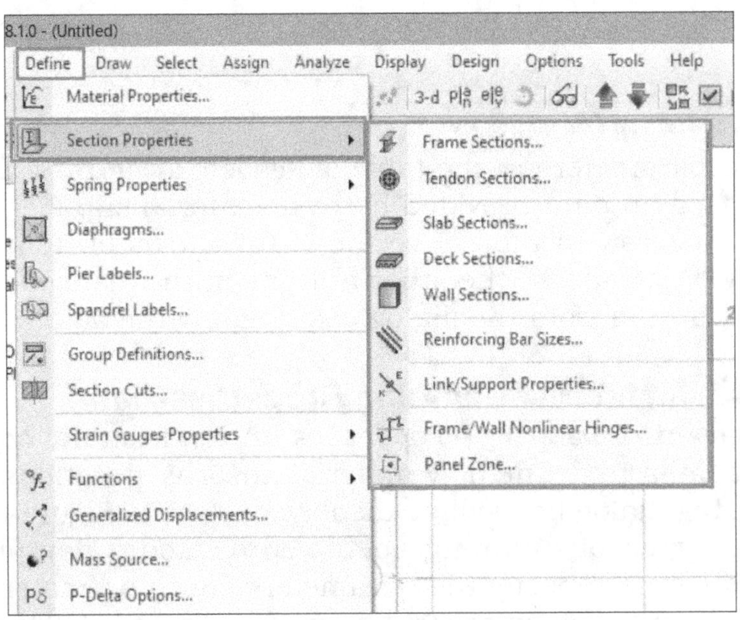

Figure-7. Section Properties cascading menu

Defining Frame Sections

The **Frame Section** tool in the **Section Properties** cascading menu of the **Define** menu is used to define the shape, size, and material of different frame sections. The procedure to define frame section is given next.

- Click on the **Frame Section** tool from the **Section Properties** cascading menu of the **Define** menu. The **Frame Properties** dialog box will be displayed; refer to Figure-8.

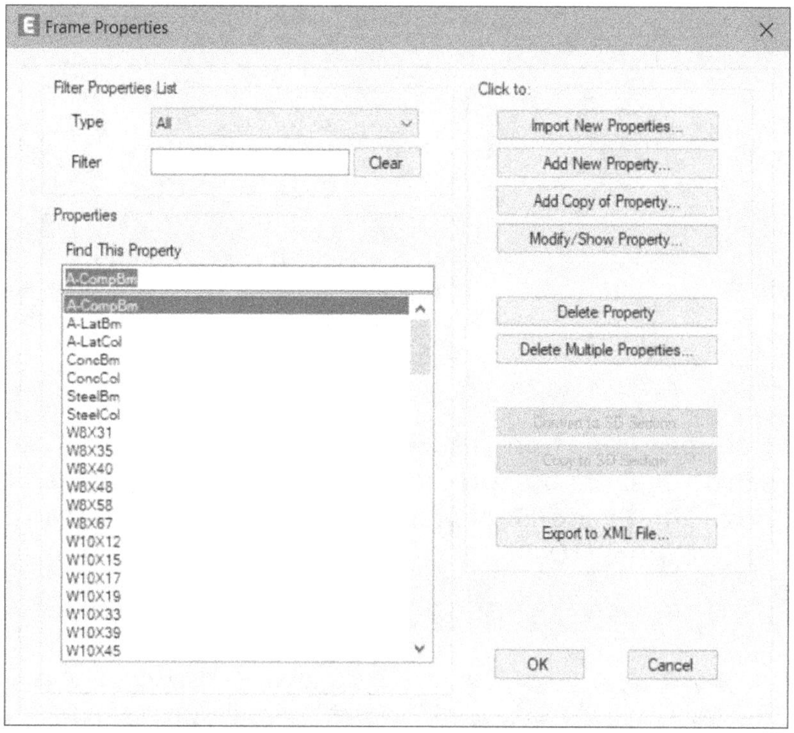

Figure-8. Frame Properties dialog box

- By default, all the frame properties are displayed in the **Properties** area of the dialog box.

Importing New Properties

- Click on the **Import New Properties** button from the dialog box. The **Frame Property Shape Type** dialog box will be displayed; refer to Figure-9.

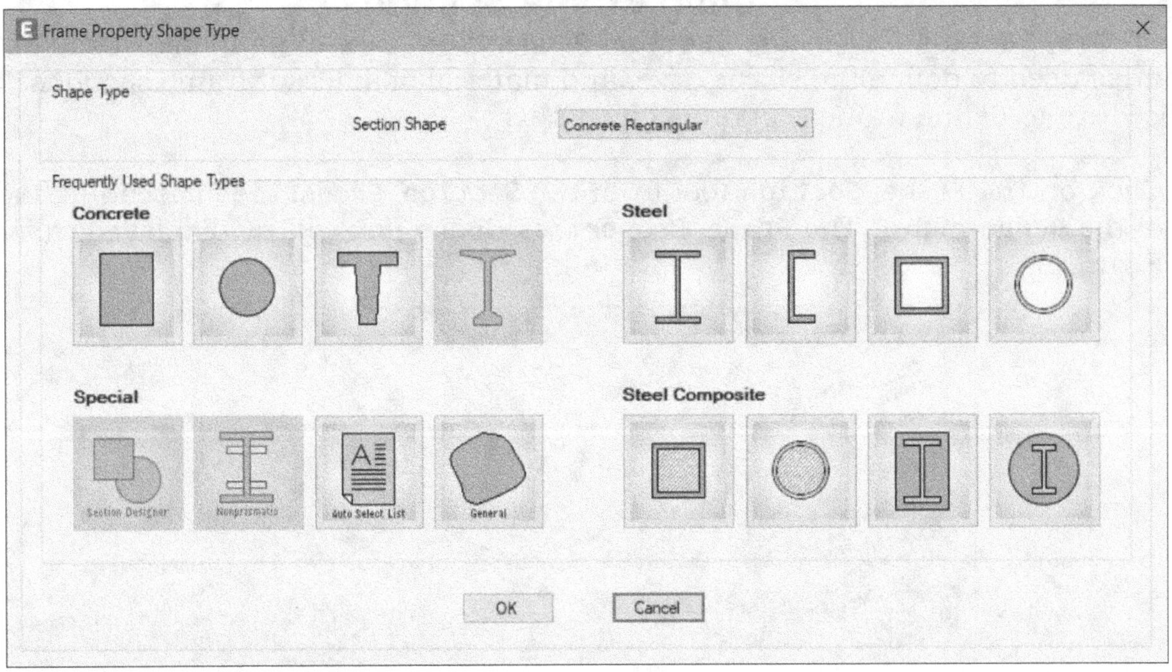

Figure-9. Frame Property Shape Type dialog box

- Select desired shape type from the **Section Shape** drop-down and click on the **OK** button or select desired button from the **Frequently Used Shape Types** area of the dialog box. The **Frame Section Property Import Data** dialog box will be displayed; refer to Figure-10.

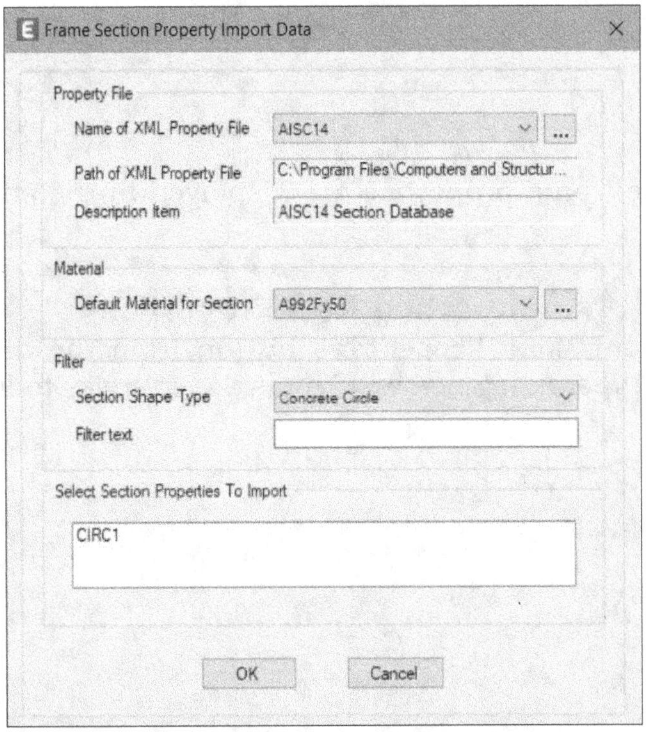

Figure-10. Frame Section Property Import Data dialog box

- Select desired XML property file from the **Name of XML Property File** drop-down to use respective properties. If you have an XML property file saved in your local drive then click on the **Add Property File** button next to the drop-down. The **Add XML Property Files** dialog box will be displayed; refer to Figure-11. Select desired file and click on the **Open** button. The options related to selected property will be displayed in the dialog box.

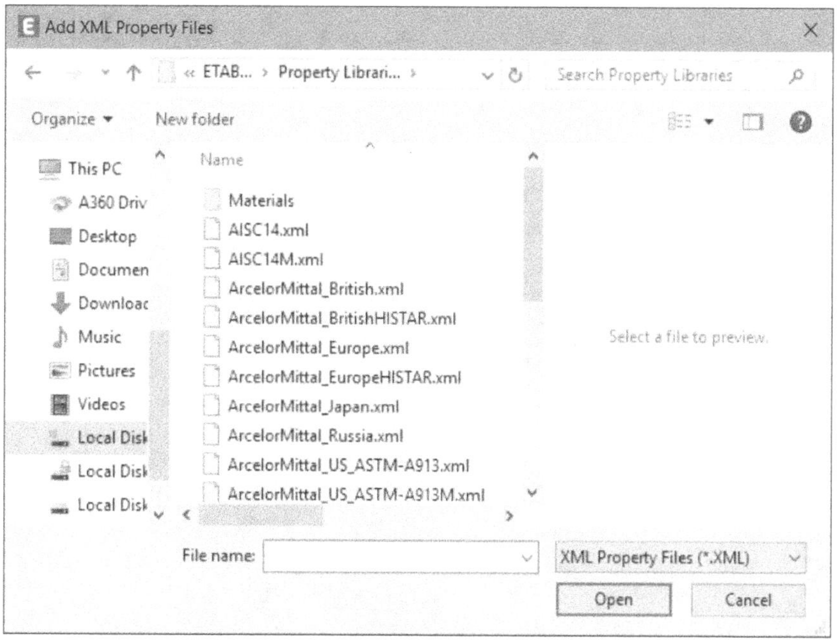

Figure-11. Add XML Property Files dialog box

- Select desired material from the **Default Material for Section** drop-down.
- Change the option in **Section Shape Type** drop-down as required. The list of section properties that can be imported, will be displayed in the **Select Section Properties to Import** area of the dialog box.
- Select the section from the list and click on the **OK** button from the dialog box. The section will be added in the list.

Adding New Section Property

The **Add New Section Property** button in the **Frame Properties** dialog box is used to add new section properties in the list. The procedure is given next.

- Click on the **Add New Property** button from the dialog box. The **Frame Property Shape Type** dialog box will be displayed as discussed earlier.
- Select desired shape type from the **Section Shape** drop-down and click on the **OK** button or select desired button from the **Frequently Used Shape Types** area of the dialog box. The **Frame Section Property Data** dialog box will be displayed; refer to Figure-12.

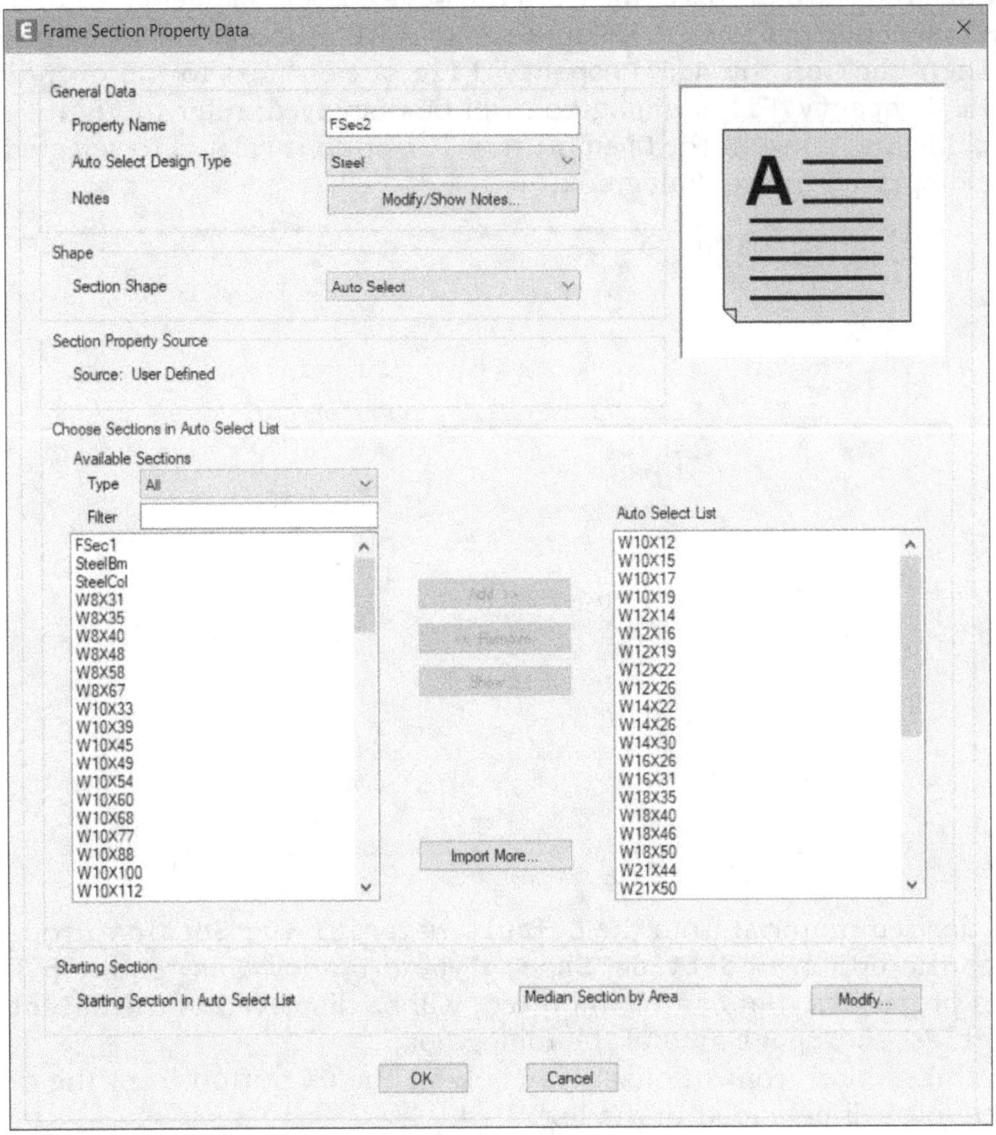

Figure-12. Frame Section Property Data dialog box

- Specify desired name of the section in **Property Name** edit box.
- Select desired material from the **Material** drop-down.
- Click on the **Change** button for **Display Color** option to change the color of the frame section.
- You can add a note related to the current section by using the **Modify/Show Notes** button in the **General Data** area of the dialog box. On clicking this button, the **Frame Property Notes** dialog box is displayed. Specify desired text note and click on the **OK** button to apply changes.
- If you want to change the section shape from what you have selected earlier, click in the **Section Shape** drop-down of the **Shape** area in the dialog box and select desired option. Note that based on your selection in this drop-down, the options of the **Section Dimensions** area get changed.
- Set the section dimensions as required in the dialog box.
- Click on the **Modify/Show Modifiers** option to check/change the parameters directly related to structural analysis. The **Property/Stiffness Modification Factors** dialog box will be displayed; refer to Figure-13. Set the multiplication factors for desired properties to modify it and click on the **OK** button.

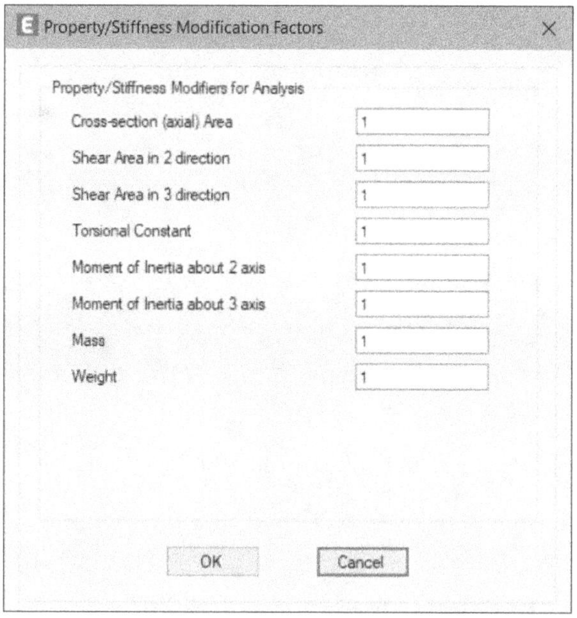

Figure-13. Property Stiffness Modification Factors dialog box

- Click on the **OK** button from **Frame Section Property Data** dialog box. The new property will be added in the list.

Adding Copy of a Property

- Select the property from the list whose copy is to be added and click on the **Add Copy of Property** button. The **Frame Section Property Data** dialog box will be displayed with all the parameters of selected property.
- Note that if you have selected a standard property then **Section Dimensions** area of the dialog box will not be active; refer to Figure-14. To make this area active for modifications, click on the **Convert to User Defined** button.
- Perform the modifications as required and click on the **OK** button from the dialog box to add the property in the list.

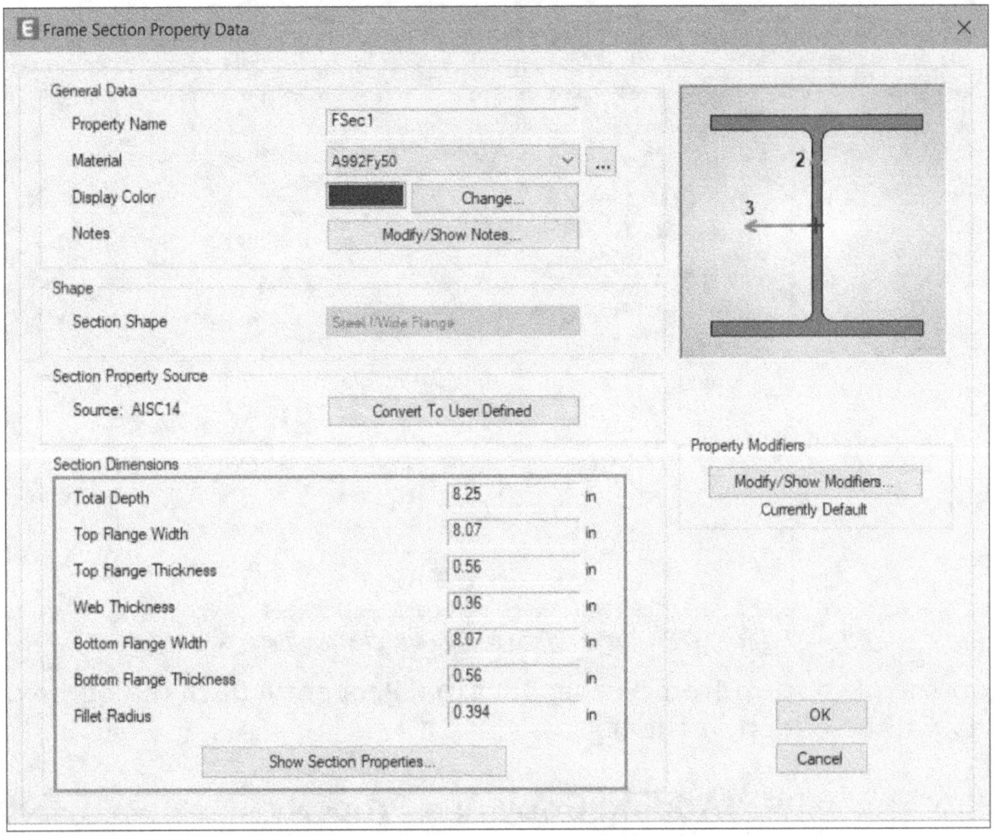

Figure-14. Section Dimensions area of Frame Section Property Data dialog box

Modifying Property Data

- To modify a section property, select it from the list and click on the **Modify/ Show Property** button from the **Frame Properties** dialog box. The **Frame Section Property Data** dialog box will be displayed as discussed earlier.
- Set desired parameters and click on the **OK** button from the dialog box.

Deleting a Property

Select the section property from the list that you want to be deleted and click on the **Delete Property** button from the **Frame Properties** dialog box. A confirmation box will be displayed. Click on the **Yes** button to delete.

Deleting Multiple Properties

The **Delete Multiple Properties** button is used to delete all the properties that can be deleted. The procedure is given next.

- Click on the **Delete Multiple Properties** button from the **Frame Properties** dialog box. The **Delete Multiple Frame Section Properties** dialog box will be displayed; refer to Figure-15.
- Select the frame sections that you want to be deleted from the list while holding the **CTRL** key and click on the **Delete Selected Frame Sections** button. The selected sections will get deleted.
- Click on the **OK** button to exit the dialog box.

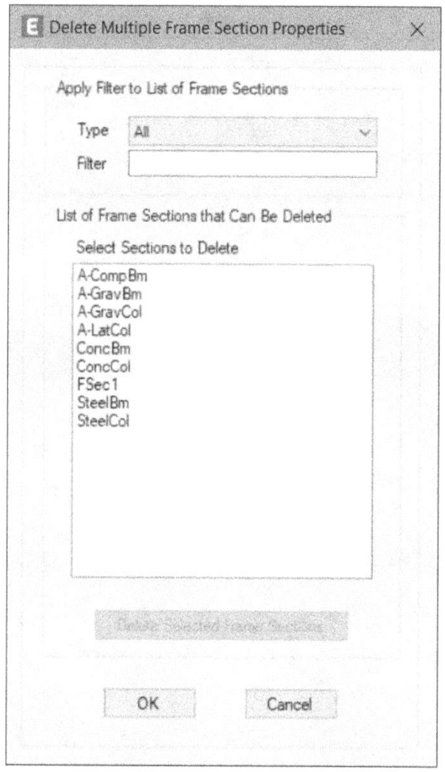

Figure-15. Delete Multiple Frame Section Properties dialog box

Converting a Standard Section to SD Section

SD stands for Section Designer. SD sections are the sections of frame member with irregular shape manually drawn by the user. So, the benefit of converting a frame member to section designer section is that you can now change its shape as required. You should always make sure that your designed section is available in market or can be produced up to your required quantity in industry. The procedure to convert a standard section to SD section is given next.

- Select the section from the list in the **Frame Properties** dialog box and click on the **Convert to SD Section** button. The **Section Designer Section Property Data** dialog box will be displayed; refer to Figure-16.
- Specify the name as desired in the **Property Name** edit box. Change the other parameters as required in the **General Data** area of the dialog box.
- In the **Design Type** area, select desired radio button. Note that radio buttons in this area are active based on the material selected in **Base Material** drop-down. On the basis of radio button selected from the **Design Type** area, the system will analyze the model using the respective post processor. For example, if you have selected the **Composite Column** radio button then system will analyze model using the **Composite Column Post Processor**.
- Selecting the **Concrete Column** or **Composite Column** radio button will activate the radio buttons available in the **Concrete Column Check/Design** area of the dialog box.
- Select desired option from the **Concrete Column Check/Design** area so that **Concrete Frame Design Post Processor** checks or designs the reinforcement in the column.
- Click on the **Section Designer** button from the **Define/Edit/Show Section** area to check or modify the section. The **Section Designer** window will be displayed; refer to Figure-17.

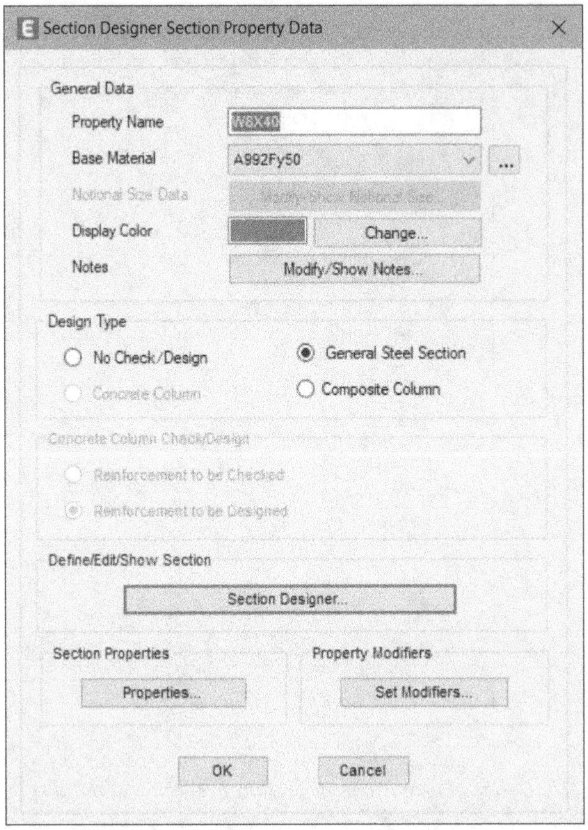

Figure-16. Section Designer Section Property Data dialog box

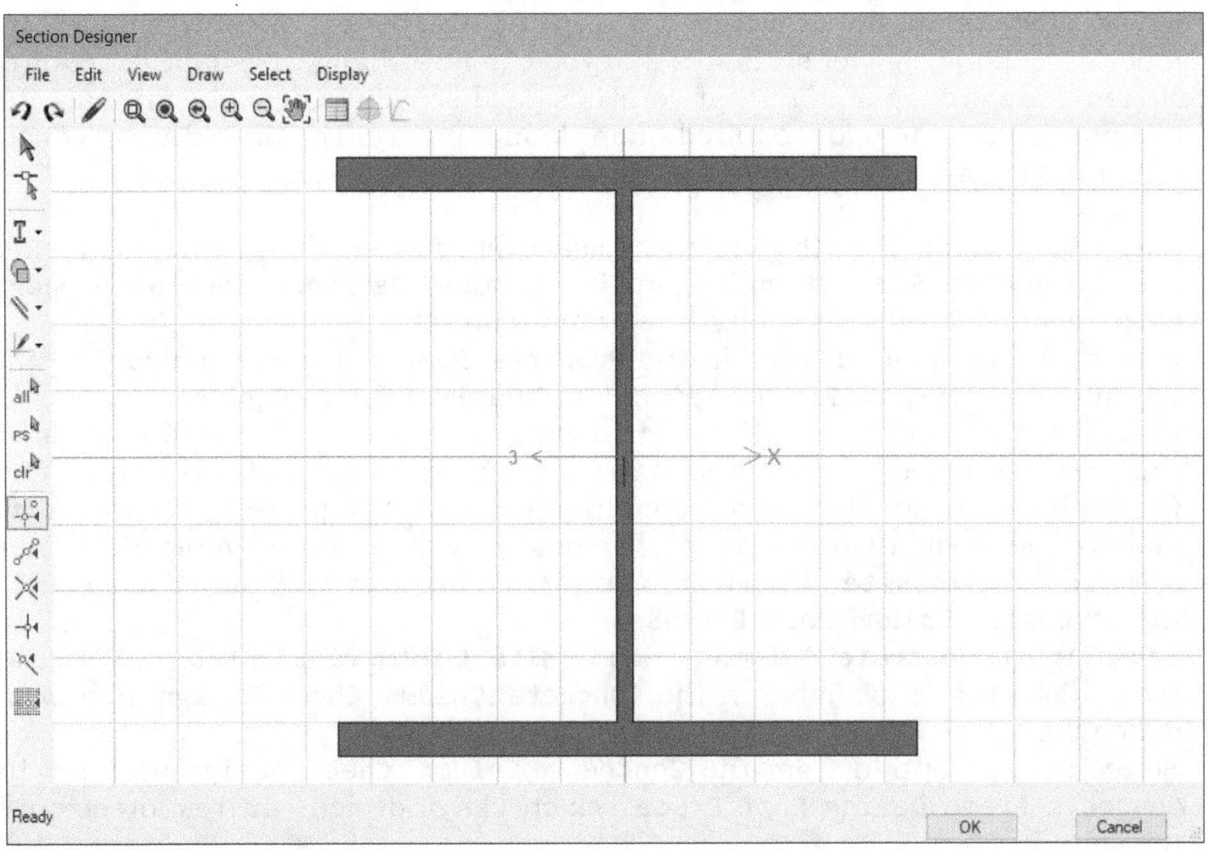

Figure-17. Section Designer window

- Create/modify the shape of section as required and click on the **OK** button from the window. The **Section Designer Section Property Data** dialog box will be displayed again. You will learn more about the tools in **Section Designer** window later in this book.
- Set the other properties as discussed earlier and click on the **OK** button. The modified section will be displayed in the list.

Similarly, you can use the **Copy to SD Section** button to create a copy of selected section property as SD section.

Exporting to XML File

The **Export to XML File** tool in the **Frame Properties** dialog box is used to export the selected section property to an XML file. The procedure is given next.

- After selecting desired property, click on the **Export to XML File** button from the dialog box. The **Frame Property Export to XML File** dialog box will be displayed; refer to Figure-18.

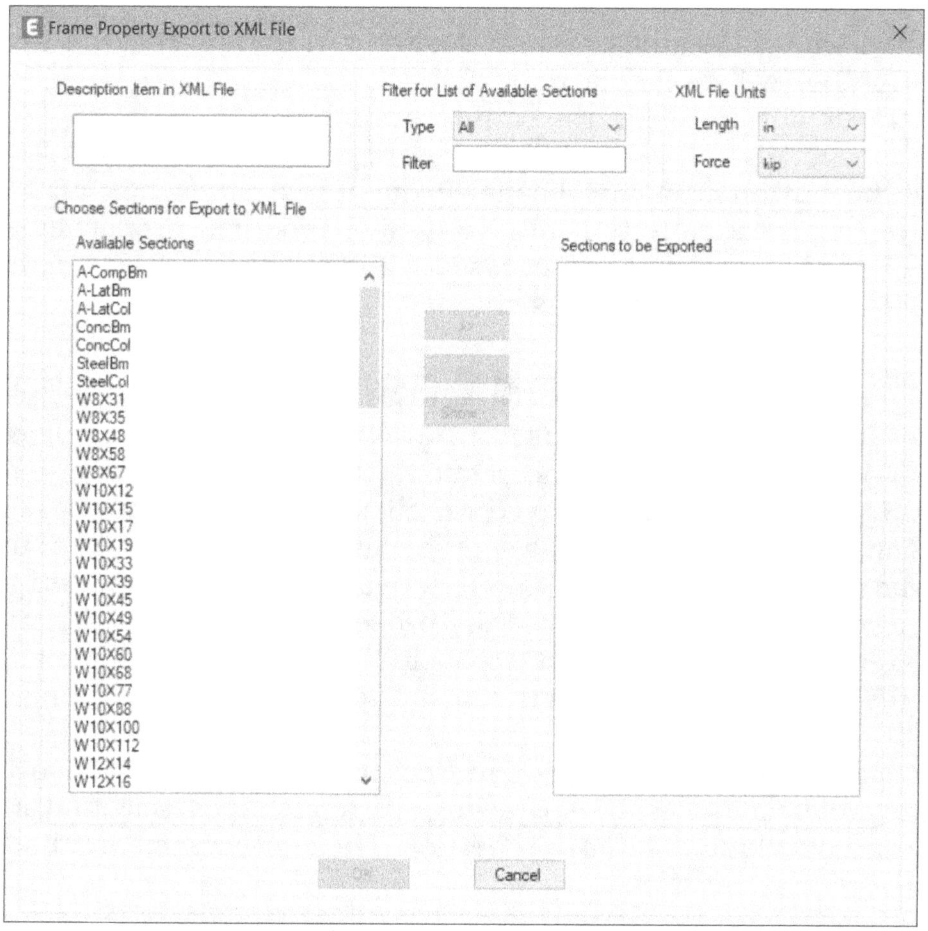

Figure-18. Frame Property Export to XML File dialog box

- Select all the properties that you want to be exported to an xml file from the left list box while holding the **CTRL** key.
- Click on the **>>** button to add selected files to export list.
- Set the unit of Length and Force in the respective drop-downs of **XML File Units** area of the dialog box.
- Click on the **OK** button after setting the parameters. The **Save Frame Property XML File As** dialog box will be displayed; refer to Figure-19.

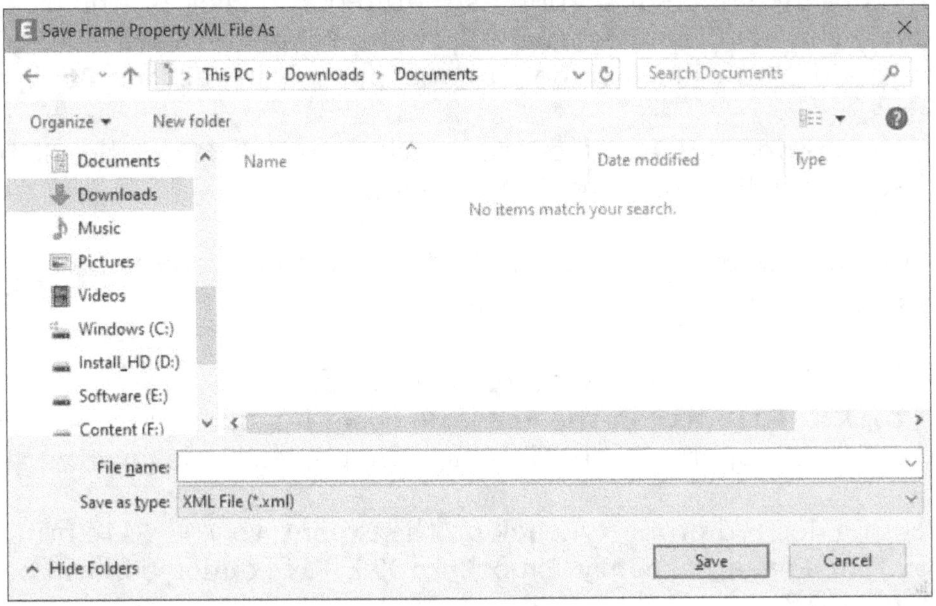

Figure-19. Save Frame Property XML File As dialog box

- Specify the name of file and save it in desired location by clicking **Save** button. A confirmation box will be displayed.
- Click on the **OK** button to exit the confirmation box.

Click on the **OK** button from the **Frame Properties** dialog box after setting the parameters as required. Note that all the new or modified properties set in **Frame Properties** dialog box can now be accessed while creating beams/columns/braces using the tools discussed in previous chapter.

SECTION DESIGNER

The **Section Designer** window is used to draw irregular designs or combine two and more standard designs of frame sections to create a new frame section. The **Section Designer** window is displayed on clicking the **Section Designer** button from the **Define/Edit/Show Section** area of the **Section Designer Section Property Data** dialog box. We have discussed about activating the **Section Designer** window in previous topic. Now, we will discuss the procedure to create different sections in **Section Designer** window. Most of the tools in this window are same as available in the ETABS interface. We will discuss the tools used for drawing sections in this topic.

Drawing a section involve many steps in Section Designer like activating required snaps, creating reference geometries, drawing different shapes, selecting and reshaping the objects, and checking different properties. These processes are discussed next.

Activating Snaps

There are six options to activate different type of snaps. These options are available in **Snap To** cascading menu of **Draw** menu in the window and at the bottom of the toolbar in the left of window; refer to Figure-20.

- Click on desired snap option from the **Snap To** cascading menu or select the button from the left toolbar. The respective snap option will be activated.
- To de-activate a snap, click on the respective option again.

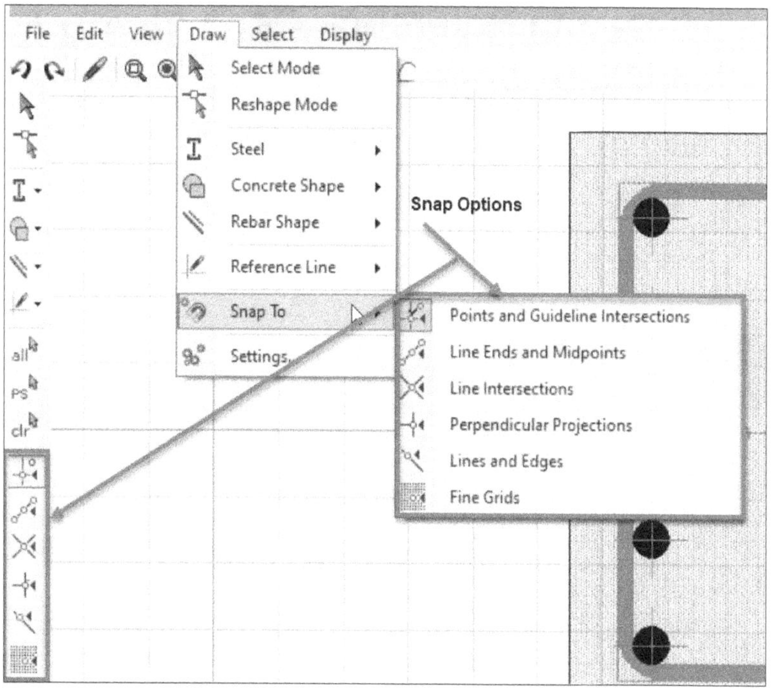

Figure-20. Snapping options

Select the **Point and Guideline intersection** option to snap cursor to the points in the object and points generated by entities at their intersection.

Select the **Line Ends and Midpoint** option to snap cursor to end points and mid points of the lines in the object.

Select the **Line Intersections** option to snap cursor to the intersection points of the lines.

Select the **Perpendicular Projections** option to snap cursor to the perpendicular projection of current geometry being created; refer to Figure-21.

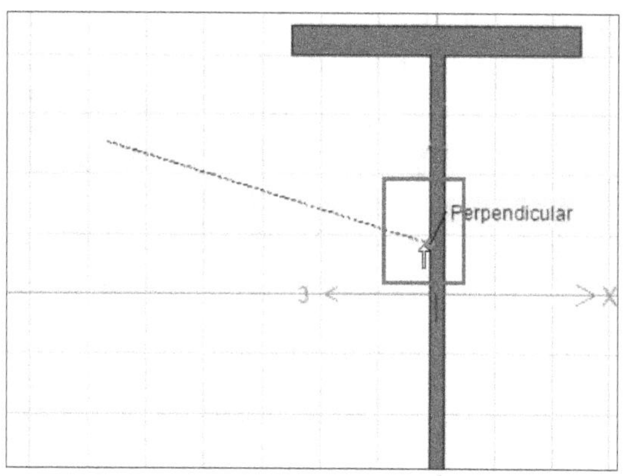

Figure-21. Snapping to perpendicular projection

Select the **Lines and Edges** option to snap cursor to lines and edges of the other objects in the section.

Select the **Fine Grids** option to snap cursor to grid lines.

Creating Reference Geometries

The tools to create reference geometries are available in the **Reference Line** cascading menu of the **Draw** menu and **Draw Reference Shape** toolbox; refer to Figure-22. The procedures to draw different reference shapes are given next.

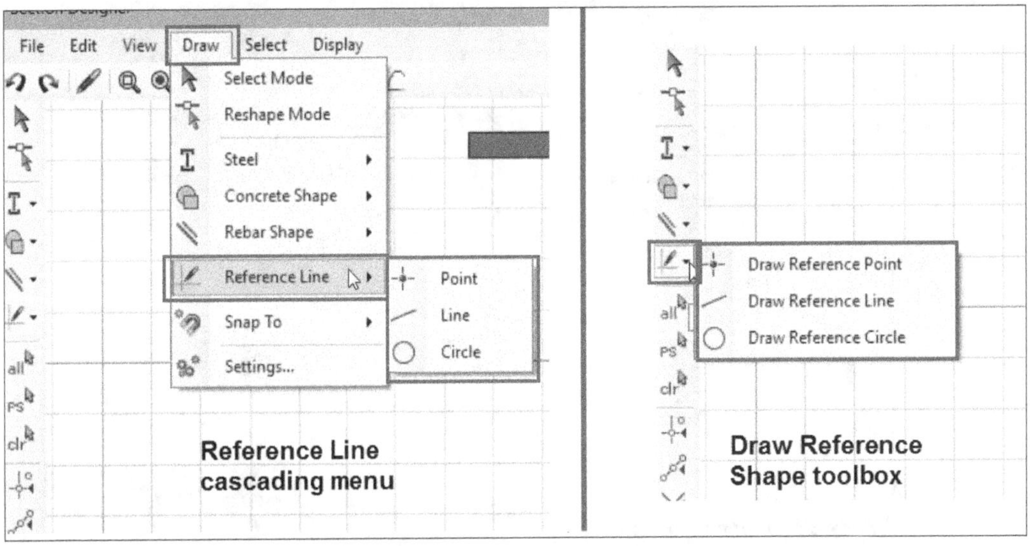

Figure-22. Draw Reference tools

Drawing Reference Point

- Click on the **Draw Reference Point** tool from the **Reference Line** cascading menu of the **Draw** menu. You will be asked to specify position of reference point.
- Click at desired location to place the reference point. The reference point will be created.

Drawing Reference Line

- Click on the **Draw Reference Line** tool from the **Reference Line** cascading menu of the **Draw** menu. You will be asked to specify the start point of the line.
- Click at desired location. A rubber band line will get attached to the cursor and you will be asked to specify the end point of the reference line.
- Click at desired location to draw the line.

Drawing Reference Circle

- Click on the **Draw Reference Circle** tool from the **Reference Line** cascading menu of the **Draw** menu. You will be asked to specify the center point of the circle.
- Click at desired location. The circle will be drawn. (What? You have not given radius/diameter!!)
- Note that the system automatically takes default diameter viz. 24 inches. To change the diameter, click on the **Select Object** tool from the left toolbar and select the circle. Now, right-click on the circle. The **Section Object Data - ReferenceCircle** dialog box will be displayed; refer to Figure-23. Note that a similar dialog box is displayed when you right-click on any object in the ETABS model. Using the options in this dialog box, you can change the parameters of the object.

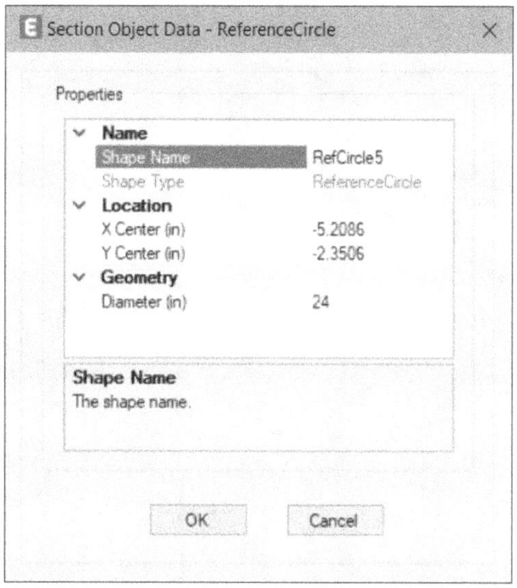

Figure-23. Section Object Data - ReferenceCircle dialog box

- Click in the **Diameter (in)** edit box of **Geometry** node and specify desired value of diameter.
- You can also change the other properties of reference circle. Click on the **OK** button from the dialog box to apply the changes.

Drawing Shapes

The tools to draw shapes are categorized in three sections: Steel Shapes, Concrete Shapes, and Rebar Shapes. In Section Designer, these tools are available in similarly named three cascading menus viz. **Steel**, **Concrete Shape**, and **Rebar Shape**; refer to Figure-24, Figure-25, and Figure-26.

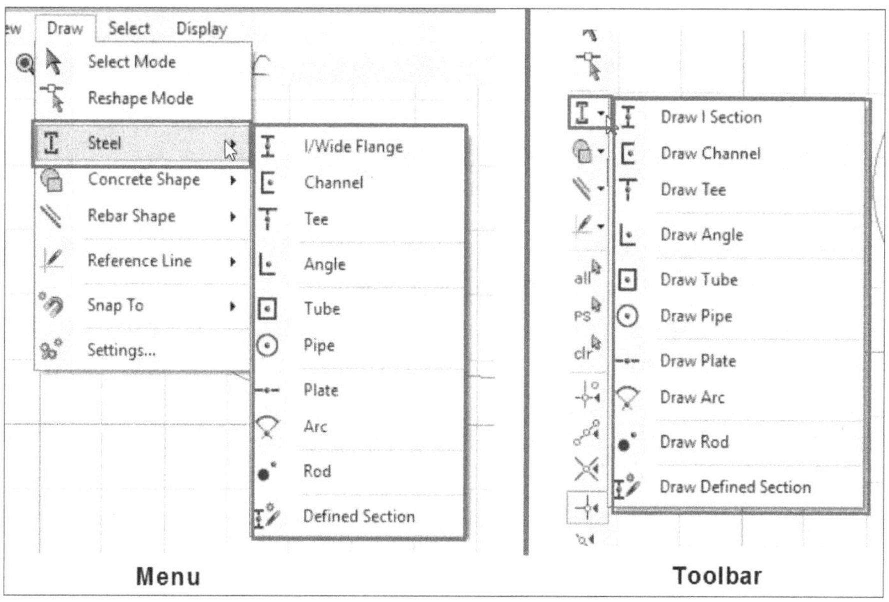

Figure-24. Steel Shapes

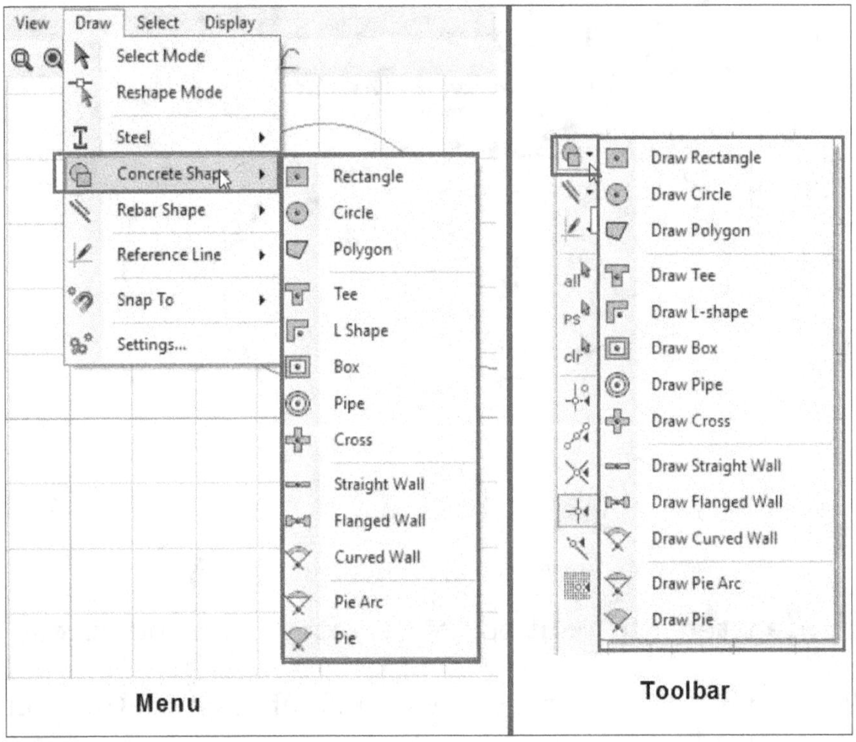

Figure-25. Concrete Shape tools

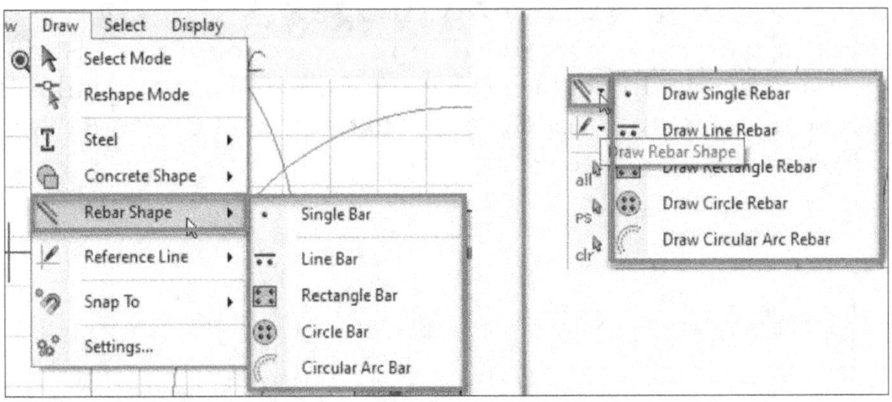

Figure-26. Rebar Shape tools

Drawing Shapes

- Click on the **Draw Channel** tool from the **Steel** cascading menu of the **Draw** menu or click on the **Draw Channel** tool from the **Draw Steel Shape** toolbox. You will be asked to specify the center location of steel channel.
- Click at desired location to place the channel. The channel will be drawn at specified location; refer to Figure-27. Select the channel and right-click on it to change its parameters.

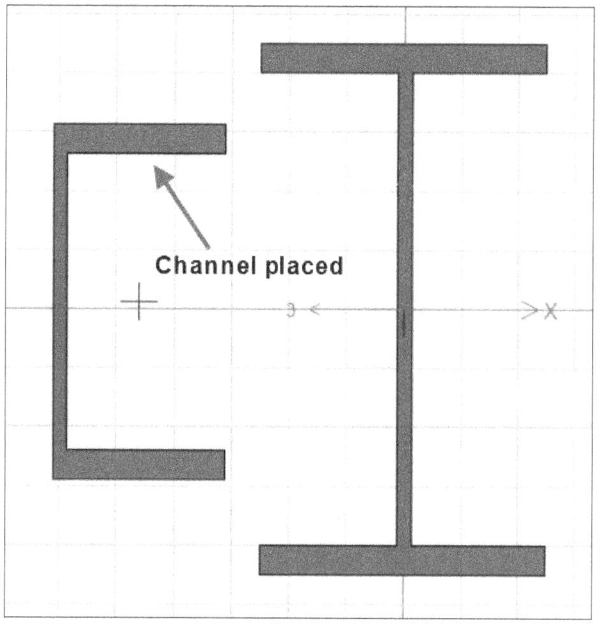

Figure-27. Channel placed

Similarly, you can draw other steel, concrete, and rebar shapes.

Drawing Defined Section

The **Draw Defined Section** tool in the **Steel** cascading menu of **Draw** menu is used to insert steel shapes already available in the library. The procedure to use this tool is given next.

- Click on the **Draw Defined Section** tool from the **Steel** cascading menu. You will be asked to specify the placement location.
- Click at desired location. The **Defined Steel Section** dialog box will be displayed; refer to Figure-28.

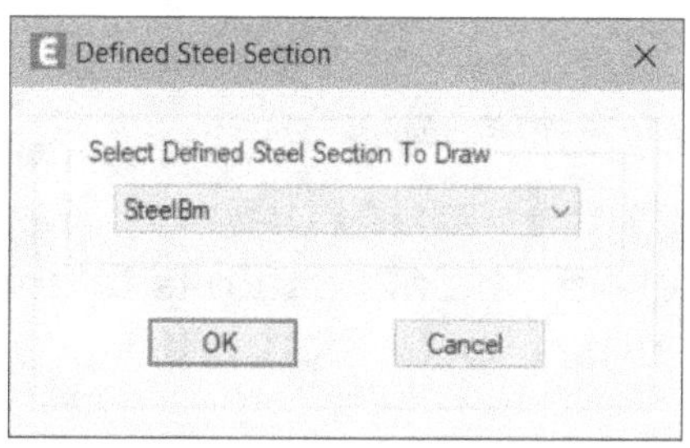

Figure-28. Defined Steel Section dialog box

- Select desired option from the drop-down and click on the **OK** button. The section will be placed.

Reshaping the Objects

The **Reshape Mode** option is used to activate the mode in which you can drag the points of the object to reshape it. The procedure is given next.

- Click on the **Reshape Mode** tool from the **Draw** menu. You will be asked to select a point on the model.
- Click at desired location on the object. Nodes will be displayed on the object to modify its shape; refer to Figure-29.

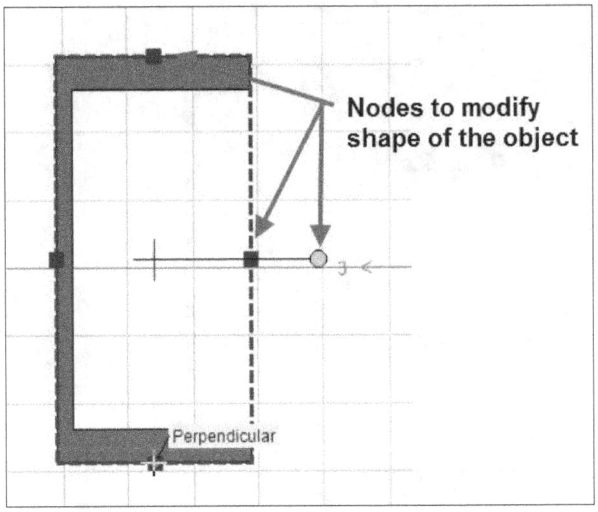

Figure-29. Nodes to modify shape of object

- Drag the nodes to modify the shape of object; refer to Figure-30.

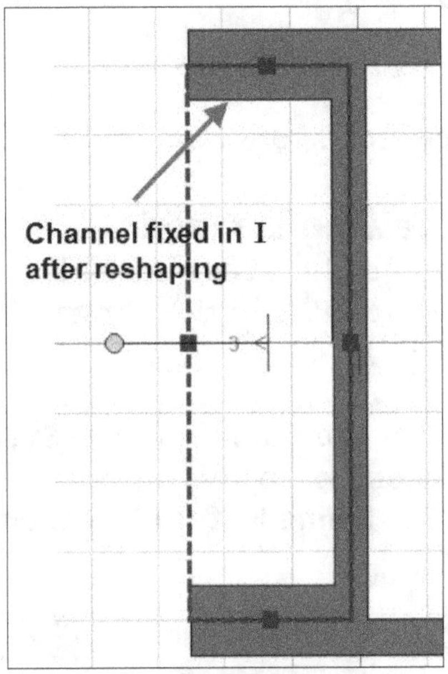

Figure-30. Reshaping objects

Checking Properties of Section

Once you have created the section, the next step is to check its properties. There are three tools in the **Display** menu; refer to Figure-31.

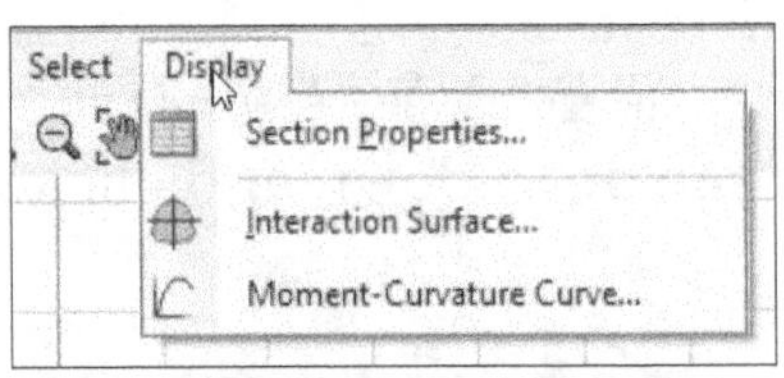

Figure-31. Display menu

The procedures to check different properties are given next.

Checking Section Properties

Section Properties are properties like center of gravity location, section modulus, moment of inertia, and so on. The procedure to check these properties for designed section is given next.

- Click on the **Section Properties** tool from the **Display** menu. The **Section Properties** dialog box will be displayed; refer to Figure-32.

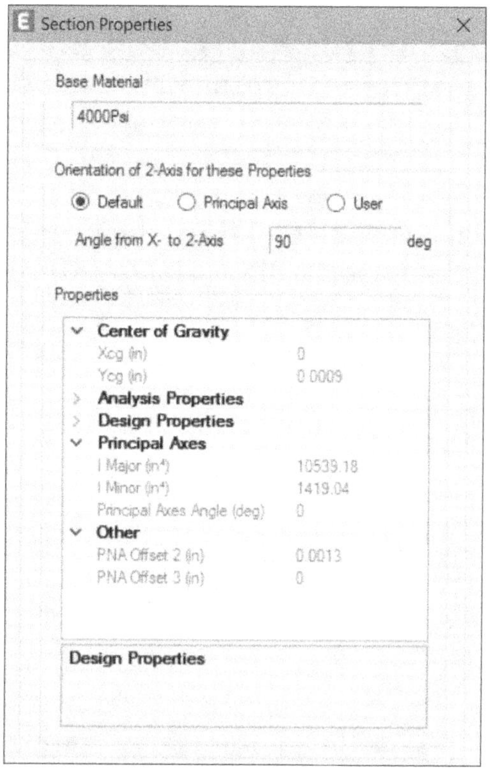

Figure-32. Section Properties dialog box

- Select desired radio button from the **Orientation of 2-Axis for these Properties** area to define orientation for checking properties.

Check the properties from the **Properties** area. Click on the **X** button at the top right corner to close the dialog box. Note that reinforced bars are not considered in section properties and following section properties are displayed in the **Properties** area of the dialog box.

X_{cg} and Y_{cg}	Location of Center of Gravity.
A	Area of Section.
AS2	Shear area for shear parallel to the 2-axis.
AS3	Shear area for shear parallel to the 3-axis.
I22 and I33	Moment of Inertia about 2-axis and 3-axis, respectively
I23	Moment of Inertia when axis 2 and 3 are not principle axis which is never the case in ETABS. So, the value of I23 is always zero.
J	Torsional Constant.
R22 and R33	Radius of Gyration about 2-axis and 3-axis.
S22(+), S22(-)	Section modulus about the 2-axis at extreme fiber of the section in the positive and negative 3-axis direction, respectively.

S33(+), S33(-)	Section modulus about the 3-axis at extreme fiber of the section in the positive and negative 2-axis direction, respectively.
Z22 and Z33	Plastic Modulus of bending for 2-axis and 3-axis respectively.
I Major	Moment of Inertia about major principle axis.
I Minor	Moment of Inertia about minor principle axis.
PNA Offset 2	Offset distance from Center of Gravity along 2-axis.
PNA Offset 3	Offset distance from Center of Gravity along 3-axis.

Checking the Interaction Surface

The **Interaction Surface** tool in the **Display** menu is used to check the interaction between reinforcement and composite material around it in concrete sections. The procedure is given next.

- Click on the **Interaction Surface** tool from the **Display** menu. The **Interaction Surface (ACI 318-14)** dialog box will be displayed; refer to Figure-33.

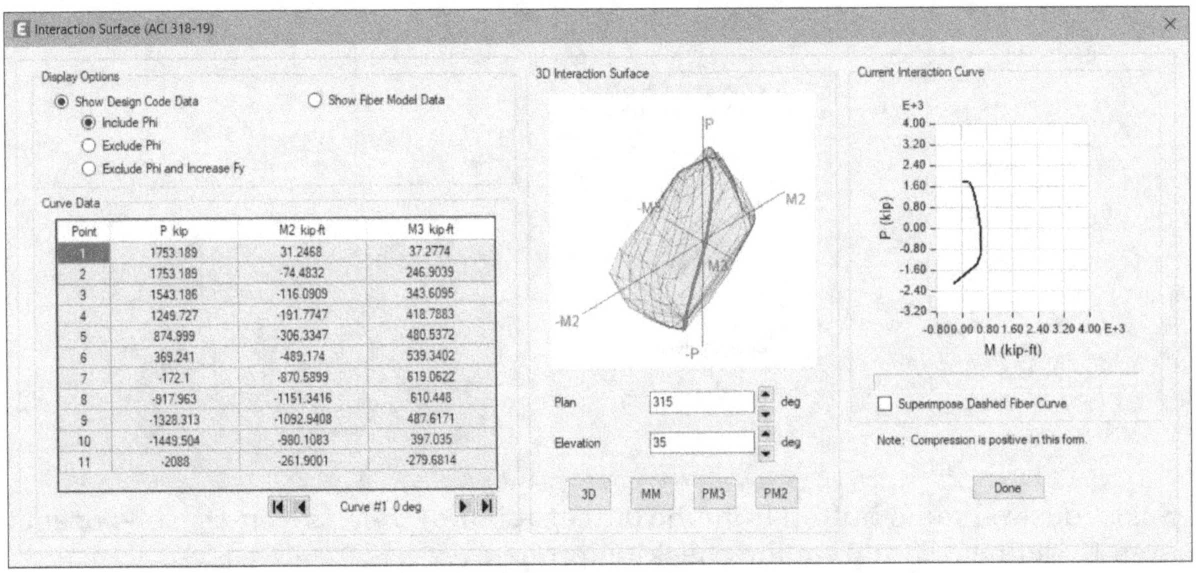

Figure-33. Interaction Surface dialog box

- Check desired modes of the interaction by selecting the button from the **3D**, **MM**, **PM3**, and **PM2** buttons.
- After checking the properties, click on the **Done** button.

Checking Moment Curvature Plot

The moment curvature plot is used to check the capability of reinforced concrete to bear load up to specified curvature. The procedure to check the plot is given next.

- Click on the **Moment-Curvature Curve** tool from the **Display** menu. The **Moment Curvature Plot** dialog box will be displayed; refer to Figure-34.

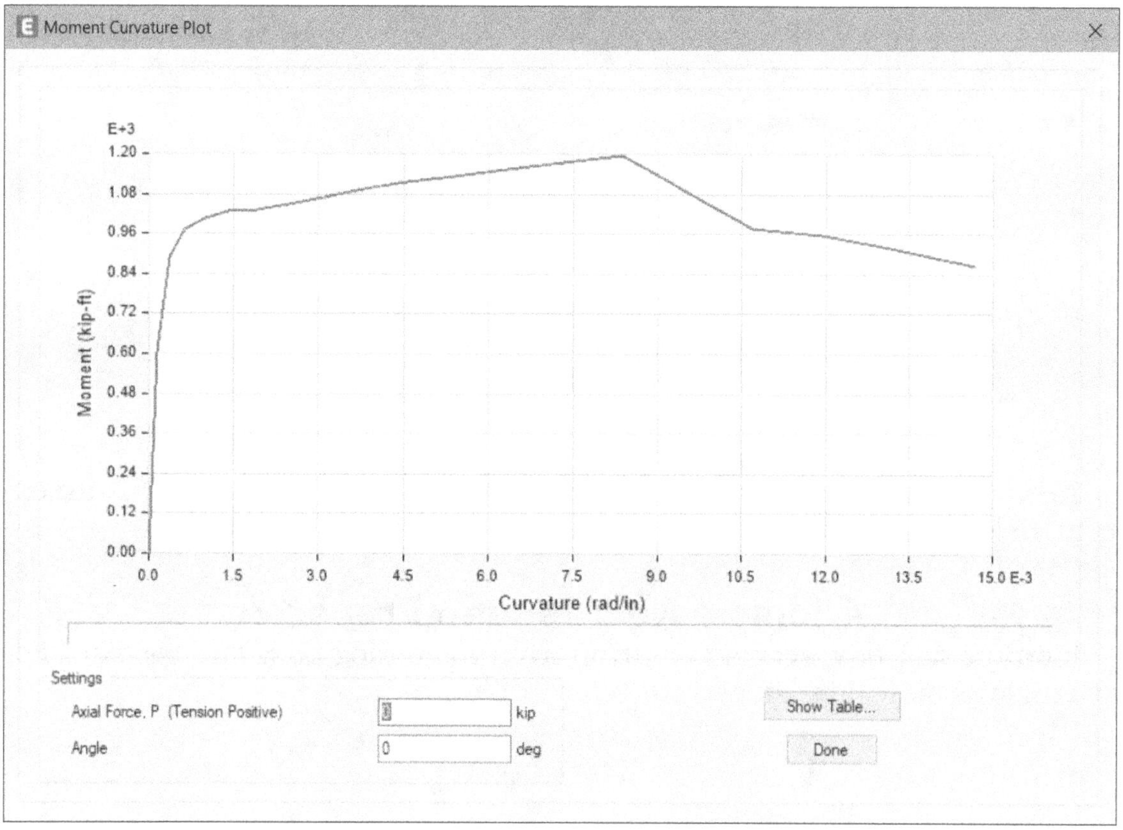

Figure-34. Moment Curvature Plot dialog box

- The vertical scale shows moment and horizontal scale shows curvature of the concrete section. Set desired values in **Axial Force** and **Angle** edit boxes in the **Settings** area to check plot in different loading conditions.
- Click on the **Show Table** button to check the data in table form.
- After checking the parameters, click on the **Done** button to exit dialog box.

Once you are satisfied with the section design, click on the **OK** button from the **Section Designer** window.

DEFINING TENDON SECTIONS

The tendons are used to simulate effect of pre-stressing and post-tensioning on the model. The procedure to define tendon section is given next.

- Click on the **Tendon Sections** tool from the **Section Properties** cascading menu of the **Define** menu. The **Tendon Properties** dialog box will be displayed; refer to Figure-35.

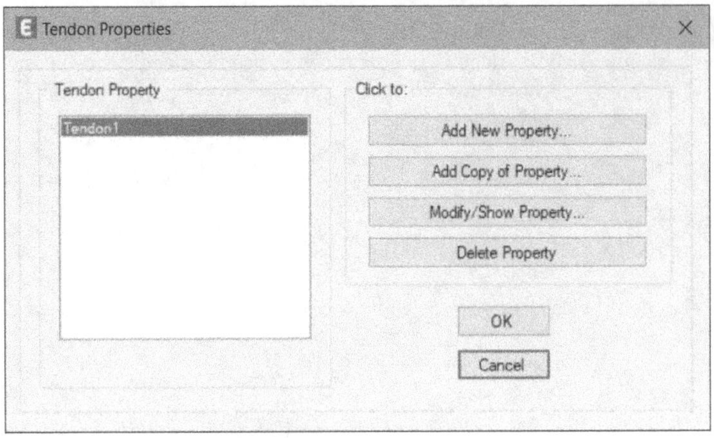

Figure-35. Tendon Properties dialog box

Using the options in this dialog box, you can add a new property, modify an existing property, delete a property, and so on. The procedures are given next.

Adding a New Tendon Property

- Click on the **Add New Property** button from the dialog box. The **Tendon Property Data** dialog box will be displayed; refer to Figure-36.

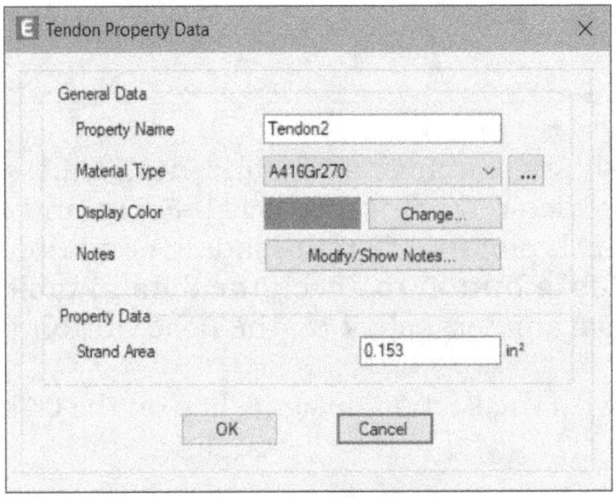

Figure-36. Tendon Property Data dialog box

- Specify desired name for the tendon property in **Property Name** edit box.
- Click on the **Add Material** button for **Material Type** and select desired material.
- Specify desired strand area value in **Strand Area** edit box and set the other options as required.
- Click on the **OK** button from the dialog box to create the new property.

Adding Copy of a Tendon Property

Click on the **Add Copy of Property** button after selecting the property that you want to be copied from the list. The **Tendon Property Data** dialog box will be displayed as discussed earlier with copied property data. Set desired parameters and click on the **OK** button to create the new property.

Modifying a Tendon Property

To modify an existing property, select it from the list and click on the **Modify/Show Property** button from the dialog box. The **Tendon Property Data** dialog box will be displayed as discussed earlier. Set desired parameters and click on the **OK** button to modify the property.

Deleting a Tendon Property

To delete a tendon, select it from the list and click on the **Delete Property** button.

After setting desired parameters, click on the **OK** button from the **Tendon Properties** dialog box to exit.

DEFINING SLAB SECTIONS

Slabs are the reinforced concrete blocks used to construct floors or ceilings. The procedure to define sections for slabs is given next.

- Click on the **Slab Sections** tool from the **Section Properties** cascading menu in the **Define** menu. The **Slab Properties** dialog box will be displayed; refer to Figure-37.

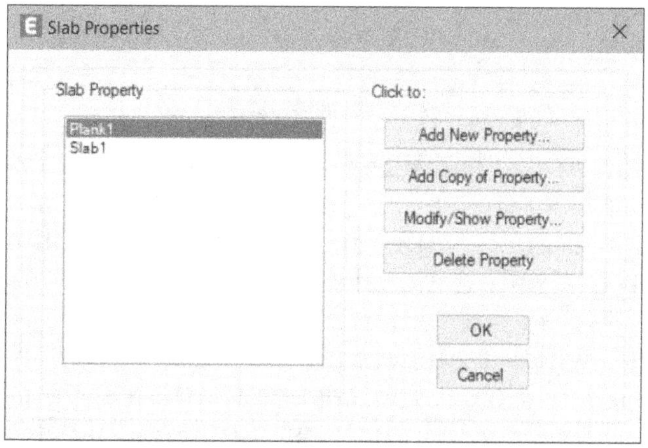

Figure-37. Slab Properties dialog box

- Click on the **Add New Property** button from the dialog box to create a new slab property. The **Slab Property Data** dialog box will be displayed; refer to Figure-38.

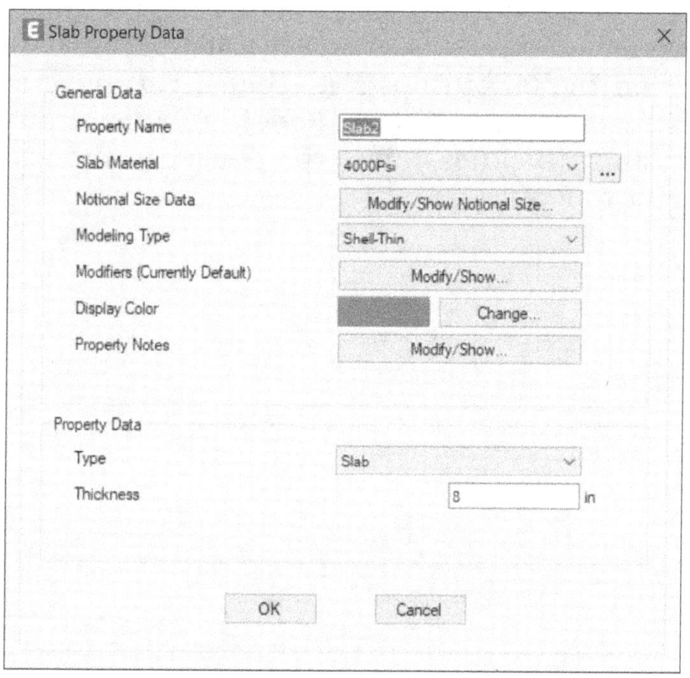

Figure-38. Slab Property Data dialog box

- Click in the **Property Name** edit box and specify desired name.

- From the **Slab Material** drop-down, select desired material for slab.
- Click on the **Modify/Show Notional Size** button from the dialog box. The **Time Dependent Parameters** dialog box will be displayed; refer to Figure-39. Specify desired value of notional size in the edit box after selecting the **User-defined** radio button. You can select the **Auto** radio button if you want this value to be specified automatically. In the **Factor** edit box, specify desired multiplication factor for notional size. Click on the **OK** button from the dialog box to apply the parameters.

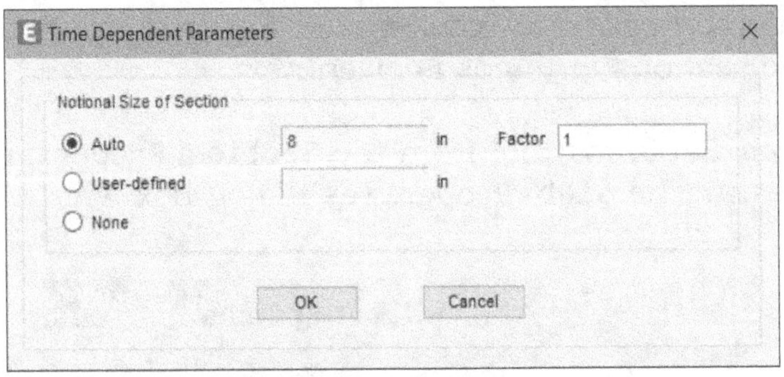

Figure-39. Time Dependent Parameters dialog box

- Select desired option in the **Modeling Type** drop-down to define how slab will be modeled in ETABS. Select the **Shell-Thin** option if you want to neglect transverse shear deformation and make your model follow Kirchhoff application. Select the **Shell-Thick** option to create a model which follows Mindlin/Reissner application and considers transverse shear deformation. Select the **Membrane** option if you want to model a slab which transfers all the load to supporting structure without taking any deformation. Select the **Layered** option if you want to model a slab with different materials layered in it. This option enables to create mixed material composite slabs. After selecting the **Layered** option, click on the **Modify/Show Layered Slab Data** button. The **Slab Property Layer Definition Data** dialog box will be displayed; refer to Figure-40. Click on the **Add** button in the dialog box. A layer will be added for the current slab. Modify the parameters in the table and add the other layers as required by using the **Add** button. After setting the other parameters, click on the **OK** button. Note that if you have selected **Shell-Thin**, **Shell-Thick**, or **Membrane** option in the **Modeling Type** drop-down then you will be asked to specify the thickness and type of slab in the **Property Data** area of the dialog box; refer to Figure-41.

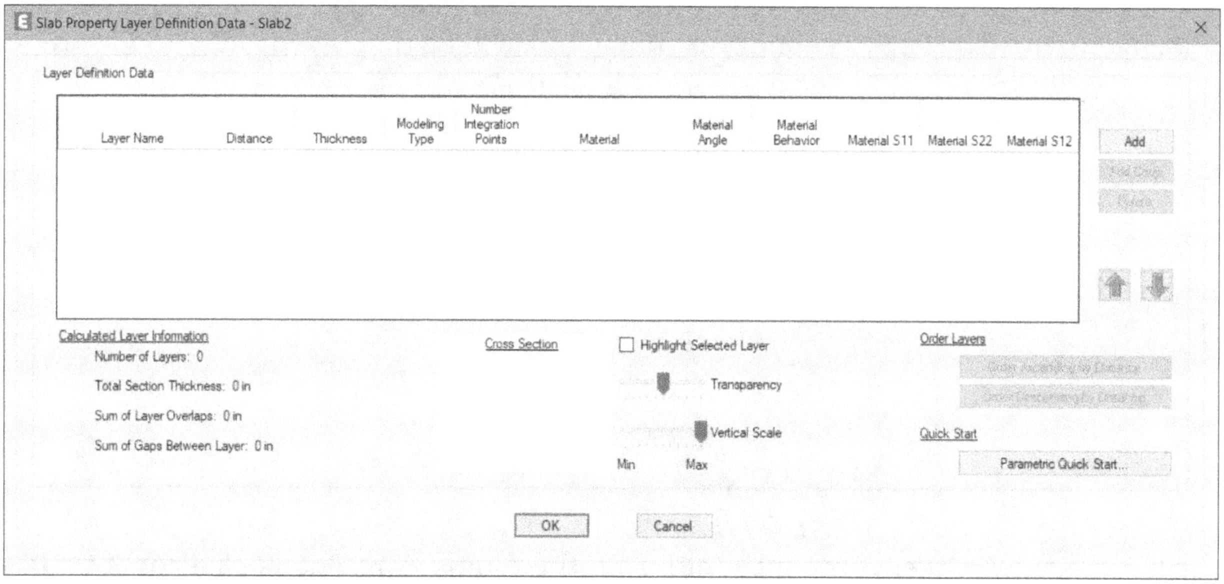

Figure-40. Slab Property Layer Definition Data dialog box

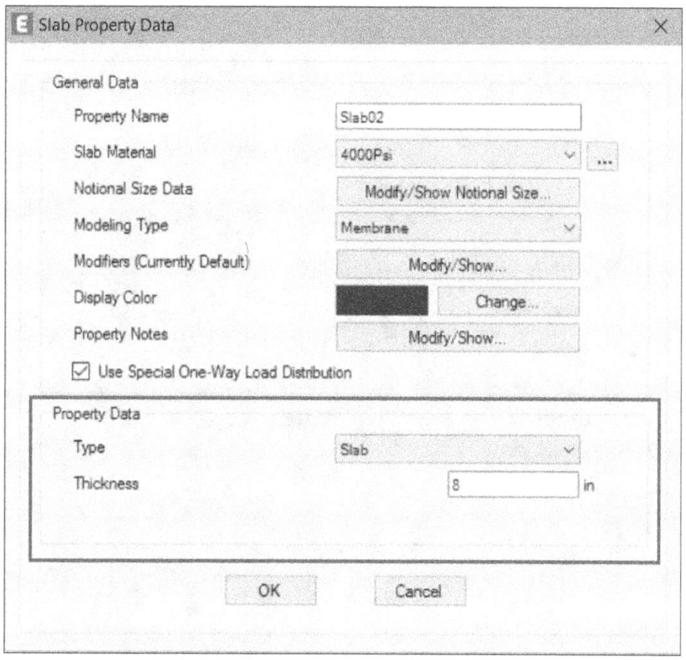

Figure-41. Property Data area

- Select desired slab type option from the **Type** drop-down and specify the related size parameters.
- Similarly, set the other parameters in the **Slab Property Data** dialog box and click on the **OK** button. The **Slab Properties** dialog box will be displayed again.

You can use the other tools in the **Slab Properties** dialog box in the same way as discussed earlier. Click on the **OK** button from the dialog box to exit.

DEFINING DECK SECTIONS

Deck is a flat surface similar to floor but constructed outdoor generally elevated from ground. The procedure to define section properties of deck is given next.

- Click on the **Deck Sections** tool from the **Section Property** cascading menu in the **Define** menu. The **Deck Properties** dialog box will be displayed; refer to Figure-42.

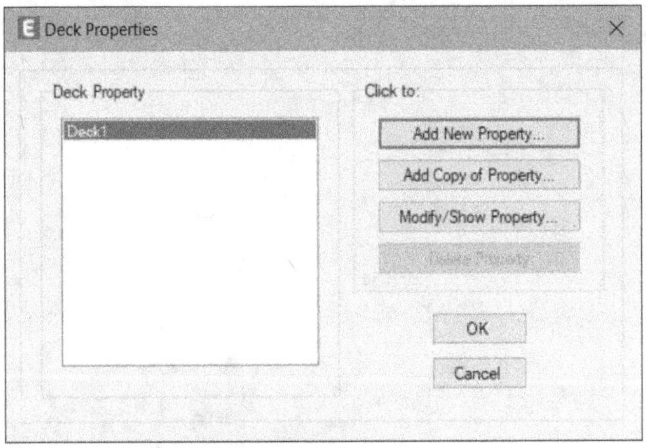

Figure-42. Deck Properties dialog box

- Click on the **Add New Property** tool from the dialog box. The **Deck Property Data** dialog box will be displayed; refer to Figure-43.
- Set desired parameters for the deck section and click on the **OK** button from the dialog box.

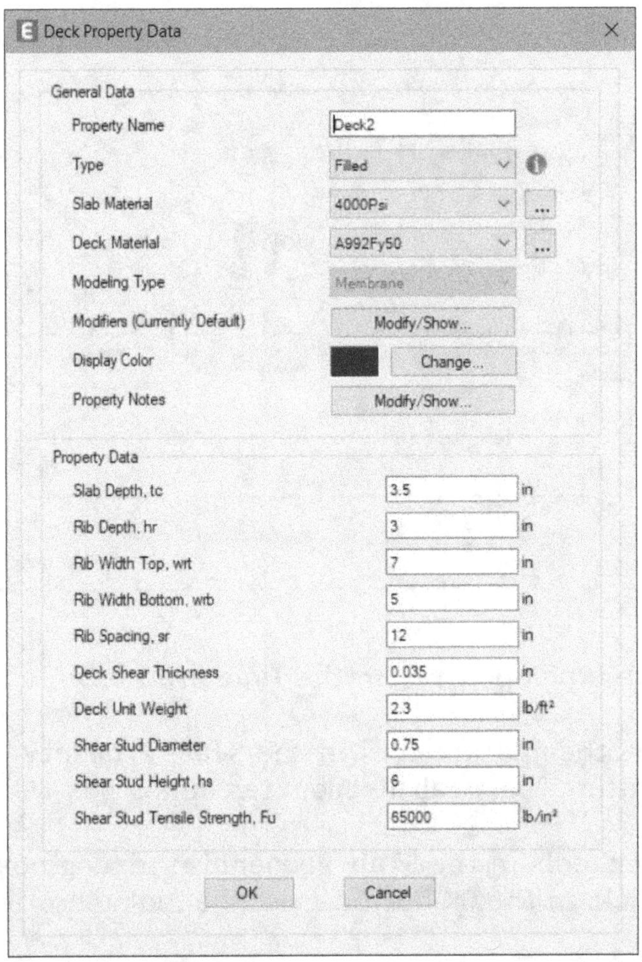

Figure-43. Deck Property Data dialog box

- You can use the other options in the **Deck Properties** dialog box in the same way as discussed earlier.

DEFINING WALL SECTION

The **Wall Sections** tool in the **Section Properties** cascading menu of the **Define** menu is used to define section properties for wall. The procedure to use this tool is same as discussed for the other tools in the **Section Properties** cascading menu.

DEFINING REINFORCING BAR SIZES

The **Reinforcing Bar Sizes** tool is used to define or modify the bar sizes used in reinforcing concrete. The procedure to use this tool is given next.

- Click on the **Reinforcing Bar Sizes** tool from the **Section Properties** cascading menu of the **Define** menu. The **Reinforcing Bar Sizes** dialog box will be displayed; refer to Figure-44.

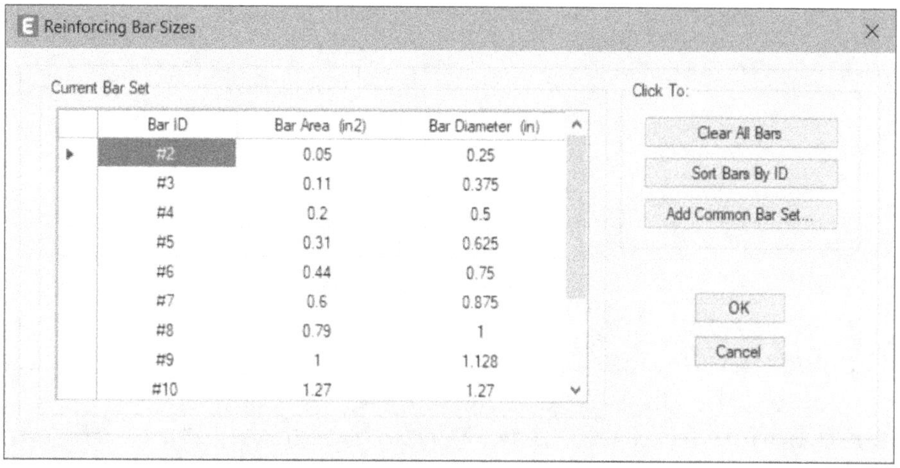

Figure-44. Reinforcing Bar Sizes dialog box

- Click on the **Add Common Bar Set** button to add the standard bar sizes in the list. The **Select Common Rebar Set** dialog box will be displayed; refer to Figure-45.

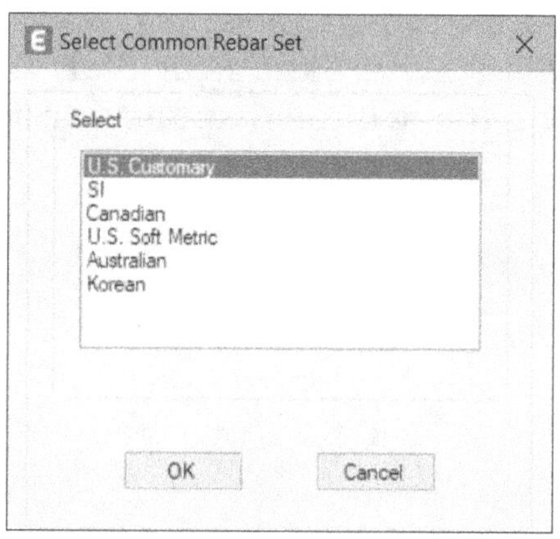

Figure-45. Select Common Rebar Set dialog box

- Select desired option from the dialog box and click on the **OK** button to add the selected rebar set in the list.
- To manually add a rebar id and size, click twice with a pause in the empty field of the table and specify desired parameters.
- To modify any field, click twice with a pause on it and set desired value.
- Click on the **Clear All Bars** button from the dialog box to clear the table.
- Click on the **OK** button from the dialog box to apply the changes.

DEFINING LINK/SUPPORT PROPERTIES

Link is used to join two points separated by some distance. The **Link/Support Properties** tool is used to define link or support properties. The procedure to use this tool is given next.

- Click on the **Link/Support Properties** tool in the **Section Properties** cascading menu of the **Define** menu. The **Define Link Properties** dialog box will be displayed; refer to Figure-46.

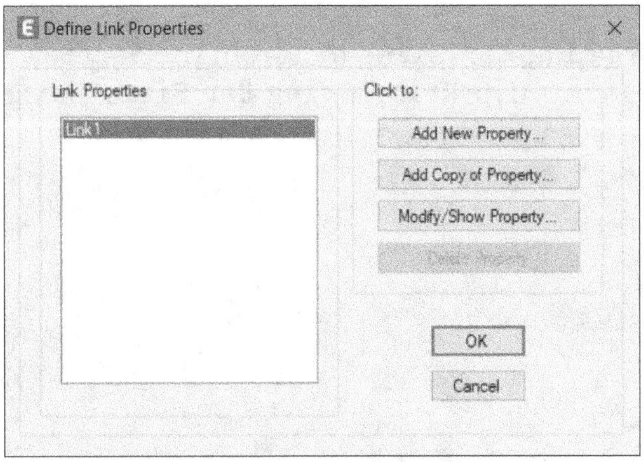

Figure-46. Define Link Properties dialog box

- Click on the **Add New Property** button from the dialog box. The **Link Property Data** dialog box will be displayed; refer to Figure-47.
- Select desired option from the **Link Type** drop-down. There are various options in this drop-down to define links for analysis purposes like Linear, Damper-Exponential, Damper-Bilinear, and so on.
- Set the other properties of link as required like mass, weight, inertia, and directional properties.
- Click on the **Modify/Show** button for **P-Delta Parameters** in the dialog box to define parameters for P-delta combinations of load on the link. The **Advanced Link P-Delta Parameters** dialog box will be displayed; refer to Figure-48.

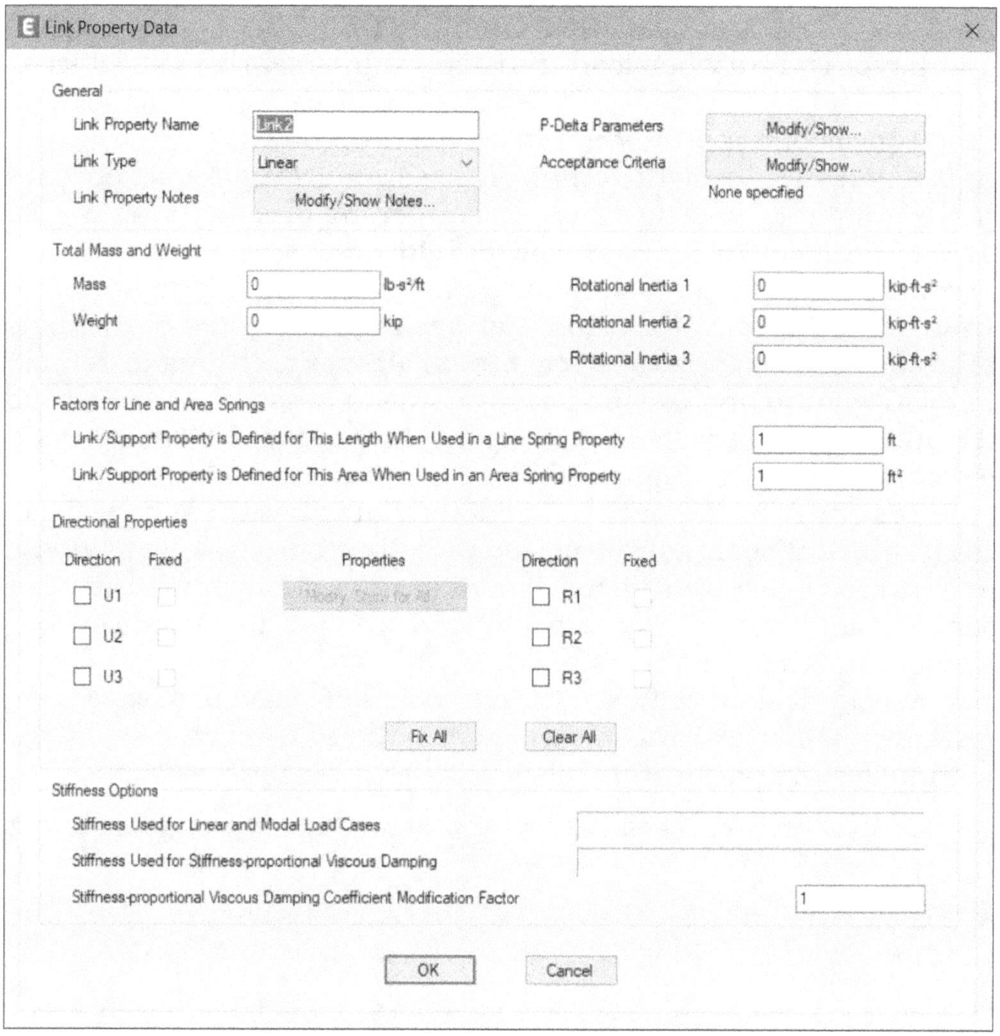

Figure-47. Link Property Data dialog box

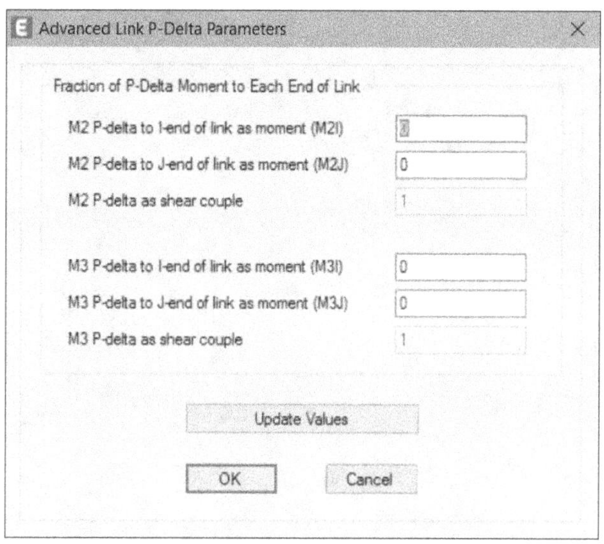

Figure-48. Advanced Link P-Delta Parameters dialog box

- Set desired parameters in the dialog box to define P-Delta load equation. Note that you should use the P-Delta iteration only when you are concerned about local buckling of structural members. Otherwise you should avoid it because it takes a lot of computing power to solve these iterations.

Linear Link Type

The linear link type is used when load transfer occurs linearly in specified directions.

- Select the **Linear** option from the **Link Type** drop-down.
- Specify the mass and weight of link in **Mass** and **Weight** edit boxes. Similarly, specify the rotational inertia values for different directions in the **Rotational Inertia** edit boxes of **Total Mass and Weight** area.
- Specify multiplication load factors for using link as line spring or area spring in edit boxes of the **Factors for Line and Area Springs** area of the dialog box.
- Select the check boxes from the **Directional Properties** area to define directions in which you want to apply load transfer using link and click on the **Modify/Show for All** button. The **Linear Link/Support Directional Properties** dialog box will be displayed; refer to Figure-49.
- Select the **Yes** check boxes for the directions that you want to have fixed stiffness.
- By default, the **Stiffness is Uncoupled** radio button is selected and hence force applied in x direction causes displacement in x direction only. Select the **Stiffness is Coupled** radio button if you want load applied in one direction to cause displacement in other directions also.
- Similarly, specify desired values in the edit boxes of **Shear Distance** area to specify location where shear deformation will occur due to loading.
- After setting desired parameters, click on the **OK** button.

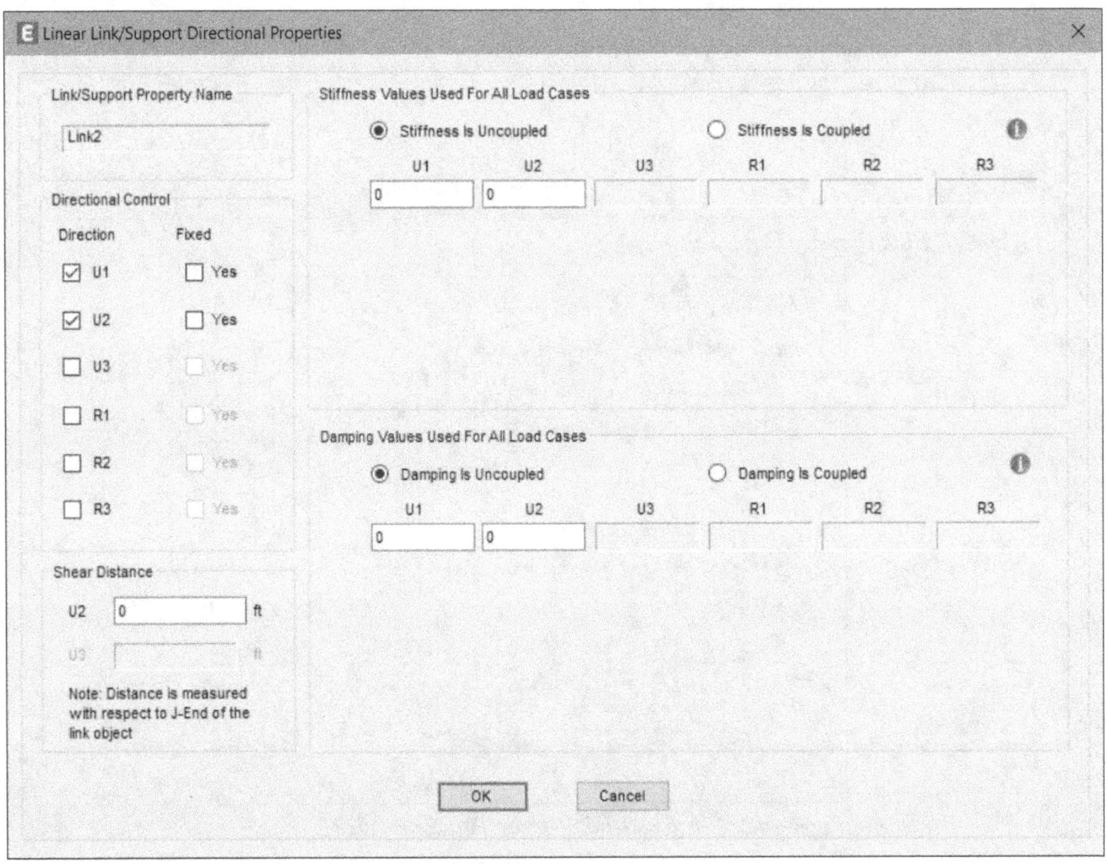

Figure-49. Linear Link/Support Directional Properties dialog box

Damper - Exponential Link Type

The damper - exponential link type is used when load transfer occurs with linear or non-linear damping.

- Select the **Damper - Exponential** option from the **Link Type** drop-down in the dialog box. The options will be displayed as shown in Figure-50.
- Select the check boxes for directions in which you want load transfer to occur in the **Directional Properties** area.
- If you want load transfer to occur non-linearly then select the **NonLinear** check boxes for desired directions.
- Click on the **Modify/Show** button in **Properties** column to specify loading parameters. The **Link/Support Directional Properties** dialog box will be displayed; refer to Figure-51.

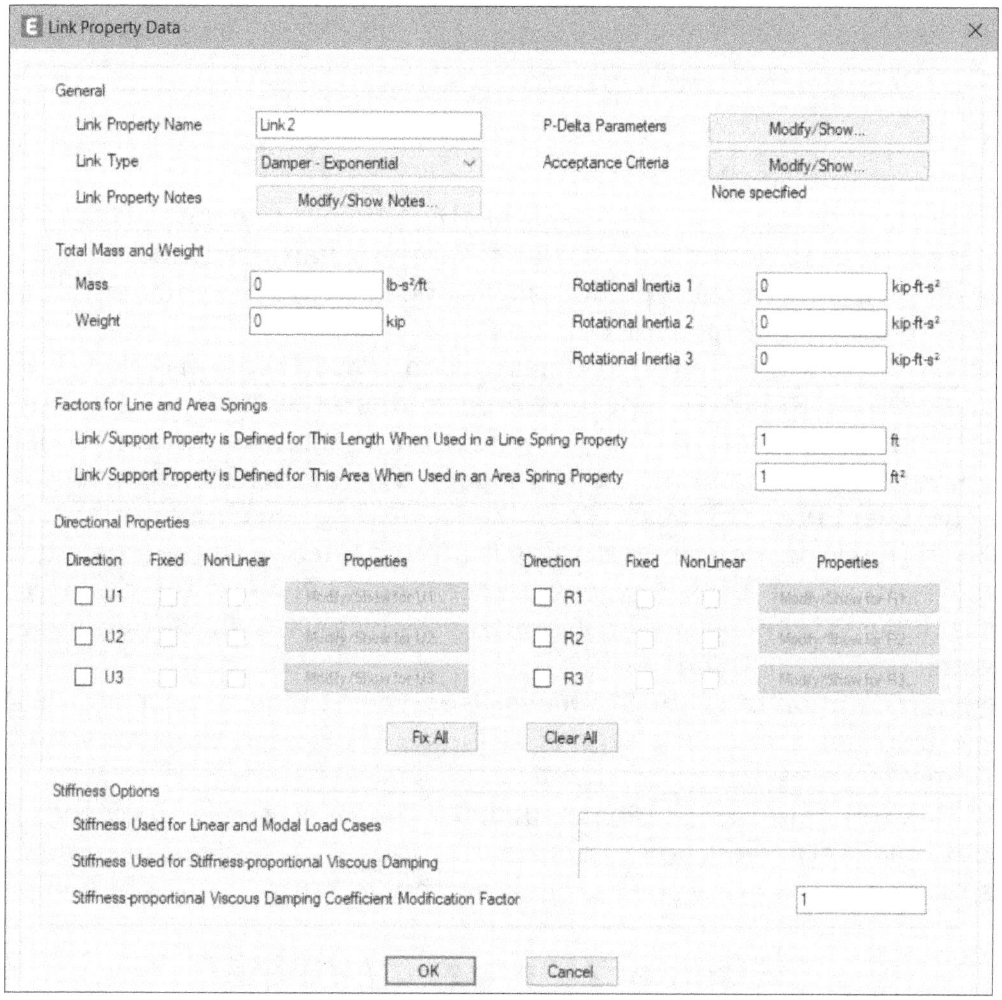

Figure-50. Link Property Data dialog box for exponential damper

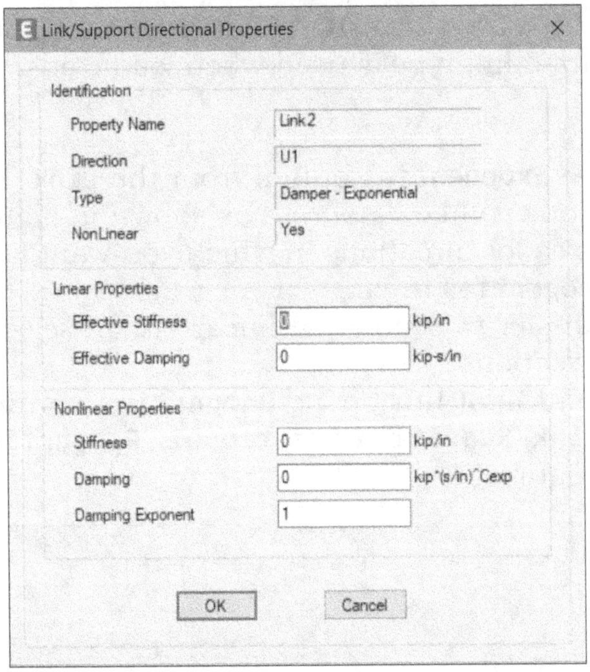

Figure-51. Link/Support Directional Properties dialog box

- Specify desired values in edit boxes of the **Linear Properties** and **Nonlinear Properties** areas to define stiffness and damping values. Since, we have selected the **Damper - Exponential** type link so you need to specify the damping exponent value in the **Damping Exponent** edit box.
- Select desired option from the **Stiffness Used for Linear and Modal Load Cases** drop-down of the **Stiffness Options** area to define what type of stiffness matrix will be used for loading. Select the **Effective Stiffness** option to use effective stiffness value specified in the **Link/Support Directional Properties** dialog box. Select the **Effective Stiffness from Zero, Else Nonlinear** option to effective stiffness in the beginning when loading is almost linear and then use the nonlinear stiffness value of the **Link/Support Directional Properties** dialog box. Select the **Nonlinear Stiffness** option from the drop-down to use nonlinear stiffness value specified in the dialog box.
- Select desired option from the **Stiffness Used for Stiffness-proportional Viscous Damping** drop-down to define type of stiffness to be used under viscous damping conditions.
- Specify desired value in **Stiffness-proportional Viscous Damping Coefficient Modification Factor** edit box to define multiplication factor damping coefficient.
- After setting desired parameters, click on the **OK** button from the dialog box.

Damper - Bilinear Link Type

The damper - bilinear link type is used when load transfer occurs with linear or non-linear damping link which has damping effect on both sides of piston movement of damper.

- Select the **Damper - Bilinear** option from the **Link Type** drop-down and select the check boxes for directions in which load/support will be applied by link.
- Select the NonLinear check boxes for link direction which will behave non-linearly under loading.

- Click on the **Modify/Show** button for the direction whose parameters are to be changed. The **Link/Support Directional Properties** dialog box will be displayed; refer to Figure-52.
- Specify desired values in the **Linear Properties** and **Nonlinear Properties** area of the dialog box to define stiffness and damping parameters of link.
- Note that you can specify the different values for initial damping and yield damping coefficients in the **Nonlinear Properties** area. After setting desired parameters, click on the **OK** button from the dialog box.

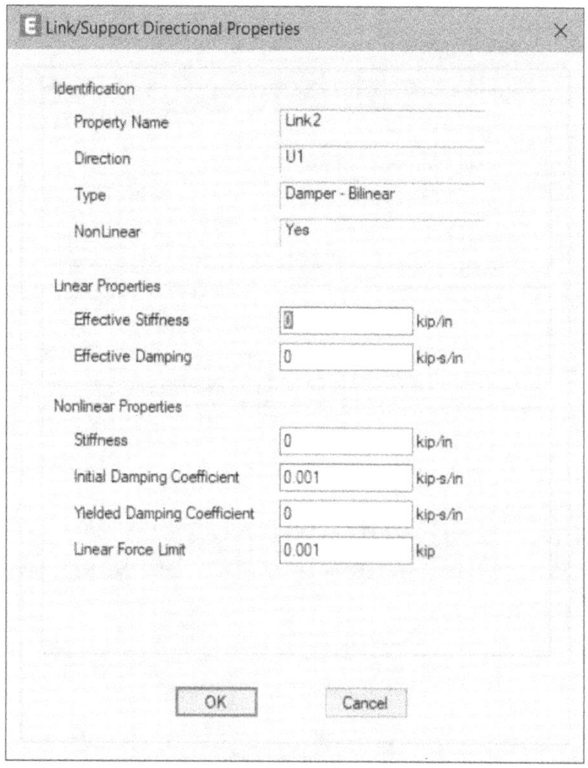

Figure-52. Link Support Directional Properties dialog box for bilinear damper

Damper - Friction Spring Link Type

The damper - friction spring link type link if the damper used in building are of friction spring type.

- Select the **Damper - Friction Spring** option from the **Link Type** drop-down.
- Select the check boxes for directions and also select the **NonLinear** check boxes to define nonlinear behavior of links.
- Click on the **Modify/Show** button for the direction whose parameters are to be changed. The **Link/Support Directional Properties** dialog box will be displayed; refer to Figure-53.
- Specify the effective stiffness and damping values in the **Linear Properties** area.
- Select desired option from the **Active Direction** drop-down to define in which direction damping will act. You can select **Compression**, **Tension**, or **Both** option to define the direction.
- Specify the nonlinear stiffness properties and displacement parameters of spring in the **Nonlinear Properties** area of the dialog box.
- Click on the **OK** button to apply the properties.

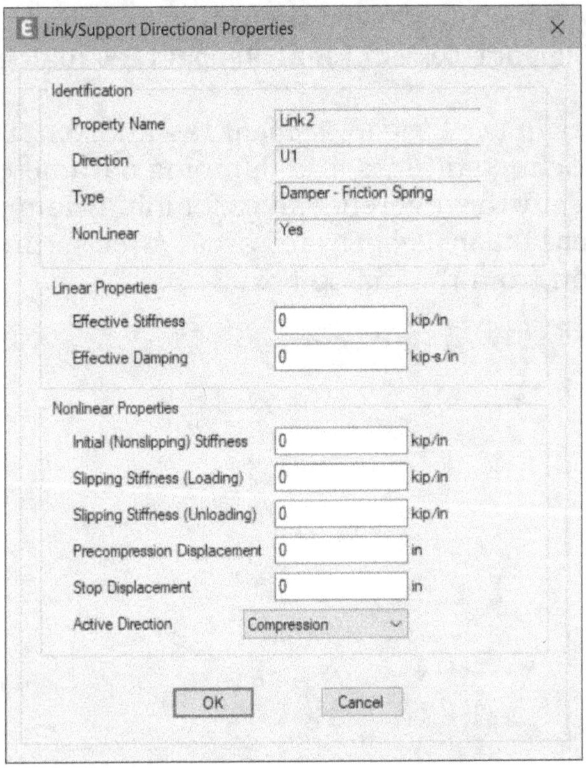

Figure-53. Link Support Directional Properties dialog box for Friction spring damper

Gap Link Type

The gap link type is used when there is a gap between two structural members of the model; refer to Figure-54.

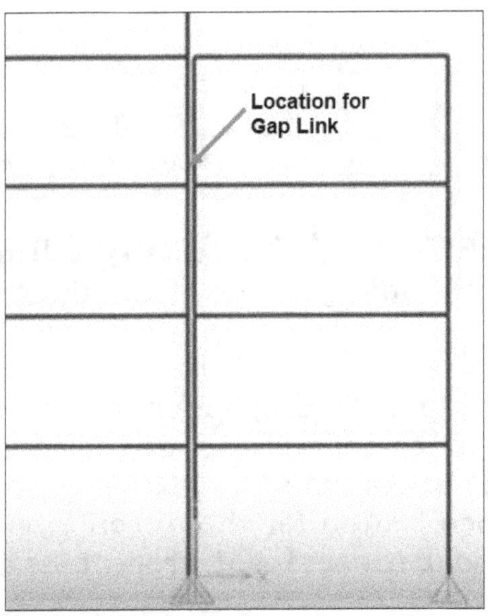

Figure-54. Gap link location

- Select the **Gap** option from the **Link Type** drop-down and set the other parameters as discussed earlier.
- If the link behaves non-linearly then select the **NonLinear** check boxes for selected directions and click on the **Modify/Show** button. The **Link/Support Directional Properties** dialog box will be displayed; refer to Figure-55.

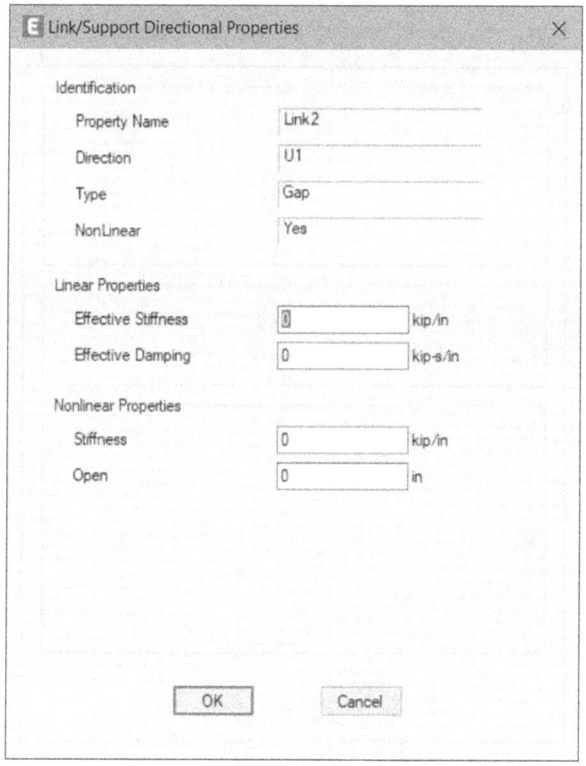

Figure-55. Link Support Directional Properties dialog box for gap link

- Specify desired value of gap between two structural members in the **Open** edit box and set the other parameters as desired.
- Click on the **OK** button from the dialog box to apply the parameters.

Similarly, you can use the other options in the **Link Type** drop-down to define the properties of the link.

- Click on the **OK** button from the **Link Property Data** dialog box to create the link property. The **Define Link Properties** dialog box will be displayed again with new link property in the list.

You can use the other tools in the **Define Link Properties** dialog box as discussed earlier. Click on the **OK** button after performing required changes.

DEFINING FRAME/WALL HINGE PROPERTIES

The Frame/wall hinges are used to provide joints in different frame/wall sections. The procedure to define these properties is given next.

- Click on the **Frame/Wall Nonlinear Hinges** tool from the **Section Properties** cascading menu of the **Define** menu. The **Define Frame/Wall Hinge Properties** dialog box will be displayed; refer to Figure-56.
- Click on the **Add New Property** button from the dialog box to create a new property. The **Default For Added Hinges** dialog box will be displayed; refer to Figure-57.

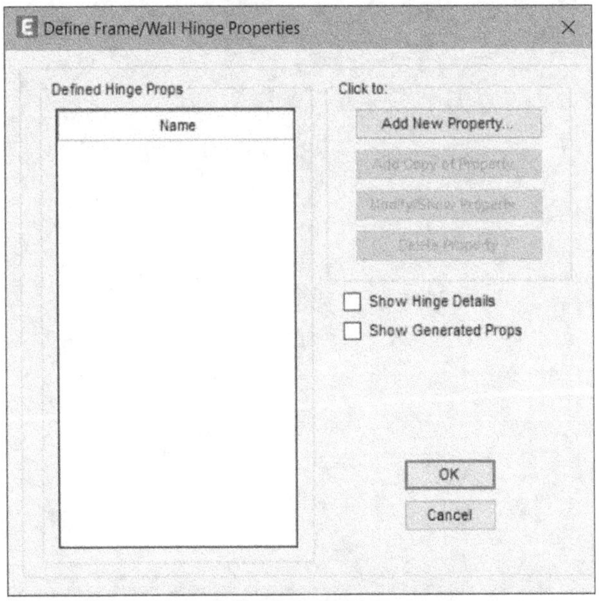

Figure-56. Define Frame/Wall Hinge Properties dialog box

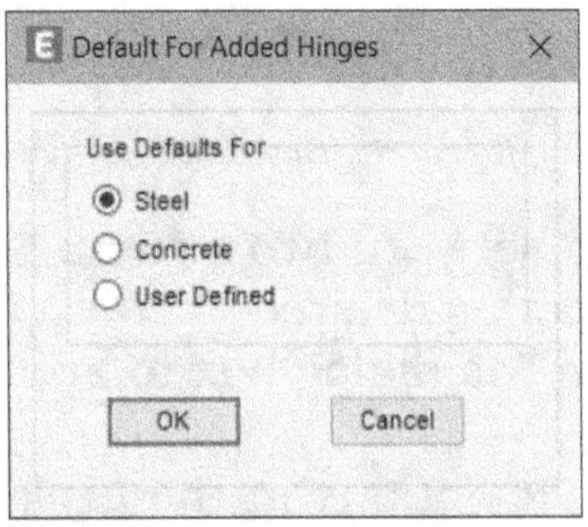

Figure-57. Default For Added Hinges dialog box

- Select desired radio button from the dialog box and click on the **OK** button. The **Hinge Property Data** dialog box will be displayed; refer to Figure-58.

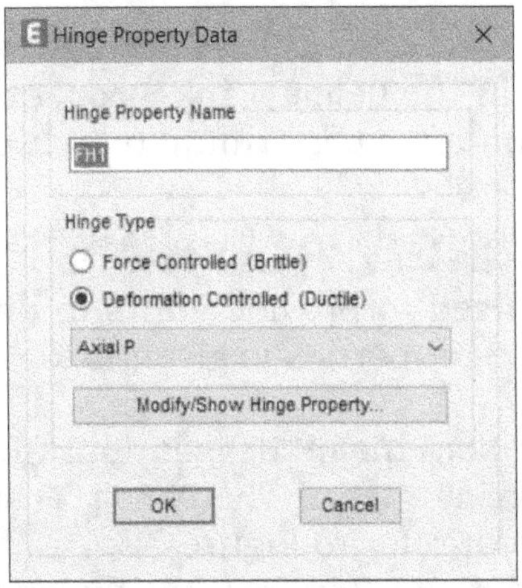

Figure-58. Hinge Property Data dialog box

- Specify the property name in the **Hinge Property Name** edit box.
- Set desired radio button from the **Hinge Type** area of the dialog box to define whether the hinge is brittle or ductile.
- Click in the drop-down and select desired option (like we have selected Axial P with ductile function). The related dialog box will be displayed; refer to Figure-59.

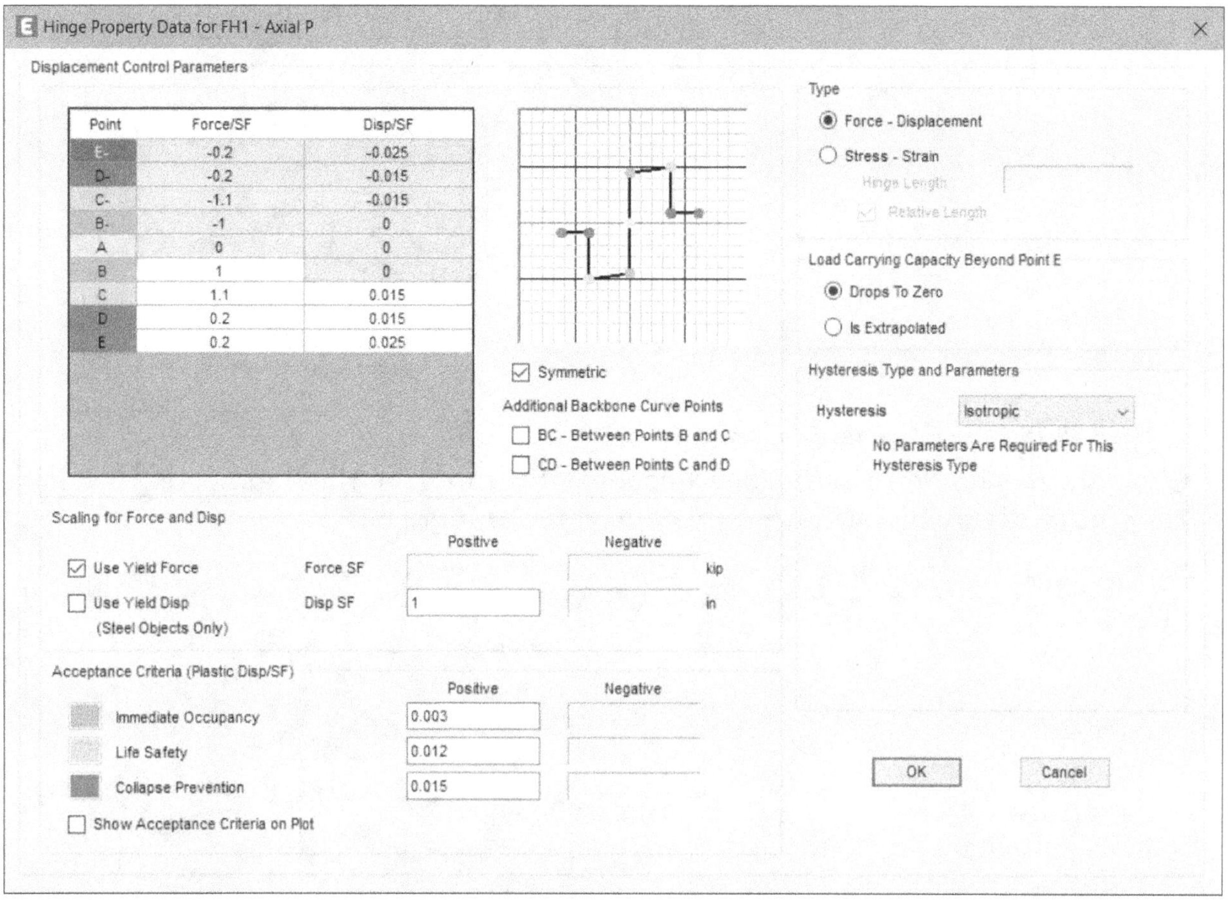

Figure-59. Hinge Property Data for FH1-Axial P dialog box

- Enter desired material parameters in the dialog box and click on the **OK** button. The **Hinge Property Data** dialog box will be displayed again.
- Click on the **OK** button from the dialog box to create it. The new property will get added in the list.

DEFINING PANEL ZONE

Panel zone in ETABS is used to define the flexibility between beam-column connection. The procedure to define panel zone properties is given next.

- Click on the **Panel Zone** tool from the **Section Properties** cascading menu in the **Define** menu. The **Panel Zone Properties** dialog box will be displayed; refer to Figure-60.

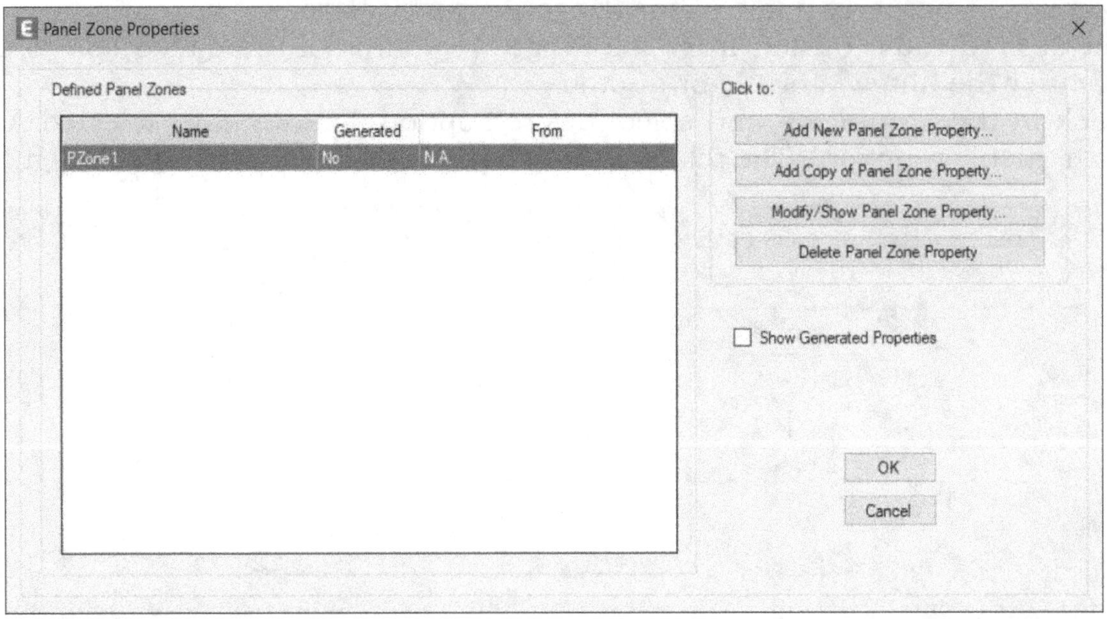

Figure-60. Panel Zone Properties dialog box

- Click on the **Add New Panel Zone Property** button from the dialog box. The **Panel Zone Data** dialog box will be displayed; refer to Figure-61.

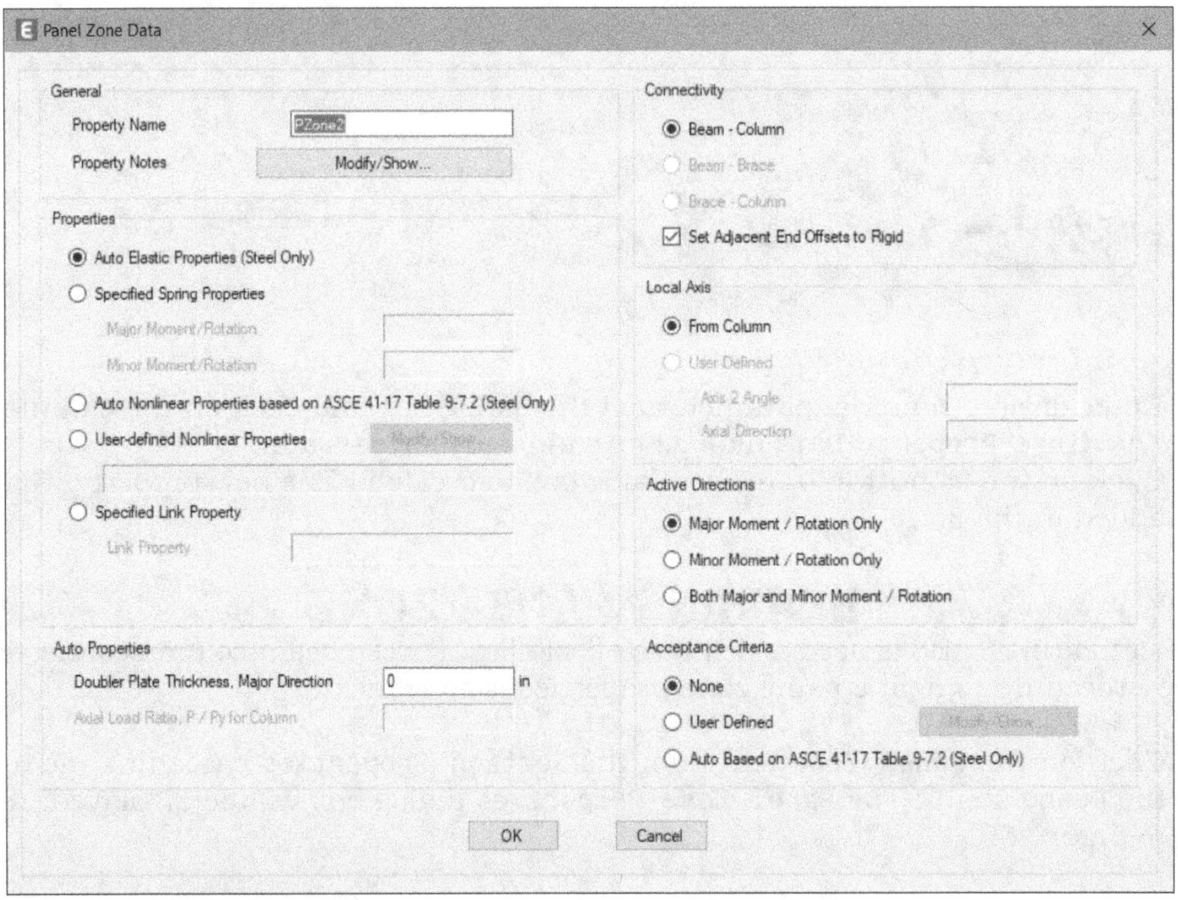

Figure-61. Panel Zone Data dialog box

- Specify desired name for the property.
- Select desired radio button from the **Properties** area and specify the related parameters.

- Click on the **OK** button to create the property. The **Panel Zone Properties** dialog box will be displayed again.
- You can use the other buttons in this dialog box as discussed earlier. Click on the **OK** button to exit the dialog box.

DEFINING SPRINGS PROPERTIES

Spring properties are applied to links for generating spring joints (nodes) when generating mesh of the model for analysis. The tools to define properties for different type of springs are available in the **Spring Properties** cascading menu of the **Define** menu; refer to Figure-62. The tools in this cascading menu are discussed next.

Defining Point Spring Properties

The procedure to define point spring properties is given next.

- Click on the **Point Springs** tool from the **Spring Properties** cascading menu. The **Point Spring Properties** dialog box will be displayed; refer to Figure-63.

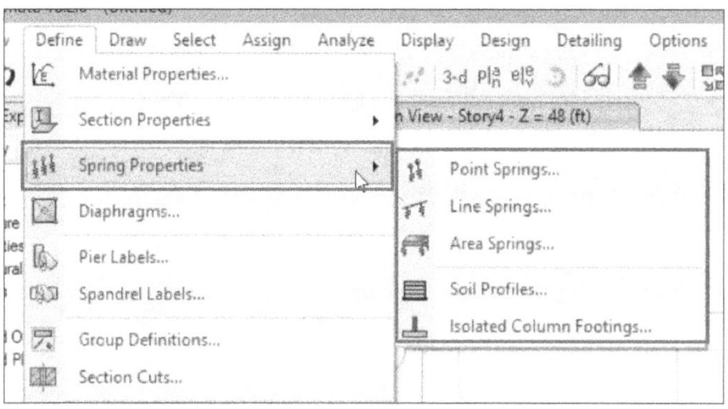

Figure-62. Spring Properties cascading menu

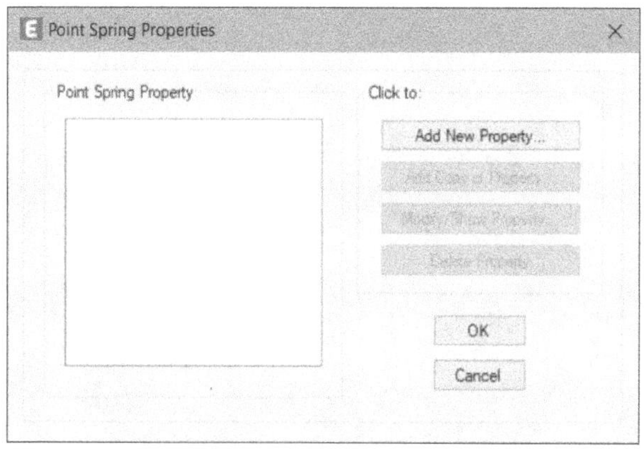

Figure-63. Point Spring Properties dialog box

- Click on the **Add New Property** button from the dialog box. The **Point Spring Property Data** dialog box will be displayed; refer to Figure-65.
- Specify desired name for the spring property and set the other general data.
- Select **User Specified/Link Properties** radio button from the **Spring Stiffness Options** area to use parameters specified in this dialog box for spring stiffness. Select the **Based on Soil Profile and Footing Dimensions** radio button to use parameters specified in Soil profile and other parameters profiles.
- Set desired stiffness parameters in different edit boxes of the **Simple Spring Stiffness in Global Directions** area for different directions. If you have defined properties of soil and footing then you can select the **Based on Soil Profile and Footing Dimensions** radio button to define spring stiffness properties.
- Select the **Quick Specification** radio button from the **Nonlinearity Specifications** area to define nonlinear behavior of spring joint by specified parameters; refer to Figure-64. Select the **None (Linear)** radio button if the link is not non-linear. Select the **Tension Only** radio button if nonlinear applies in stretching direction of spring. Select the **Compression Only** radio button if nonlinearity applies in compression direction. Select the **Elasto-plastic** radio button to apply nonlinearity to both compression and tension directions; and set desired parameters in edit boxes below it.

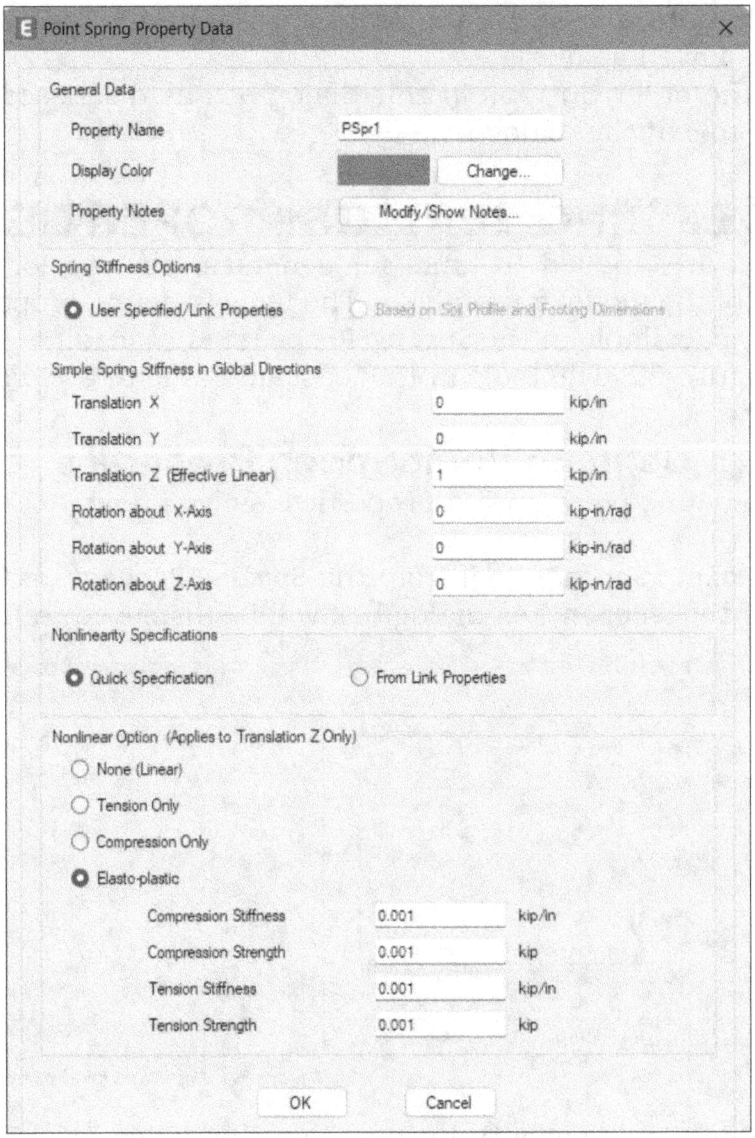

Figure-64. Point Spring Properties Data dialog box

- Select the **From Link Properties** radio button to use link property define in the table to specify nonlinearity.
- Click on the **Add** button from the **Single Joint Links at Point** area. A link property will be added.
- Set desired parameter for link property and click on the **OK** button.
- You can use the other parameters of the **Point Spring Properties** dialog box in the same way as discussed earlier.

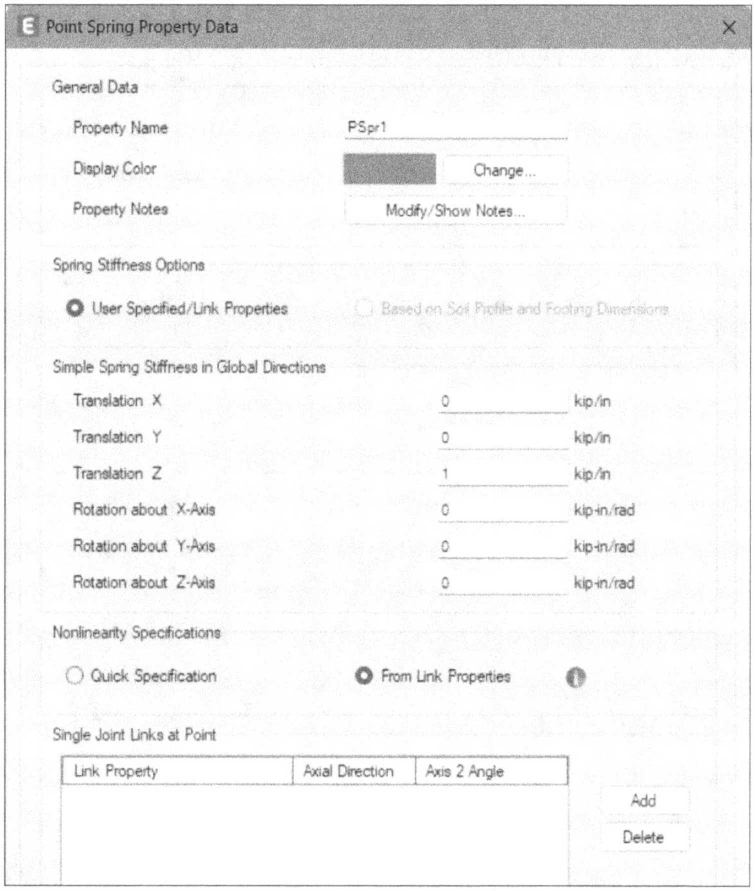

Figure-65. Point Spring Property Data dialog box

Defining Line Spring and Area Spring Properties

The procedures to create line spring and area spring properties are similar to Point springs. In case of Line spring properties, you can create properties for an array of spring in straight line. In case of Area spring properties, you can create properties of a full area to be used as spring.

Soil Profile

The soil properties are important factors for construction of a building. The procedure to define soil profile in ETABS is given next.

- Click on the **Soil Profiles** tool from the **Spring Properties** cascading menu of the **Define** menu. The **Soil Profiles** dialog box will be displayed; refer to Figure-66.

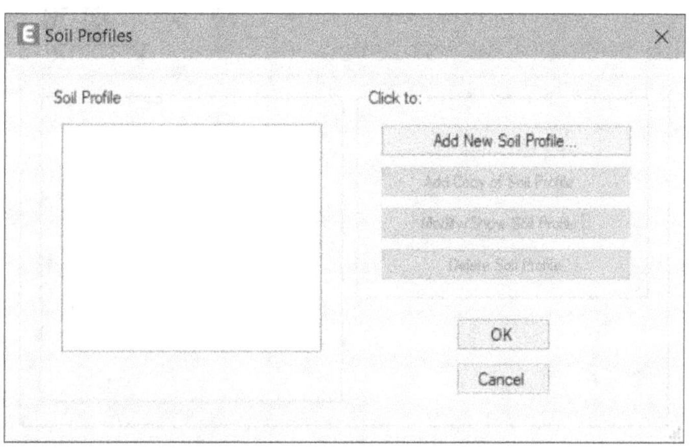

Figure-66. Soil Profiles dialog box

- Click on the **Add New Soil Profile** button from the dialog box. The **Soil Profile Data** dialog box will be displayed; refer to Figure-67.

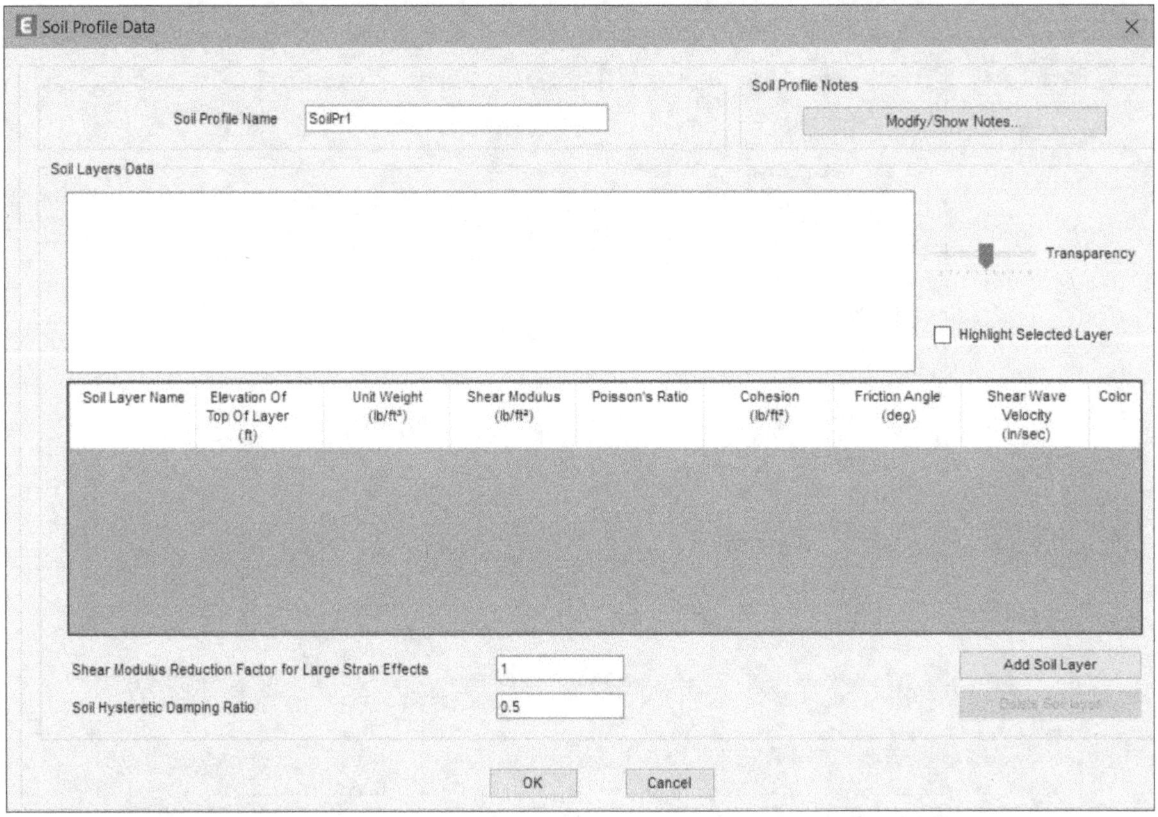

Figure-67. Soil Profile Data dialog box

- Set desired name for the property in the **Soil Profile Name** edit box.
- Click on the **Add Soil Layer** button. A new layer of soil will be added in the table.
- Triple-click in the field of table to edit it. You can add as many soil layers as needed and edit their parameters.
- After setting desired properties, click on the **OK** button.

The other tools in the **Soil Profiles** dialog box are same as discussed earlier.

Properties of Isolated Column Footing

The Isolated column footing is used to support single column in the building. The procedure to define properties of isolated column footing is given next.

- Click on the **Isolated Column Footings** tool from the **Spring Properties** cascading menu of the **Define** menu. The **Isolated Column Footing Specifications** dialog box will be displayed; refer to Figure-68.

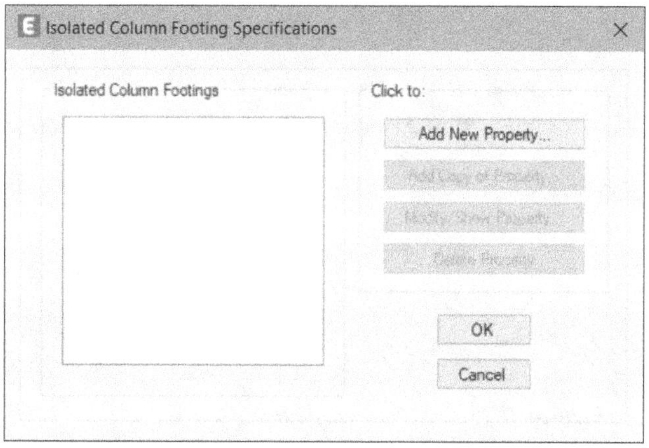

Figure-68. Isolated Column Footing Specifications dialog box

- Click on the **Add New Property** button. The **Isolated Column Footing Data** dialog box will be displayed; refer to Figure-69.

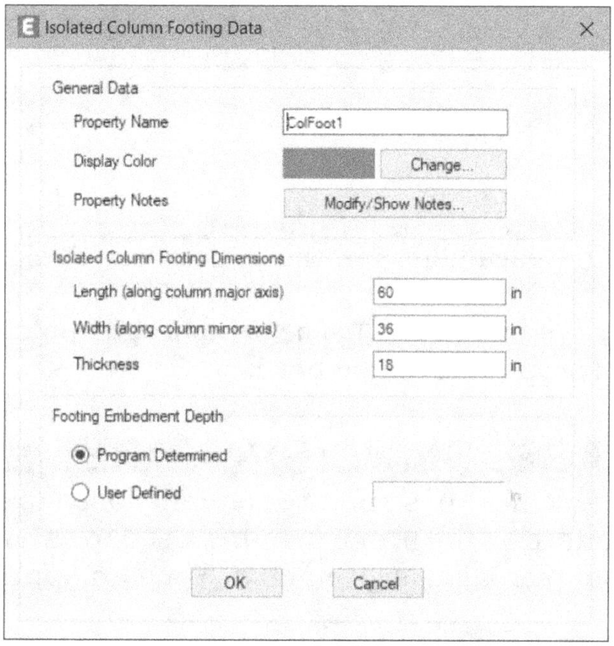

Figure-69. Isolated Column Footing Data dialog box

- Specify desired parameters like length, width, and thickness of column footing and click on the **OK** button to create the property.

You can use the other tools in the **Isolated Column Footing Specifications** dialog box as discussed for other properties.

DIAPHRAGMS PROPERTIES

Diaphragm is used to transfer lateral loads to vertical resisting members of structure. The procedure to create diaphragm property is given next.

- Click on the **Diaphragms** tool from the **Define** menu. The **Define Diaphragm** dialog box will be displayed; refer to Figure-70.
- Click on the **Add New Diaphragm** button from the dialog box. The **Diaphragm Data** dialog box will be displayed; refer to Figure-71.

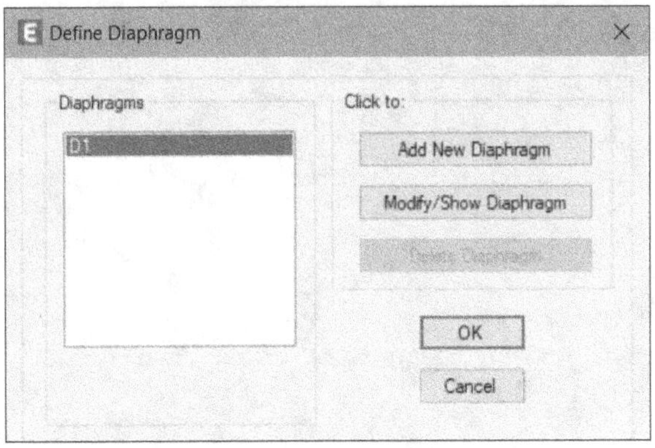

Figure-70. Define Diaphragm dialog box

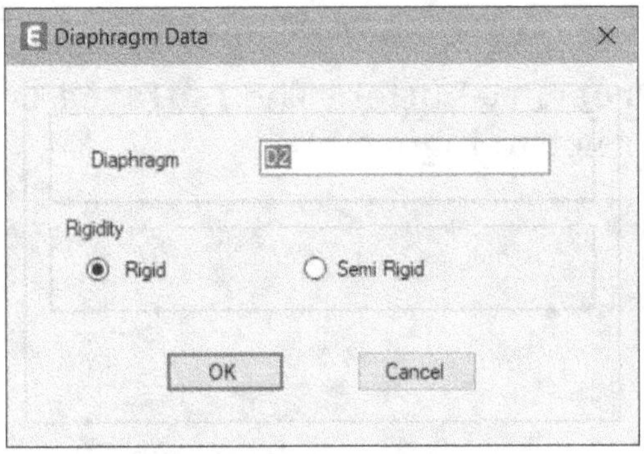

Figure-71. Diaphragm Data dialog box

- Set desired name and parameters in the dialog box. Click on the **OK** button.

You can use the other options in the **Define Diaphragm** dialog box in the same way as discussed for other properties dialog boxes.

PIER LABELS AND SPANDREL LABELS

Pier is the load bearing member of structure generally used to support arcs or middle section of bridges. Spandrel is the frame member used to carry load of exterior walls. The procedure to create pier label is given next. You can apply similar procedure for spandrel labels.

- Click on the **Pier Labels** tool from the **Define** menu. The **Pier Labels** dialog box will be displayed; refer to Figure-72.

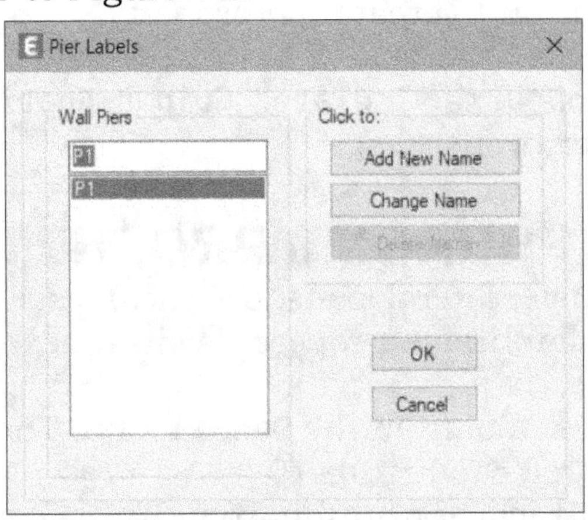

Figure-72. Pier Labels dialog box

- Specify desired name in the **Wall Piers** edit box and click on the **Add New Name** button. A new pier label will be added.

- If you want to change the name of a pier label, select it from the list and click on the **Change Name** button.
- To delete any label, select it from the list and click on the **Delete Name** button.
- Click on the **OK** button from the dialog box to apply the modifications.

CREATING GROUPS

Groups are used to combine desired objects to form one type of system. The procedure to create group is given next.

- Click on the **Group Definitions** tool from the **Define** menu. The **Define Groups** dialog box will be displayed; refer to Figure-73.

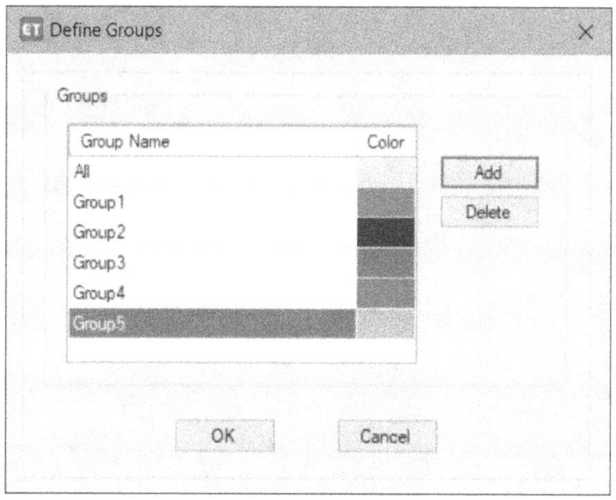

Figure-73. Define Groups dialog box

- Click on the **Add** button to define a new group.
- Triple-click on the name of group to be renamed. Click on the color block next to group name to change it.
- Click on the **OK** button from the **Define Groups** dialog box to create/modify the groups.

DEFINING SECTION CUT PROPERTIES

The **Section Cuts** tool is used to define properties of section cuts for checking internal layers of the structures. The procedure to define these properties is given next.

- Click on the **Section Cuts** tool from the **Define** menu. The **Section Cuts** dialog box will be displayed; refer to Figure-74.

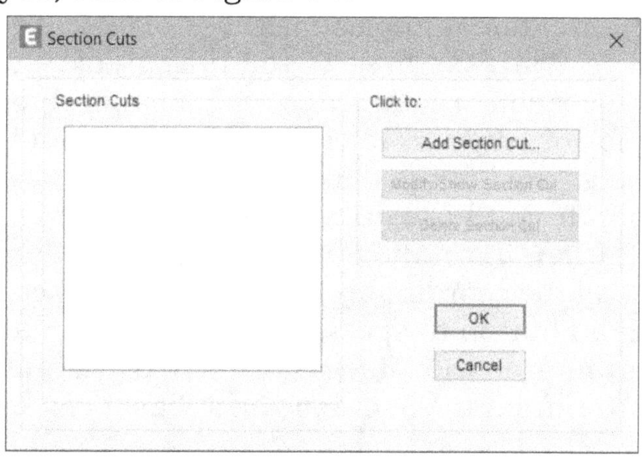

Figure-74. Section Cuts dialog box

- Click on the **Add Section Cut** button from the dialog box. The **Section Cut Data** dialog box will be displayed; refer to Figure-75.

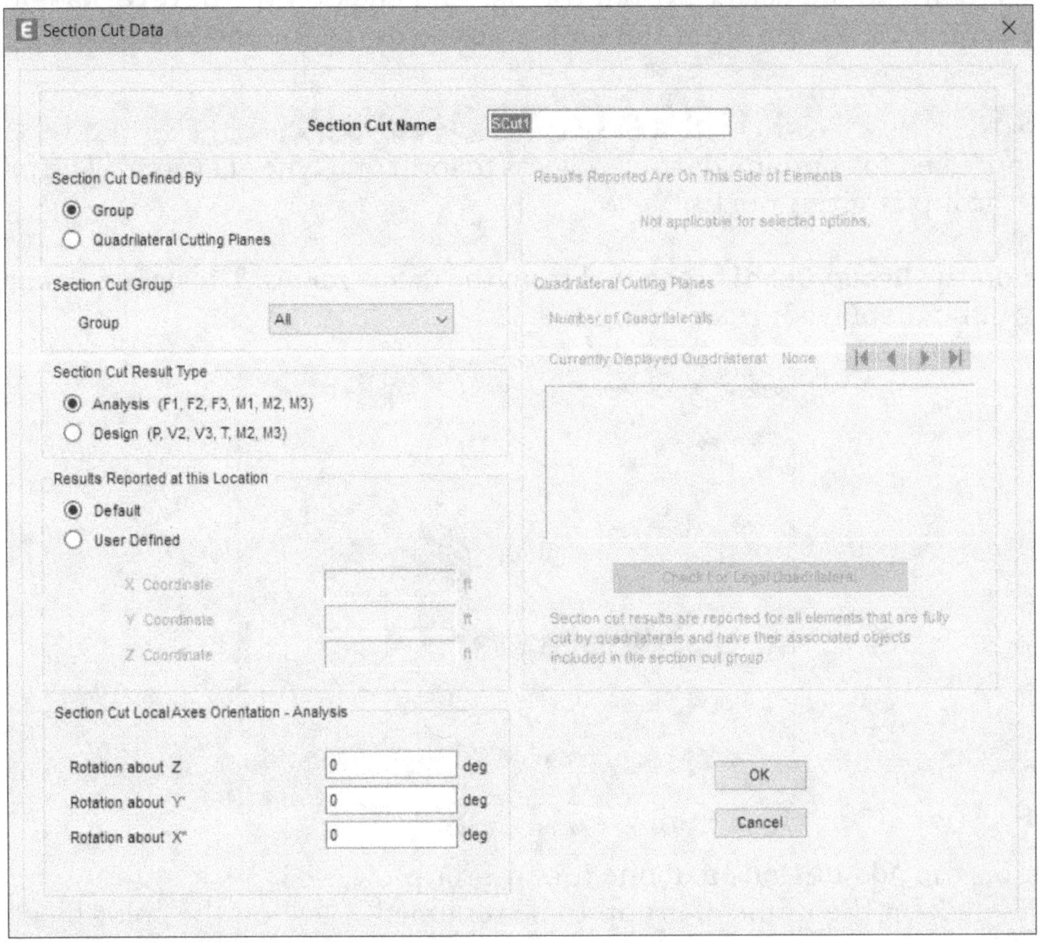

Figure-75. Section Cut Data dialog box

- Specify desired name for the property in the **Section Cut Name** edit box.
- Select desired option from the **Section Cut Defined By** area to define the scope of generating section cut results. Select the **Group** radio button to generate section cut results on all the elements of selected group. Select the **Quadrilateral Cutting Planes** radio button to use table in **Quadrilateral Cutting Planes** area of the dialog box for defining points where section cut results will be generated.
- Select desired option from the **Section Cut Result Type** area of the dialog box to define the type of results to be generated for the section cuts. Select the **Analysis** radio button to make local 1, 2, and 3 axes correspond to global X, Y, and Z axes. You can specify the rotational angles for these axes in the **Section Cut Local Axes Orientation - Analysis** area of the dialog box. Select the **Design** radio button to orient the section cut axes to a wall, spandrel, or slab. You can select the radio buttons for orientation references from the **Section Cut Local Axes Orientation - Design** area of the dialog box.
- Select desired option from the **Results Reported at this Location** area of the dialog box. Select the **Default** radio button to report section cut forces at location which is average of all the coordinates of all the point objects in the group. Select the **User Defined** radio button to specify the coordinate manually.
- Click on the **OK** button from the dialog box to create parameters.
- Click on the **OK** button from the **Section Cuts** dialog box to apply settings.

DEFINING PROPERTIES OF STRAIN GAUGES

The tools in **Strain Gauges Properties** cascading menu of the **Define** menu are used to specify the properties for strain measurement gauges. The tools in this cascading menu are discussed next.

Defining Line Gauge Properties

The **Line Gauges** tool in **Strain Gauges Properties** cascading menu is used to define properties for strain measurement gauges. The procedure to use this tool is given next.

- Click on the **Line Gauges** tool from the **Strain Gauges Properties** cascading menu of the **Define** menu. The **Line Gauge Properties** dialog box will be displayed; refer to Figure-76.

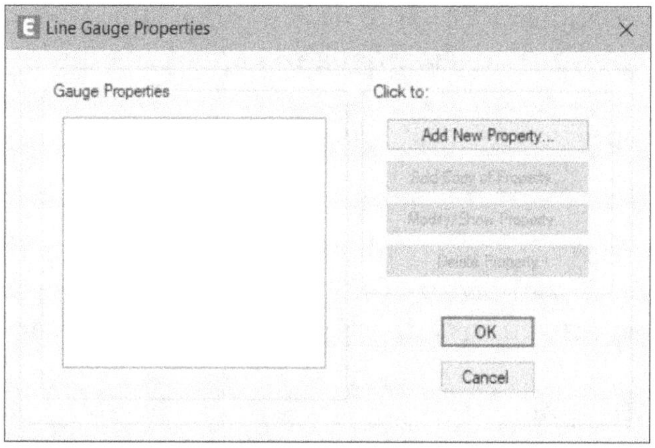

Figure-76. Line Gauge Properties dialog box

- Click on the **Add New Property** button from the dialog box. The **Line Gauge Property Data** dialog box will be displayed; refer to Figure-77.

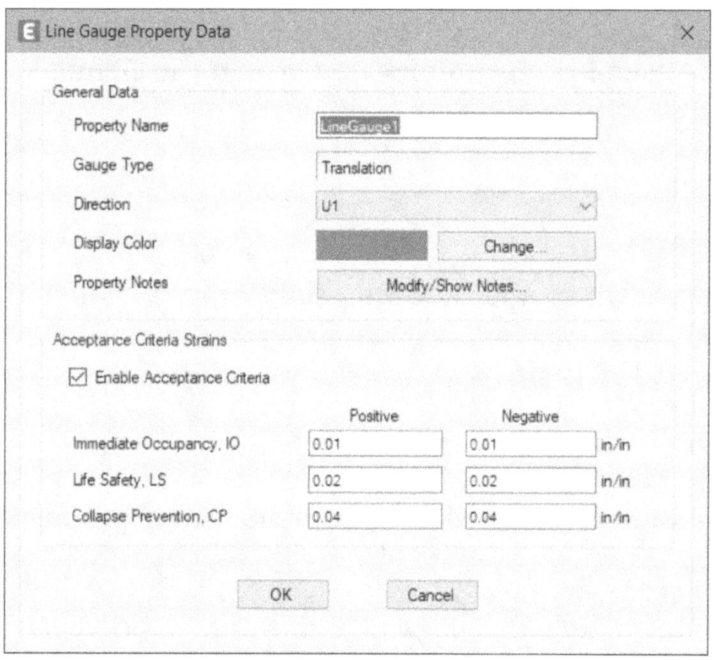

Figure-77. Line Gauge Properties Data dialog box

- Specify desired name of the property in the **Property Name** edit box.

- Select desired option from the **Direction** drop-down to select the direction in which you want to check the strain.
- Set desired color and property notes as discussed earlier.
- Select the **Enable Acceptance Criteria** check box to define range within which strain will be measured.
- After setting desired parameters, click on the **OK** button. The **Line Gauge Properties** dialog box will be displayed again. The other options are same as discussed earlier.

Defining Quad Gauge Properties

The **Quad Gauges** option in the **Strain Gauges Properties** cascading menu of the **Define** menu is used to create properties for strain measurement of a quad link in the structure.

- Click on the **Quad Gauges** tool from the **Strain Gauges Properties** cascading menu of the **Define** menu. The **Quad Gauge Properties** dialog box will be displayed.

You can create the properties for quad gauges in the same way as discussed for line gauge.

This concludes the properties generally defined regarding the objects in ETABS. But this does not end the tool array of ETABS. In the next chapter, you will learn to assign these properties and many others to the objects in the model.

SELF-ASSESSMENT

Q1. Which tool in the Define menu is used to create and manage material properties?
A. Frame Section Tool
B. Material Properties Tool
C. Section Designer Tool
D. Snap Tool

Q2. Which button is used to create a new material in the Define Materials dialog box?
A. Modify Material
B. Import Material
C. Add New Material
D. Create Section

Q3. The Region drop-down in Add New Material Property dialog box is used to define:
A. Color of material
B. Country/region for the material
C. Thickness of material
D. Shape of section

Q4. Which option should be selected if material has same properties in all directions?
A. Orthotropic
B. Anisotropic
C. Isotropic
D. Composite

Q5. Which directional symmetry type is used when material properties differ in X, Y, and Z directions?
A. Linear
B. Orthotropic
C. Isotropic
D. Elastic

Q6. Which radio button is selected to define density as mass per unit volume?
A. Specify Weight Density
B. Specify Linear Density
C. Specify Mass Density
D. Specify Area Density

Q7. Which properties are specified in the Mechanical Property Data area for isotropic materials?
A. Load combinations
B. Modulus of elasticity and Poisson's ratio
C. Story heights
D. Grid spacing

Q8. Which button is used to modify properties separately for each direction in orthotropic materials?
A. Define Material
B. Show Density Data
C. Modify/Show Mechanical Property Data
D. Section Designer

Q9. Nonlinear material data is mainly used for:
A. Linear static analysis
B. Non-linear analysis
C. Grid editing
D. Printing reports

Q10. Which metals are mentioned as having nonlinear material behavior?
A. Copper and zinc
B. Steel and aluminum
C. Iron and brass
D. Titanium and silver

Q11. Which levels are set in nonlinear analysis using Acceptance Criteria Strains?
A. Load, Stress, and Force
B. IO, LS, and CP
C. X, Y, and Z
D. Tension, Compression, and Shear

Q12. Which check box ignores tension levels in nonlinear analysis?
A. Ignore Compression Criteria
B. Ignore Load Criteria
C. Ignore Tension Acceptance Criteria
D. Ignore Shear Criteria

Q13. Material Damping Properties are used to define how material:
A. Expands
B. Conducts electricity
C. Absorbs vibrations
D. Changes color

Q14. Which two damping coefficients are mentioned in ETABS?
A. Linear and Nonlinear
B. Modal and Viscous Proportional
C. Static and Dynamic
D. Elastic and Plastic

Q15. The Frame Section tool is available in which menu?
A. Analyze Menu
B. Draw Menu
C. Define Menu
D. File Menu

Q16. Which button is used to import frame section properties from XML files?
A. Add New Property
B. Import New Properties
C. Convert to SD Section
D. Export to XML File

Q17. Which dialog box appears after selecting a shape type for imported frame properties?
A. Material Property Data
B. Frame Section Property Import Data
C. Section Designer Window
D. Frame Property Notes

Q18. Which drop-down is used to define material for imported sections?
A. Shape Material
B. Default Material for Section
C. Property Material
D. Material Shape Type

Q19. Which button allows changing the display color of a frame section?
A. Change Color
B. Modify Color
C. Display Color
D. Change button for Display Color

Q20. Which dialog box is used to modify stiffness modification factors?
A. Material Damping Dialog Box
B. Property/Stiffness Modification Factors
C. Frame Notes Dialog Box
D. Nonlinear Material Data Dialog Box

Q21. SD in SD Section stands for:
A. Structural Design
B. Section Definition
C. Section Designer
D. Standard Design

Q22. What is the main advantage of converting a standard section into an SD section?
A. It reduces file size
B. It allows shape modification
C. It changes material automatically
D. It creates reports automatically

Q23. Which button is used to open the Section Designer window?
A. Modify/Show Section
B. Open Shape Tool
C. Section Designer
D. Draw Shape

Q24. Which snap option snaps the cursor to line intersections?
A. Fine Grids
B. Line Intersections
C. Line Ends and Midpoint
D. Perpendicular Projections

Q25. Which tool is used to modify an object shape by dragging nodes?
A. Select Object Tool
B. Draw Defined Section Tool
C. Reshape Mode Tool
D. Reference Line Tool

Q26. Which tool is used to check properties like center of gravity and moment of inertia of a section?
A. Interaction Surface Tool
B. Section Properties Tool
C. Moment Curvature Tool
D. Panel Zone Tool

Q27. Which property represents the area of section in the Section Properties dialog box?
A. J
B. AS2
C. A
D. I23

Q28. What does Xcg and Ycg represent in section properties?
A. Radius of Gyration
B. Location of Center of Gravity
C. Plastic Modulus
D. Torsional Constant

Q29. In ETABS, the value of I23 is always:
A. 1
B. 100
C. Zero
D. Infinite

Q30. Which tool is used to check interaction between reinforcement and composite material?
A. Section Properties Tool
B. Interaction Surface Tool
C. Tendon Section Tool
D. Link Property Tool

Q31. Which button is used to view interaction modes in the Interaction Surface dialog box?
A. Shell
B. PM2
C. Deck
D. Grid

Q32. In the Moment Curvature Plot, the vertical axis represents:
A. Curvature
B. Thickness
C. Moment
D. Area

Q33. Which tool is used to define tendon sections in ETABS?
A. Slab Sections Tool
B. Tendon Sections Tool
C. Deck Sections Tool
D. Wall Sections Tool

Q34. Which edit box is used to specify strand area in tendon properties?
A. Strand Area
B. Tendon Width
C. Section Area
D. Gap Area

Q35. Which tool is used to define slab sections?
A. Deck Sections Tool
B. Slab Sections Tool
C. Wall Sections Tool
D. Tendon Sections Tool

Q36. Which slab modeling type neglects transverse shear deformation?
A. Shell-Thick
B. Membrane
C. Layered
D. Shell-Thin

Q37. Which slab modeling type follows Mindlin/Reissner application?
A. Membrane
B. Shell-Thick
C. Shell-Thin
D. Layered

Q38. Which slab modeling type is used to create mixed material composite slabs?
A. Membrane
B. Shell-Thin
C. Layered
D. Shell-Thick

Q39. Which tool is used to define deck section properties?
A. Deck Sections Tool
B. Slab Sections Tool
C. Wall Section Tool
D. Link Tool

Q40. Which tool is used to define reinforcing bar sizes?
A. Panel Zone Tool
B. Reinforcing Bar Sizes Tool
C. Diaphragm Tool
D. Tendon Tool

Q41. Which button adds standard rebar sizes in the Reinforcing Bar Sizes dialog box?
A. Add Layer Button
B. Add Material Button
C. Add Common Bar Set Button
D. Add Grid Button

Q42. Which tool is used to define link or support properties?
A. Link/Support Properties Tool
B. Tendon Tool
C. Wall Tool
D. Deck Tool

Q43. Which link type is used when load transfer occurs linearly?
A. Gap
B. Linear
C. Damper-Bilinear
D. Damper-Exponential

Q44. Which radio button allows force in one direction to cause displacement in other directions?
A. Stiffness is Uncoupled
B. Stiffness is Coupled
C. NonLinear
D. Effective Stiffness

Q45. Which link type is used when there is a gap between two structural members?
A. Linear
B. Damper-Bilinear
C. Gap
D. Friction Spring

Q46. Which edit box specifies the gap between structural members in Gap Link Type?
A. Width
B. Open
C. Distance
D. Thickness

Q47. Which tool is used to define frame/wall hinge properties?
A. Point Springs Tool
B. Panel Zone Tool
C. Frame/Wall Nonlinear Hinges Tool
D. Deck Sections Tool

Q48. Panel zones in ETABS are used to define flexibility between:
A. Wall and Slab
B. Beam and Column
C. Soil and Footing
D. Deck and Wall

Q49. Which tool is used to define point spring properties?
A. Soil Profiles Tool
B. Point Springs Tool
C. Deck Sections Tool
D. Diaphragms Tool

Q50. Which nonlinearity specification applies behavior only in stretching direction?
A. Compression Only
B. Elasto-plastic
C. Tension Only
D. None (Linear)

Q51. Which tool is used to define soil properties in ETABS?
A. Soil Profiles Tool
B. Point Springs Tool
C. Deck Sections Tool
D. Wall Sections Tool

Q52. Which footing type supports a single column in a building?
A. Combined Footing
B. Raft Footing
C. Isolated Column Footing
D. Strip Footing

Q53. Diaphragms are mainly used to transfer:
A. Gravity Loads
B. Wind Loads Only
C. Lateral Loads
D. Thermal Loads

Q54. Which tool is used to create pier labels?
A. Group Definitions Tool
B. Pier Labels Tool
C. Panel Zone Tool
D. Soil Profiles Tool

Q55. Which tool is used to combine objects into one system in ETABS?
A. Diaphragm Tool
B. Group Definitions Tool
C. Link Tool
D. Slab Sections Tool

FOR STUDENTS NOTES

Chapter 4

Assigning Properties and Applying Loads

Topics Covered

The major topics covered in this chapter are:

- ***Introduction***
- ***Assigning Properties to Joints***
- ***Assigning Frame Properties***
- ***Assigning Shell Properties***
- ***Assigning Link Properties***
- ***Assigning Properties to Tendons***
- ***Assigning Joint Loads***
- ***Assigning Frame Loads***
- ***Assigning Shell Loads***
- ***Assigning Loads to Tendons***
- ***Copying and Pasting Assigns***

INTRODUCTION

In previous chapter, you have learned to define various properties of frame sections, springs, diaphragms, and so on. In this chapter, you will learn to assign these properties and many other to the frame members in the model. You will also learn to apply load and prepare model for different analyses.

ASSIGNING PROPERTIES TO JOINTS

Joints are the places where two or more frame members join together. Depending on the joint made at this place, you need to assign the related properties for the analysis. The tools to assign joint properties are available in the **Joint** cascading menu of the **Assign** menu; refer to Figure-1. The tools in this cascading menu are discussed next.

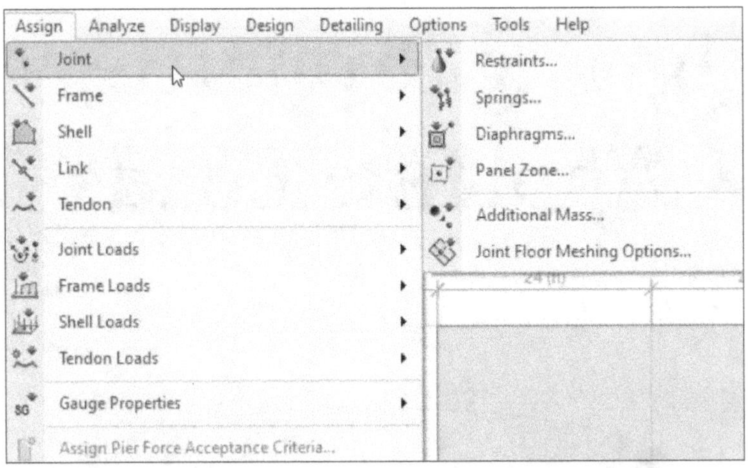

Figure-1. Joint cascading menu

Assigning Restraint Joint Properties

The **Restraint** tool in **Joint** cascading menu is used to restrict the motion of selected joint in specified directions. The procedure to use this tool is given next.

- Select desired joints from the model to which you want to assign restraints; refer to Figure-2 and click on the **Restraints** tool from the **Joint** cascading menu of the **Assign** menu. The **Joint Assignment - Restraints** dialog box will be displayed; refer to Figure-3.
- Select the check boxes for the directions which you want to be restrained from the **Restraints in Global Directions** area of the dialog box.
- There are four buttons in the **Fast Restraints** area to quickly apply restraints viz. **Fixed Base** button, **Pinned Base** button, **Roller Support** button, and **No Support** button. Select the **Fixed Base** button if you want to restrict motion in all six directions for selected joints. Select the **Pinned Base** button if you want to restrict the translation of joints but they are free to rotate. Select the **Roller Support** button if you want to restrict the movement of joint along Z axis only. Select the **No Support** button if you want the joints to be free to move in all directions.

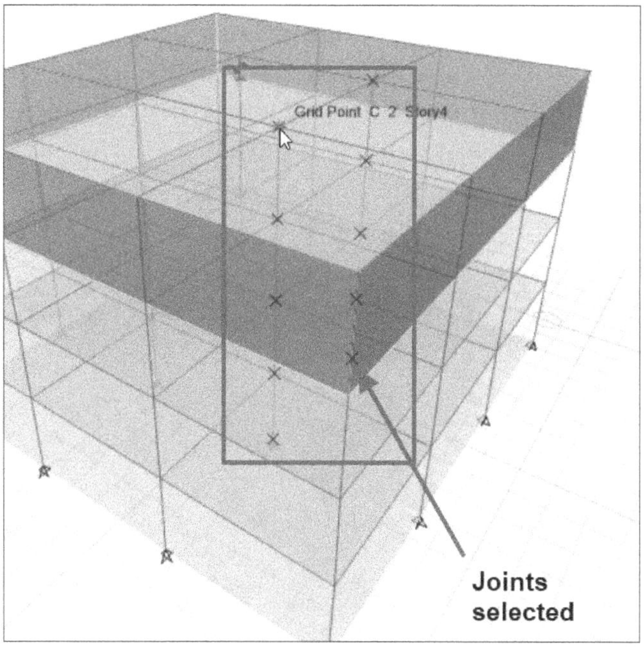

Figure-2. Selecting joints

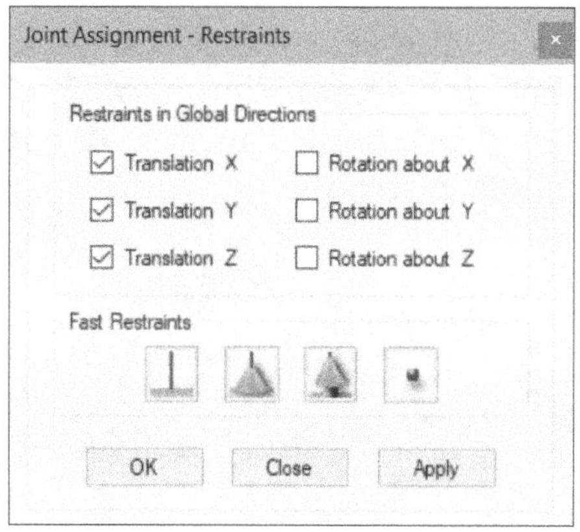

Figure-3. Joint Assignment-Restraints dialog box

- After setting desired restraints, click on the **OK** button from the dialog box to apply changes.

Assigning Spring Joint Properties

The **Springs** tool in **Joint** cascading menu is used to assign spring properties to the selected joints. Assigning spring joint is also useful when the beams/columns are joined by soft materials to absorb shocks. The procedure to use this tool is given next.

- Select the joints to which you want to assign spring properties and click on the **Springs** tool from the **Joint** cascading menu of the **Assign** menu. The **Joint Assignment - Spring** dialog box will be displayed with list of spring properties earlier created; refer to Figure-4. The procedure to create spring properties has been discussed in previous chapter.

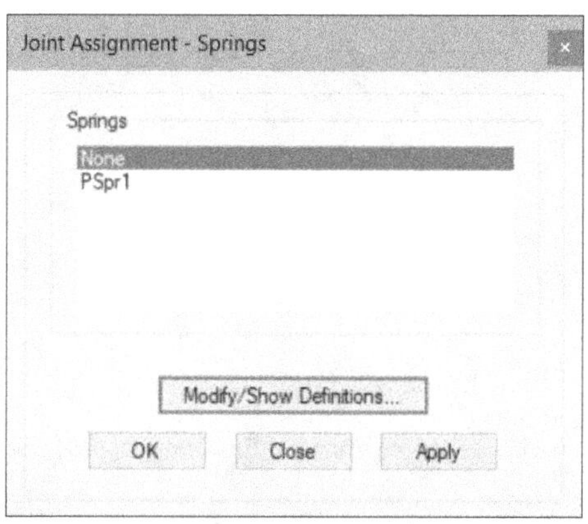

Figure-4. Joint Assignment-Springs dialog box

- Select desired property and click on the **Modify/Show Definitions** button to check or modify the selected properties. The **Point Spring Properties** dialog box will be displayed; refer to Figure-5. The options in this dialog box have been discussed earlier in previous chapter. Create or modify the properties as required and click on the **OK** button.

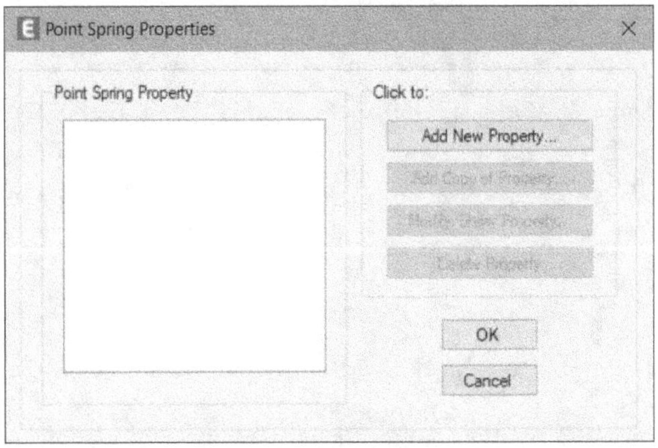

Figure-5. Point Spring Properties dialog box

- Select desired property from the list in **Joint Assignment - Spring** dialog box and click on the **OK** button to assign properties.

Assigning Diaphragms Joint Properties

A diaphragm joint creates links between joints located within a plane such that they move together as a planar diaphragm, rigid against membrane (in-plane) deformation, but susceptible to plate (out-of-plane) deformation and associated effects. The **Diaphragms** tool in the **Joint** cascading menu is used to assign diaphragms properties to the selected joints. The procedure to use this tool is given next.

- After selecting the joints, click on the **Diaphragms** tool from the **Joint** cascading menu of the **Assign** menu. The **Joint Assignment - Diaphragms** dialog box will be displayed; refer to Figure-6.

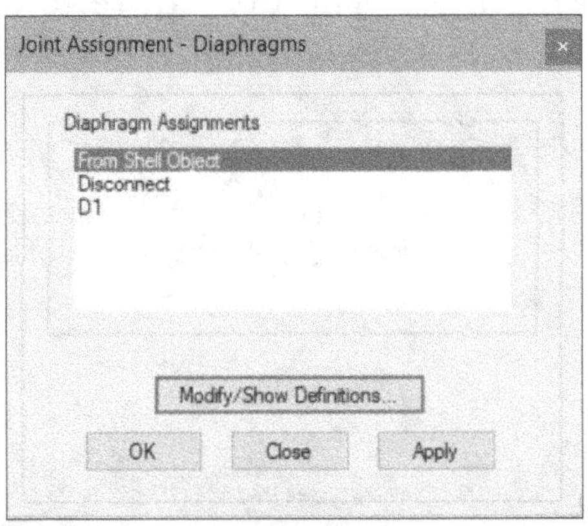

Figure-6. Joint Assignment-Diaphragms dialog box

- List of all the diaphragm properties earlier created will be displayed in the list box. The procedure to create and modify these properties has already been discussed in previous chapter. Select desired property and click on the **OK** button to assign the property to selected joints.

Assigning Panel Zone Properties to Joints

Panel zones are assigned to joints to model the flexibility of beam-column connections. Like, you can define the translation and rotation values for the selected connections. The procedure to assign panel zone to selected joints is given next.

- Select desired joints and click on the **Panel Zone** tool from the **Joint** cascading menu of the **Assign** menu. The **Joint Assignment-Panel Zone Property** dialog box will be displayed; refer to Figure-7.

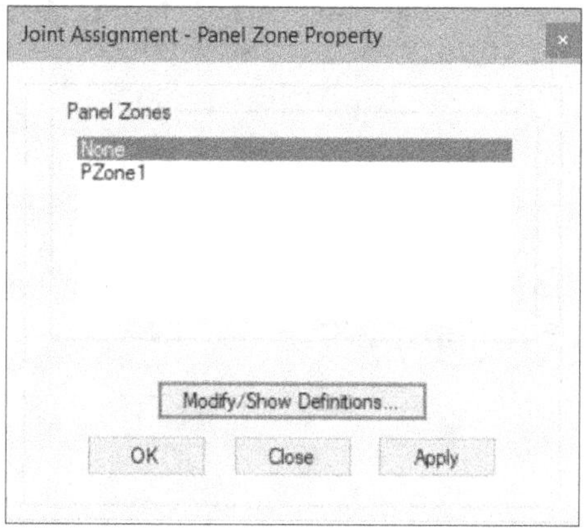

Figure-7. Joint Assignment-Panel Zone Property dialog box

- Select desired panel zone from the list and click on the **OK** button. The selected panel zone will be assigned.

Assigning Additional Mass to Selected Joints

The **Additional Mass** tool in the **Joint** cascading menu is used to assign additional mass to the selected joints. This additional mass can be representation of any physical object like a machine or furniture etc. The procedure to use this tool is given next.

- Select the joints to which you want to assign additional mass and click on the **Additional Mass** tool from the **Joint** cascading menu of the **Assign** menu. The **Joint Assignment-Additional Mass** dialog box will be displayed; refer to Figure-8.

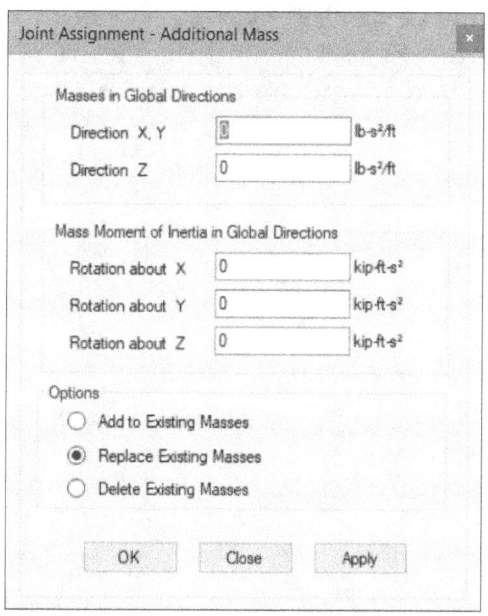

Figure-8. Joint Assignment-Additional Mass dialog box

- Specify desired values of mass and mass moment of inertia in respective edit boxes.
- Select desired option from the **Options** area of the dialog box. Select the **Add to Existing Masses** radio button to add the specified value of mass to existing masses. Select the **Replace Existing Masses** radio button to replace the values of masses for selected joints. Select the **Delete Existing Masses** radio button to delete the masses of selected joints for analysis.
- After setting desired parameters, click on the **OK** button from the dialog box.

Setting Joint Floor Meshing Options

The **Joint Floor Meshing Options** tool is used to set the parameters for meshing of joints. Using this option, you can include or exclude joint objects in mesh. The procedure to use this tool is given next.

- Click on the **Joint Floor Meshing Options** tool from the **Joint** cascading menu of the **Assign** menu. The **Joint Assignment-Joint Floor Meshing Option** dialog box will be displayed; refer to Figure-9.

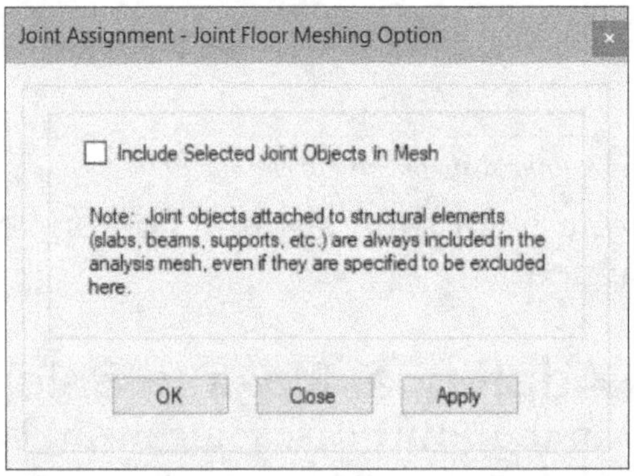

Figure-9. Joint Assignment-Joint Floor Meshing Option dialog box

- Select the **Include Selected Joint Objects In Mesh** check box if you want to include the selected joints in meshing.
- Click on the **OK** button from the dialog box to apply the changes.

ASSIGNING FRAME PROPERTIES

There are various tools in the **Frame** cascading menu of the **Assign** menu to assign different properties to selected frame members; refer to Figure-10. These tools are discussed next.

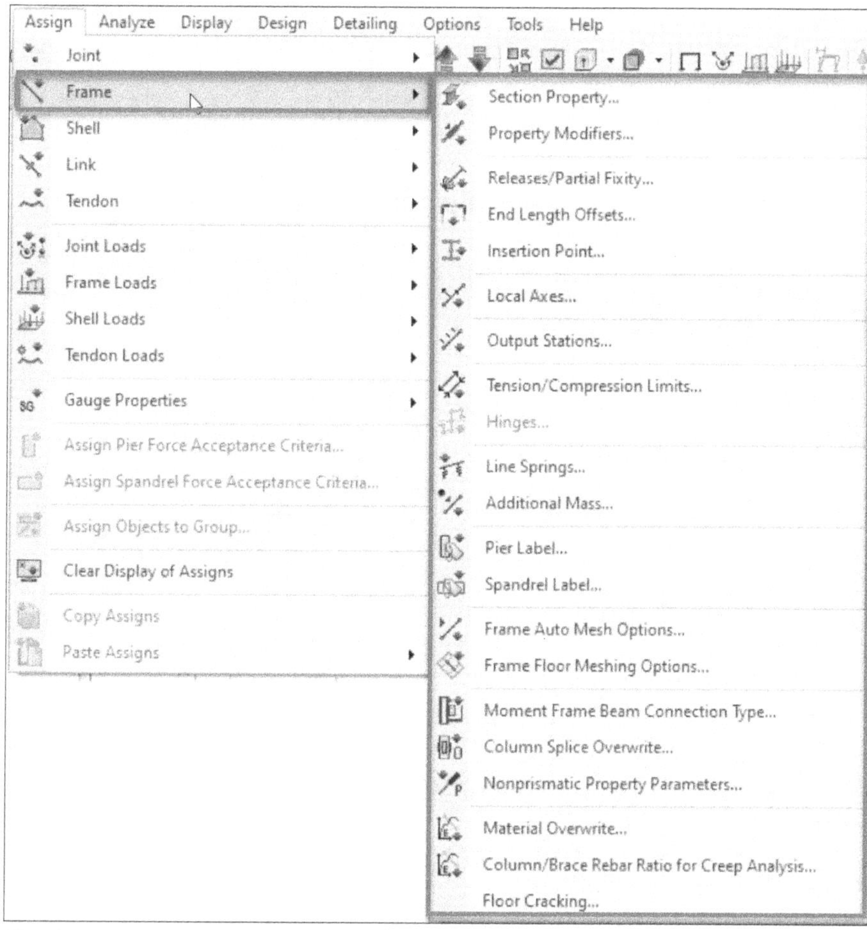

Figure-10. Frame cascading menu

Assigning Section Properties

Section properties are the properties of frame member to define shape, size, and so on. The procedure to assign section property is given next.

- Select desired frame member and click on the **Section Property** tool from the **Frame** cascading menu of the **Assign** menu. The **Frame Assignment - Section Property** dialog box will be displayed; refer to Figure-11.

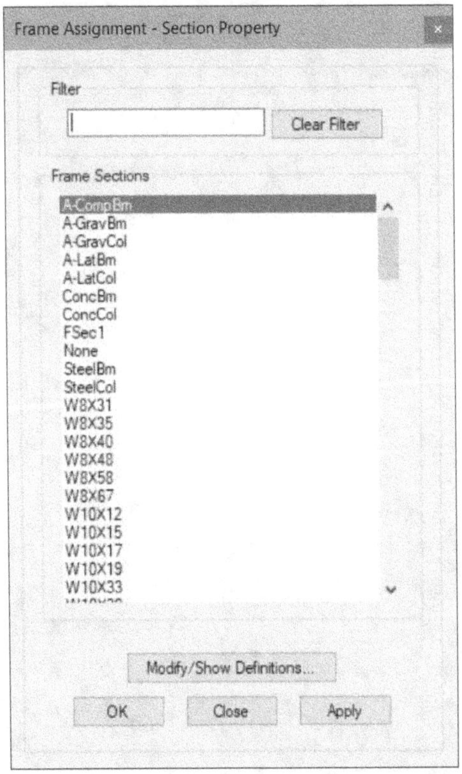

Figure-11. Frame Assignment-Section Property dialog box

- Select desired option from the list to assign respective property to selected members.
- Click on the **Modify/Show Definitions** tool from the dialog box to check or modify the property of selected section. The procedure to modify or create section properties has been discussed earlier in previous chapter.
- After selecting desired section property, click on the **OK** button to assign the selected property.

Assigning Property Modifier to Frame

The property modifier is used to modify properties of selected sections exclusively for analysis. The procedure to apply property modifier is given next.

- Select the section whose properties are to be modified for analysis and click on the **Property Modifiers** tool from the **Frame** cascading menu of the **Assign** menu. The **Frame Assignment-Property Modifiers** dialog box will be displayed; refer to Figure-12.

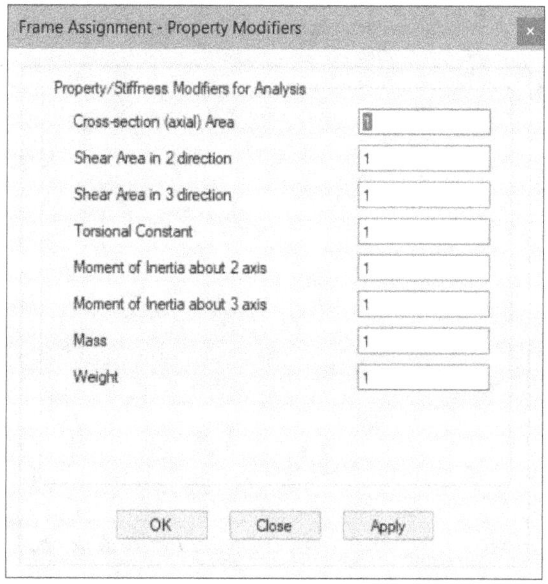

Figure-12. Frame Assignment-Property Modifiers dialog box

- Specify desired values in the edit boxes of the dialog box and click on the **OK** button. The specified values are multiplier assigned to selected frame members. For example, if you specify 2 in **Mass** edit box then mass of selected frame members will be multiplied by 2.

Releasing or Partially Fixing Ends of Frame Members

The **Releases/Partially Fixity** tool is used to release or partially fix frame members for analysis. The benefit of doing so is to make selected beams/columns free to deform practically under load. Assume you have fixed both ends of the beam and applied a large load on it in analysis. Once you run the study, the beam will be stretched like a rubber band while both ends retain their original position as well as line trajectory. But if you want the beams to be simply supported then you need to release the moment at end points which make it work like pin connection; refer to Figure-13. The procedure to use this tool is given next.

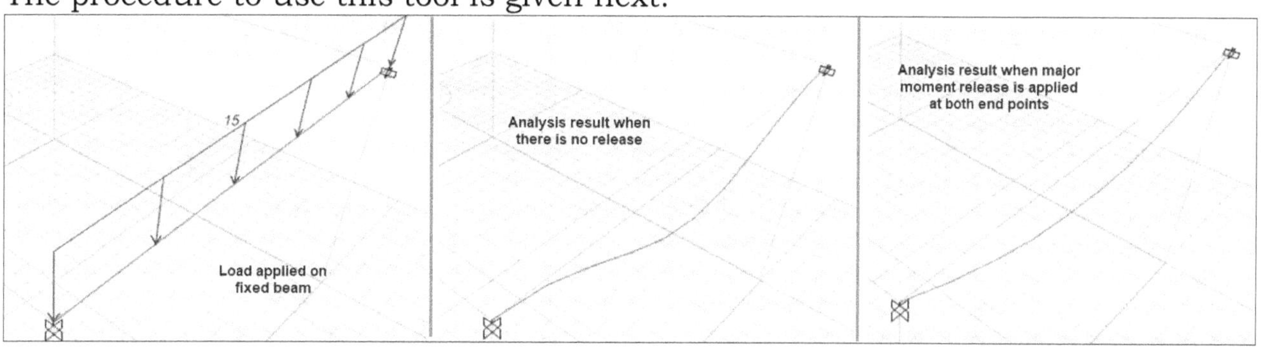

Figure-13. Applying moment release

- Select the frame members to which you want to apply releasing or partially fixing and click on the **Releases/Partially Fixity** tool from the **Frame** cascading menu of the **Assign** menu. The **Frame Assignment-Releases/Partial Fixity** dialog box will be displayed; refer to Figure-14.

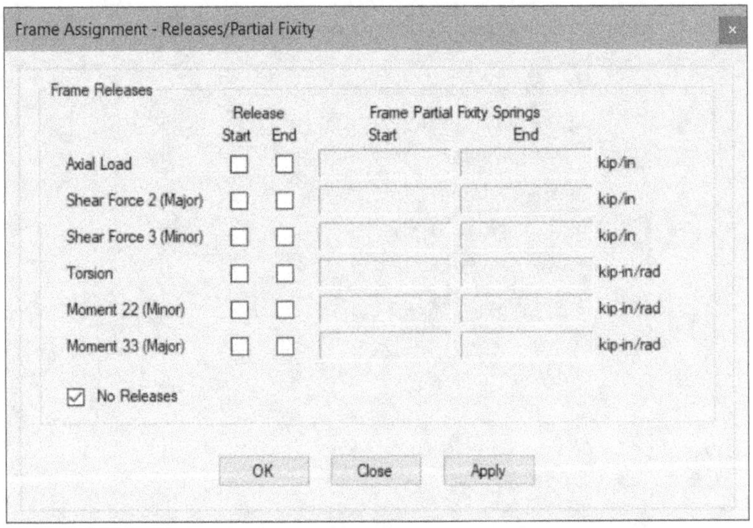

Figure-14. Frame Assignment-Releases Partial Fixity dialog box

- Select the **Start** or **End** check boxes for desired type of loads to be released. On selecting the **Start** check box, you release the respective load from start of the frame member and on selecting the **End** check box, you release the load from the end of the frame member. Note that a **0** value means full release.
- If you want to apply partial load on the selected end of frame then specify desired value in the edit box next to selected check box.
- After setting desired parameters, click on the **OK** button from the dialog box.

Assigning End Length Offset

The **End Length Offsets** tool is used to offset extra length of beam/column at an end point on the joint so that the beam/columns extends at the joint like it happens in real world. Otherwise, the software will join the beams/columns at the centroid. The procedure to use this tool is given next.

- Select the beams/columns to which you want to assign offset and click on the **End Length Offsets** tool from the **Frame** cascading menu of the **Assign** menu. The **Frame Assignment-End Length Offsets** dialog box will be displayed; refer to Figure-15.

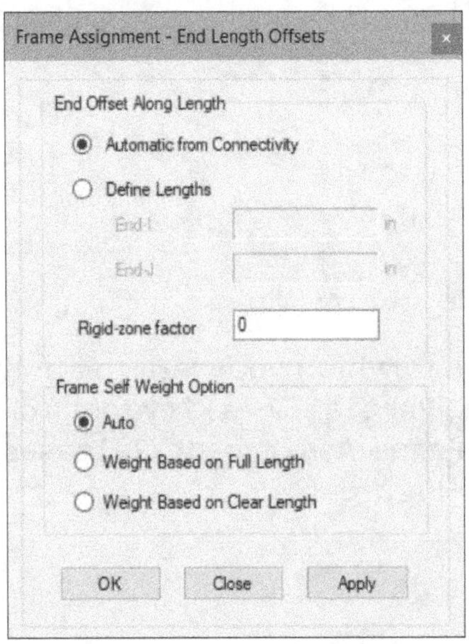

Figure-15. Frame Assignment-End Length Offsets dialog box

- Select the **Automatic from Connectivity** radio button to automatically offset one beam/column to accommodate other beam/column in the joint.
- Select the **Define Lengths** radio button to specify desired values of offset along End-I and End-J of the beam/column. End-I is the start point of beam/column and End-J is the end point of the beam/column.
- Specify desired value in **Rigid-zone factor** edit box. The rigid-zone factor specifies the fraction of each end offset assumed to be rigid for bending and shear deformations. When a fraction of the end offset is specified rigid, the outside portion of the end offset is assumed rigid, that is, the portion at the end of the frame member.
- Select desired option from the **Frame Self Weight Option** area to define how the weight of the frame member will be calculated after assigning the offset.
- After setting desired parameters, click on the **OK** button.

Assigning Insertion Points for Frame Members

The insertion point is the location of beam/column face at which two more beams/ columns join together. The **Insertion Point** tool in the **Frame** cascading menu of the **Assign** menu is used to specify insertion points for selected beams/columns. The procedure to use this tool is given next.

- After selecting desired beams/columns, click on the **Insertion Point** tool from the **Frame** cascading menu of the **Assign** menu. The **Frame Assignment-Insertion Point** dialog box will be displayed; refer to Figure-16.

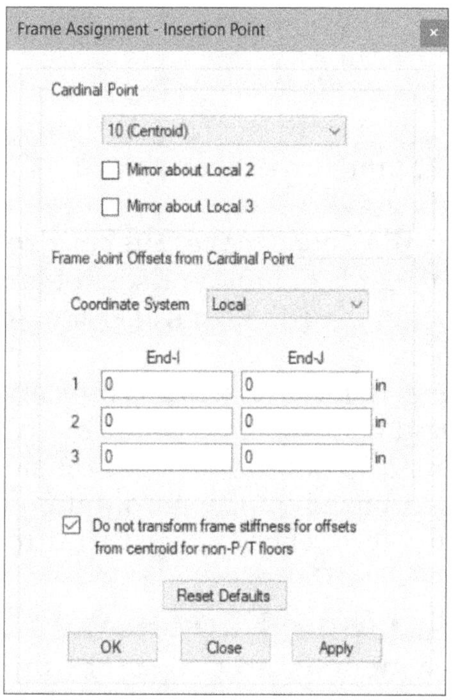

Figure-16. Frame Assignment-Insertion Point dialog box

- Select desired option from the **Cardinal Point** drop-down to define the location of insertion point. Location of various points on the beam is shown in Figure-17.

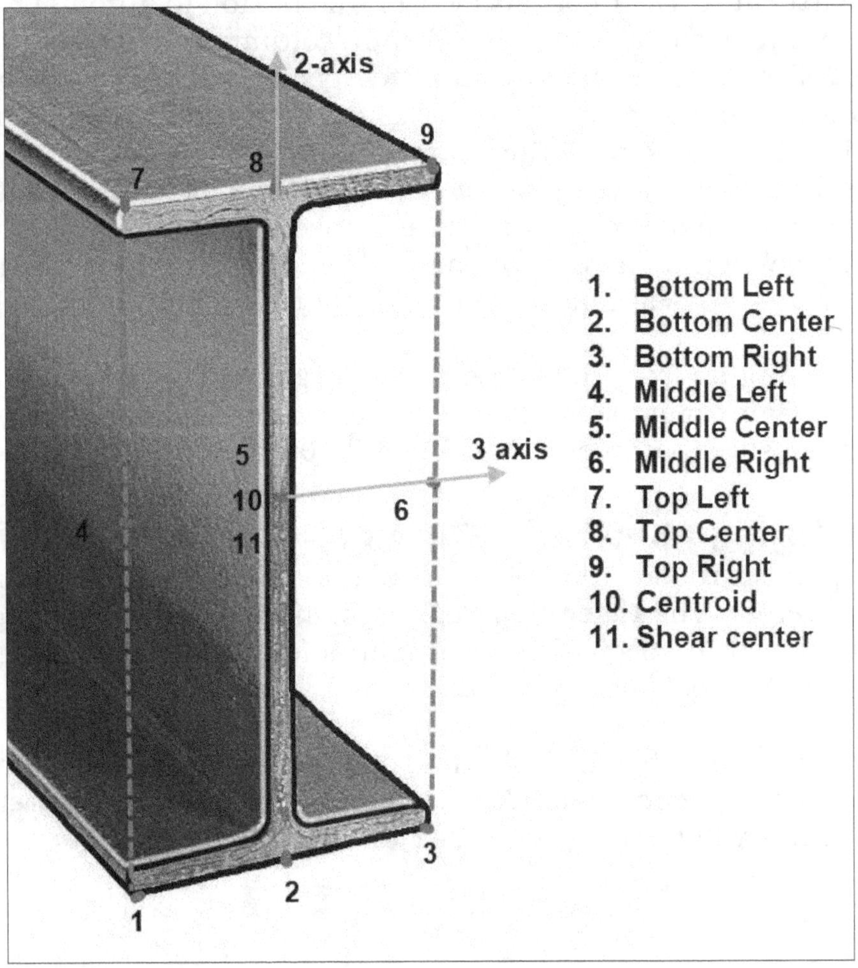

Figure-17. I beam

- Select the **Mirror about 2** and/or **Mirror about 3** check boxes to mirror the cardinal point about 2 axis and/or 3 axis.
- Select desired option from the **Coordinate System** drop-down to define the user coordinate system to be used for defining offsets of end points.
- Specify desired offset values in the edit boxes of **Frame Joint Offsets from Cardinal Point** area of the dialog box.
- Select the **Do not transform frame stiffness for offsets from centroid for non-P/T floors** check box to avoid transforming frame stiffness for non-post tensioning floors based on offsets. Note that floors with post-tensioning will always have their frame stiffness transformed based on the cardinal point and joint offsets.
- After setting desired parameters, click on the **OK** button from the dialog box.

Assigning Local Axes

The **Local Axes** tool in **Assign** menu is used to set the orientation of axis of selected frame member. The procedure to use this tool is given next.

- Click on the **Local Axes** tool from the **Frame** cascading menu of the **Assign** menu. The **Frame Assignment - Local Axes** dialog box will be displayed; refer to Figure-18.
- Select the **Angle** radio button to specify the orientation of 2nd axis of the selected frame member. Select the **Rotate by Angle** radio button to specify the angle value by which the 2nd axis will be rotated with respect to 1st axis.

- Select the **Orient with Grid System - Applies to Vertical Frame Objects Only** radio button to orient the axes by using grid system. Note that these options are applicable for vertical frame objects only like columns.

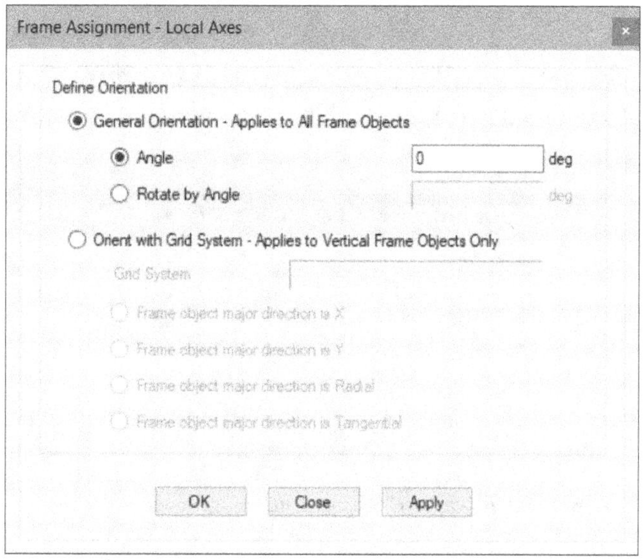

Figure-18. Frame Assignment-Local Axes dialog box

- Select desired option below the selected radio button to specify the orientation of selected frame members with respect to grid system selected.
- After setting desired parameters, click on the **OK** button. The marking of axes will be displayed on the model; refer to Figure-19.

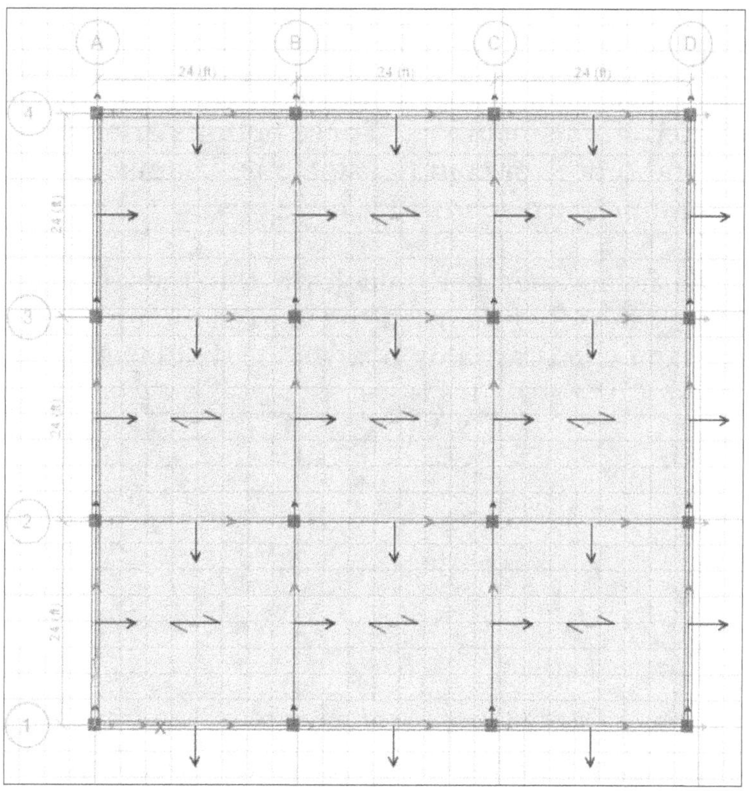

Figure-19. Model with axes denotation

Assigning Output Stations

Output stations are designated locations along a frame element used to output forces, perform design, and plot points used for graphic display of force diagrams. The procedure to assign output stations to selected frame members is given next.

- After selecting the frame members, click on the **Output Stations** tool from **Frame** cascading menu of the **Assign** menu. The **Frame Assignment - Output Stations** dialog box will be displayed; refer to Figure-20.

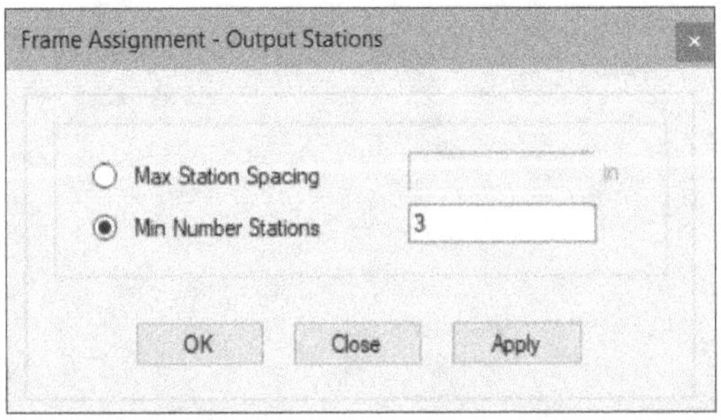

Figure-20. Frame Assignment-Output Stations dialog box

- Select the **Max Station Spacing** radio button to specify the maximum distance between two consecutive stations being placed on selected frame members. Specify desired distance value in the adjacent edit box.
- Select the **Min Number Stations** radio button and specify the minimum number of stations to be assigned on selected frame members.
- Click on the **OK** button to apply changes.

Assigning Tension/Compression Limits

The tension and compression limits are the maximum allowed tension and compression stresses on the selected frame members. These options are used primarily to model tension-only cables and braces or compression only beams. The procedure to assign tension/compression limits is given next.

- After selecting the frame members, click on the **Tension/Compression Limits** tool from the **Frame** cascading menu of the **Assign** menu. The **Frame Assignment-Tension/Compression Limits** dialog box will be displayed; refer to Figure-21.

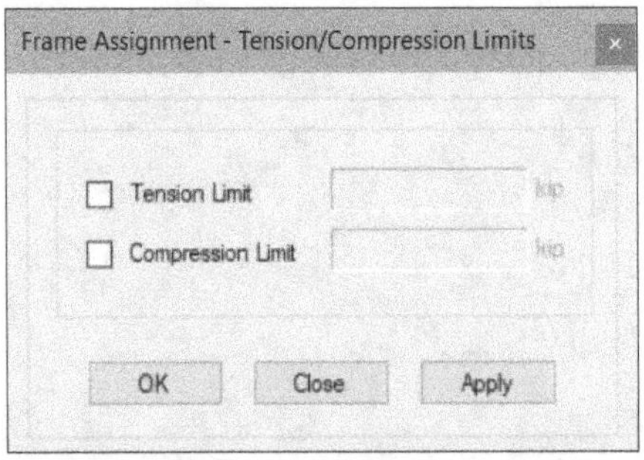

Figure-21. Frame Assignment-Tension Compression Limits dialog box

- Select the **Tension Limit** check box to specify tension limit in selected frame members and specify the value in adjacent edit box. Note that the unit for tension limit is kip which is 1000 pound-force.

- Select the **Compression Limit** check box to specify compression limit in selected frame members and specify the value in adjacent edit box. Note that the unit for compression limit is kip which is 1000 pound-force.
- After setting desired values, click on the **OK** button to apply limits.

Assigning Hinges

The **Hinges** tool of **Frame** cascading menu in **Assign** menu is used to assign hinge properties to the selected frame member. The procedure to use this tool is given next.

- Select the frame member to which you want to assign hinge properties and click on the **Hinges** tool from the **Frame** cascading menu of the **Assign** menu. The **Frame Assignment - Hinges** dialog box will be displayed.

Nonlinear Beam/Column

- Select **Nonlinear Beam/Column** radio button to assign concentrated nonlinear hinges at specific locations in a beam or column. The options of the dialog box will be displayed as shown Figure-22.

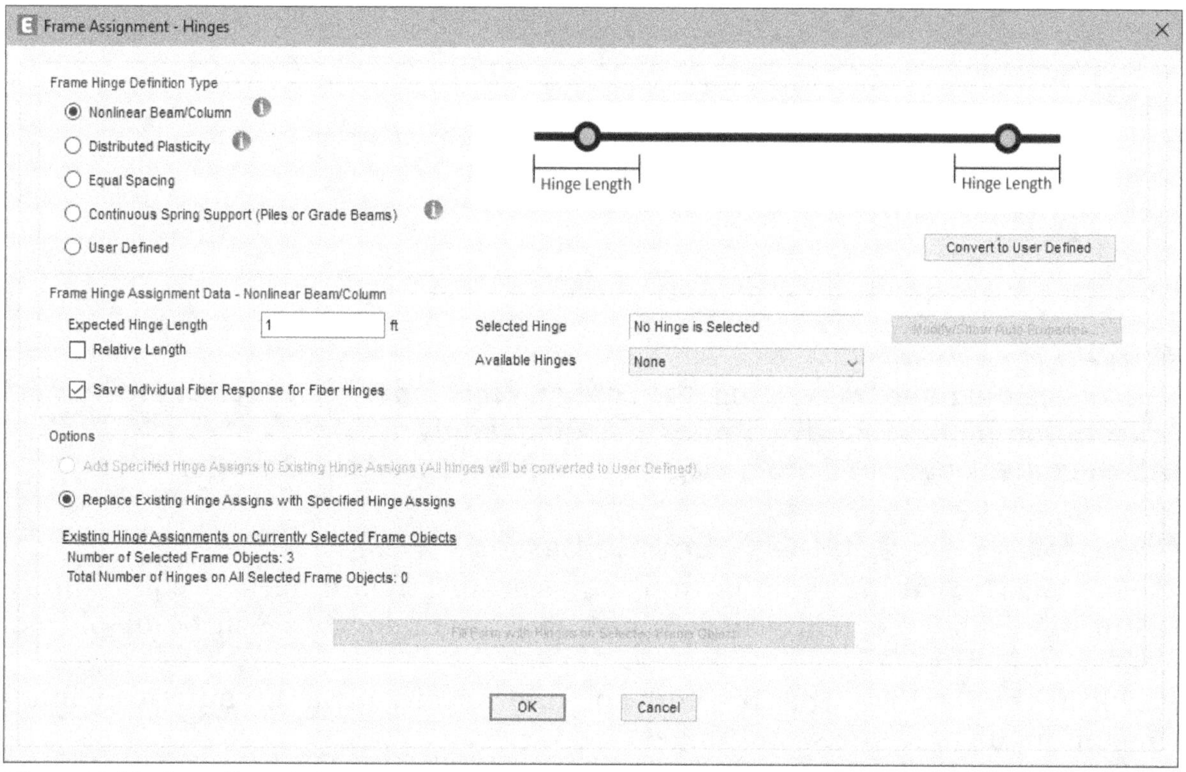

Figure-22. Nonlinear Beam Column dialog box

- Click in the **Expected Hinge Length** edit box and specify desired value to define the length of the hinge region.
- Select **Relative Length** check box to define the hinge length as a fraction of the total element length.
- Select **Save Individual Fibre Response for Fibre Hinges** check box to store the response of each fibre in the hinge at every step.
- Select the **Auto** option from the **Available Hinges** drop-down, the **Auto Hinge Assignment Data** dialog box will be displayed. The options of the dialog box have been discussed earlier.

Distributed Plasticity

- Select **Distributed Plasticity** radio button to model plastic behavior along the entire length of the frame element. These options of the dialog box will be displayed as shown in Figure-23.

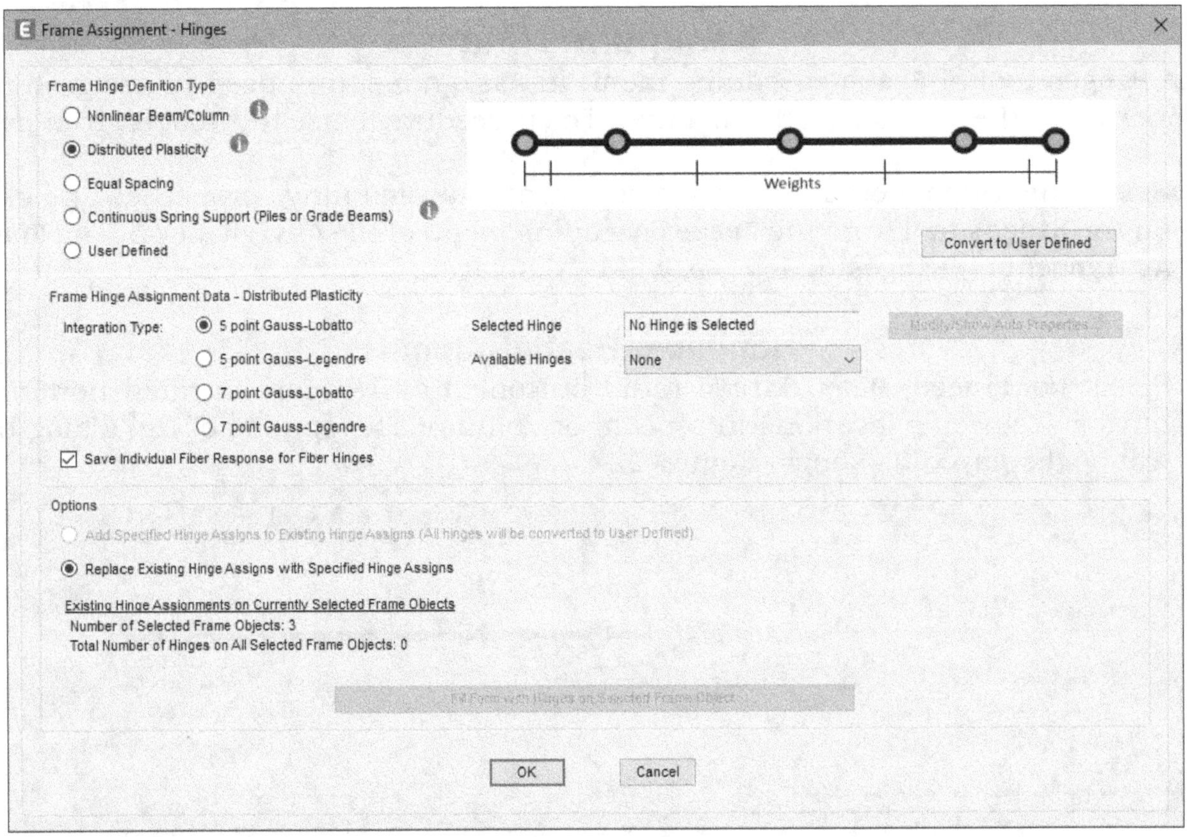

Figure-23. Distributed Plasticity dialog box

- Select desired radio button from the **Integration Type** area to select the numerical integration method for distributed plasticity analysis.
- All the other options of the dialog box have been discussed earlier.

Equal Spacing

- Select **Equal Spacing** radio button to assign multiple hinges at equal intervals along the frame element. These options of the dialog box will be displayed as shown in Figure-24.
- Click in the **Maximum Spacing** edit box and specify desired value to define the maximum distance between hinges.
- Select **Place Hinges at the Two Ends** check box to assign hinge at both ends of the member.
- All the other options of the dialog box have been discussed earlier.

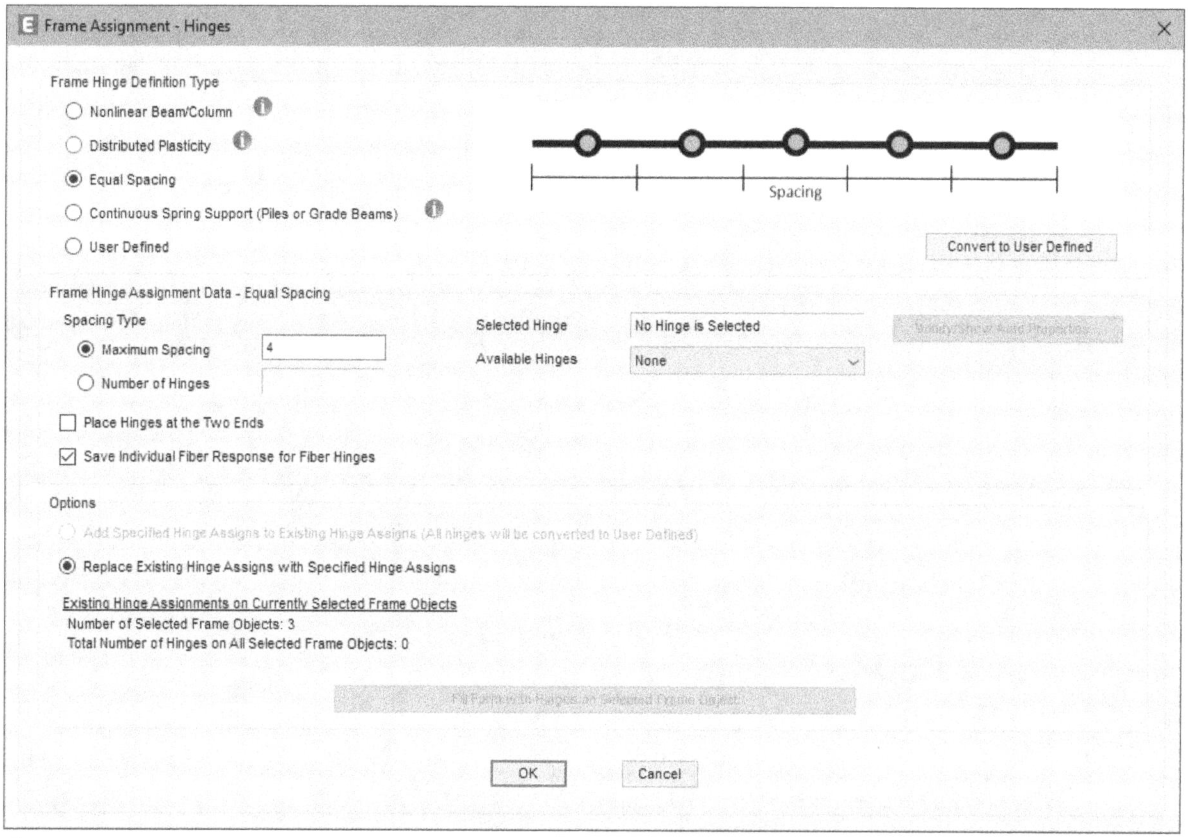

Figure-24. Equal Spacing dialog box

Continuous Spring Support

- Select **Continuous Spring Support** radio button to represent continuous soil or spring support along the length of the element.
- The options of the dialog box have been discussed earlier.

User Defined

- Select **User Defined** radio button to manually define hinge properties and locations. These options of the dialog box will be displayed as shown in Figure-25.

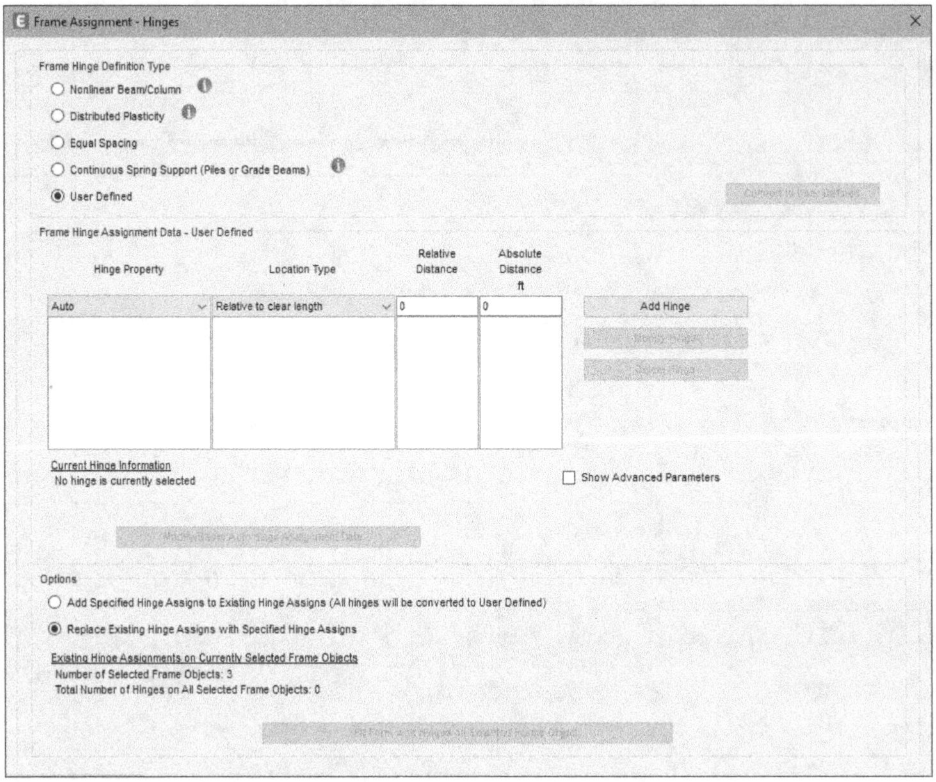

Figure-25. Frame Assignment-Hinges dialog box

- Specify the relative distance of hinge at the selected frame member in **Relative Distance** edit box of the dialog box. The value specified here should be between 0 to 1. Here, 0 and 1 represent the end points of selected frame member and 0.5 represents the middle of frame member. After setting desired value, click on the **Add Hinge** button. The **Auto Hinge Assignment Data** dialog box will be displayed; refer to Figure-26.

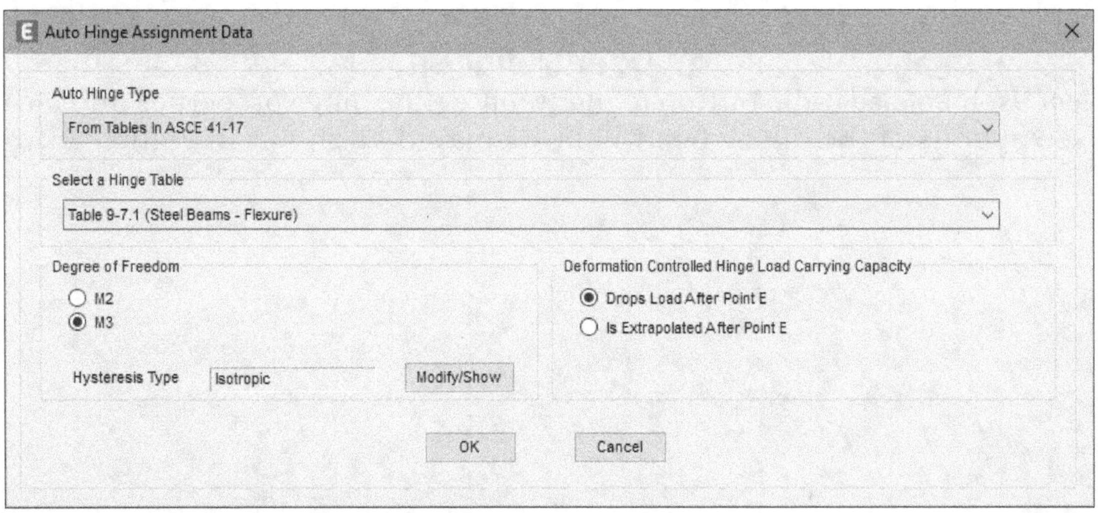

Figure-26. Auto Hinge Assignment Data dialog box

- Select desired option from **Auto Hinge Type** drop-down to define the source of list from where hinges will be selected. If you have selected the **Buckling Restrained Brace** option from the drop-down then hinges will be selected automatically based on braces created in the model.
- Select desired table from the **Select a Hinge Table** drop-down. Note that the selection of table depends on type and material of selected frame member.

- Select desired radio button from **Degree of Freedom** area to define the degree of freedom to be free for hinging.
- Click on the **Modify/Show** button for the **Hysteresis Type** parameter to view or edit the selected hysteresis properties. Hysteresis parameter defines energy dissipation for cyclic loads. On clicking the **Modify/Show** button, the **Hysteresis Type** dialog box will be displayed; refer to Figure-27.

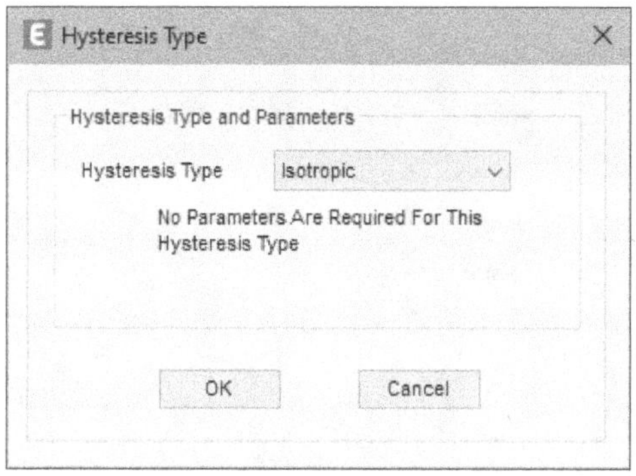

Figure-27. Hysteresis Type dialog box

- Select desired option from the **Hysteresis Type** drop-down to define the cyclic load behavior of the hinge. Selecting **Isotropic** option assumes that if strength is decreasing in one direction due to loading then it will be increasing in opposite direction. **Degrading** option is used when energy dissipation decreases with more plastic deformation in frame member. Select **Kinematic** option to use Bauschinger Effect in ductile materials like steel which means stretching in one direction will reduce compressive yield point of frame member. Select **Takeda** option if you need RC modeling that focuses on stiffness degradation without needing complex extra parameters. Select the **Pivot** option if you want to define specific pivot points at which loading and unloading lines of frame member are forced to pivot. Similarly, you can select other options to define hysteresis settings.
- After specifying desired parameters, click on the **OK** button from the dialog box.
- Select desired radio button from the **Deformation Controlled Hinge Load Carrying Capacity** area to define whether load carrying capacity of beam/frame member will be 0 or load carrying capacity is extended after point E.
- After specifying parameters, click on the **OK** button to apply automatic hinge. Based on material and loading requirements, the hinge will be created automatically.

Assigning Hinge Overwrites

The **Hinge Overwrites** tool is used to assign overwrite to the selected frame member. These overwrites include the sub-division of hinge and no load drop on hinges. Note that this tool is displayed only after you have assigned hinge properties of some frame members in the model. The procedure to use this tool is given next.

- After selecting the frame members, click on the **Hinge Overwrites** tool from the **Frame** cascading menu of the **Assign** menu. The **Frame Assignment-Hinge Overwrites** dialog box will be displayed; refer to Figure-28.

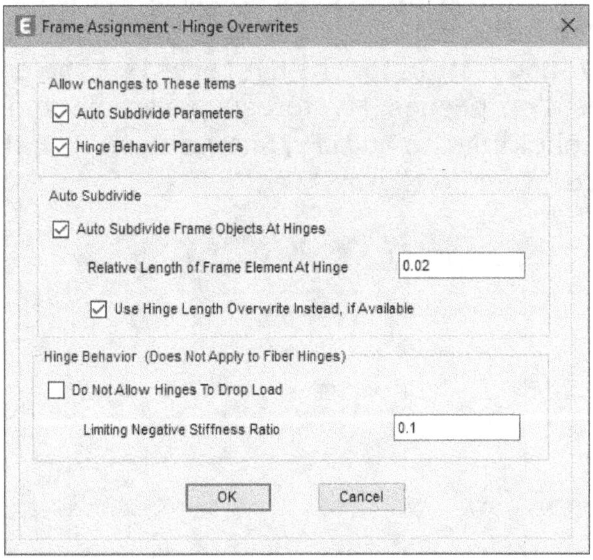

Figure-28. Frame Assignment-Hinge Overwrites dialog box

- Select the **Auto Subdivide Parameters** check box to allow changes in sub-divisions of hinge.
- Select the **Hinge Behavior Parameters** check box to allow modification of hinge behavior parameters.
- Select the **Auto Subdivide Frame Objects At Hinges** check box to divide the frame objects on which hinge point exist. You can specify the value of relative length in the edit box as discussed for previous tool.
- Select **Use Hinge Length Overwrite Instead, if Available** check box to automatically use the predefined hinge length.
- Select the **Do Not Allow Hinges To Drop Load** check box to make sure the hinged frame members deform as per their stiffness and do not fail suddenly.
- After setting desired parameters, click on the **OK** button.

Assigning Line Spring

The **Line Spring** tool in the **Frame** cascading menu of the **Assign** menu is used to assign line spring properties to selected frame members. The procedure to use this tool is given next.

- Select desired frame members and click on the **Line Spring** tool from the **Frame** cascading menu of the **Assign** menu. The **Frame Assignment-Line Springs** dialog box will be displayed; refer to Figure-29.

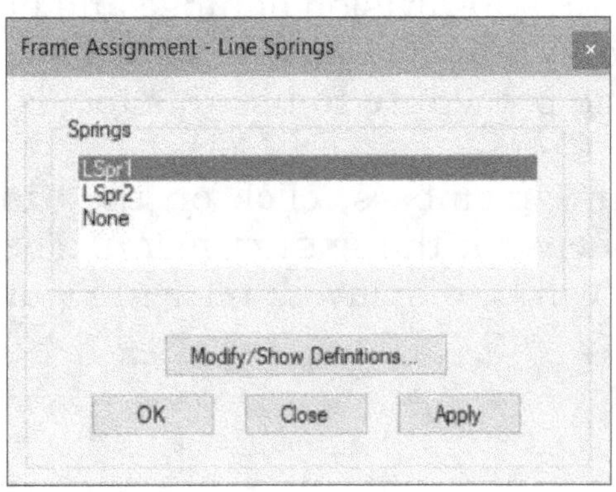

Figure-29. Frame Assignment-Line Springs dialog box

- Select desired property to be applied.
- Click on the **OK** button to apply the settings.

Assigning Additional Mass to Frame Members

The **Additional Mass** tool in the **Frame** cascading menu of the **Assign** menu is used to assign additional mass to the selected frame members. The procedure to use this tool is given next.

- After selecting the frame members, click on the **Additional Mass** tool from the **Frame** cascading menu of the **Assign** menu. The **Frame Assignment-Additional Mass** dialog box will be displayed; refer to Figure-30.

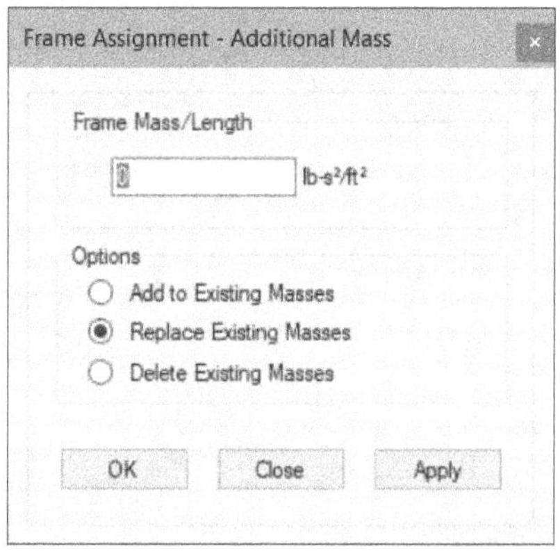

Figure-30. Frame Assignment - Additional Mass dialog box

- Specify desired value of mass per length in the **Frame Mass/Length** edit box.
- Select desired radio button from the **Options** area to specify whether the mass should be added, replaced, or deleted.
- After setting desired parameters, click on the **OK** button from the dialog box.

Assigning Pier Labels

The **Pier Label** tool in **Frame** cascading menu of the **Assign** menu is used to assign earlier created pier label to selected frame member. The procedure to use **Pier Label** tool is given next.

- After selecting desired frame members, click on the **Pier Label** tool from the **Frame** cascading menu of the **Assign** menu. The **Frame Assignment-Pier Label** dialog box will be displayed; refer to Figure-31.

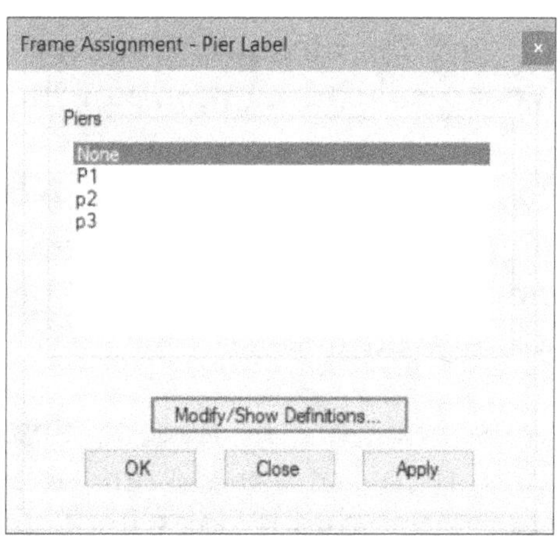

Figure-31. Frame Assignment-Pier Label dialog box

- Select desired pier label for selected frame member and click on the **OK** button.

You can assign spandrel labels in the same way.

Frame Auto Mesh Options

The **Frame Auto Mesh Options** tool in the **Frame** cascading menu of the **Assign** menu is used to specify whether the auto meshing will be applied to selected frames or not during the analysis. The procedure to use this tool is given next.

- Select desired frame members and click on the **Frame Auto Mesh Options** tool from the **Frame** cascading menu of the **Assign** menu. The **Frame Assignment-Frame Auto Mesh Options** dialog box will be displayed; refer to Figure-32.

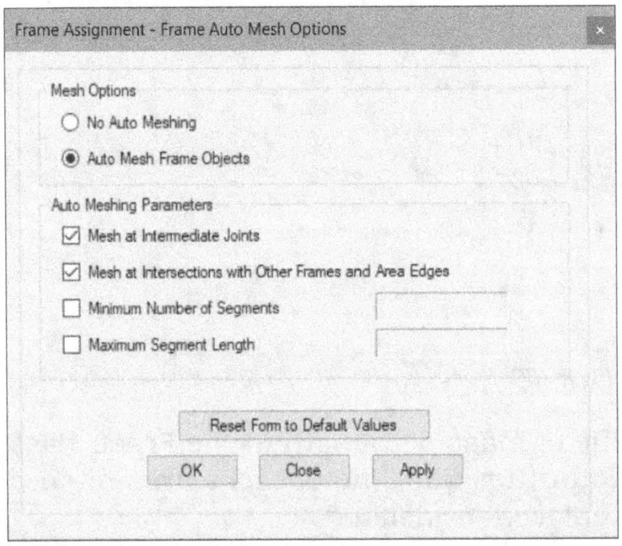

Figure-32. Frame Assignment-Frame Auto Mesh Options dialog box

- Select desired option from the dialog box. Select the **Auto Mesh at Intermediate Joints** radio button to assign auto meshing at the joints. Select the **Auto Mesh at Intermediate Joints and Intersecting Frames/Edges** radio button to assign auto meshing at joints as well as at intersections of frame members and edges. After selecting this radio button, select desired check boxes from the **Auto Meshing Parameters** area of the dialog box. Select the **No Auto Meshing** radio button to exclude selected frame members from auto meshing.
- After setting desired parameters, click on the **OK** button.

Similarly, you can set the frame floor meshing options by using the **Frame Floor Meshing Options** tool in the **Frame** cascading menu of the **Assign** menu.

Defining Moment Connections of Frame Beams

The **Moment Frame Beam Connection Type** tool in the **Frame** cascading menu of the **Assign** menu is used to define the connection type of the beam related to moment. The procedure to use this tool is given next.

- After selecting desired beams, click on the **Moment Frame Beam Connection Type** tool from the **Frame** cascading menu of the **Assign** menu. The **Frame Assignment-Moment Frame Beam Connection Type** dialog box will be displayed; refer to Figure-33.

- Select the **Standard Moment Connection** radio button if you want to use standard connections without any section reduction or seismic loading conditions.

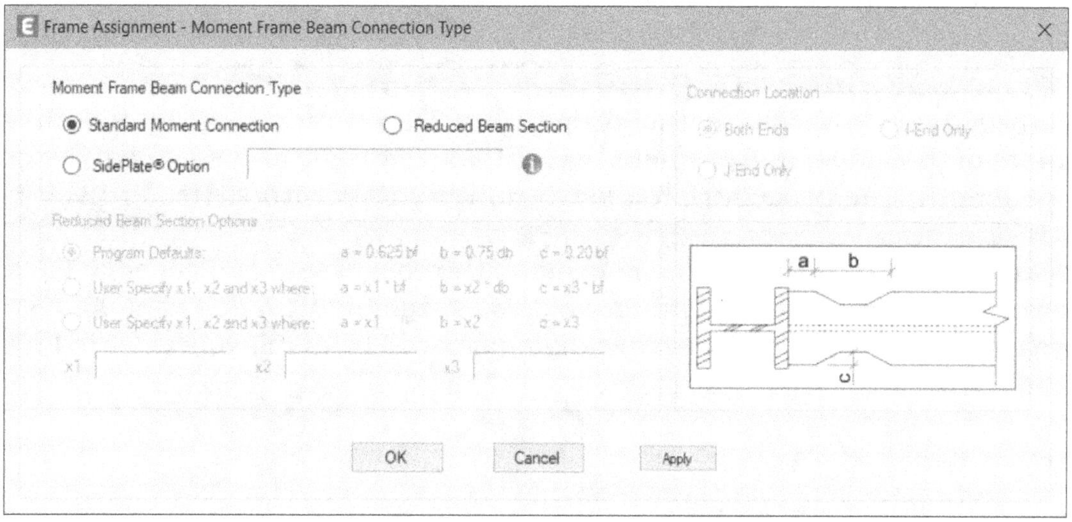

Figure-33. Frame Assignment-Moment Frame Beam Connection Type dialog box

- Select the **Reduced Beam Section** radio button if you want reduction in size of the beam near connection point. The options to specify parameters for reduction in beam section will be displayed in the **Reduced Beam Section Options** area of the dialog box. Specify desired values of a, b, and c using these options.
- Select the **SidePlate Option** radio button if you want to update the connection as per the seismic loading of the joint. In this type of connection, a series of plates are welded to the beam end and the column to strengthen the beam-column connection and force hinging to occur in the beam away from the connection. On selecting this radio button, a drop-down is displayed at the right in the dialog box. Select desired option from the drop-down to define seismic loading.
- After setting desired parameters, click on the **OK** button.

Assigning Column Splice Overwrites

The **Column Splice Overwrite** tool is used to apply overwrites to splicing of columns. The procedure to do so is given next.

- Select desired columns and click on the **Column Splice Overwrite** tool from the **Frame** cascading menu of the **Assign** menu. The **Assign Column Splice Overwrites to Column Objects** dialog box will be displayed; refer to Figure-34.

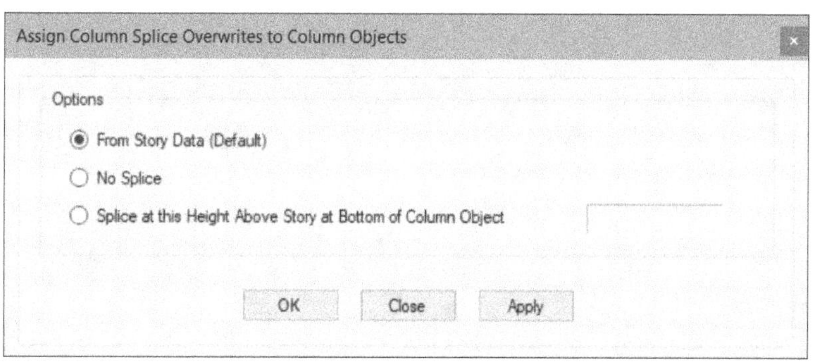

Figure-34. Assign Column Splice Overwrites to Column Objects dialog box

- Select the **From Story Data (Default)** radio button to splice columns based on the story data.

- Select the **No Splice** radio button if you do not want to splice columns.
- Select the **Splice at this Height Above Story at Bottom of Column Object** radio button to specify height above story bottom at which columns will be spliced.

Assigning Nonprismatic Property Parameters

The Nonprismatic Property is the property of frame members by which various parameters of the model like stiffness, weight, axial stress, and so on can be changed along the length. The procedure to assign nonprismatic properties is given next.

- Select desired frame members and click on the **Nonprismatic Property Parameters** tool from the **Frame** cascading menu of the **Assign** menu. The **Frame Assignment - Nonprismatic Property Parameters** dialog box will be displayed; refer to Figure-35.

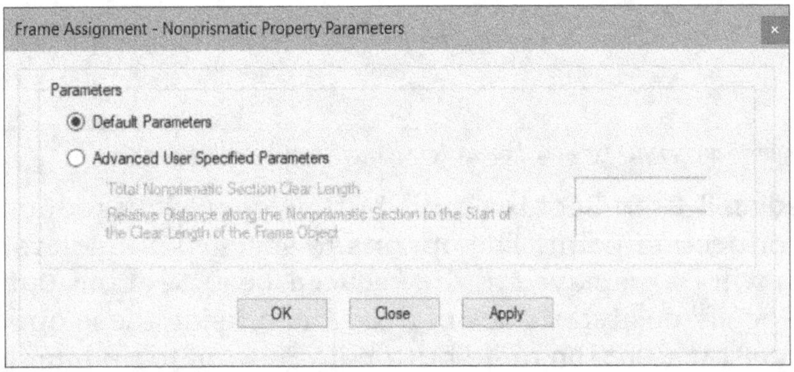

Figure-35. Frame Assignment-Nonprismatic Property Parameters dialog box

- Select the **Default Parameters** radio button to keep the frame members as they are by default.
- Select the **Advanced User Specified Parameters** radio button to specify the relative length of frame members till which the frame members will not demonstrate nonprismatic properties and enter desired values in adjacent edit boxes.
- After setting desired parameters, click on the **OK** button.

Assigning Material Overwrites

The material overwrites are assigned to modify the material properties of selected frame members. The procedure to do so is given next.

- After selecting desired frame members, click on the **Material Overwrites** tool from the **Frame** cascading menu of the **Assign** menu. The **Frame Assignment - Material Overwrite** dialog box will be displayed; refer to Figure-36.

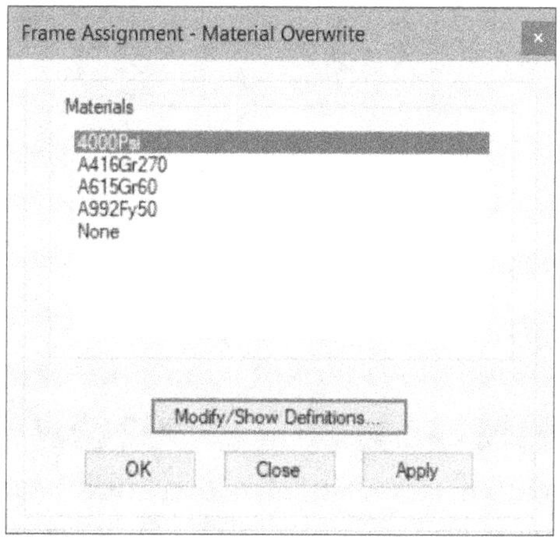

Figure-36. Frame Assignment-Material Overwrite dialog box

- Select desired material to be assigned to selected frame members and click on the **OK** button.

Assigning Column/Brace Rebar Ratio for Creep Analysis

The rebar ratio is the ratio of steel bars to concrete mixture per volume. The procedure to assign this ratio to selected columns/beams is given next.

- Click on the **Column/Brace Rebar Ratio for Creep Analysis** tool from the **Frame** cascading menu of the **Assign** menu. The **Frame Assignment-Rebar Ratio for Creep Analysis** dialog box will be displayed; refer to Figure-37.

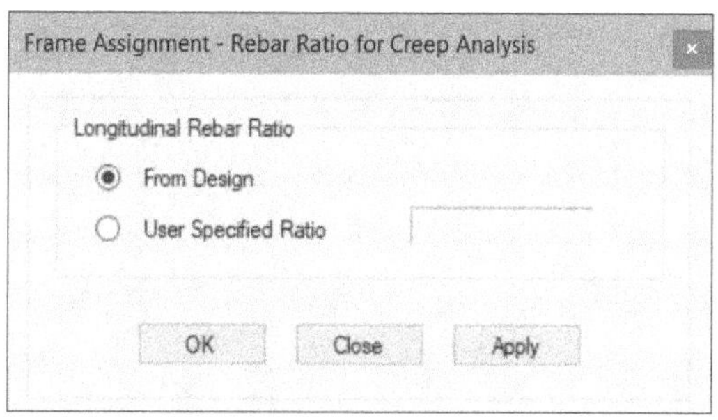

Figure-37. Frame Assignment-Rebar Ratio for Creep Analysis dialog box

- Select the **From Design** radio button if you want to apply rebar ratios as per the design codes automatically.
- Select the **User Specified Ratio** radio button to specify the rebar ratio manually in the adjacent edit box.
- After setting desired parameters, click on the **OK** button.

Assigning Floor Cracking

The **Floor Cracking** tool is used to define whether beams will be included in the floor cracking analysis or not. On clicking this tool, the **Assign Floor Cracking Option to Beam Objects** dialog box will be displayed; refer to Figure-38. Select the **Consider Selected Beam Objects in Floor Cracking Analysis** check box to include selected beams to analysis and click on the **OK** button from the dialog box.

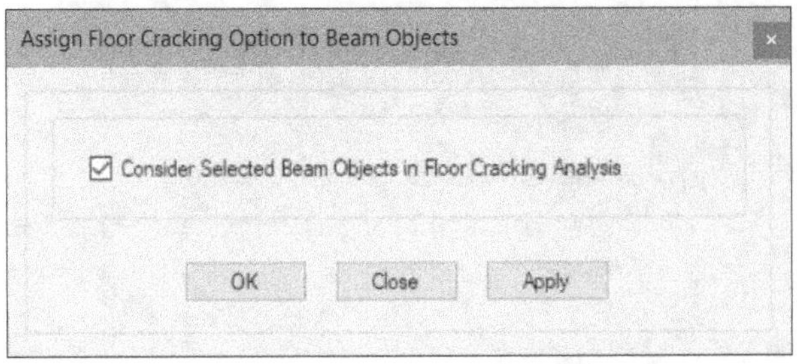

Figure-38. Assign Floor Cracking Option to Beam Objects dialog box

ASSIGNING SHELL PROPERTIES

Shell objects are those made of three or four nodes and are used to create floor, wall, bridge deck, and so on in the model. The tools to assign various properties to shell objects are available in the **Shell** cascading menu of the **Assign** menu; refer to Figure-39. Various tools in this menu are discussed next.

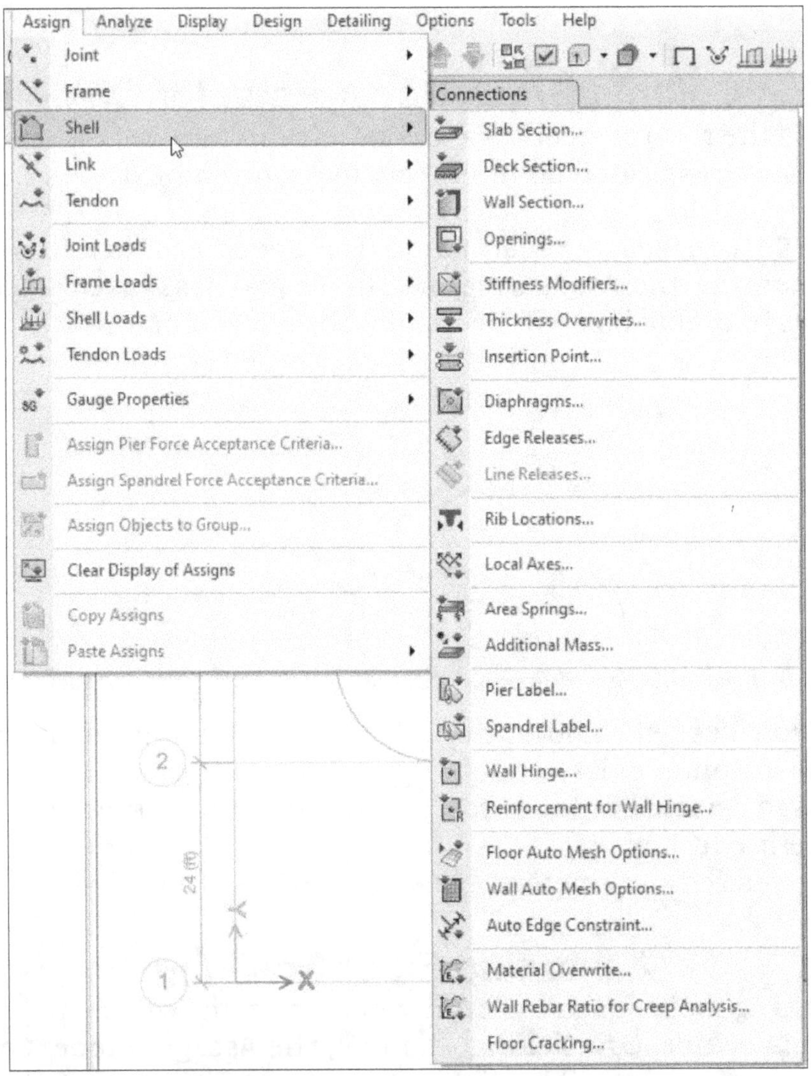

Figure-39. Shell cascading menu

Assigning Slab Section

The **Slab Section** tool in **Shell** cascading menu of the **Assign** menu is used to assign slab section properties to selected shell objects. The procedure to use this tool is given next.

- Select desired shell objects and click on the **Slab Section** tool from the **Shell** cascading menu of the **Assign** menu. The **Shell Assignment-Slab Section** dialog box will be displayed; refer to Figure-40.

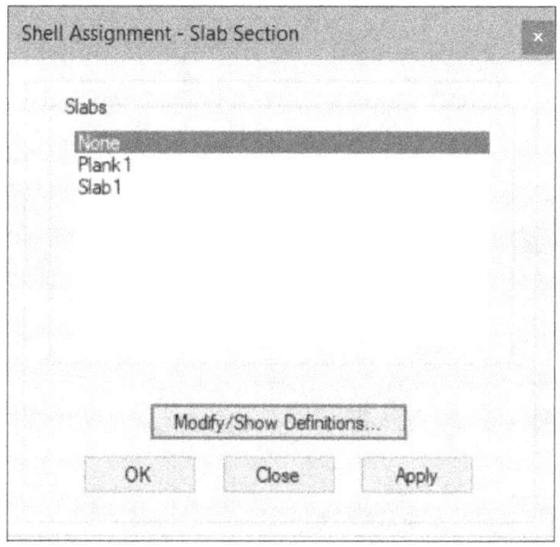

Figure-40. Shell Assignment-Slab Section dialog box

- Select desired slab section property from the list and click on the **OK** button from the dialog box. Note that the procedure to create and modify slab section properties has been discussed in previous chapter.

Assigning Deck Section Properties

The **Deck Section** tool in the **Shell** cascading menu of the **Assign** menu is used to assign deck properties to selected shell objects. The procedure to use this tool is given next.

- After selecting the shell object, click on the **Deck Section** tool in the **Shell** cascading menu of the **Assign** menu. The **Shell Assignment - Deck Section** dialog box will be displayed; refer to Figure-41.

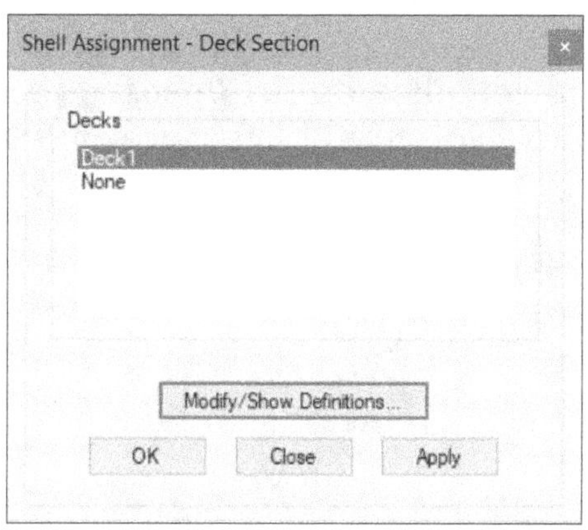

Figure-41. Shell Assignment-Deck Section dialog box

- Select desired deck property and click on the **OK** button. The selected deck property will be assigned shell objects. Note that the procedure to create and modify deck section properties has been discussed in previous chapter.

Assigning Wall Section

The **Wall Section** tool in the **Shell** cascading menu of the **Assign** menu is used to assign wall sections to selected shell objects. The procedure to use this tool is given next.

- After selecting the shell objects, click on the **Wall Section** tool in the **Shell** cascading menu of the **Assign** menu. The **Shell Assignment-Wall Section** dialog box will be displayed; refer to Figure-42.
- Select desired wall section from the dialog box and click on the **OK** button. Note that the procedure to create and modify wall section properties has been discussed in previous chapter.

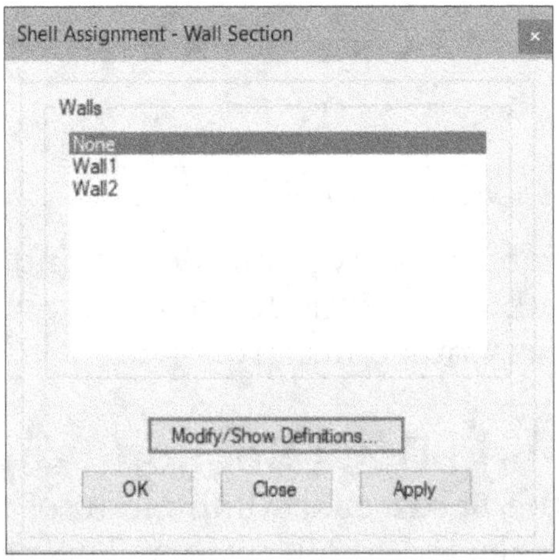

Figure-42. Shell Assignment-Wall Section dialog box

Assigning Openings

The **Openings** tool in the **Shell** cascading menu of the **Assign** menu is used to designate selected shell objects as opening. Openings are assumed to carry no load. The procedure to use this tool is given next.

- After selecting the shell object, click on the **Openings** tool from the **Shell** cascading menu of the **Assign** menu. The **Shell Assignment-Openings** dialog box will be displayed; refer to Figure-43.

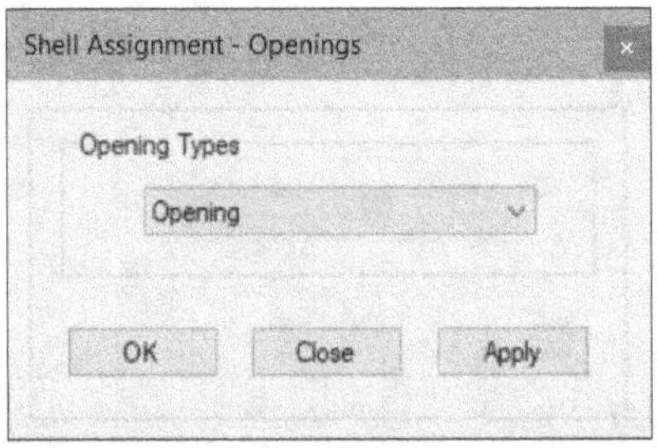

Figure-43. Shell Assignment-Openings dialog box

- Select the **Opening** option from the drop-down and click on the **OK** button. The selected shell object will be converted to an opening. If you want to close an opening then select it and use the **Not an Opening** option from the drop-down.

Assigning Stiffness Modifiers

The **Stiffness Modifiers** tool is used to apply modifiers to stiffness properties of the selected shell object. The procedure to use this tool is given next.

- After selecting desired shell objects, click on the **Stiffness Modifiers** tool from the **Shell** cascading menu of the **Assign** menu. The **Shell Assignment-Stiffness Modifiers** dialog box will be displayed; refer to Figure-44.

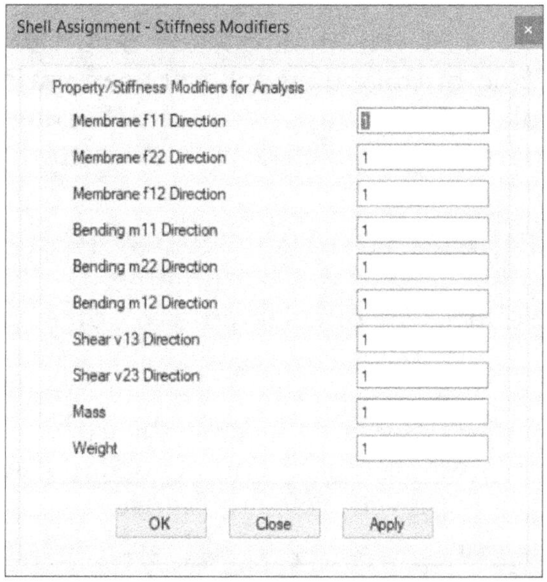

Figure-44. Shell Assignment-Stiffness Modifiers dialog box

- Specify desired values for modifier and click on the **OK** button to apply values.

Assigning Thickness Overwrites

The **Thickness Overwrites** tool is used to modify the thickness of selected shell objects. The procedure to use this tool is given next.

- Select desired shell objects and click on the **Thickness Overwrites** tool from the **Shell** cascading menu of the **Assign** menu. The **Shell Assignment - Thickness Overwrites** dialog box will be displayed; refer to Figure-45.

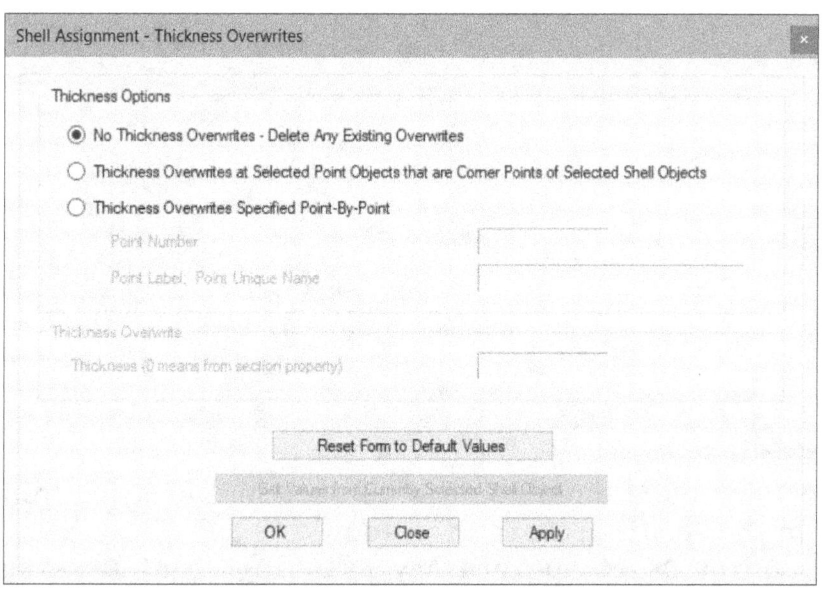

Figure-45. Shell Assignment-Thickness Overwrites dialog box

- Select the **No Thickness Overwrites - Delete Any Existing Overwrites** radio button if you want to delete overwrites of selected shell objects.
- Select the **Thickness Overwrites at Selected Point Objects that are Corner Points of Selected Shell Objects** radio button to overwrite thickness of shell objects at selected points. This option is available only when you have selected shell object and points on it. Specify the value of the thickness in **Thickness** edit box of **Thickness Overwrite** area of the dialog box.
- Select the **Thickness Overwrites Specified Point-By-Point** radio button to specify thickness overwrites for selected shell objects point by point. Use the buttons adjacent to selected radio button to switch between various buttons. Specify the thickness overwrite values for each point in the **Thickness** edit box.
- After setting desired parameters, click on the **OK** button.

Assigning Insertion Point to Shell Objects

The **Insertion Point** tool in the **Shell** cascading menu of the **Assign** menu is used to specify offset for insertion point of selected shell objects. The procedure to use this tool is given next.

- After selecting desired shell objects, click on the **Insertion Point** tool from the **Shell** cascading menu of the **Assign** menu. The **Shell Assignment - Insertion Point** dialog box will be displayed; refer to Figure-46.

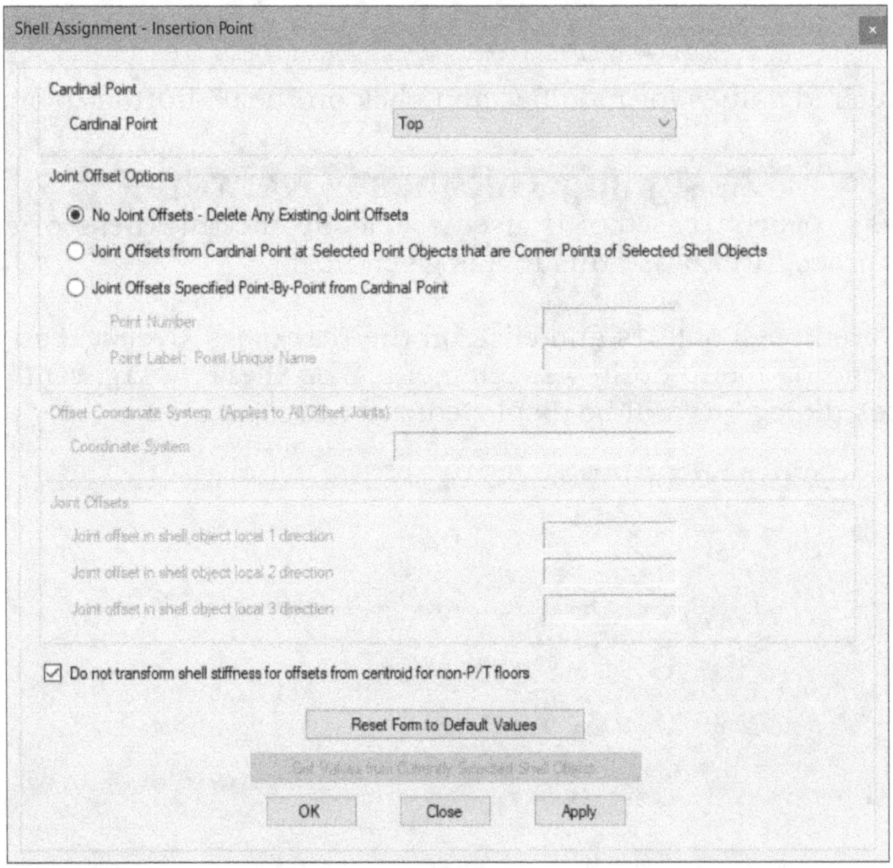

Figure-46. Shell Assignment-Insertion Point dialog box

- Select the location of cardinal point to be used for offsetting from each selected shell object in the **Cardinal Point** drop-down.
- Select the **No Joint Offsets-Delete Any Existing Joint Offsets** radio button to remove any offset applied to selected points of the selected shell objects.

- Select the **Joint Offsets from Cardinal Point at Selected Point Objects that are Corner Points of Selected Shell Objects** radio button to specify the value of offset in each direction using the edit boxes of **Joint Offsets** area of the dialog box. Note that by default, **Local** option is selected in the **Coordinate System** drop-down of the dialog box. If you want to specify the offset along the global X, Y, and Z axes then select the **Global** option from the **Coordinate System** drop-down.
- Select the **Joint Offsets Specified Point-By-Point from Cardinal Point** radio button to specify different offset values for each selected point. To switch between the points, click on the navigation buttons in the dialog box; refer to Figure-47.

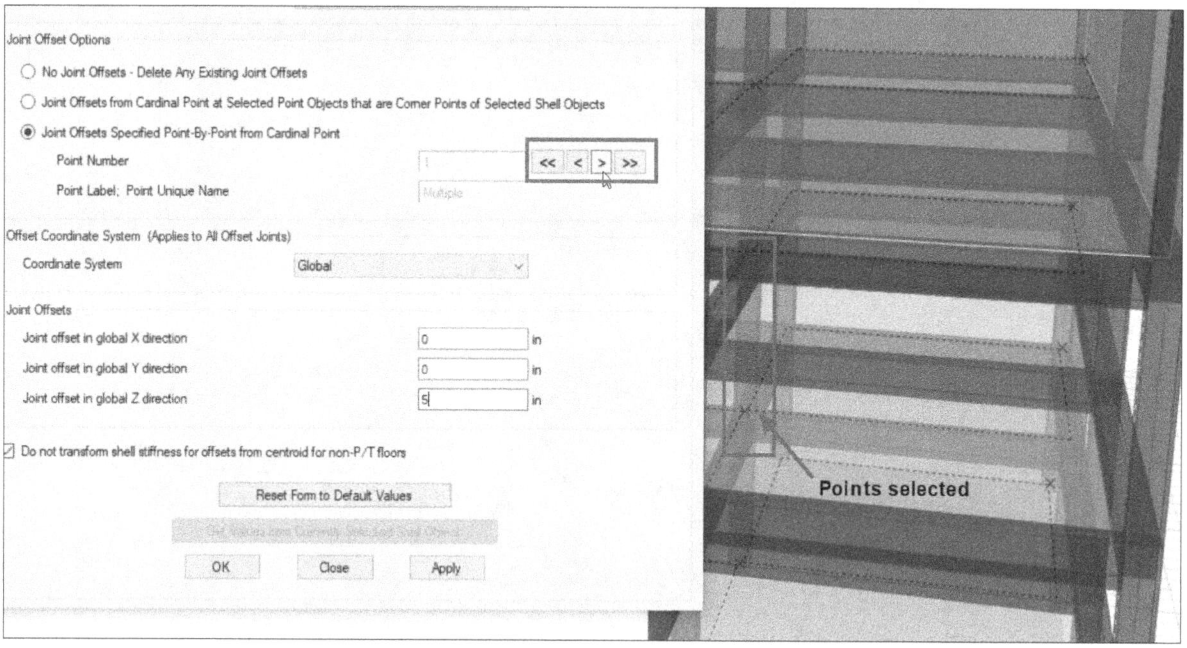

Figure-47. Navigation buttons

- After setting desired parameters, click on the **OK** button.

Assigning Diaphragms to Shell Objects

The **Diaphragms** tool in the **Shell** cascading menu of the **Assign** menu is used to assign diaphragm property to selected shell objects. The procedure to use this tool is given next.

- Select desired shell objects and click on the **Diaphragms** tool from the **Shell** cascading menu of the **Assign** menu. The **Shell Assignment - Diaphragms** dialog box will be displayed; refer to Figure-48.

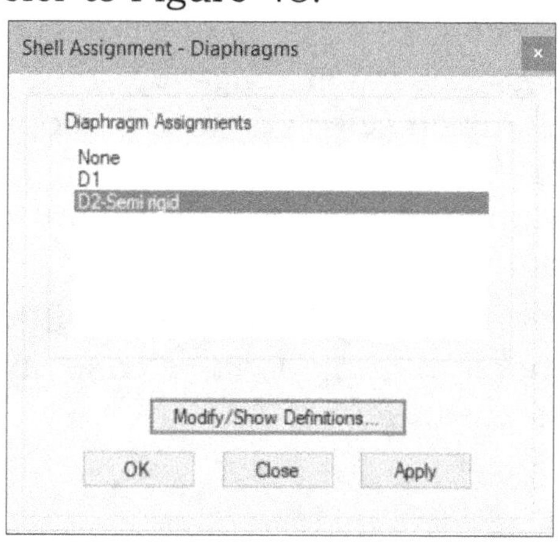

Figure-48. Shell Assignment-Diaphragms dialog box

- Select desired property from the dialog box and click on the **OK** button.

Assigning Edge Releases

The **Edge Releases** tool is used to release load from edges of selected shell objects. The procedure to use this tool is given next.

- After selecting desired shell objects, click on the **Edge Releases** tool from the **Shell** cascading menu of the **Assign** menu. The **Shell Assignment - Edge Releases** dialog box will be displayed; refer to Figure-49.

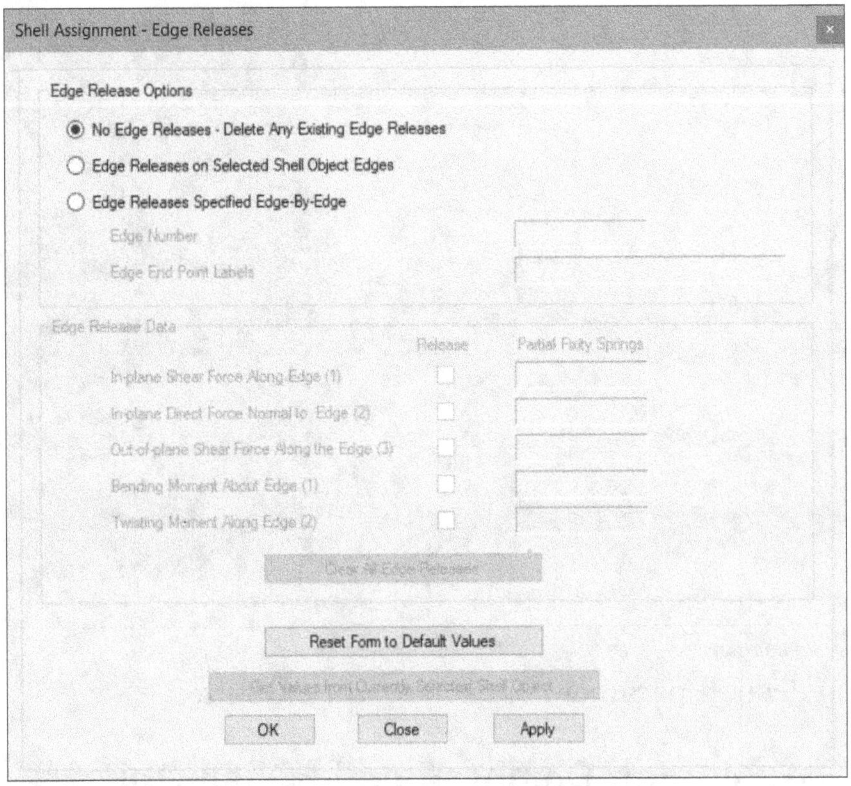

Figure-49. Shell Assignment-Edge Releases dialog box

- Select the **No Edge Releases - Delete Any Existing Edge Releases** radio button to delete any release earlier assigned to selected shell objects.
- Select the **Edge Releases on Selected Shell Object Edges** radio button to apply releases to selected objects. The options in the **Edge Release Data** area of the dialog box will become active. Select the check boxes for loads which you want to be removed. If you want to apply partial load then specify the value in respective edit box of selected check box.
- Select the **Edge Releases Specified Edge-By-Edge** radio button to select the individually set releases for each selected edge. The options are same as discussed earlier.
- After setting desired parameters, click on the **OK** button from the dialog box.

Assigning Line Releases

The **Line Releases** tool is used to apply load release on selected line element. The procedure to use this tool is given next.

- Click on the **Line Releases** tool from **Shell** cascading menu of **Assign** menu. The **Shell Assignment - Slab Line Releases** dialog box will be displayed; refer to Figure-50.

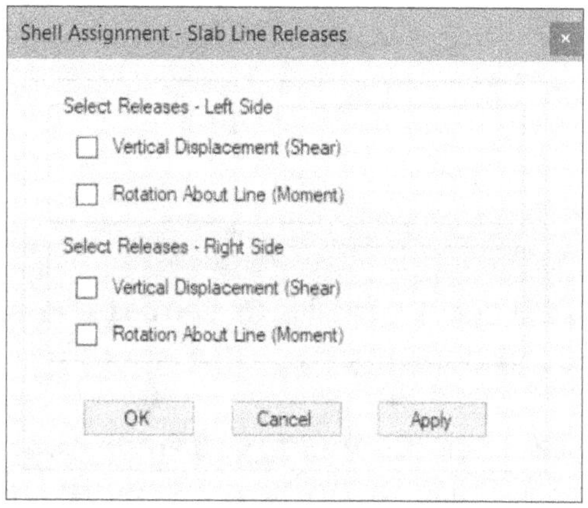

Figure-50. Shell Assignment-Slab Line Releases dialog box

- Select desired check boxes from the dialog box to release load from the selected line frame member.
- Click on the **OK** button from the dialog box to apply shell assignment.

Assigning Rib Locations

The **Rib Locations** tool in **Shell** cascading menu of **Assign** menu is used to set point location for ribs. On clicking this tool, the **Shell Assignment - Local Axes** dialog box will be displayed; refer to Figure-51. Specify desired values in the edit boxes to define X and Y locations for point, and click on the **OK** button.

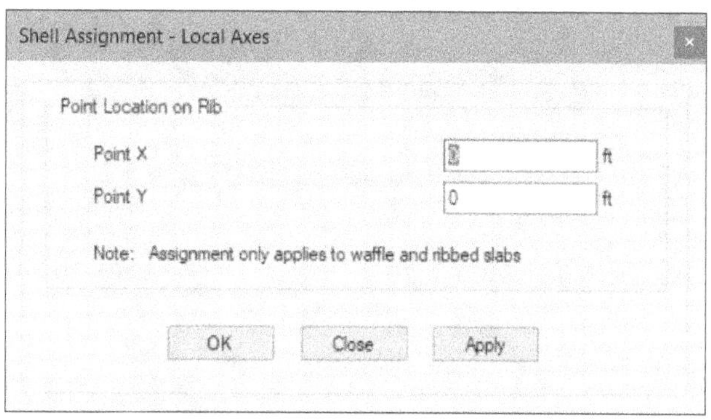

Figure-51. Shell Assignment-Local Axes dialog box

Assigning Local Axes

The **Local Axes** tool in the **Shell** cascading menu of the **Assign** menu is used to reorient the local axes of selected shell object. The procedure to use this tool is given next.

- Select desired shell objects and click on the **Local Axes** tool from the **Shell** cascading menu of the **Assign** menu. The **Shell Assignment - Local Axes** dialog box will be displayed; refer to Figure-52.

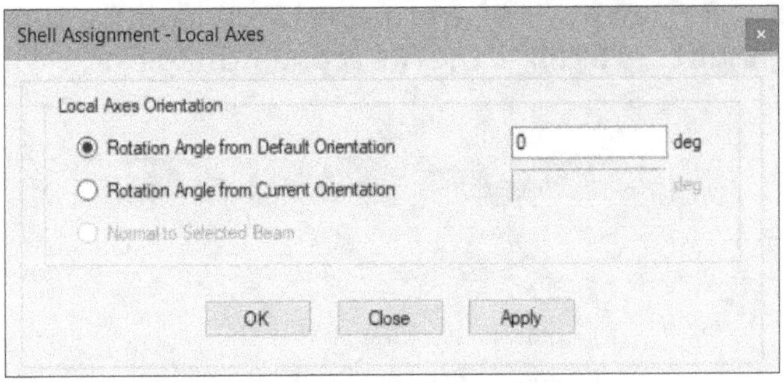

Figure-52. Shell Assignment-Local Axes dialog box

- Select the **Rotation Angle from Default Orientation** radio button if you want to specify angle with respect to default axes and specify the angle value in adjacent edit box.
- Select the **Rotation Angle from Current Orientation** radio button to rotate the axes with respect to current orientation and specify the angle value in adjacent edit box.
- Select the **Normal to Selected Beam** radio button to rotate the local 1 axis perpendicular to selected beam.
- After setting desired parameters, click on the **OK** button.

Assigning Area Springs

The **Area Spring** tool is used to assign area springs to selected shell objects. The procedure to use this tool is given next.

- Select desired shell objects and click on the **Area Spring** tool from the **Shell** cascading menu of the **Assign** menu. The **Shell Assignment - Area Spring** dialog box will be displayed; refer to Figure-53.

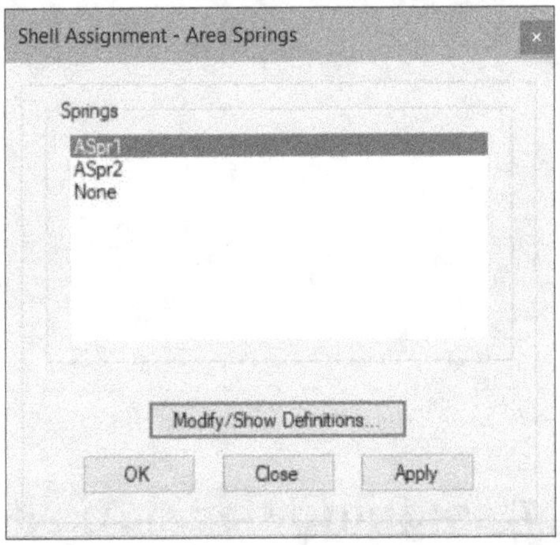

Figure-53. Shell Assignment-Area Springs dialog box

- Select desired area spring property from the dialog box and click on the **OK** button. The procedure to create and modify area spring properties has been discussed in previous chapter.

Assigning Additional Mass to Shell Objects

The **Additional Mass** tool is used to assign additional mass to selected shell objects apart from their dead weight. The procedure to assign additional mass is given next.

- Select the shell objects and click on the **Additional Mass** tool from the **Shell** cascading menu of the **Assign** menu. The **Shell Assignment - Additional Mass** dialog box will be displayed; refer to Figure-54.
- Specify the value of mass per area value in the **Shell Mass/Area** edit box.
- Select desired option from the **Options** area of the dialog box and click on the **OK** button.

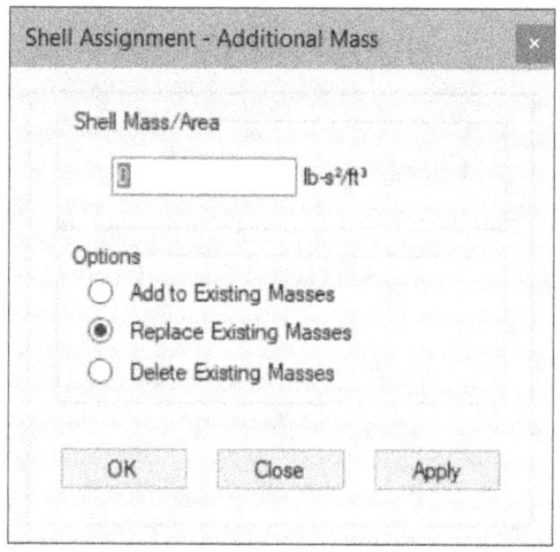

Figure-54. Shell Assignment - Additional Mass dialog box

Similarly, you can assign pier label and spandrel label using respective tools in the **Shell** cascading menu of the **Assign** menu.

Assigning Shell Hinge Data

The wall hinge properties are assigned to define shell data for hinges in the model. The procedure to do so is given next.

- Click on the **Wall Hinge** tool from the **Shell** cascading menu of the **Assign** menu. The **Shell Assignment - Hinges** dialog box will be displayed; refer to Figure-55.

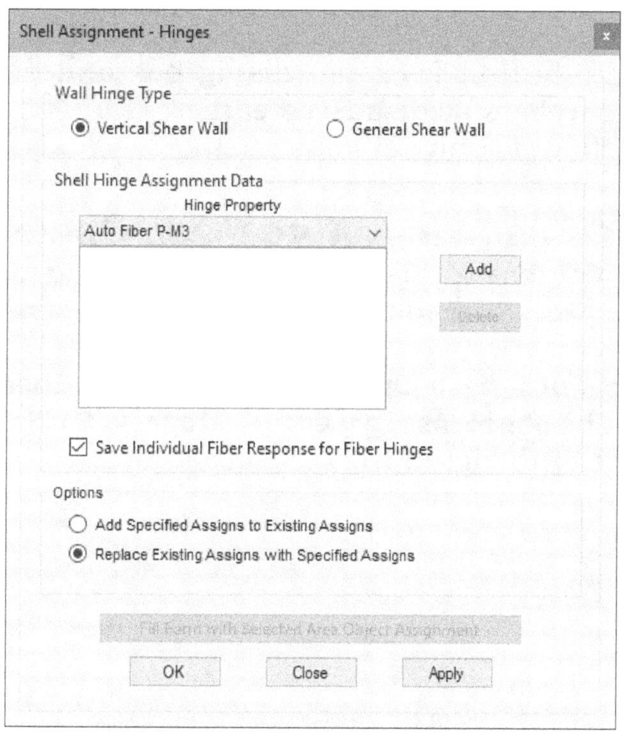

Figure-55. Shell Assignment-Hinges dialog box

- Select **Save Individual fibre response for Fibre Hinges** check box to enable saving the response of each individual fiber in a fiber hinge at every step.
- Select the **Add Specified Assigns to Existing Assigns** radio button to add hinge properties currently selected in top drop-down of dialog box. If you want to replace current selected assignment then select the **Replace Existing Assigns with Specified Assigns** radio button.
- Select desired hinge property from the top drop-down and click on the **Add** button.
- Click on the **OK** button to apply hinge properties to selected shell object.

Applying Reinforcement for Wall Hinge

The **Reinforcement for Wall Hinge** tool is used to apply reinforcement to selected wall hinges. The procedure to use this tool is given next.

- Click on the **Reinforcement for Wall Hinge** tool from the **Shell** cascading menu of the **Assign** menu. The **Wall Hinge Reinforcement Selection** dialog box will be displayed; refer to Figure-56.

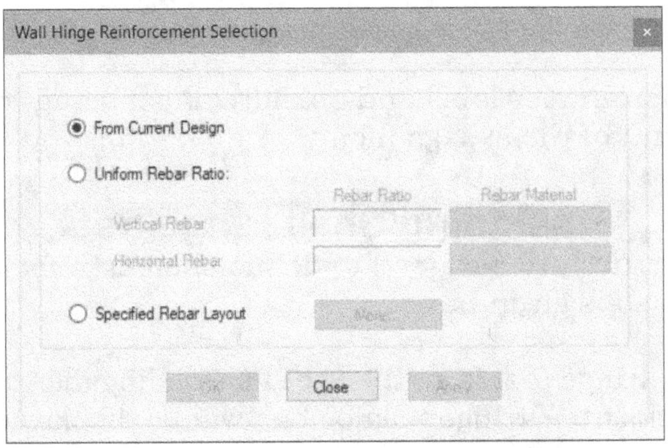

Figure-56. Wall Hinge Reinforcement Selection dialog box

- Select desired radio button from the dialog box and set the parameters for reinforcement using rebars as discussed earlier.
- Click on the **OK** button to apply reinforcement.

Defining Floor Auto Mesh Options

The **Floor Auto Mesh Options** tool is used to set automatic meshing parameters for analysis of floors. The procedure to use this tool is given next.

- Click on the **Floor Auto Mesh Options** tool from the **Shell** cascading menu in the **Assign** menu. The **Shell Assignment - Floor Auto Mesh Options** dialog box will be displayed; refer to Figure-57.

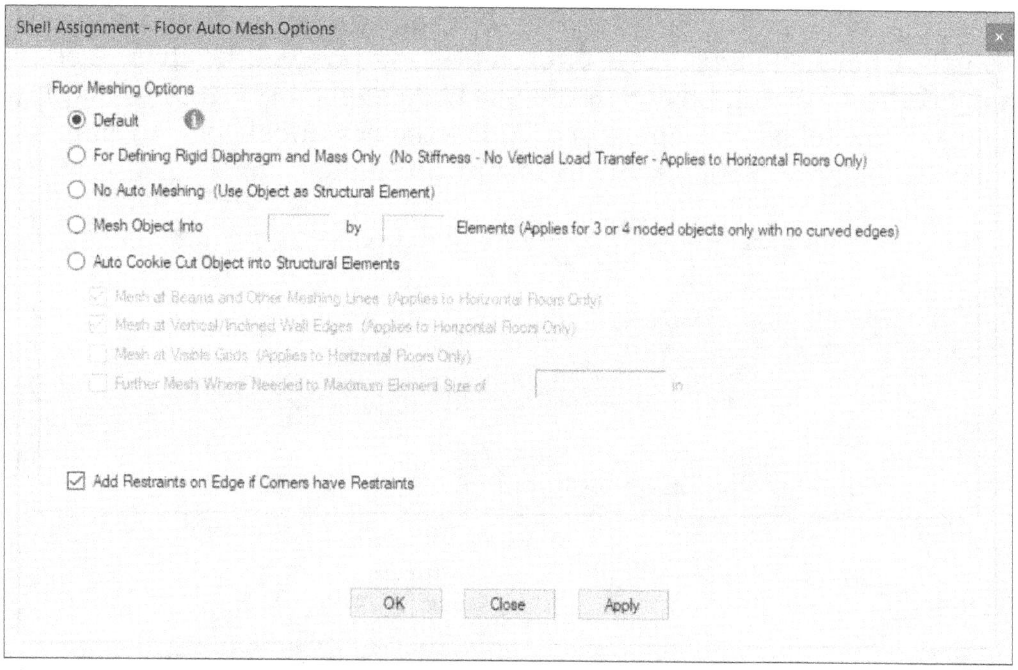

Figure-57. Shell Assignment-Floor Auto Mesh Options dialog box

- Selecting the **Default** radio button uses one of the three options for automatic meshing of shell floor objects based on their properties:

1. Horizontal decks and planks use the auto cookie cut at beams and walls meshing option. Because these types of floor objects do not have out-of-plane stiffness, vertical loads are transformed from the floors onto the beams and walls using tributary area algorithms.
2. Horizontal slabs and layered slabs use the automatic general meshing or rectangular meshing as specified on the Automatic Mesh Options (for Floors) form. For slabs with membrane behavior, vertical loads are transformed to vertical supports using tributary area algorithms.
3. Inclined floors use the no auto meshing option, i.e., they are not internally meshed and remain as a single element equivalent in size to the floor object.

- Select the **For Defining Rigid Diaphragm and Mass Only** radio button if you want to mesh only horizontal floors having rigid diaphragm and mass defined. There will be no vertical load transfer in this case so this option is only suited for shear deformation.
- Select the **No Auto Meshing** radio button if you want to use individually created structural elements for shell objects in analysis.
- Select the **Mesh Object Into** radio button to create mesh elements of specified a x b size. Note that you cannot use this option if floors have curved edges.
- Select the **Auto Cookie Cut Object into Structural Elements** radio button to apply meshing based on categories of objects in the model. For example, after selecting this radio button, you can clear the **Mesh at Beams and Other Meshing Lines** check box to avoid meshing of beams and lines in the model. Select/clear the check boxes based on your requirements.
- Select the **Add Restraints on Edge if Corners have Restraints** check box to automatically apply restraints on edges.
- Click on the **OK** button from the dialog box to apply meshing.

To view meshing, click on the **Set Display Options** tool from the **View** menu or press **CTRL+W** shortcut key. The **Set View Options** dialog box will be displayed. Select the **Shell Analysis Mesh** check box from the **Shell Assignments** area to display meshes; refer to Figure-58. Select the **Apply to All Windows** check box to display mesh in all viewports and click on the **OK** button from the dialog box.

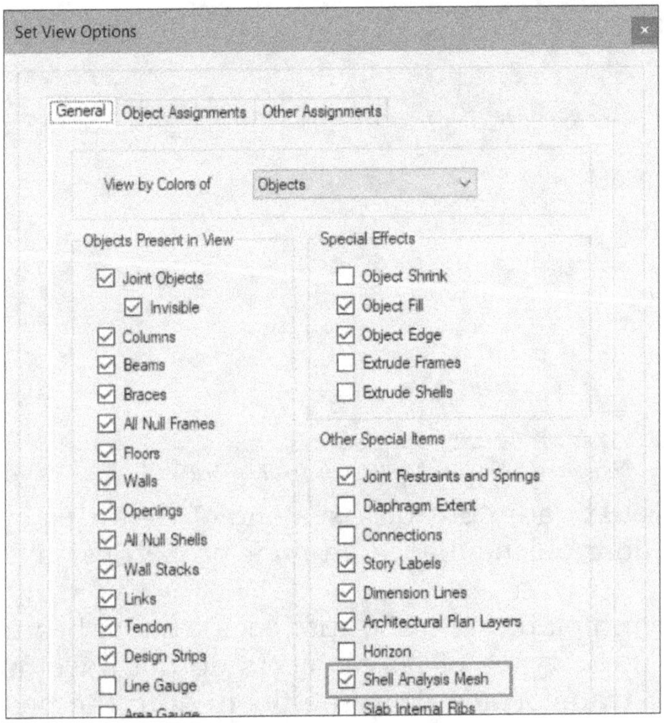

Figure-58. Shell Analysis Mesh check box

Defining Wall Auto Mesh Options

The **Wall Auto Mesh Options** tool is used to define meshing parameters for walls. The procedure to use this tool is given next.

- Click on the **Wall Auto Mesh Options** tool from the **Shell** cascading menu of the **Assign** menu. The **Shell Assignment - Wall Auto Mesh Options** dialog box will be displayed; refer to Figure-59.

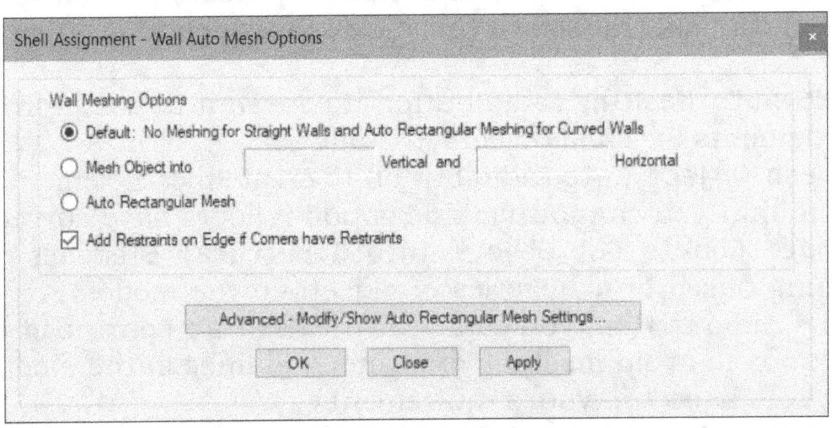

Figure-59. Shell Assignment-Wall Auto Mesh Options dialog box

- Select the **Default** radio button from the dialog box if you do not want to mesh straight walls but use their structural elements for analysis. The Curved walls will be automatically meshed using rectangular elements.

- Select the **Mesh Objects into** radio button to define a x b size of elements.
- Select the **Auto Rectangular Mesh** radio button to create rectangular mesh of automatic size to walls.
- Click on the **Advanced** button from the dialog box. The **Automatic Rectangular Mesh Options** dialog box will be displayed; refer to Figure-60.

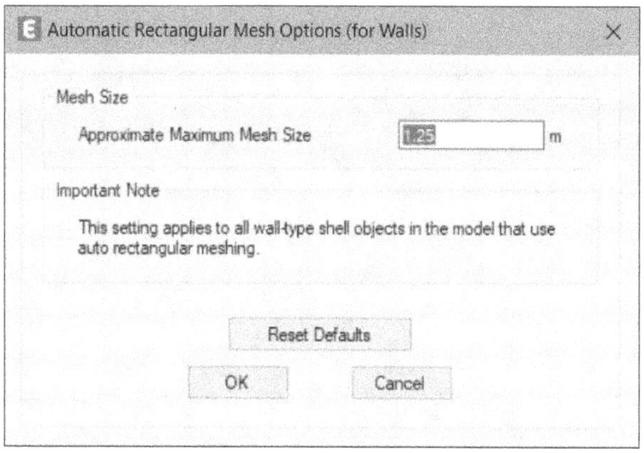

Figure-60. Automatic Rectangular Mesh Options (for Walls) dialog box

- Specify desired size of mesh in the **Approximate Maximum Mesh Size** edit box and click on the **OK** button. The **Shell Assignment - Wall Auto Mesh Options** dialog box will be displayed again.
- Click on the **OK** button from the dialog box to generate mesh.

Defining Auto Edge Constraints

The **Auto Edge Constraint** tool is used to automatically define constraints for the model. The procedure to use this tool is given next.

- Click on the **Auto Edge Constraint** tool from the **Shell** cascading menu of the **Assign** menu. The **Shell Assignment - Auto Edge Constraints** dialog box will be displayed; refer to Figure-61.

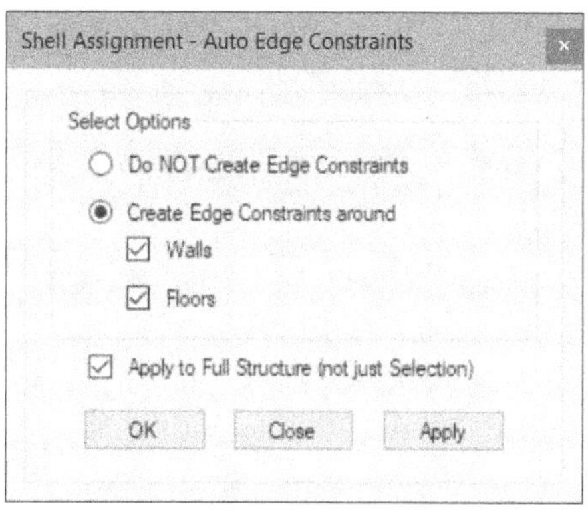

Figure-61. Shell Assignment-Auto Edge Constraints dialog box

- Set the options as discussed earlier and click on the **OK** button to generate edge constraints.

The other options in **Shell** cascading menu are same as discussed earlier.

ASSIGNING LINK PROPERTIES

The links are used to join two or more joints. The tools in **Link** cascading menu are used to assign link properties to selected links and manage local axes; refer to Figure-62. The tools in this menu are discussed next.

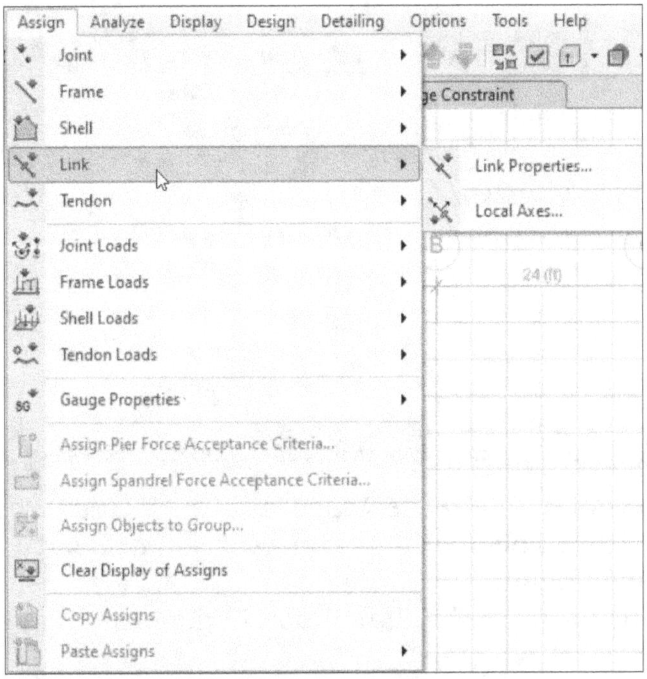

Figure-62. Link cascading menu

Assigning Link Properties

The **Link Properties** tool in the **Link** cascading menu of the **Assign** menu is used to assign link properties earlier created to the selected links. The procedure to use this tool is given next.

- After selecting desired links, click on the **Link Properties** tool from the **Link** cascading menu of the **Assign** menu. The **Link Assignment-Link Property** dialog box will be displayed; refer to Figure-63.

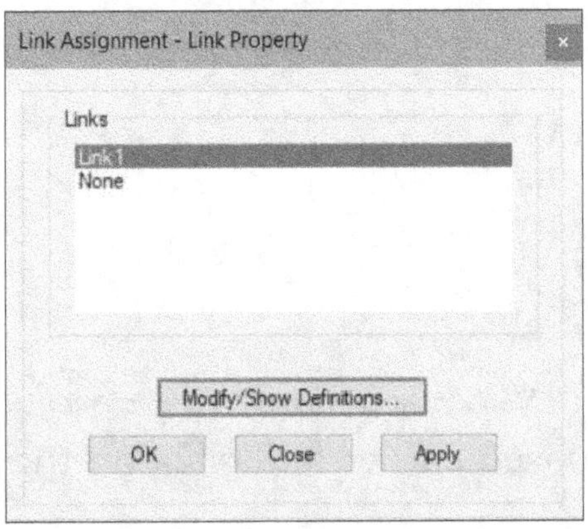

Figure-63. Link Assignment-Link Property dialog box

- Select desired option from the dialog box and click on the **OK** button. The procedure to create and modify link properties has been discussed earlier.

Orienting Local Axes for the Links

The **Local Axes** tool in the **Link** cascading menu of the **Assign** menu is used to orient axes of selected links. The procedure to use this tool is given next.

- Click on the **Local Axes** tool from the **Link** cascading menu of the **Assign** menu after selecting the links. The **Link Assignment-Local Axes** dialog box will be displayed; refer to Figure-64.

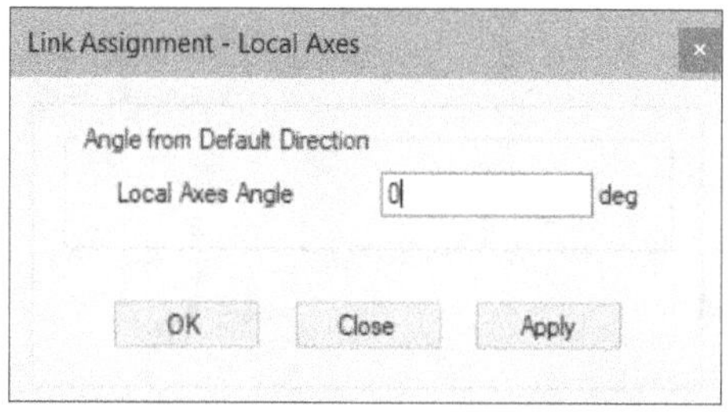

Figure-64. Link Assignment-Local Axes dialog box

- Specify desired angle value to rotate 1st axis of the selected links.
- Click on the **OK** button from the dialog box to apply specified rotation.

ASSIGNING PROPERTIES TO TENDONS

The tendons are used to prestress other objects in the structure. The procedure to assign properties to tendons has been discussed earlier.

- Select the tendons earlier created in the model and click on the **Tendon Properties** tool from the **Tendon** cascading menu of the **Assign** menu. The **Tendon Property Assign** dialog box will be displayed; refer to Figure-65.

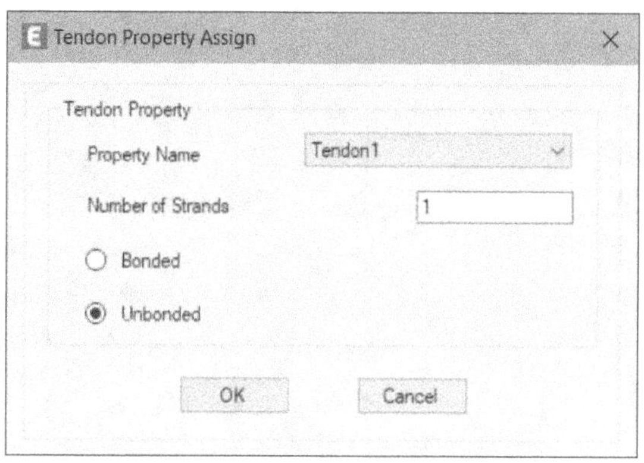

Figure-65. Tendon Property Assign dialog box

- Select desired tendon property from the **Property Name** drop-down.
- Specify the number of strands of tendon in the **Number of Strands** edit box.
- Select the **Bonded** radio button to bond tendons as if they are modeled with the structure. Select the **Unbonded** radio button to make tendons unbonded to the model and created as a separate feature.
- After setting desired parameters, click on the **OK** button.

ASSIGNING JOINT LOADS

The tools in the **Joint Loads** cascading menu are used to assign different types of loads on joints in the model; refer to Figure-66. The tools in this menu are discussed next.

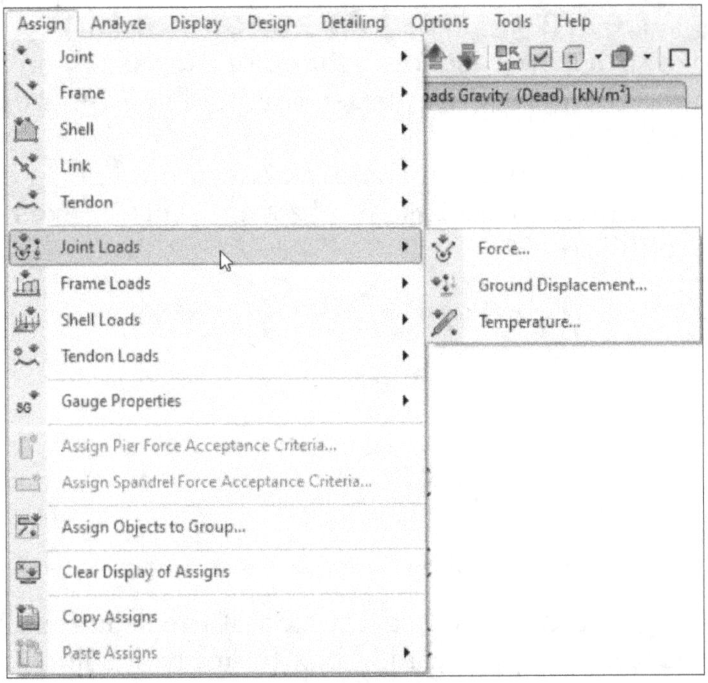

Figure-66. Joint Loads cascading menu

Applying Force at the Joint

The **Force** tool in the **Joint Loads** cascading menu is used to apply force at the joints. The procedure to use this tool is given next.

- Select the joints on which you want to apply loads and click on the **Force** tool from the **Joint Loads** cascading menu of the **Assign** menu. The **Joint Load Assignment - Force** dialog box will be displayed; refer to Figure-67.

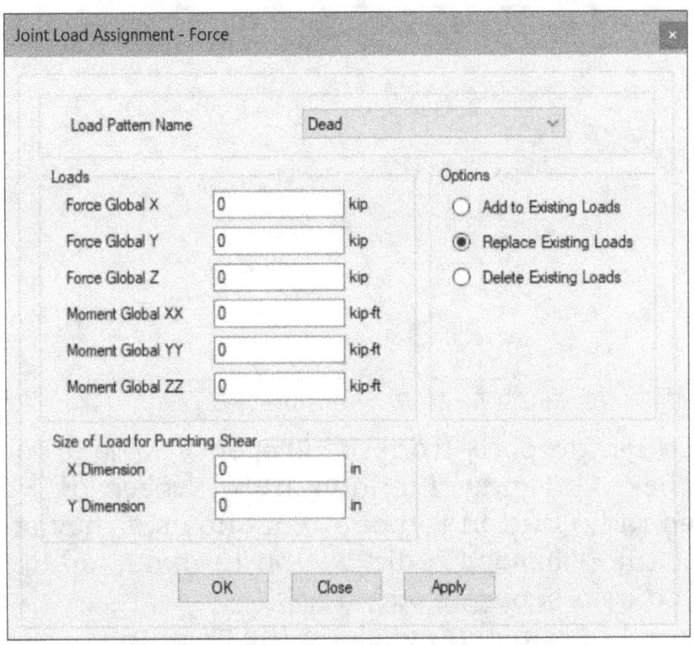

Figure-67. Joint Load Assignment-Force dialog box

- Click in the **Load Pattern Name** drop-down and select desired type of load. The **Dead** option is selected to apply fixed load (mostly exerted by fixed assets of building) on the joint. Select the **Live** option if you want to apply maximum load of moving objects in the building (like people, moving machines etc.). Select the **PT-FINAL** option to define post tension final load. Select the **PT-TRANSFER** option to define post tension transferred load on selected joints.

- Specify the values of different components of force and moment in respective edit boxes of the **Loads** area.
- Specify the size of punching load in the edit boxes of **Size of Load for Punching Shear** area of the dialog box.
- Select desired radio button from the **Options** area of the dialog box to add, replace, or delete loads.
- After setting desired parameters, click on the **OK** button from the dialog box.

Applying Ground Displacement at the Joints

The **Ground Displacement** tool is used to apply desired translation or rotation to selected joints. The procedure to use this tool is given next.

- Select the joints to which you want to apply translation or rotation and click on the **Ground Displacement** tool from the **Joint Loads** cascading menu of the **Assign** menu. The **Joint Load Assignment-Ground Displacement** dialog box will be displayed; refer to Figure-68.

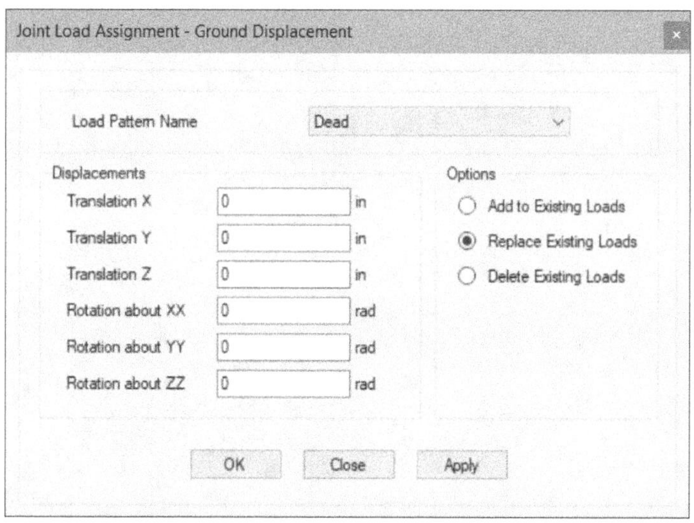

Figure-68. Joint Load Assignment-Ground Displacement dialog box

- Specify desired values of translation and rotation in respective edit boxes.
- After setting the parameters, click on the **OK** button.

Assigning Temperature to Joints

The **Temperature** tool in the **Joint** cascading menu is used to assign temperature to selected joints. The procedure to use this tool is given next.

- Select desired joints and click on the **Temperature** tool from the **Joint** cascading menu of the **Assign** menu. The **Joint Load Assignment- Temperature** dialog box will be displayed; refer to Figure-69.

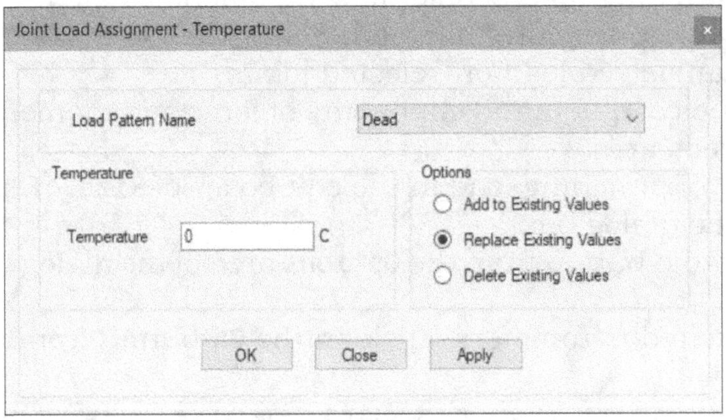

Figure-69. Joint Load Assignment-Temperature dialog box

- Specify desired temperature value in the **Temperature** edit box. The other options in the dialog box are same as discussed earlier.
- After setting desired parameters, click on the **OK** button.

ASSIGNING FRAME LOADS

The tools in the **Frame Loads** cascading menu are used to assign loads to frame members; refer to Figure-70. Various tools in this cascading menu are discussed next.

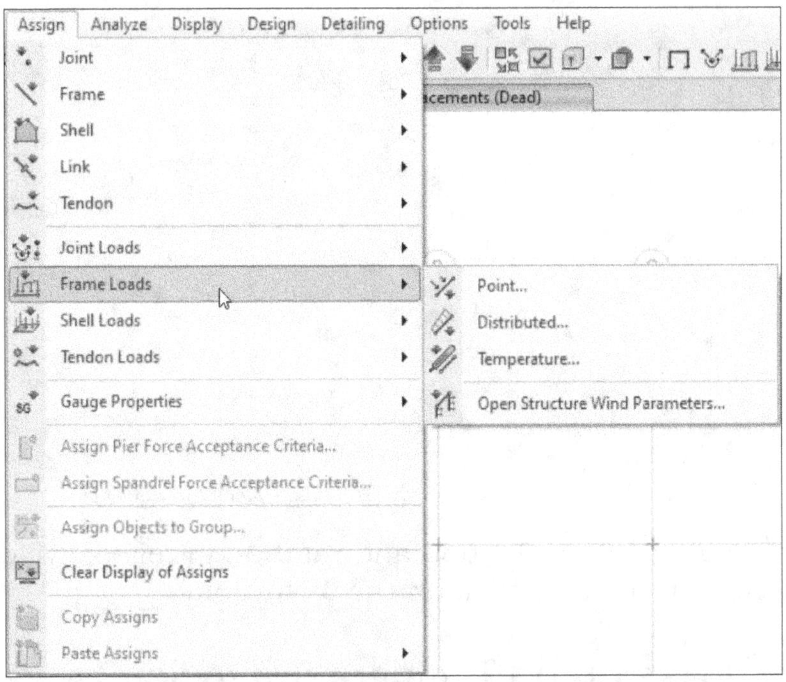

Figure-70. Frame Loads cascading menu

Assigning Point load on Frame Members

The **Point** tool in the **Frame** cascading menu is used to apply point load at frame members. The procedure to use this tool is given next.

- Select desired frame members and click on the **Point** tool from the **Frame** cascading menu of the **Assign** menu. The **Frame Load Assignment-Point** dialog box will be displayed; refer to Figure-71.

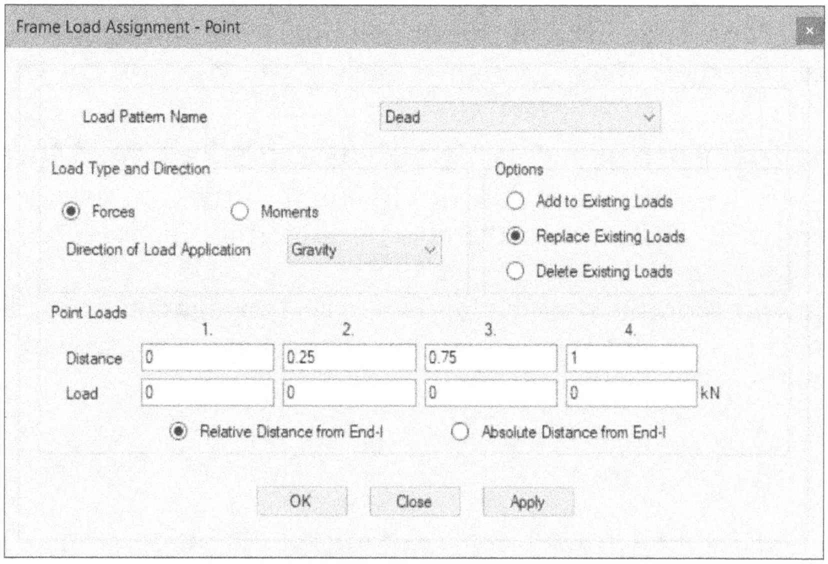

Figure-71. Frame Load Assignment-Point dialog box

- Select desired radio button to define type of load from the **Load Type and Direction** area of the dialog box.
- Select desired option from the **Direction of Load Application** drop-down to define direction of load.
- Specify the relative distance of load points and loads on them in the edit boxes of the **Point Loads** area.
- Set the other options as discussed earlier and click on the **OK** button to apply loads.

Assigning Distributed Load on Frame Members

The **Distributed** tool in the **Frame Loads** cascading menu is used to apply uniform load on selected frame members. The procedure to use this tool is given next.

- Select desired frame members and click on the **Distributed** tool from the **Frame Loads** cascading menu of the **Assign** menu. The **Frame Load Assignment - Distributed** dialog box will be displayed; refer to Figure-72.

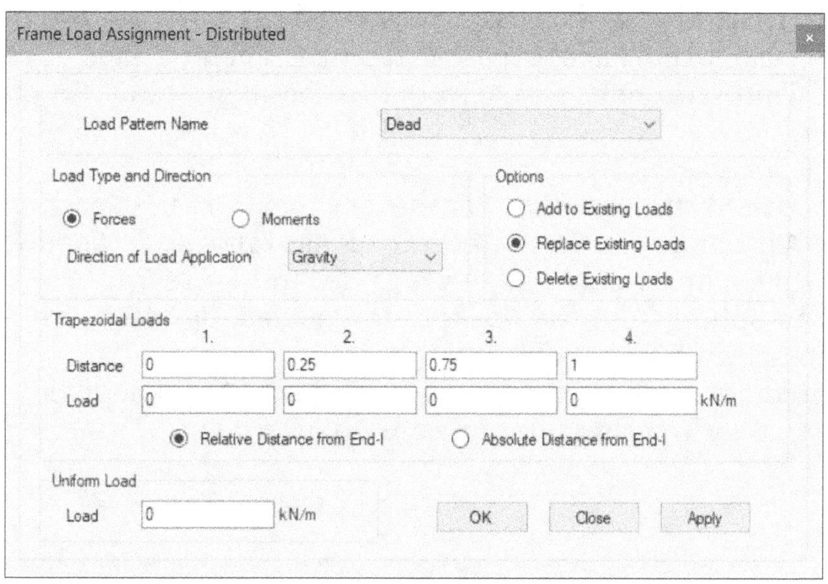

Figure-72. Frame Load Assignment-Distributed dialog box

- Specify desired value of uniform load over selected frame members in the **Load** edit box of **Uniform Load** area of the dialog box.
- Set the other parameters as discussed earlier and click on the **OK** button.

Assigning Temperature to Selected Frame Members

The **Temperature** tool in the **Frame** cascading menu is used to assign temperature to selected frame members. The procedure to use this tool is given next.

- Select the frame members and click on the **Temperature** tool from the **Frame** cascading menu of the **Assign** menu. The **Frame Load Assignment - Temperature** dialog box will be displayed; refer to Figure-73.

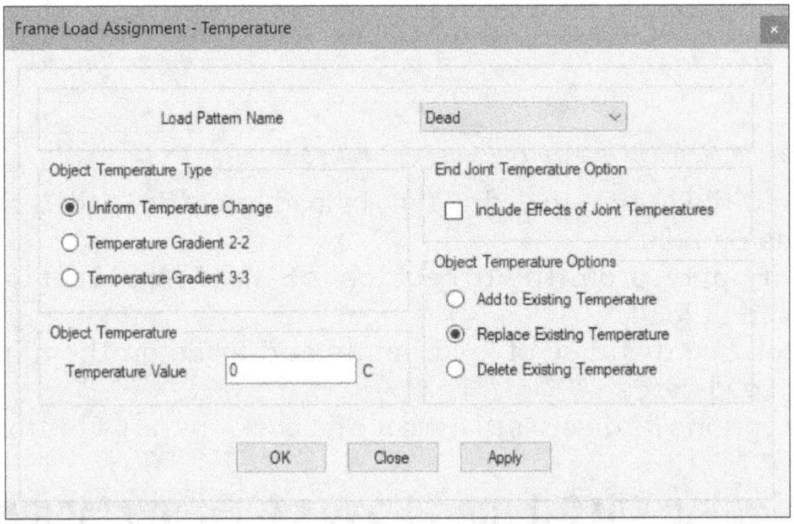

Figure-73. Frame Load Assignment-Temperature dialog box

- Specify desired temperature value in the **Uniform Temperature Change** edit box.
- Select the **Include Effects of Joint Temperatures** check box to include the effect of temperatures applied to joints of the frame members.
- Set the other parameters as discussed earlier and click on the **OK** button.

Assigning Open Structure Wind to Frame Members

The **Open Structure Wind Parameters** tool in the **Frame Loads** cascading menu is used to assign wind parameters to selected frame members. The procedure to use this tool is given next.

- Click on the **Open Structure Wind Parameters** tool from the **Frame Loads** cascading menu of the **Assign** menu. The **Frame Load Assignment - Open Structure Wind Parameters** dialog box will be displayed; refer to Figure-74.
- Select the **Yes** option from the **Object is Loaded by Wind** drop-down to apply wind load on selected frame objects.
- Specify desired value of ice thickness in the **Ice Thickness** edit box if you are designing frame for freezing temperature environment.

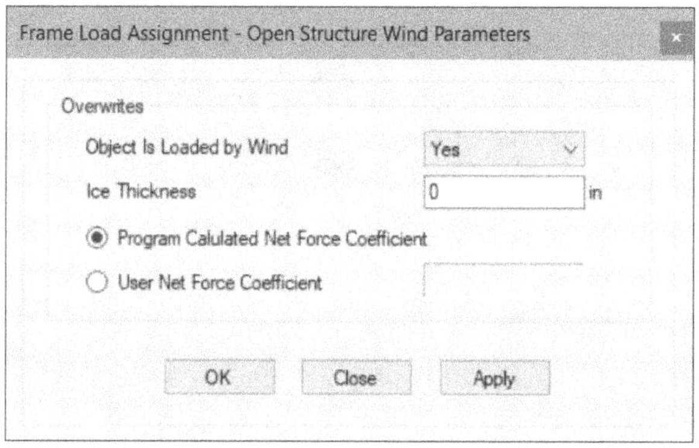

Figure-74. Frame Load Assignment-Open Structure Wind Parameters dialog box

- Select the **Program Calculated Net Force Coefficient** radio button to apply default net force coefficient based on design code of your project.
- Select the **User Net Force Coefficient** radio button to specify the value of force coefficient manually.
- After setting desired parameters, click on the **OK** button.

ASSIGNING SHELL LOADS

The tools in the **Shell Loads** cascading menu of the **Assign** menu are used to assign loads to selected shell objects; refer to Figure-75. The tools in this cascading menu are discussed next.

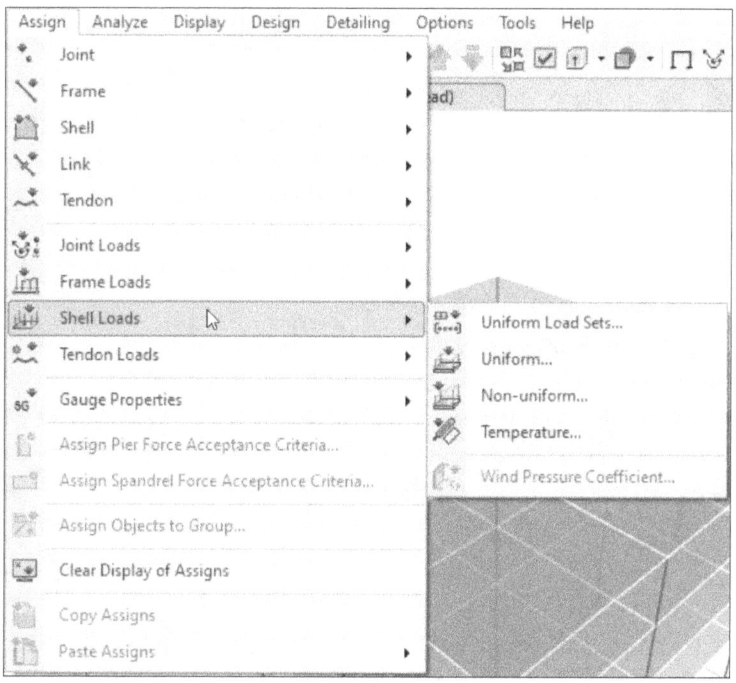

Figure-75. Shell Loads cascading menu

Assigning Uniform Load Sets

The **Uniform Load Sets** tool in the **Shell Loads** cascading menu is used to assign a combination of dead, live, post-tension transfer, and post-tension final loads. The procedure to use this tool is given next.

- Select desired shell objects and click on the **Uniform Load Sets** tool from the **Shell Loads** cascading menu of the **Assign** menu. The **Shell Assignment - Uniform Load Set** dialog box will be displayed; refer to Figure-76.

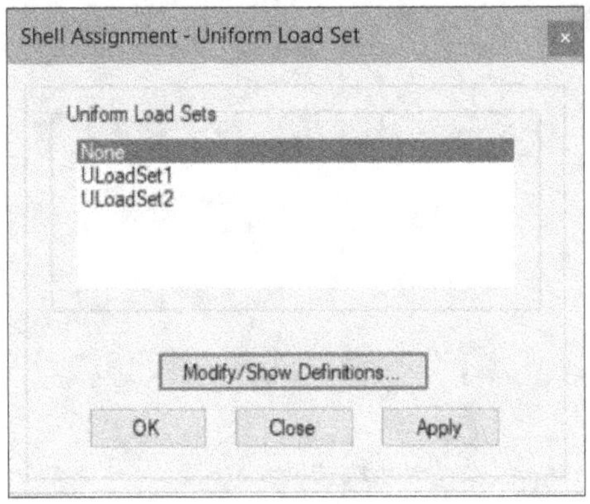

Figure-76. Shell Assignment-Uniform Load Set dialog box

- Select desired load set from the list and click on the **OK** button.

Defining Uniform Load Set

- If you have not created uniform load sets earlier then click on the **Modify/Show Definitions** button from the dialog box. The **Shell Uniform Load Sets** dialog box will be displayed; refer to Figure-77. Note that this dialog box can also be invoked by **Shell Uniform Load Sets** tool in the **Define** menu.
- Click on the **Add New Load Set** button from the dialog box to add a new load set. The **Shell Uniform Load Set Data** dialog box will be displayed; refer to Figure-78.

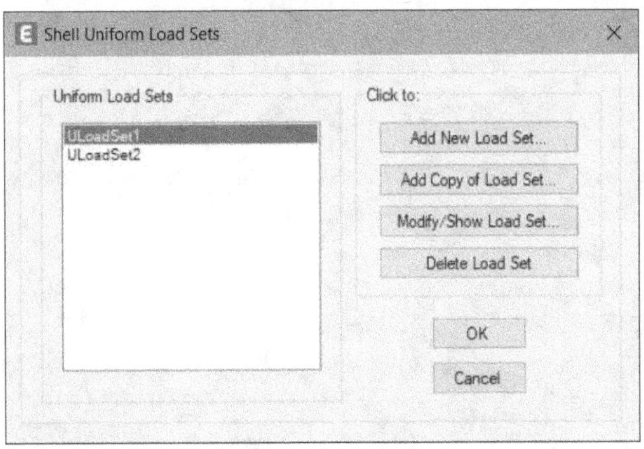

Figure-77. Shell Uniform Load Sets dialog box

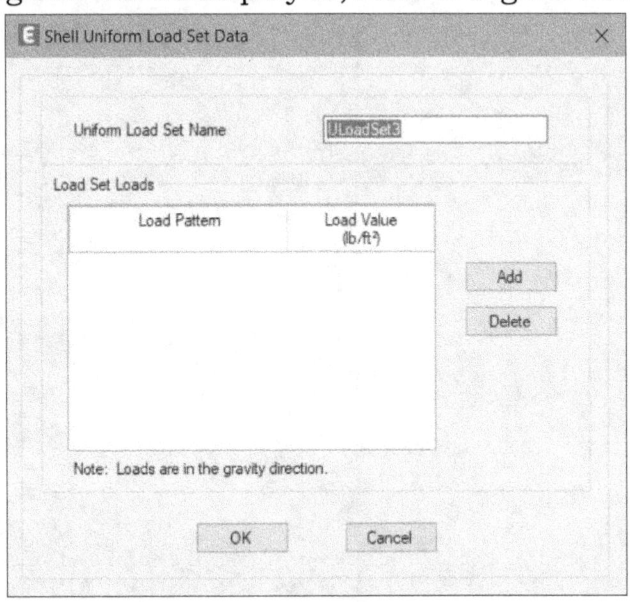

Figure-78. Shell Uniform Load Set Data dialog box

- Specify desired name of load set in the **Uniform Load Set Name** edit box.
- Click on the **Add** button from the dialog box to add a load. A dead load of **0** value will be added. Click on the **Dead** option under **Load Pattern** column in the table and select desired option like Live, Dead, PT-FINAL, and so on. Specify the value of load in adjacent edit box.
- Repeat the above step to add as many loads as you want in current load set. After adding desired load, click on the **OK** button.

- The **Shell Uniform Load Sets** dialog box will be displayed again with newly added load sets in the list. The other options in this dialog box are same as discussed in previous chapter.
- Click on the **OK** button from the dialog box. The **Shell Assignment - Uniform Load Set** dialog box will be displayed again. Select desired load set and click on the **OK** button.

Assigning Uniform Shell Load

The **Uniform** Shell Load tool is used to assign a uniform load all over the selected shell objects. The procedure to use this tool is given next.

- Click on the **Uniform** Shell Load tool from the **Shell Loads** cascading menu of the **Assign** menu. The **Shell Load Assignment - Uniform** dialog box will be displayed; refer to Figure-79.

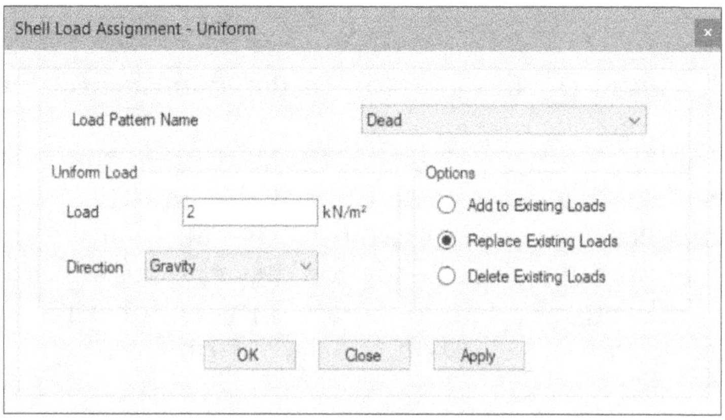

Figure-79. Shell Load Assignment-Uniform dialog box

- Select desired type of load from the **Load Pattern Name** drop-down to define the type of load.
- Specify desired value and direction of load in the **Uniform Load** area of the dialog box.
- Set the other parameters as discussed earlier and click on the **OK** button.

Assigning Non-uniform Load to Shell Objects

The **Non-uniform** tool is used to assign a varying load on selected shell objects. The procedure to use this tool is given next.

- Select desired shell objects and click on the **Non-uniform** tool from the **Shell Loads** cascading menu of the **Assign** menu. The **Shell Load Assignment - Non-uniform** dialog box will be displayed; refer to Figure-80.

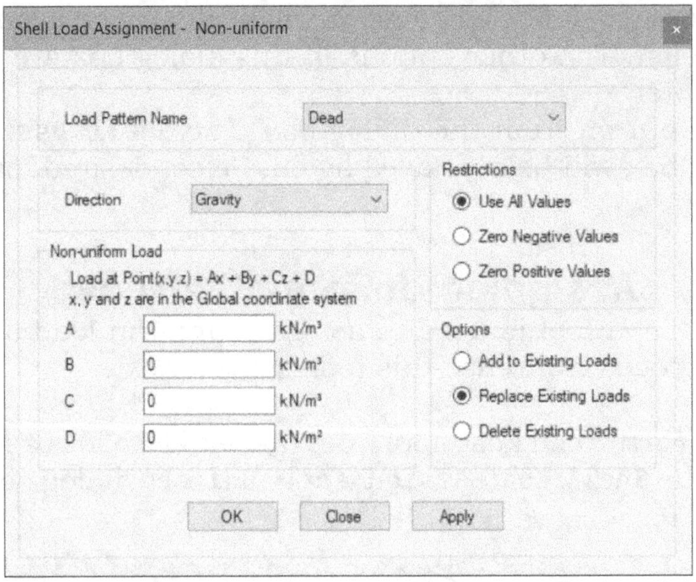

Figure-80. Shell Load Assignment-Non-uniform dialog box

- Specify the A, B, C, and D values of non-uniform load equation in respective edit boxes.
- Set desired restriction from the **Restrictions** area of the dialog box. Select the **Use All Values** radio button to use all A, B, C, and D values specified. Select the **Zero Negative Values** radio button to use all positive values specified and set negative values 0. Select the **Zero Positive Values** radio button to use all negative values and set the positive values 0 in the formula.
- After setting desired values, click on the **OK** button.

Applying Temperature to Shell Objects

The **Temperature** tool in the **Shell Loads** cascading menu is used to assign temperature to selected shell objects. The procedure to use this tool is same as discussed earlier.

ASSIGNING LOADS TO TENDONS

The tools to assign loads to tendons are available in the **Tendon Loads** cascading menu of the **Assign** menu; refer to Figure-81. These tools are discussed next.

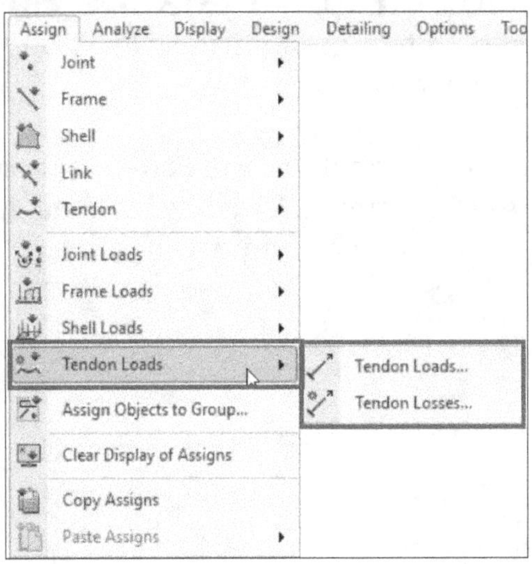

Figure-81. Tendon Loads cascading menu

Assigning Tendon Loads

The **Tendon Loads** tool is used to apply loads on selected tendons for analysis. The procedure to use this tool is given next.

- Select the tendons on which you want to apply the loads and click on the **Tendon Loads** tool from the **Tendon Loads** cascading menu of the **Assign** menu. The **Tendon Load** dialog box will be displayed; refer to Figure-82.

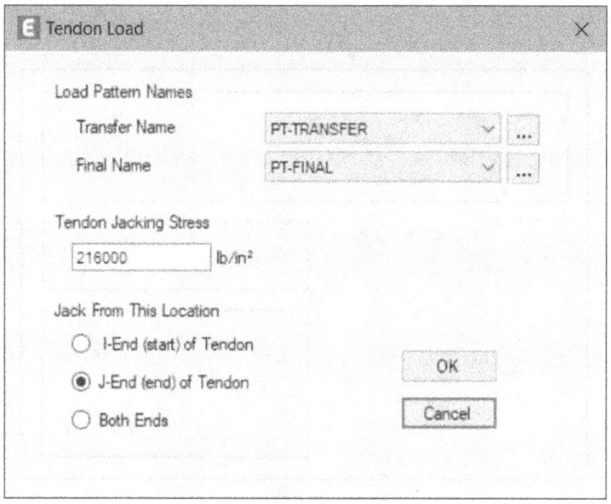

Figure-82. Tendon Load dialog box

- Select desired options from the **Load Pattern Names** area. You will learn about creating load patterns in the next chapter.
- Specify desired value in the **Tendon Jacking Stress** edit box to define jacking stress.
- From the **Jack From This Location** area select the radio button to define the location at which jacking stress will be applied.
- After setting desired parameters, click on the **OK** button.

Assigning Tendon Losses

The **Tendon Losses** tool is used to apply load/stress losses to selected tendons. The procedure to use this tool is given next.

- Select the tendons and click on the **Tendon Losses** tool from the **Tendon Loads** cascading menu in the **Assign** menu. The **Tendon Loss Options** dialog box will be displayed; refer to Figure-83.
- Select desired radio button to specify loss calculation method. Select the **Based on Force Percentage (%)** radio button to specify the percentage of force that will be lost after post tensioning. Select the **Based on Fixed Stress Value** radio button to specify fix value of load that will be lost after tensioning tendons. Select the **Based on Detailed Calculations** radio button to specify various coefficients for short term losses and values for long term losses.
- After selecting desired radio button, specify the related values in respective edit boxes.
- Click on the **OK** button. The specified losses will be applied to selected tendon objects.

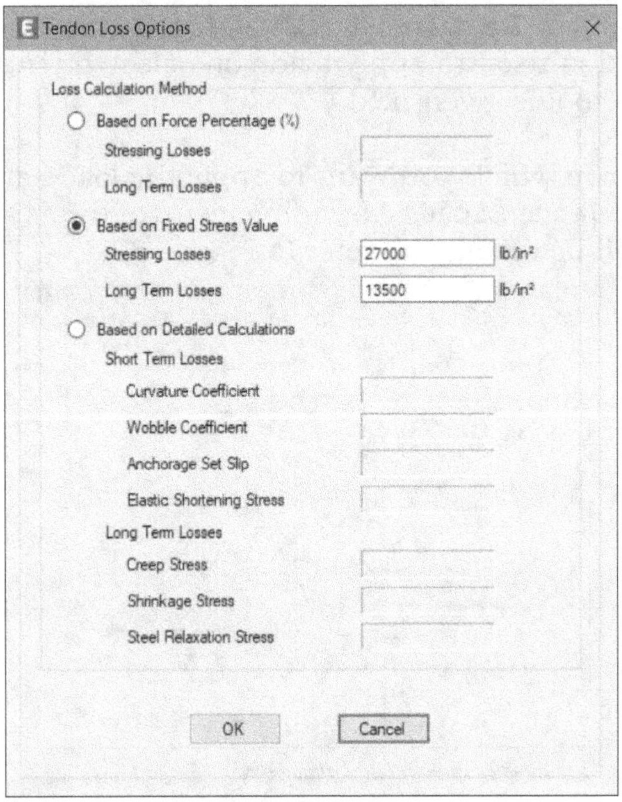

Figure-83. Tendon Loss Options dialog box

ASSIGNING GAUGE PROPERTIES

The options in the **Gauge Properties** cascading menu are used to assign earlier created properties to the strain measurement gauges; refer to Figure-84.

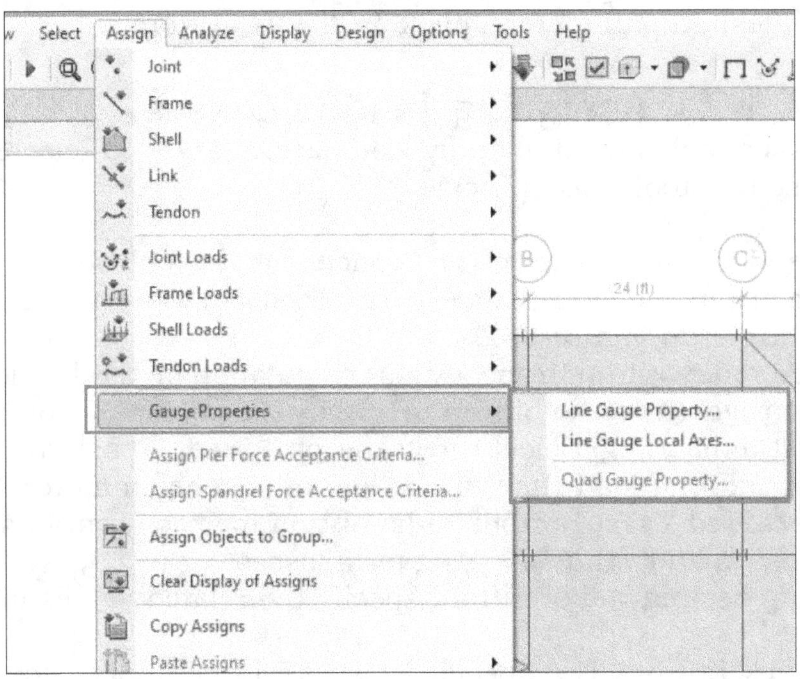

Figure-84. Gauge Properties options

The options in this cascading menu are discussed next.

Assigning Line Gauge Property

The **Line Gauge Property** tool is used to assign line gauge properties to selected gauge. The procedure is given next.

- Click on the **Line Gauge Property** tool from the **Gauge Properties** cascading menu of the **Assign** menu. The **Line Gauge Assignment - Properties** dialog box will be displayed; refer to Figure-85.

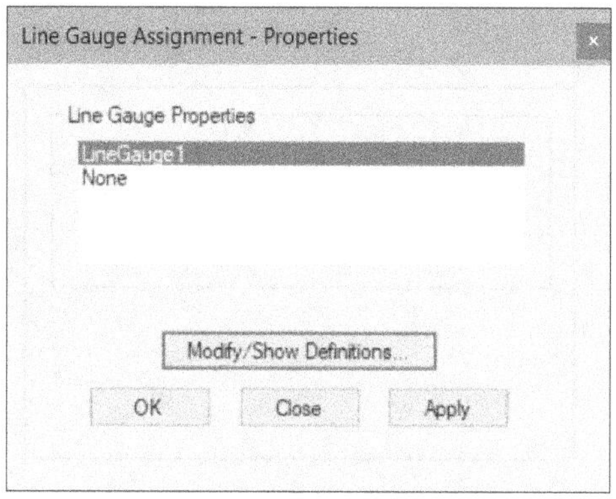

Figure-85. Line Gauge Assignment - Properties dialog box

- The procedure to modify the line gauge properties have been discussed earlier. Select desired line gauge and click on the **OK** button.

Assigning Line Gauge Local Axes

The **Line Gauge Local Axes** tool is used to modify orientation of local axes of line gauges. The procedure is given next.

- Click on the **Line Gauge Local Axes** tool from the **Gauge Properties** cascading menu of the **Assign** menu. The **Line Gauge Assignment - Local Axes** dialog box will be displayed; refer to Figure-86.

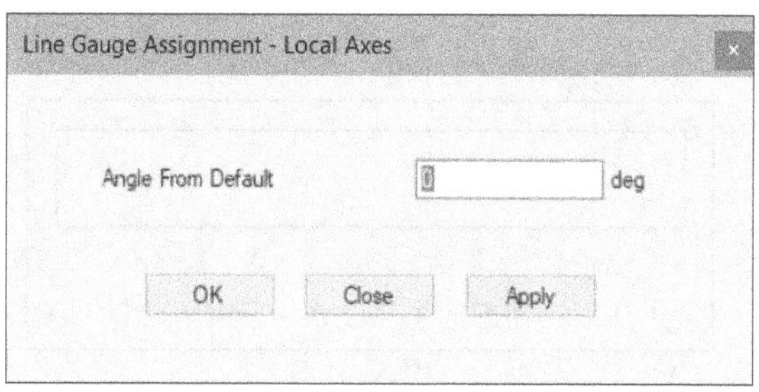

Figure-86. Line Gauge Assignment-Local Axes dialog box

- Specify desired value in the **Angle From Default** edit box and click on the **OK** button.

ASSIGNING PIER FORCE ACCEPTANCE CRITERIA

The **Assign Pier Force Acceptance Criteria** tool is used to assign load acceptance criteria to pier labels. The procedure to use this tool is given next.

- Click on the **Assign Pier Force Acceptance Criteria** tool in **Assign** menu. The **Assign Pier Force Acceptance Criteria** dialog box will be displayed; refer to Figure-87.

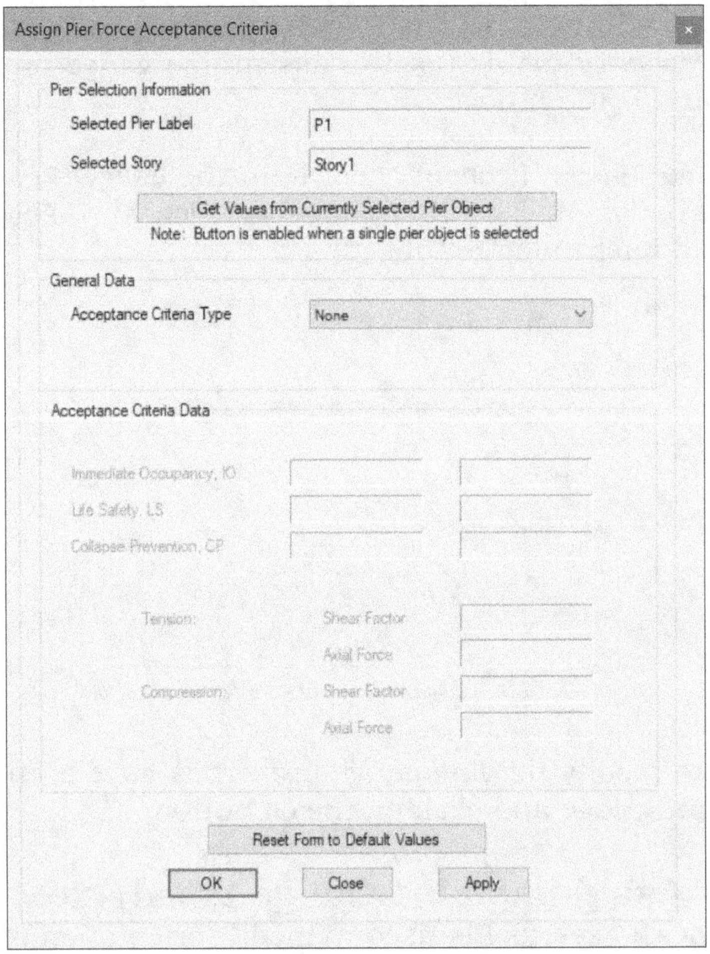

Figure-87. Assign Pier Force Acceptance Criteria dialog box

- Select desired option from the **Acceptance Criteria Type** drop-down to define criteria. After selecting desired option, you can specify the related parameters in **Acceptance Criteria Data** area of the dialog box.
- After setting desired parameters, click on the **OK** button from the dialog box.

You can similarly assign the spandrel force assessment criteria using the **Assign Spandrel Force Acceptance Criteria** tool in **Assign** menu.

ASSIGNING OBJECTS TO GROUPS

The **Assign Objects to Group** tool is used to add selected objects to specified group. The procedure to use this tool is given next.

- Select all the objects that you want to assign to a group and click on the **Assign Objects to Group** tool from the **Assign** menu. The **Assign Objects to Group** dialog box will be displayed; refer to Figure-88.

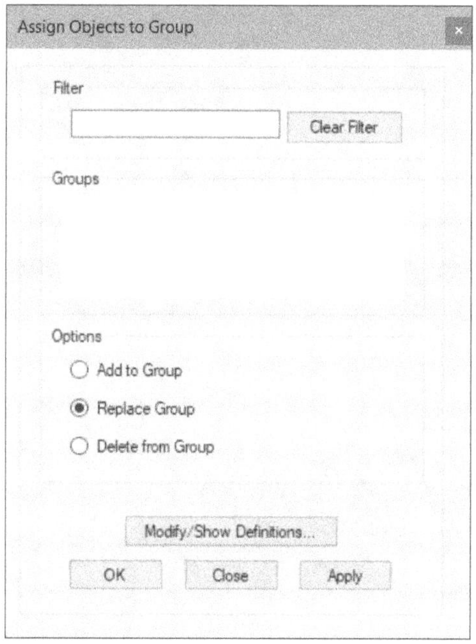

Figure-88. Assign Objects to Group dialog box

- If you have not created groups earlier, click on the **Modify/Show Definitions** button from the dialog box and create the groups as discussed earlier.
- Select desired group from the list and select the **Add to Group** radio button if you want to add all selected objects to the group.
- If you want to replace the group with newly selected objects then select the **Replace Group** radio button.
- If you want to remove selected objects from a group then select the **Delete from Group** radio button.
- Click on the **OK** button from the dialog box to apply changes.

If you want to clear all the assigns of selected object then select the objects and click on the **Clear Display of Assigns** tool from the **Assign** menu.

COPYING AND PASTING ASSIGNS

The **Copy Assigns** tool in the **Assign** menu is used to copy assigned properties and parameters of selected objects so that you can apply the same to other similar objects. The **Paste Assigns** tools in the **Assign** menu are used to assign copied properties to selected objects. The procedure to copy and paste assigns is discussed next.

- Select the object whose assigns are to be copied and click on the **Copy Assigns** tool from the **Assign** menu; refer to Figure-89.

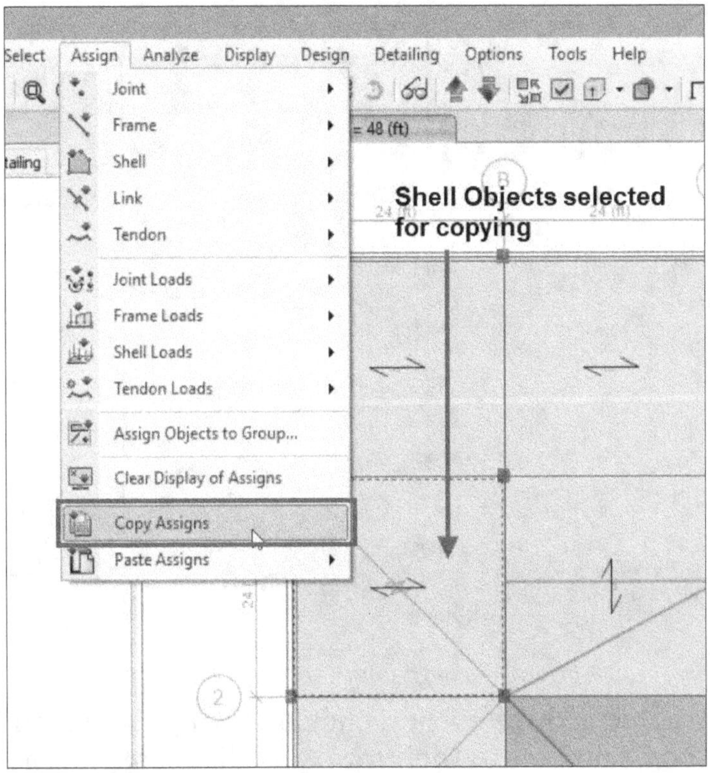

Figure-89. Copying assigns

- Now, select the same type of objects to which you want to assign copied properties and click on the respective button from the **Paste Assigns** cascading menu. For example, to paste assigns copied from a shell object, select the **Shell** tool from the **Paste** cascading menu and apply them to another shell objects.

SELF-ASSESSMENT

Q1. What are joints in structural modeling?
A. Places where two or more frame members join together
B. Only the supports of structure
C. Areas where loads are applied
D. Mesh generation points

Q2. In which menu are the tools for assigning joint properties available?
A. Analyze menu
B. Draw menu
C. Joint cascading menu of Assign menu
D. Design menu

Q3. What is the purpose of the Restraint tool?
A. To assign loads
B. To restrict motion of selected joints
C. To create mesh
D. To modify materials

Q4. Which button restricts motion in all six directions?
A. Roller Support
B. No Support
C. Pinned Base
D. Fixed Base

Q5. Which restraint option allows translation restriction but free rotation?
A. Roller Support
B. Fixed Base
C. Pinned Base
D. No Support

Q6. Roller Support restricts movement along which axis only?
A. X-axis
B. Y-axis
C. Z-axis
D. All axes

Q7. Which restraint option allows joints to move freely in all directions?
A. Fixed Base
B. No Support
C. Roller Support
D. Pinned Base

Q8. The Springs tool is used to assign what type of properties?
A. Load properties
B. Material properties
C. Spring properties
D. Mesh properties

Q9. Spring joints are useful when beams or columns are joined by ________.
A. Concrete blocks
B. Soft materials to absorb shocks
C. Steel plates only
D. Fixed hinges

Q10. Which dialog box appears after clicking the Springs tool?
A. Joint Assignment - Restraints
B. Joint Assignment - Spring
C. Frame Assignment - Spring
D. Joint Assignment - Mesh

Q11. Which button is used to check or modify spring properties?
A. Add Properties
B. Assign Spring
C. Modify/Show Definitions
D. Update Values

Q12. A diaphragm joint is rigid against which deformation?
A. Out-of-plane deformation
B. Membrane in-plane deformation
C. Plastic deformation
D. Thermal deformation

Q13. Which tool is used to assign diaphragm properties?
A. Springs tool
B. Restraints tool
C. Diaphragms tool
D. Panel Zone tool

Q14. Panel zones are assigned to model the flexibility of _______.
A. Slab openings
B. Beam-column connections
C. Foundations
D. Mesh joints

Q15. Which tool is used to assign panel zones?
A. Hinges tool
B. Panel Zone tool
C. Local Axes tool
D. Section Property tool

Q16. Additional mass assigned to joints may represent objects like _______.
A. Soil layers
B. Furniture or machines
C. Steel bolts
D. Hinges

Q17. Which option adds specified mass to existing masses?
A. Delete Existing Masses
B. Replace Existing Masses
C. Add to Existing Masses
D. New Mass Option

Q18. Which option replaces existing masses at selected joints?
A. Replace Existing Masses
B. Add to Existing Masses
C. Delete Existing Masses
D. Fixed Masses

Q19. The Joint Floor Meshing Options tool is used for _______.
A. Material design
B. Meshing of joints
C. Assigning loads
D. Creating hinges

Q20. Which check box includes selected joints in meshing?
A. Enable Joint Loads
B. Include Selected Joint Objects In Mesh
C. Mesh All Frames
D. Assign Mesh Supports

Q21. Section properties define the _______ of frame members.
A. Loads only
B. Shape and size
C. Meshing options
D. Rebar ratio

Q22. Which tool is used to assign section properties?
A. Hinges tool
B. Output Stations tool
C. Section Property tool
D. Local Axes tool

Q23. Property modifiers are used to modify properties exclusively for _______.
A. Drawing
B. Printing
C. Analysis
D. Rendering

Q24. If value 2 is entered in the Mass edit box of Property Modifiers, the mass will be _______.
A. Halved
B. Removed
C. Multiplied by 2
D. Divided by 2

Q25. The Releases/Partially Fixity tool is used to _______.
A. Assign loads
B. Release or partially fix frame members
C. Define materials
D. Create joints

Q26. A value of 0 in release options means _______.
A. Partial release
B. Full release
C. No release
D. Rigid connection

Q27. End Length Offsets are used so beams or columns extend at the joint like in the _______.
A. Drawing sheet
B. Real world
C. Mesh view
D. Analysis graph

Q28. Which radio button automatically offsets one beam or column to accommodate another?
A. Define Lengths
B. Automatic from Connectivity
C. Offset from Axis
D. Relative Length

Q29. End-I of a beam or column represents the _______.
A. Midpoint
B. End support
C. Start point
D. Offset point

Q30. The rigid-zone factor specifies the fraction of end offset assumed rigid for _______.
A. Temperature loads
B. Bending and shear deformations
C. Tension only
D. Compression only

Q31. Which tool is used to define whether beams will be included in floor cracking analysis?
A. Wall Section Tool
B. Floor Cracking Tool
C. Deck Section Tool
D. Thickness Overwrites Tool

Q32. Which check box is selected to include beams in floor cracking analysis?
A. Include Beam Loads
B. Mesh Beam Objects
C. Consider Selected Beam Objects in Floor Cracking Analysis
D. Enable Beam Cracking

Q33. Shell objects are generally made of how many nodes?
A. 1 or 2
B. 2 or 3
C. 3 or 4
D. 5 or 6

Q34. Shell objects are commonly used to create which of the following?
A. Pipes only
B. Floors and walls
C. Bolts only
D. Bearings only

Q35. Which tool is used to assign slab section properties to shell objects?
A. Deck Section Tool
B. Wall Section Tool
C. Slab Section Tool
D. Openings Tool

Q36. Which dialog box appears after using the Slab Section tool?
A. Frame Assignment-Slab Section
B. Shell Assignment-Slab Section
C. Joint Assignment-Slab Section
D. Load Assignment-Slab Section

Q37. Which tool is used to assign deck properties to shell objects?
A. Wall Section Tool
B. Slab Section Tool
C. Deck Section Tool
D. Area Spring Tool

Q38. Which tool is used to assign wall sections to shell objects?
A. Wall Section Tool
B. Edge Releases Tool
C. Thickness Overwrites Tool
D. Local Axes Tool

Q39. Which tool converts a shell object into an opening?
A. Deck Section Tool
B. Openings Tool
C. Stiffness Modifiers Tool
D. Diaphragms Tool

Q40. Openings in shell objects are assumed to carry:
A. Full load
B. Half load
C. No load
D. Dynamic load only

Q41. Which option is selected to close an opening?
A. Remove Opening
B. Close Shell
C. Not an Opening
D. Fill Opening

Q42. Which tool is used to apply modifiers to shell stiffness properties?
A. Area Spring Tool
B. Stiffness Modifiers Tool
C. Deck Section Tool
D. Openings Tool

Q43. Which tool is used to modify thickness of shell objects?
A. Thickness Overwrites Tool
B. Insertion Point Tool
C. Diaphragms Tool
D. Local Axes Tool

Q44. Which radio button deletes existing thickness overwrites?
A. Clear Thickness Data
B. Remove Thickness Values
C. No Thickness Overwrites - Delete Any Existing Overwrites
D. Reset Thickness Option

Q45. Which tool is used to specify insertion point offsets for shell objects?
A. Insertion Point Tool
B. Slab Section Tool
C. Deck Section Tool
D. Openings Tool

Q46. Which coordinate system option is selected to define offsets along global X, Y, and Z axes?
A. User
B. Local
C. Grid
D. Global

Q47. Which tool assigns diaphragm properties to shell objects?
A. Wall Section Tool
B. Diaphragms Tool
C. Thickness Overwrites Tool
D. Area Spring Tool

Q48. Which tool is used to release loads from edges of shell objects?
A. Edge Releases Tool
B. Line Releases Tool
C. Stiffness Modifiers Tool
D. Additional Mass Tool

Q49. Which radio button deletes earlier assigned edge releases?
A. Delete Edge Loads
B. No Edge Releases - Delete Any Existing Edge Releases
C. Remove Edge Supports
D. Clear Boundary Loads

Q50. Which tool is used to apply load release on selected line elements?
A. Edge Releases Tool
B. Line Releases Tool
C. Rib Locations Tool
D. Local Axes Tool

Q51. Which tool is used to define X and Y locations for ribs?
A. Local Axes Tool
B. Area Spring Tool
C. Rib Locations Tool
D. Deck Section Tool

Q52. Which tool is used to reorient local axes of shell objects?
A. Local Axes Tool
B. Wall Section Tool
C. Openings Tool
D. Slab Section Tool

Q53. Which radio button rotates axes with respect to default axes?
A. Normal to Selected Beam
B. Rotate Along Beam
C. Rotation Angle from Default Orientation
D. Rotation from Current Grid

Q54. Which radio button rotates axes perpendicular to the selected beam?
A. Rotation Angle from Current Orientation
B. Rotation Angle from Default Orientation
C. Normal to Selected Beam
D. Parallel to Selected Beam

Q55. Which tool assigns area springs to shell objects?
A. Area Spring Tool
B. Additional Mass Tool
C. Diaphragms Tool
D. Openings Tool

Q56. Which tool assigns additional mass to shell objects apart from dead weight?
A. Area Spring Tool
B. Additional Mass Tool
C. Local Axes Tool
D. Edge Releases Tool

Q57. In the Shell Assignment - Additional Mass dialog box, mass is specified per:
A. Volume
B. Point
C. Area
D. Length

Q58. Which tool is used to assign wall hinge properties?
A. Reinforcement for Wall Hinge Tool
B. Wall Hinge Tool
C. Local Axes Tool
D. Area Spring Tool

Q59. Which check box stores response of each individual fiber in a fiber hinge?
A. Save Individual fibre response for Fibre Hinges
B. Enable Fibre Storage
C. Store Load Response
D. Save Shell Stress Data

Q60. Which tool is used to apply reinforcement to wall hinges?
A. Wall Section Tool
B. Reinforcement for Wall Hinge Tool
C. Area Spring Tool
D. Diaphragms Tool

FOR STUDENT NOTES

Chapter 5

Creating Load Cases and Performing Analysis

Topics Covered

The major topics covered in this chapter are:

- ***Introduction***
- ***Defining Load Patterns***
- ***Creating Modal Cases***
- ***Setting Load Cases***
- ***Setting Load Combinations***
- ***Creating Auto Construction Sequence Load Case***
- ***Creating Walking Vibrations***
- ***Checking Model for Analysis***
- ***Setting Degree of Freedom for Analysis***
- ***Setting Load Cases to Run Analysis***
- ***Setting SAPFire Solver Options***
- ***Mesh Setting for Floors and Walls***
- ***Setting Hinges and Cracking Analysis***
- ***Running Analysis***
- ***Applying Modal Alive***
- ***Modifying Undeformed Geometry***
- ***Checking Analysis Log***
- ***Unlocking Model***
- ***Setting Display Options***

INTRODUCTION

In previous chapters, you have learned to assign properties and loads to various frame members. In this chapter, you will learn to set load cases and combinations for analysis. You will also learn to perform analysis under specified load conditions. Later in this chapter, you will learn to analyze the results of analysis.

DEFINING LOAD PATTERNS

The **Load Pattern** tool is used to define the distribution of automatic loads like self weight, wind, earth quake, and other phenomenons for design studies. You will learn more about design studies in next chapter. Note that Load pattern is the first step for applying load in the model. After creating load pattern, you will assign different loads in the model using the **Assign** menu options. These assigned loads will fall under different categories of load pattern. Once you have assigned different loads to respective patterns then you can create load cases to run analysis. The procedure to create load pattern is given next.

- Click on the **Load Pattern** tool from the **Define** menu. The **Define Load Patterns** dialog box will be displayed; refer to Figure-1.

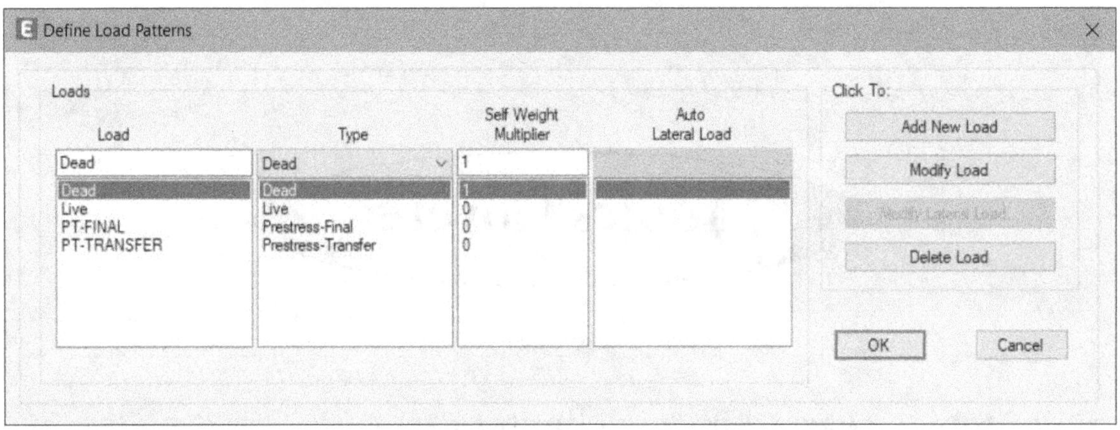

Figure-1. Define Load Patterns dialog box

- Specify desired name of load in the **Load** edit box.
- Click in the **Type** drop-down and select desired type of load. Like to apply wind load, select the **Wind** option from the **Type** drop-down.
- Specify desired value in the **Self-weight Multiplier** edit box to define the amount of weight to be added to the current load.
- Set desired auto lateral loading code in the **Auto Lateral Load** drop-down if you are defining seismic or wind load.
- Click on the **Add New Load** button to add it in the pattern.
- To modify a load, select it from the list and change the parameters as required from the top of the table. After setting desired changes, click on the **Modify Load** button.
- If you want to modify lateral loading of a seismic or wind load then select it from the table and click on the **Modify Lateral Load** button. The **Wind Load Pattern** or **Seismic Loading** dialog box will be displayed based on the type of load; refer to Figure-2 and Figure-3.

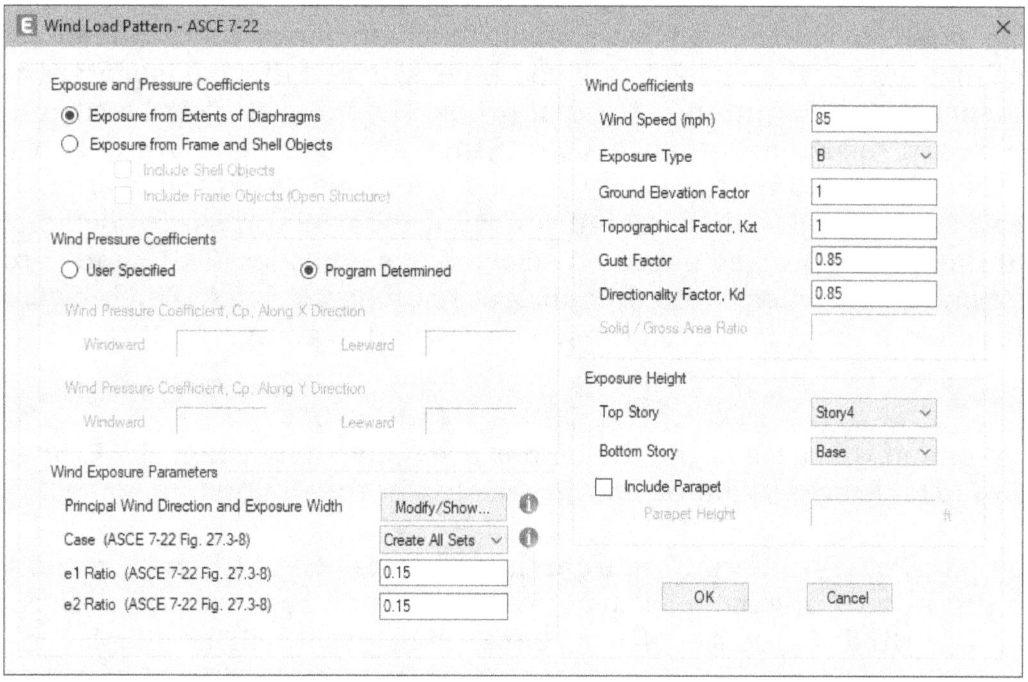

Figure-2. Wind Load Pattern dialog box

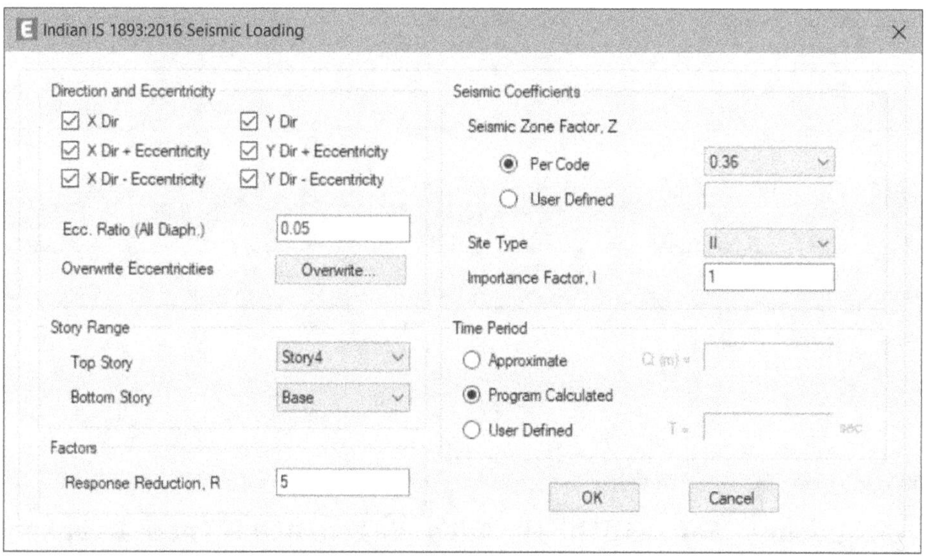

Figure-3. Seismic Loading dialog box

Modifying Lateral load of wind

- Select the **Exposure from Extents of Diaphragms** radio button to specify wind load individually to each of the diaphragms. A separate lateral load is created for each diaphragm present at a story level. The wind loads calculated at any story level are based on the story level elevation, the story height above and below that level, the assumed exposure width for the diaphragm(s) at the story level, and various code-dependent wind coefficients.
- Select the **Exposure from Frame and Shell Objects** radio button to specify wind load for each frame and shell object as per the wind pressure coefficient, C_p. Select the **Include Shell Objects** check box to include shell objects for wind loading. Select the **Include Frame Objects (Open Structure)** check box to include open frames for wind loading.

Wind Pressure Coefficients

- Specify desired values of wind pressure coefficients in the **Wind Pressure Coefficients** area of the dialog box if you have selected the **Exposure from Extents of Diaphragms** radio button in the dialog box. Note that the **Windward** and **Leeward** edit boxes in the **Wind Pressure Coefficients** area will be active only if you have **User Specified** radio button from the **Wind Pressure Coefficients** area. Select the **Program Determined** radio button if you want the wind pressure coefficients to be determined automatically based on selected design codes. The Windward defines the direction from where wind is coming at point of reference. The Leeward defines the direction from point of reference where wind is going.

Wind Exposure Parameters

All the options in this area of the dialog box are active only when the **Exposure from Extents of Diaphragms** radio button is selected in the dialog box.

- Click on the **Modify/Show** button from the **Principal Wind Direction and Exposure Width** drop-down in **Wind Exposure Parameters area** to set wind direction and width. The **Wind Exposure Width Data** dialog box will be displayed; refer to Figure-4.

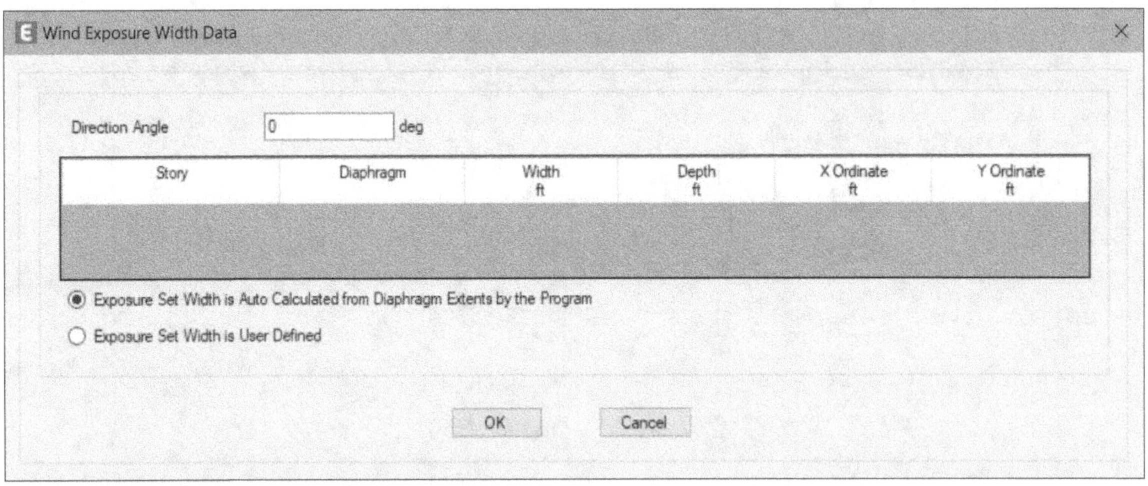

Figure-4. Wind Exposure Width Data dialog box

- Select the **Exposure Set Width is Auto Calculated from Diaphragm Extents by the Program** radio button to automatically set the width of wind exposure automatically. Specify the angle value in the **Direction Angle** edit box.
- Select the **Exposure Set Width is User Defined** radio button to also set the width of exposure manually. To add wind exposure, click on the **Add Row** button. A new row will be added in the list. Set desired parameters in the table.
- Select desired case from the **Case** drop-down. Hover the cursor on information button ( )to check the details of cases in the drop-down.
- Set desired values of **e1 Ratio** and **e2 Ratio** in respective edit boxes.

Wind Coefficients

The options in the **Wind Coefficients** area is used to specify basic parameters of wind.

- Specify desired value of wind speed in miles per hour in the **Wind Speed (mph)** edit box.

- Select desired option from the **Exposure Type** drop-down. The **B** type is used for urban and suburban areas, wooded areas, or other terrain with numerous closely spaced obstructions having the size of single-family dwellings or larger. The **C** type is used for open terrain with scattered obstructions having heights generally less than 30 feet. (Commonly associated with flat open country and grasslands). Select the **D** type for structures at a close distance (typically within 600 feet) from an "open waterway" one mile or more across. This category is readily distinguishable, where the locally enforced code very likely has considered this in their requirements.
- Specify desired value in the **Ground Elevation Factor** edit box to define density of wind at height.
- Set desired value in **Topographical Factor, Kzt** edit box. There are different values of Topographical factor for different type of terrains. Like for plain terrain value of Kzt is 1. For other terrains, height of ridges, escarpments, and hills are considered to derive the factor.
- Specify desired value of Gust factor in the **Gust Factor** edit box. A gust factor, defined as the ratio between a peak wind speed and mean wind speed over a period of time, can be used along with other statistics to examine the structure against the wind.
- Specify desired value of directionality factor in the **Directionality Factor, Kd** edit box. The directionality factor (Kd) used in the ASCE 7 wind load provisions for components and cladding is a load reduction factor intended to take into account the less than 100% probability that the design event wind direction aligns with the worst case building aerodynamics. The Directionality Factor Kd is defined as a parameter that makes the design more rational by considering the dependencies of the wind speed, the frequency of occurrence of extreme wind and the aerodynamic property on wind direction. The wind directionality factor Kd is affected by the frequency of occurrence and the routes of typhoons, climatological factors, large-scale topographic effects, and so on.

Exposure Height

- Set desired upper and lower limit in the **Top Story** and **Bottom Story** edit boxes, respectively in the **Exposure Height** area of the dialog box.
- If you have selected the **Exposure from Extents of Diaphragms** radio button then the **Include Parapet** check box will be active. Select the check box to include parapets while considering wind loads. A parapet is a barrier which is an extension of the wall at the edge of a roof, terrace, balcony, walkway, or other structure. Specify desired value of height (ft) in the **Parapet Height** edit box.

- After setting desired parameters, click on the **OK** button to apply modifications to lateral loads.

You can modify seismic loading in the same way.

SETTING LOAD CASES

A load case defines how load patterns are applied (statically or dynamically), how the structure responds (linearly or nonlinearly), and how analysis is performed (through modal analysis, direct integration, etc.). For each analysis to be performed, a load case is defined. Each load case may apply a single load pattern or a combination of load patterns. The procedure to create and manage load cases is given next.

- Click on the **Load Cases** tool from the **Define** menu. The **Load Cases** dialog box will be displayed with a list of load patterns earlier created; refer to Figure-5.

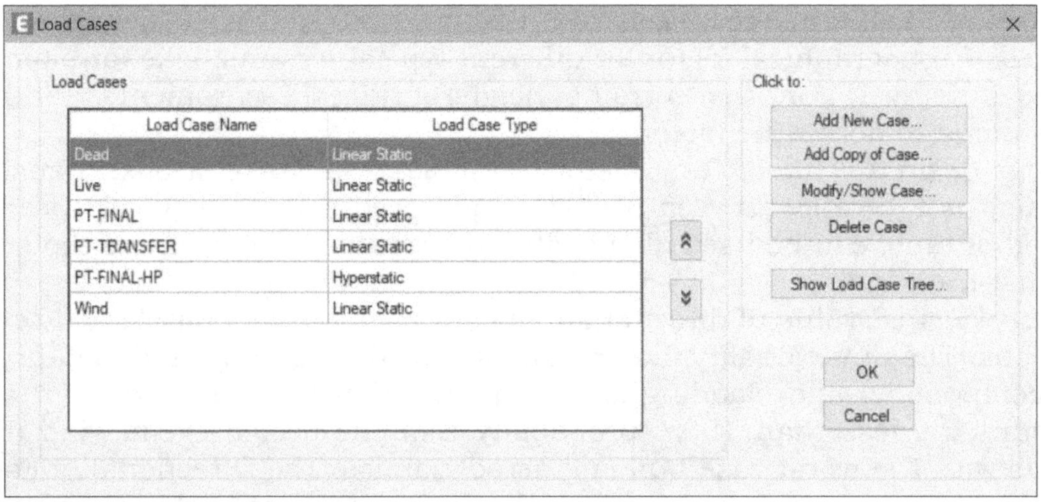

Figure-5. Load Cases dialog box

- To add a new load case, click on the **Add New Case** button. The **Load Case Data** dialog box will be displayed; refer to Figure-6.
- Set desired name for load case in the **Load Case Name** edit box.
- Select desired load case type from the **Load Case Type** drop-down. Select the **Linear Static** option if you want to apply fix loads without any dynamic effects. Select the **Nonlinear Static** option if you want to apply time changing force and check the large displacement effects in the model. Select the **Nonlinear Staged Construction** option to perform analysis at different stages. Select the **Response Spectrum** option to check the plot of the peak or steady-state response (displacement, velocity, or acceleration) of a series of oscillators of varying natural frequency, that are forced into motion by the same base vibration or shock. Select the **Time History** option from the drop-down if Time-varying loads are applied. The solution may be by modal superposition or direct integration methods. Select the **Buckling** option from the drop-down if you are concerned about the buckling failure of the structure. Select the **Hyperstatic** option to apply hyperstatic load. A hyperstatic load case is required in certain codes when performing design with post-tensioning.

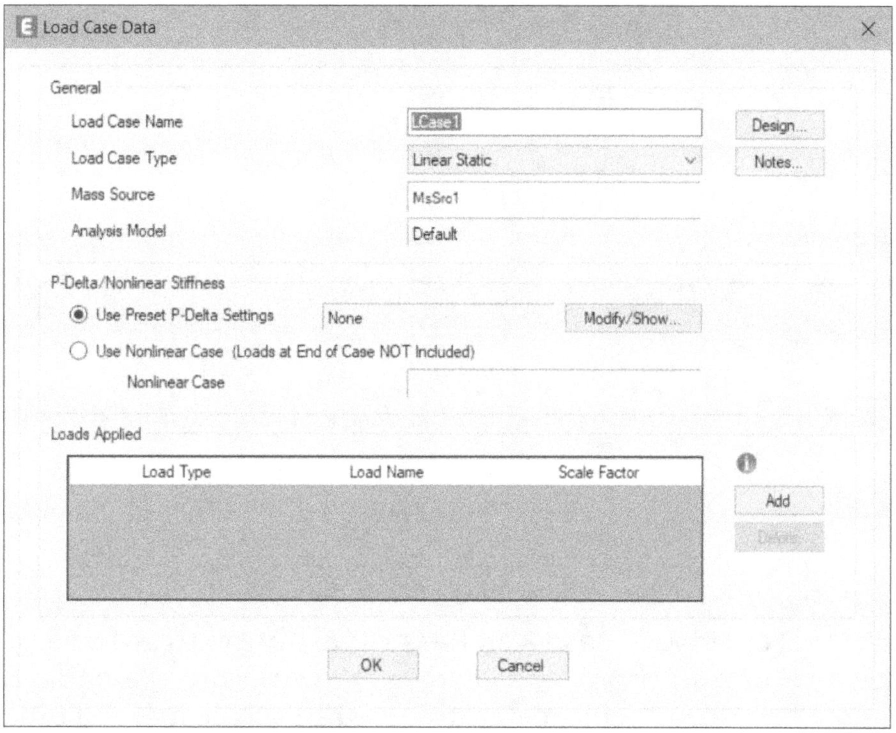

Figure-6. Load Case Data dialog box

- To move the load cases up or down, select the load case from the list and click on the **Up** or **Down** button at the right in the dialog box. Note that moving the load cases up or down does not define their position while performing analysis. All the load cases will run as per their dependency.
- The options in this dialog box will be modified accordingly. Set the parameters as required and click on the **OK** button. You will learn more about these options later.
- The **Add Copy of Case**, **Modify/Show Case**, and **Delete Case** buttons in the dialog box work in the same way as they have been discussed in previous chapters.
- Click on the **OK** button from the **Load Cases** dialog box to apply the load cases.

Creating Linear Static Load Case

The Linear Static load case is created to perform analysis with static load on the model. The procedure to create linear static load case is given next.

- Click on the **Add New Case** button from the **Load Cases** dialog box. The **Load Case Data** dialog box will be displayed as discussed earlier.
- Specify desired name for the case in the **Load Case Name** edit box.
- Select the **Linear Static** option from the **Load Case Type** drop-down to create linear static case.
- Select the **Use Preset P-Delta Settings** radio button to apply P-Delta settings to the current linear static analysis. For most tall building structures, the P-Delta effect is of most concern in the columns because of gravity load, including dead and live load.
- Click on the **Modify/Show** button to specify parameters for P-Delta. The **Preset P-Delta Options** dialog box will be displayed; refer to Figure-7.

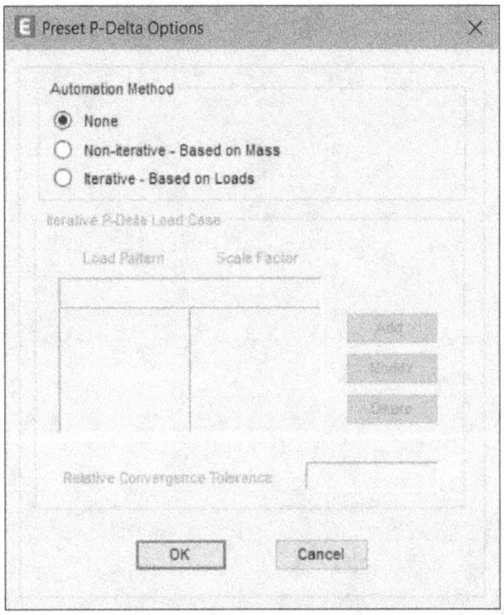

Figure-7. Preset P-Delta Options dialog box

- Select the **Non-iterative - Based on Mass** radio button to automatically compute load at each level as a story-by-story load on the structure. Note that this approach is approximate and do not require iterative solution. This method essentially treats the building as a simplified stick model to consider the P-Delta effect. It does not capture local buckling as well as the iterative method. This method works best if the model has a single diaphragm at each floor level, although it also works for other cases as well. Select the **Iterative - Based on Loads** radio button to compute load cases by specified combination of static load patterns. For example, the load case may be the sum of a dead load case plus a fraction of a live load case. This approach requires an iterative solution to determine the P-Delta effect upon the structure. This method considers the P-Delta effect on an element-by-element basis. It captures local buckling effects better than the non-iterative method. We recommend the use of this iterative method in all cases except those for which no gravity load is specified in the model. On selecting this radio button, the options in the **Iterative P-Delta Load Case** area will become active. Select desired load pattern, respective scale factor in the table, and click on the **Add** button. Click on the **OK** button to set preset of P-Delta.
- Click in the field under **Load Type** column and select desired load type. Select **Load Pattern** option to add desired load. Select the **Acceleration** option to apply acceleration load on the model. Note that Static, modal, and buckling load cases permit only acceleration loads, UX, UY, UZ, RX, RY, and RZ. Response-spectrum and time-history load cases permit application of ground acceleration in a local coordinate system. Specify an angle of rotation about the Z axis. The acceleration load names are U1, U2, U3, R1, R2, and R3.
- Click in the field under **Load Name** column and select desired load.
- Specify desired scale factor in the **Scale Factor** column. Click on the **Add** button to add the current load in load case data. Add as many loads as desired and click on the **OK** button from the dialog box to create load case.

Creating Nonlinear Static Load Case

The Nonlinear Static load case is created to perform analysis with specified initial conditions and static load on the model. The procedure to create nonlinear static load case is given next.

- Click on the **Add New Case** button from the **Load Cases Type** dialog box. The **Load Case Data** dialog box will be displayed. Select the **Nonlinear Static** option from the **Load Case Type** drop-down. The dialog box will be displayed as shown in Figure-8.

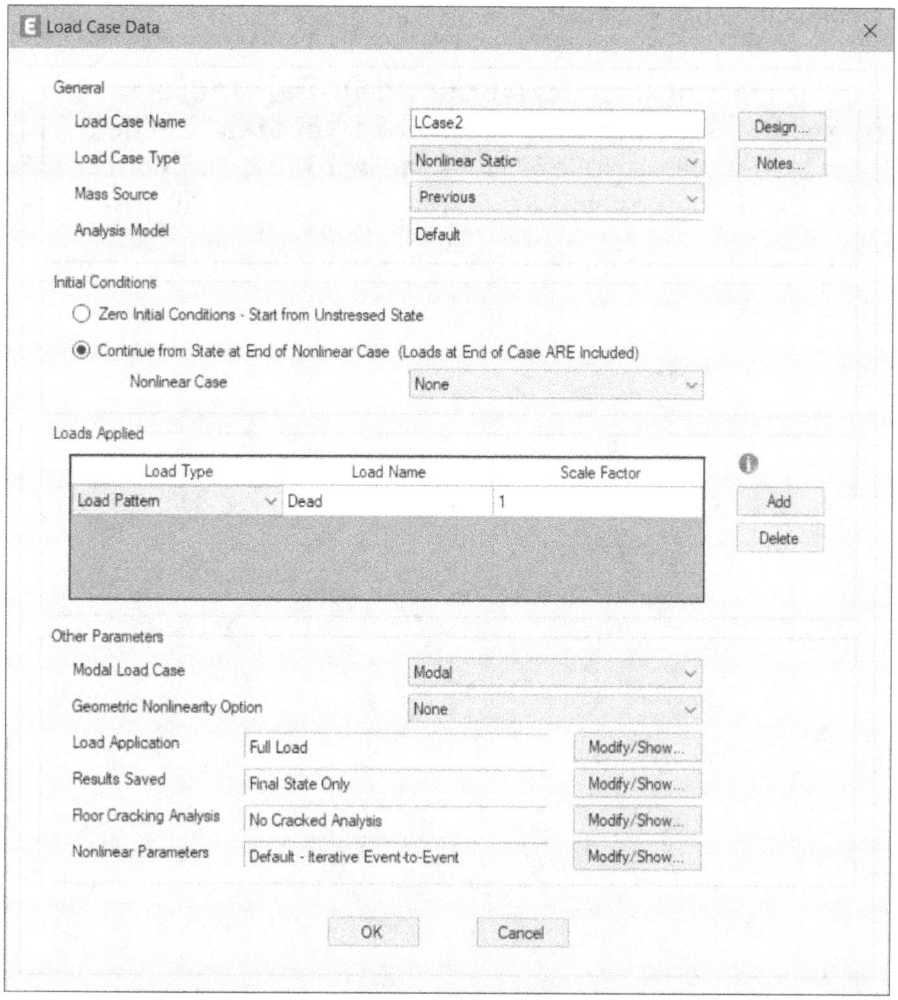

Figure-8. Load Case Data dialog box for Nonlinear Static case

- Select desired mass source from the **Mass Source** drop-down.
- Select the **Zero Initial Conditions - Start from Unstressed State** radio button to set zero displacement and velocity. All elements are unstressed and there is no history of nonlinear deformation.
- Select the **Continue from State at End of Nonlinear Case (Loads at End of Case ARE Included)** radio button to select desired nonlinear load whose end results will be used as starting point for current analysis. Select desired load case from the **Nonlinear Case** drop-down.
- Add desired loads in the table in **Loads Applied** area as discussed earlier.
- Select desired option from the **Modal Load Case** drop-down. All the modal load cases created earlier are displayed in this drop-down.
- Select desired option to define non-linearity of geometry in the **Geometric Nonlinearity Option** drop-down. If you have selected **None** option then the All equilibrium equations are considered in the undeformed configuration of the structure. If you have selected **P-Delta** option then the equilibrium equations take into account the deformed configuration of the structure, partially. Tensile forces tend to resist the rotation of elements and stiffen the structure, and compressive forces tend to enhance the rotation of elements and destabilize the structure.

This may require a moderate amount of iteration. If you have selected **P-Delta plus Large Displacements** option then all equilibrium equations are written in the deformed configuration of the structure. This may require a large amount of iteration. Although large displacement and large rotation effects are modeled, all strains are assumed to be small.

Defining Load Application Parameters

- Click on the **Modify/Show** button next to **Load Application** field in the dialog box to specify how load will be applied. The **Load Application Control for Nonlinear Static Analysis** dialog box will be displayed; refer to Figure-9.

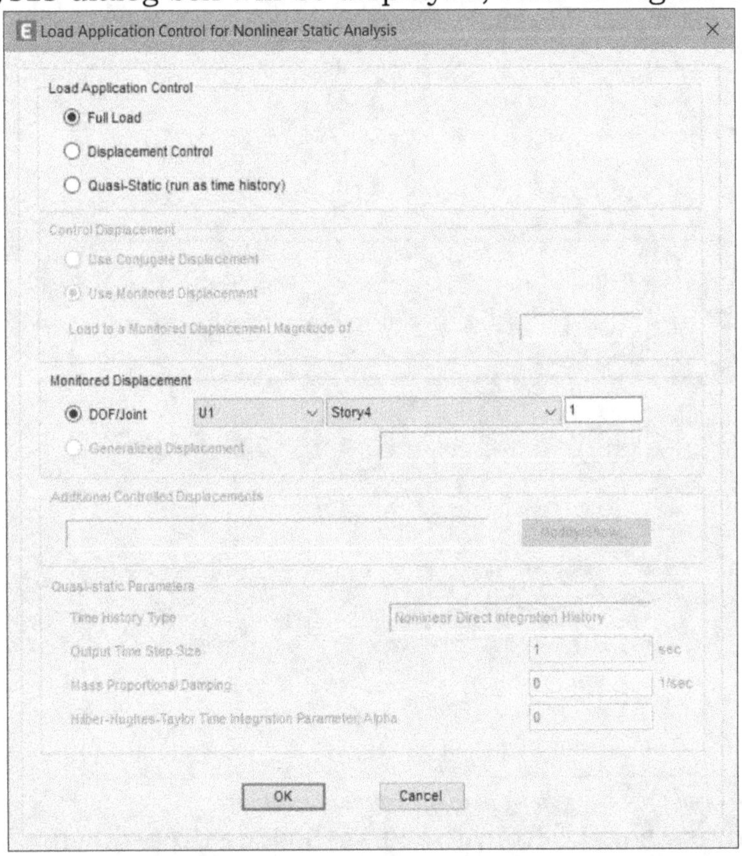

Figure-9. Load Application Control for Nonlinear Static Analysis dialog box

- Select the **Full Load** radio button if you want to apply full load at once. Select the **Displacement Control** radio button if you know the distance moved by the structure but the amount of load is unknown. This is most useful for structures that become unstable and may lose load-carrying capacity during the course of the analysis. Typical applications include static pushover analysis and snap-through buckling analysis. Select the **Quasi-Static** radio button when the distance moved by the structure is known. However, the solution is found by applying a specified amount of load incrementally, and solving using dynamic analysis rather than static analysis. This is most useful for structures that become unstable and may lose load-carrying capacity during the course of the analysis, and for which a solution could not be found using displacement control. Typical applications include static pushover analysis and snap-through buckling analysis. Set the other parameters as required and click on the **OK** button.
- Click on the **Modify/Show** button next to **Results Saved** field. The **Results Saved for Nonlinear Static Case** dialog box will be displayed; refer to Figure-10. Select the **Final State Only** radio button to save the result of analysis only once and

that too at the end of analysis. Select the **Multiple States** radio button to save the results of non-linear analysis at regular intervals as specified in the dialog box. Click on the **OK** button to apply the settings.

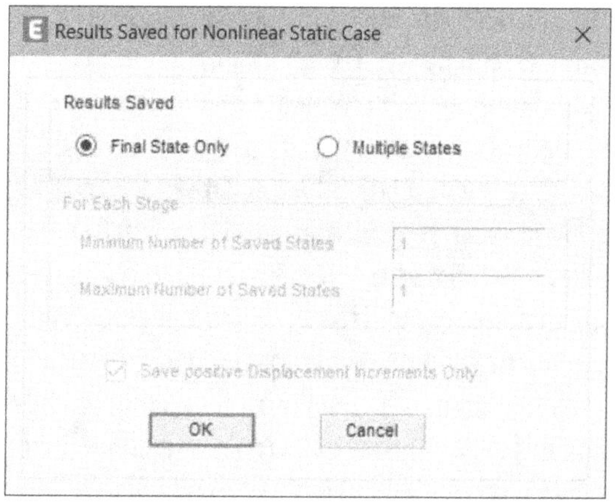

Figure-10. Results Saved for Nonlinear Static Case dialog box

- Click on the **Modify/Show** button next to **Nonlinear Parameters** field. The **Nonlinear Parameters** dialog box will be displayed; refer to Figure-11. Set desired parameters and click on the **OK** button.

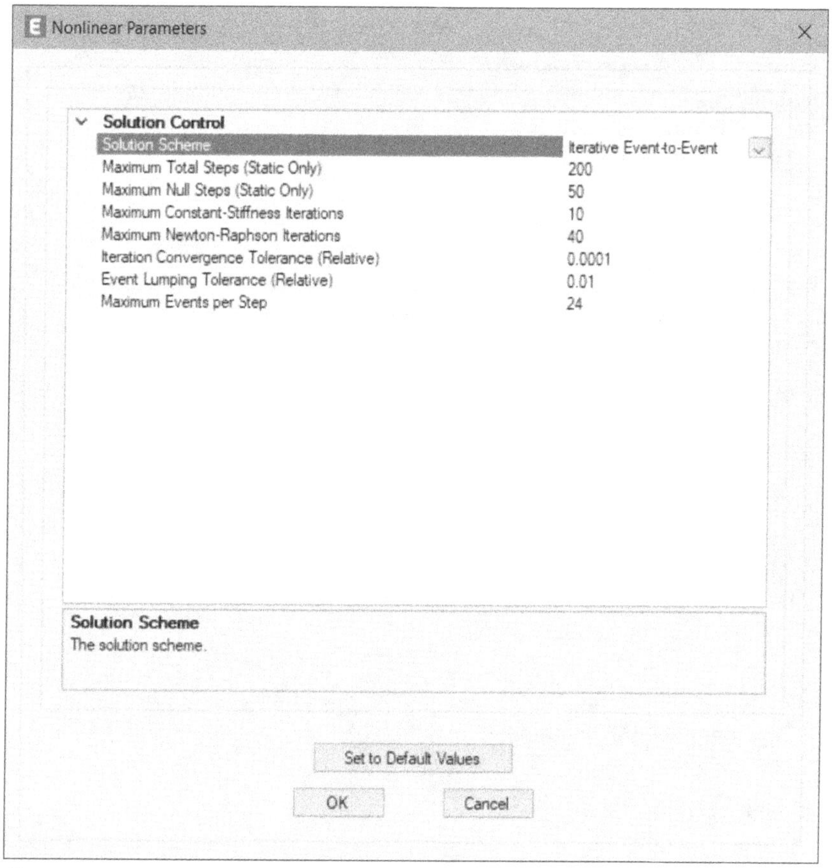

Figure-11. Nonlinear Parameters dialog box

- After setting desired parameters, click on the **OK** button to create load case.

Creating Nonlinear Staged Construction Load Case

The Nonlinear staged construction load case is created to perform non-linear analysis at different stages generally when creating bridges. The procedure to create this load case is given next.

- Click on the **Add New Case** button from the **Load Cases** dialog box and select the **Nonlinear Staged Construction** option from the **Load Case Type** drop-down. The **Load Case Data** dialog box will be displayed as shown in Figure-12.
- Most of the parameters are same as discussed earlier.
- Click on the **Add** button in the **Stage Definition** area to create a new stage at which analysis will be performed. A new row will be added in the table. Double-click in the field of table to change the parameters.
- Select desired stage and click on the **Stage Operations** button to add operations. The **Stage Data** dialog box will be displayed; refer to Figure-13.
- Select desired stage from the **Stage Name** drop-down. Click on the **Add** button. A new operation will be added in the list.
- Click in the field under **Operation** column and select desired type of operation like adding structure, removing structure, loading objects, and so on.
- Click in the field under **Object Type** column and select desired type of object on which you want to perform the selected operation.

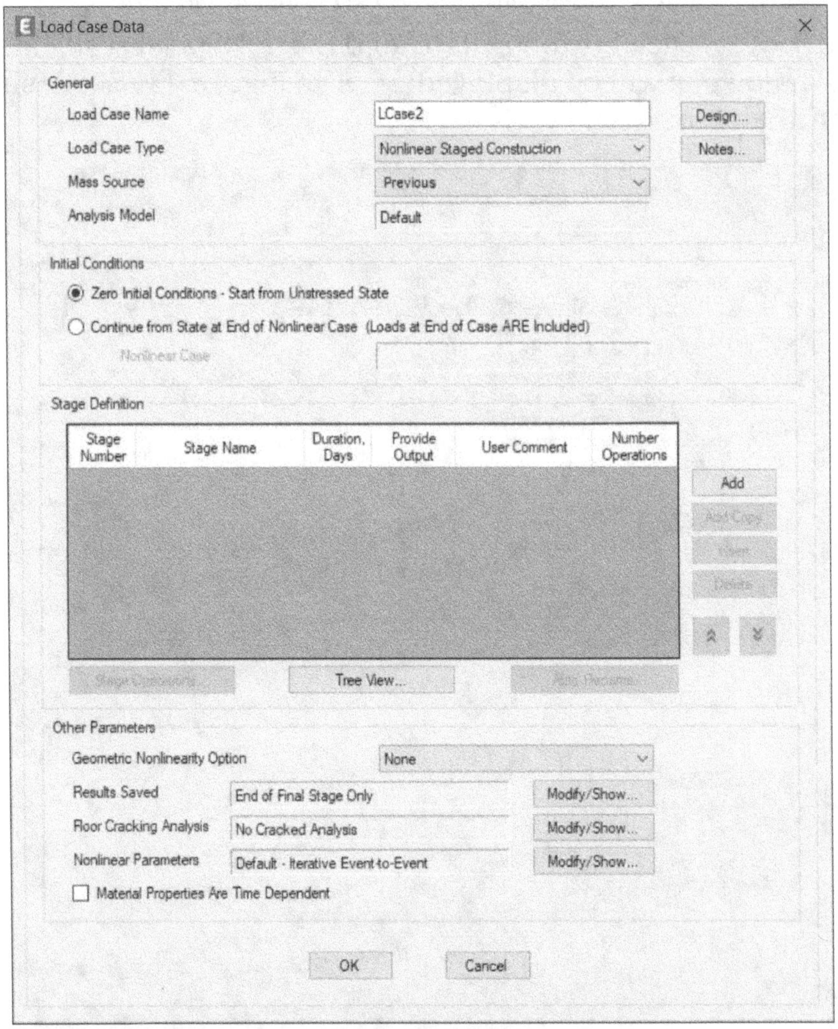

Figure-12. Load Case Data dialog box for Nonlinear Staged Construction case

- Specify the age of structure being added in current stage in the **Age at Add, days** edit box. This column is active only when **Add Structure** option selected in the **Operation** column.
- If you have selected the **Load Objects** or **Load Objects if Added** in the **Operation** column then **Type**, **Name**, and **Scale Factor** columns will be active. Specify desired load type, load name, and scale factor under respective columns.

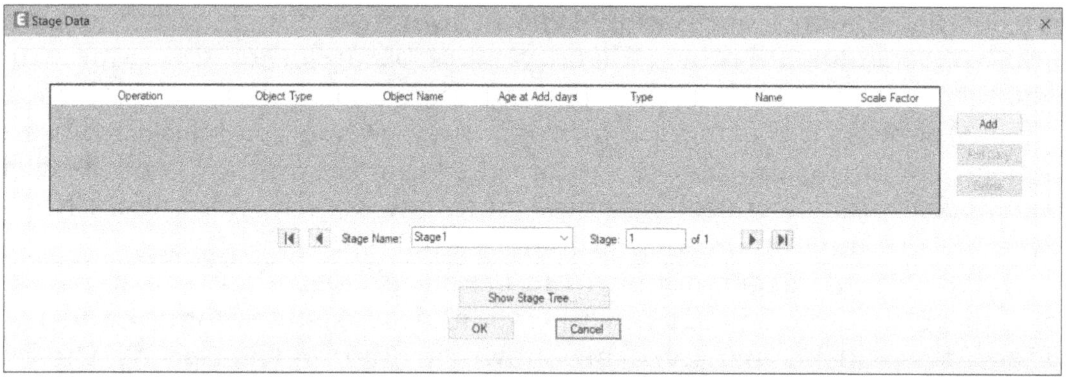

Figure-13. Stage Data dialog box

- If you have selected the **Change Sections** option under **Operation** column then specify the section type and size in the **Type** and **Name** columns, respectively.
- After setting desired data, click on the **OK** button. The **Load Case Data** dialog box will be displayed again.
- Set the other parameters as discussed earlier and click on the **OK** button to create the load case.

Creating Modal Load Case

The modal case is used to find natural frequencies of the model. The procedure to create modal case is given next.

- Select the **Modal** option from the **Load Case Type** drop-down. The **Modal Cases** dialog box will be displayed with list of modal cases earlier created; refer to Figure-14.

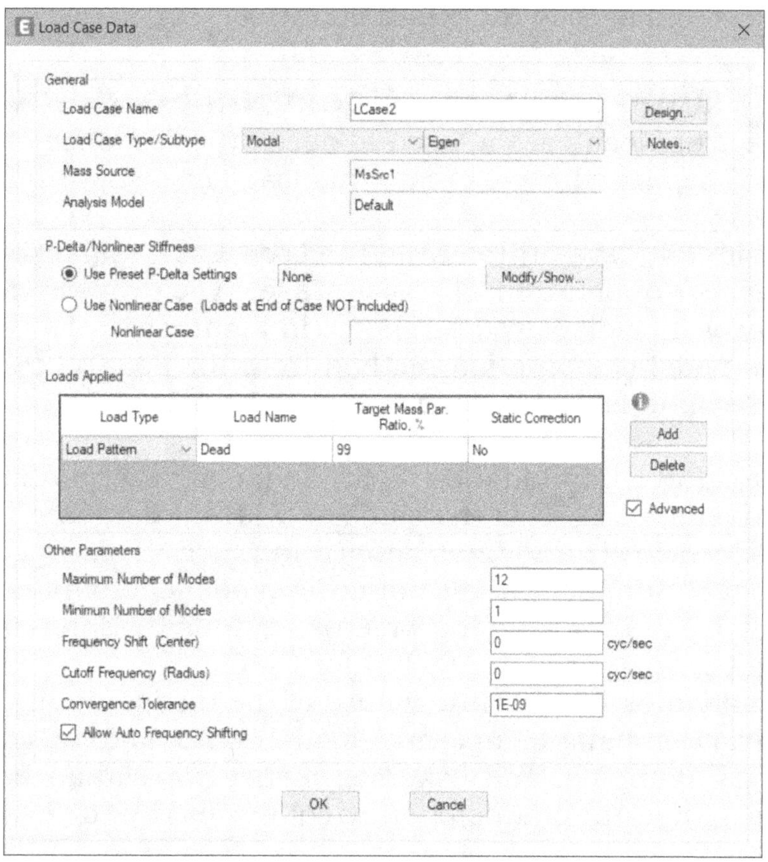

Figure-14. Modal load case data

- Select desired option from the **Subtype** drop-down. For dynamic analysis, Ritz vectors are recommended over Eigen vectors because, for the same number of modes, Ritz vectors provide a better participation factor, which enables the analysis to run faster with the same level of accuracy. When analysis involves ground motion in the horizontal plane, the benefit is not as pronounced. However, for vertical acceleration, localized machine vibration, and the nonlinear FNA method, Ritz vectors are much more well-suited for analysis.
- If you have selected **Eigen** option from the drop-down then select the **Advanced** check box to define loads along with modal analysis. The options in the **Loads Applied** area will be displayed as shown in Figure-15. Specify the loads as discussed earlier. Similarly, you can add load for **Ritz** load case sub-type.

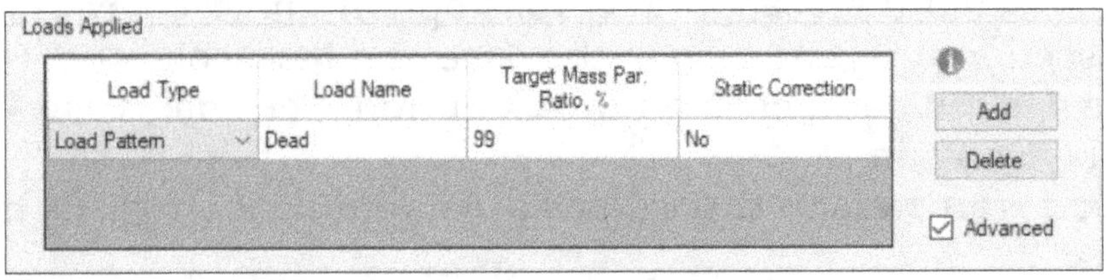

Figure-15. Loads Applied area with Eigen load case subtype

- If you have selected Ritz Modal Case Sub Type then specify the maximum and minimum modes in **Maximum Number of Modes** and **Minimum Number of Modes** edit boxes, respectively.
- Specify desired values for frequency shift, cutoff frequency, and convergence tolerance in respective edit boxes if you have selected **Eigen** option in **Modal Case SubType** drop-down and click on the **OK** button.
- Note that if you want to define a design load case then click on the **Design** button in the dialog box and specify the parameters.

Creating Response Spectrum Load Case

The response spectrum load case is used to check the response of structure under vibrational loads. The procedure to create this load case is given next.

- Click on the **Add New Case** button from the **Load Cases** dialog box and select the **Response Spectrum** option from the **Load Case Type** drop-down. The **Load Case Data** dialog box will be displayed as shown in Figure-16.
- Select desired modal combination method from the **Modal Combination Method** drop-down. The options in this drop-down are as below:

CQC - Complete Quadratic Combination option. A modal combination technique that accounts for modal damping. It is the same as SRSS if damping is zero. Enter the characteristic frequencies f1 and f2 as defined in ASCE 4 for GMC.
SRSS - Square Root of Sum of Squares option. A modal combination technique that does not account for modal damping or cross coupling. Enter the characteristic frequencies f1 and f2 as defined in ASCE 4 for GMC.
Absolute - Summation of the absolute values of the modal results.
GMC - General Modal Combination option. A modal combination technique that takes into account modal damping, and assumes higher correlation between modes at higher frequencies. Enter the characteristic frequencies f1 and f2 as defined in ASCE 4 for GMC.

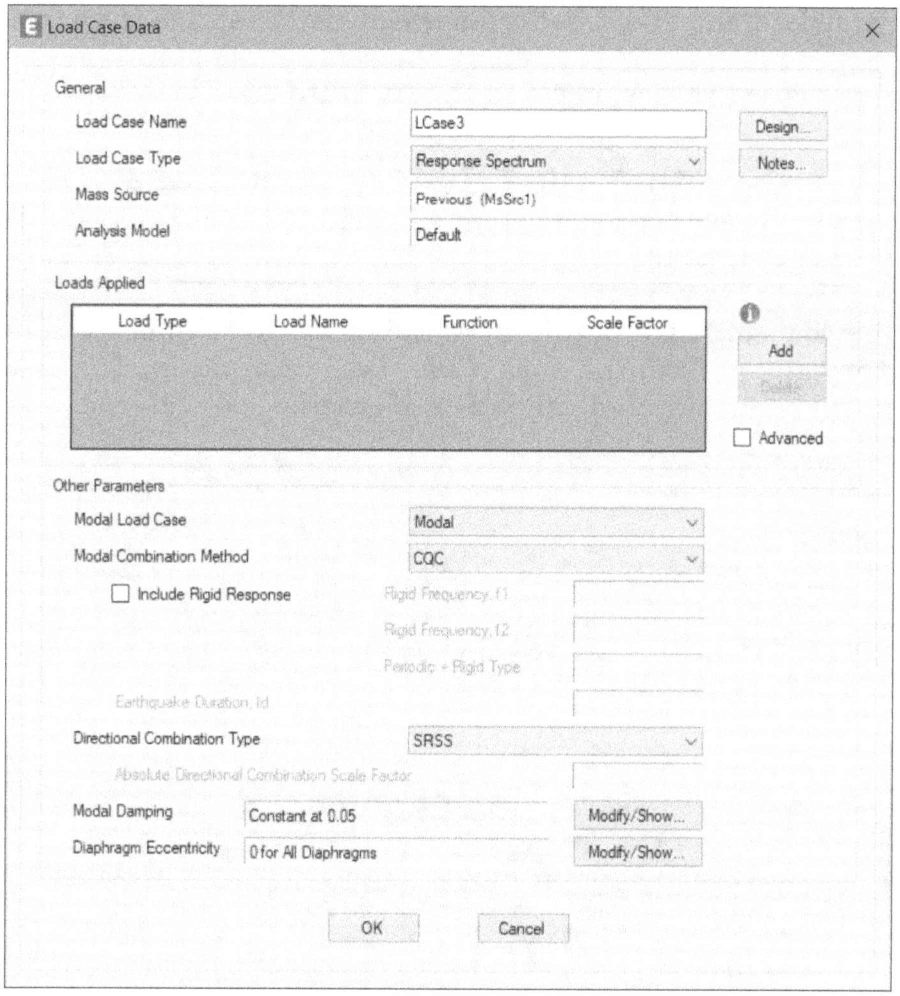

Figure-16. Load Case Data dialog box for Response Spectrum case

NRC 10 Pct - the Ten Percent method of the U.S. Nuclear Regulatory Commission Regulatory Guide 1.92 option. The Ten Percent method assumes full, positive coupling between all modes whose frequencies differ from each other by 10% or less of the smaller of the two frequencies. Modal damping does not affect the coupling. Enter the characteristic frequencies f1 and f2 as defined in ASCE 4 for GMC.

Double Sum - the Double Sum method of the U.S. Nuclear Regulatory Commission Regulatory Guide 1.92 option. The Double Sum method assumes a positive coupling between all modes, with correlation coefficients that depend upon damping in a fashion similar to the CQC and GMC methods, and that also depend upon the duration of the earthquake. Specify this duration as parameter t_d, part of the load cases definition.

CQC3 - An extension of the SRSS method for finding the maximum response when the horizontal (U1 and U2) directions of loading use the same response spectrum function but have different scale factors. The critical angle of loading is determined automatically independent of the angle specified for the loading. The vertical response is combined with the maximum horizontal response using the SRSS method. If different response-spectrum functions are used for U1 and U2, the results must be interpreted carefully by the engineer.

- Select the **Include Rigid Response** check box to include rigid frequencies. Specify desired parameters in edit boxes adjacent to the check box.

- Set the modal damping and diaphragm eccentricity values using the **Modify/Show** buttons.
- After setting desired values, click on the **OK** button.

Creating Time History Load Case

The Time History load case is used when time varying loads are applied. The procedure to use this tool is given next.

- Click on the **Add New Case** button from the **Load Cases** dialog box and select the **Time History** option from the **Load Case Type** drop-down. The **Load Case Data** dialog box will be displayed as shown in Figure-17.

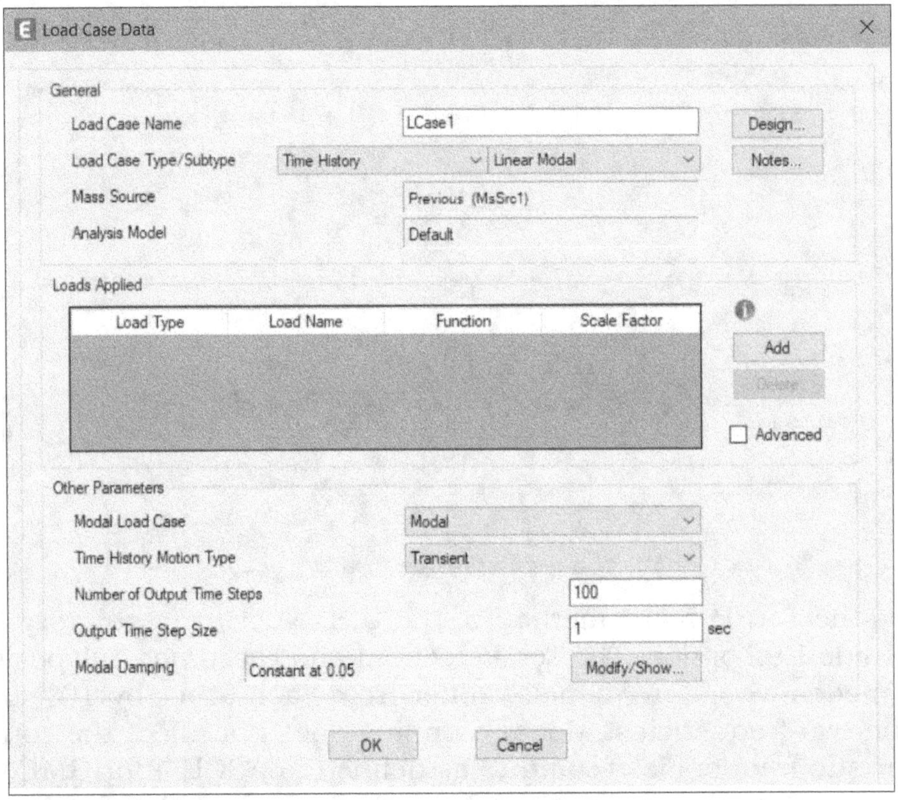

Figure-17. Load Case Data dialog box for Time History case

- Select desired subtype of Time History load case from the **Subtype** drop-down. The options will be modified in the dialog box accordingly.

Setting Linear Modal Time History Load Case

- Select the **Time History** option from the **Load Case Type** drop-down and select the **Linear Modal** option from the **Subtype** drop-down. The options will be displayed as shown in Figure-17.
- Set the loads applied and Modal Load Case as discussed earlier.
- Select the **Transient** option from the **Time History Motion Type** drop-down when structure starts at rest and is subjected to the specified loads only during the time period specified for the analysis. Select the **Periodic** option from the drop-down when specified loads are assumed to be periodic i.e. they repeat indefinitely with a period given by the length of the analysis.
- Specify desired number of steps in the **Number of Output Time Steps** edit box.

- Set desired value of time in the **Output Time Step Size** edit box.
- After setting desired parameters, click on the **OK** button.

Setting Nonlinear Modal Time History Load Case

- Select the **Time History** option from the **Load Case Type** drop-down and select the **Nonlinear Modal (FNA)** option from the **Subtype** drop-down. The options will be displayed as shown in Figure-18.

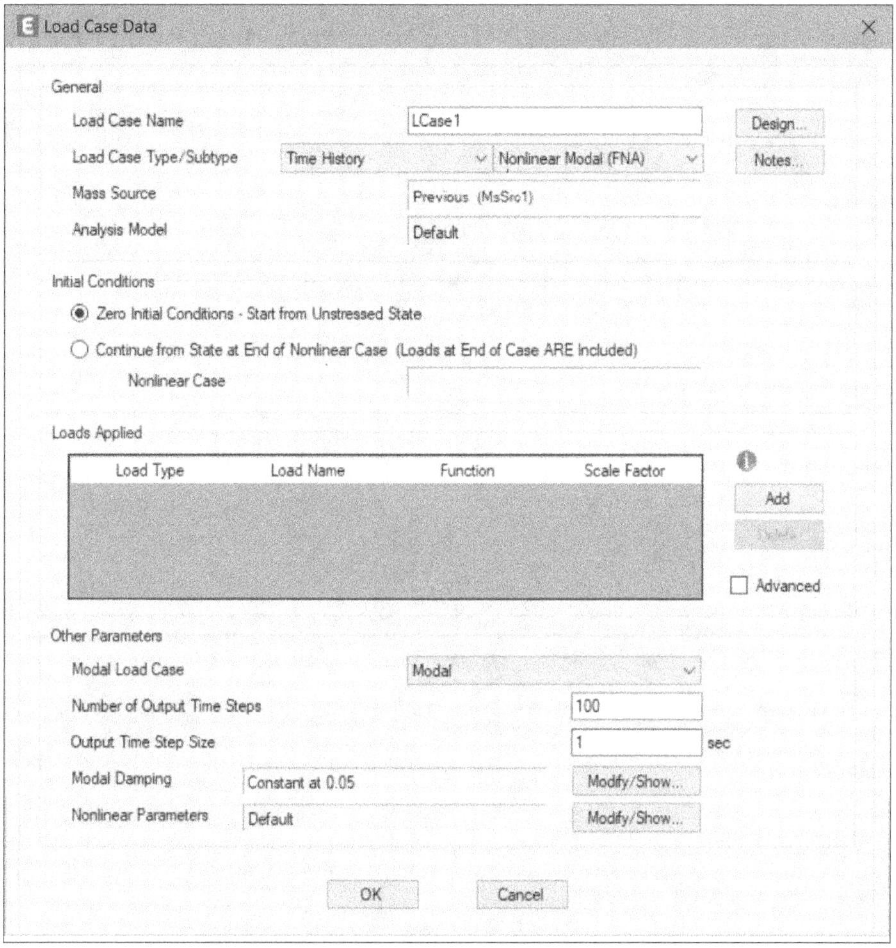

Figure-18. Load Case Data dialog box for Time History Nonlinear Modal case

- Set all the parameters as discussed earlier and click on the **OK** button to create load case.

Setting Linear Direct Integration Time History Load Case

- Select the **Time History** option from the **Load Case Type** drop-down and select the **Linear Direct Integration** option from the **Subtype** drop-down. The options will be displayed as shown in Figure-19.
- Click on the **Modify/Show** button next to **Damping** field in the dialog box. The **Direct Integration Damping** dialog box will be displayed; refer to Figure-20. The options in this dialog box are used to specify the method by which damping will be applied.

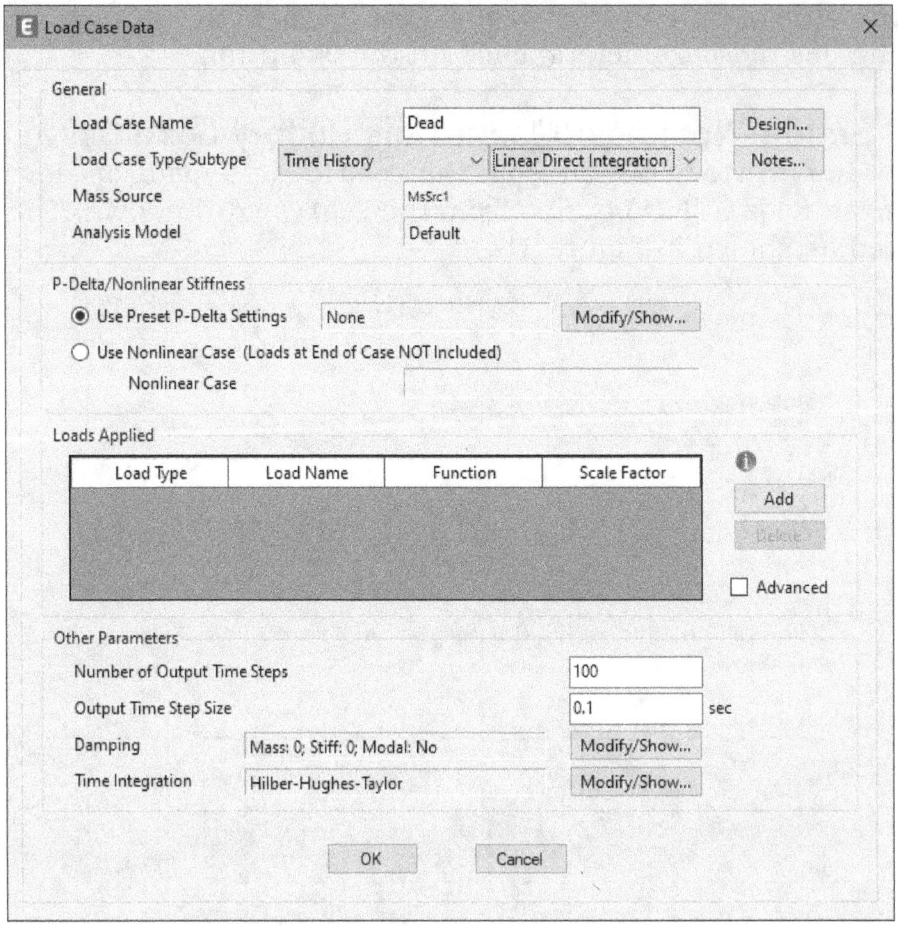

Figure-19. Load Case Data dialog box for Time History Linear Direct Integration load case

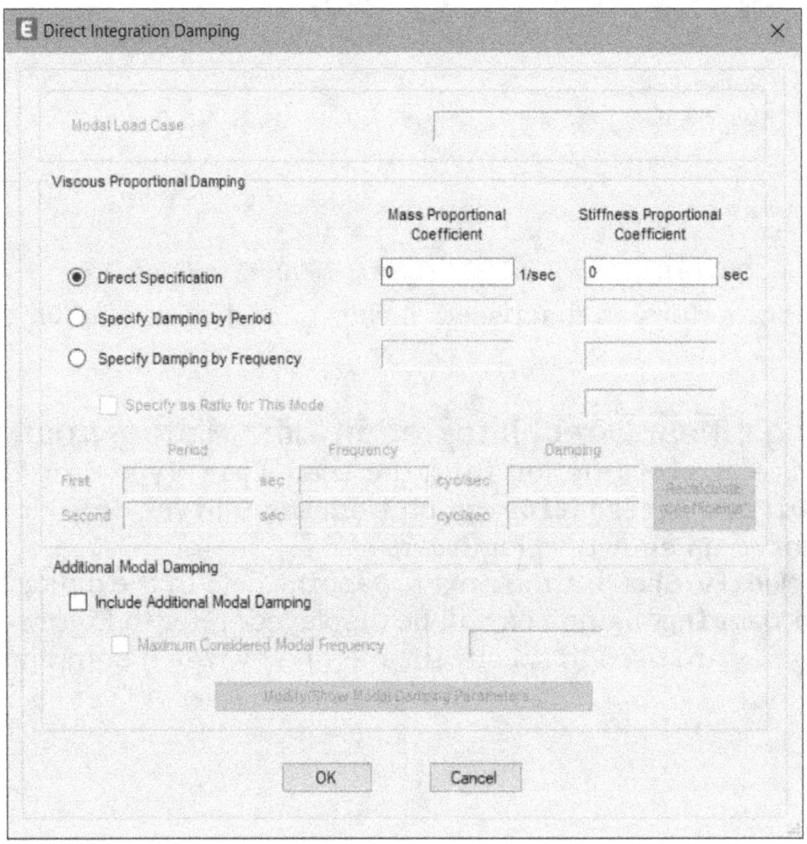

Figure-20. Direct Integration Damping dialog box

Setting Direct Integration Damping

- Select the **Direct Specification** radio button to specify the mass and stiffness coefficient of damping.
- Select the **Specify Damping by Period** radio button to specify the time period and relative damping in the table. After specifying the time period and damping value, click on the **Recalculate Coefficient** button.
- If you want to add modal damping also to the damping coefficient then select the **Include Additional Modal Damping** check box. The options below it will become active. Set desired parameters.
- Select the **Specify Damping by Frequency** radio button to specify frequency and damping in the table. After setting parameters, click on the **Recalculate Coefficients** button.
- Click on the **OK** button to apply direct integration damping.

Specifying Time Integration Method

The Time Integration methods are used to define how the time integration will be performed during calculations.

- Click on the **Modify/Show** button next to **Time Integration** field. The **Time Integration Parameters** dialog box will be displayed; refer to Figure-21.

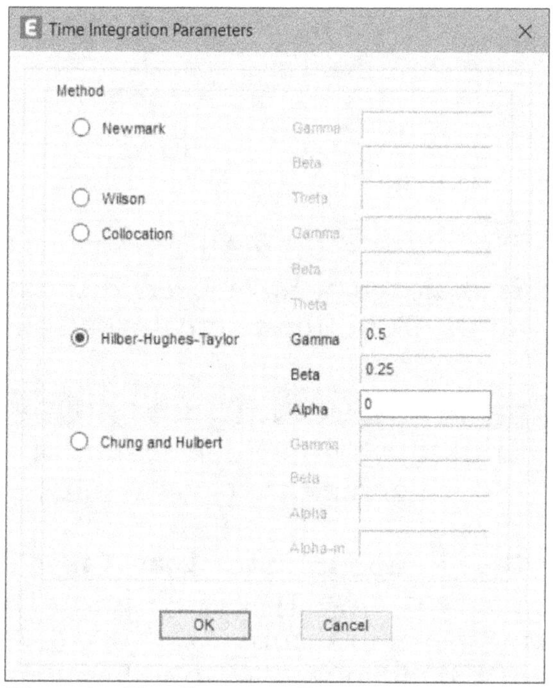

Figure-21. Time Integration Parameters dialog box

- Select desired radio button to specify time integration method. Set the respective parameters and click on the **OK** button. The **Load Case Data** dialog box will be displayed again.
- Set the other parameters in the **Load Case Data** dialog box as discussed earlier.

Setting Nonlinear Direct Integration Time History Load Case

- Select the **Time History** option from the **Load Case Type** drop-down and select the **Nonlinear Direct Integration** option from the **Subtype** drop-down. The options will be displayed as shown in Figure-22.

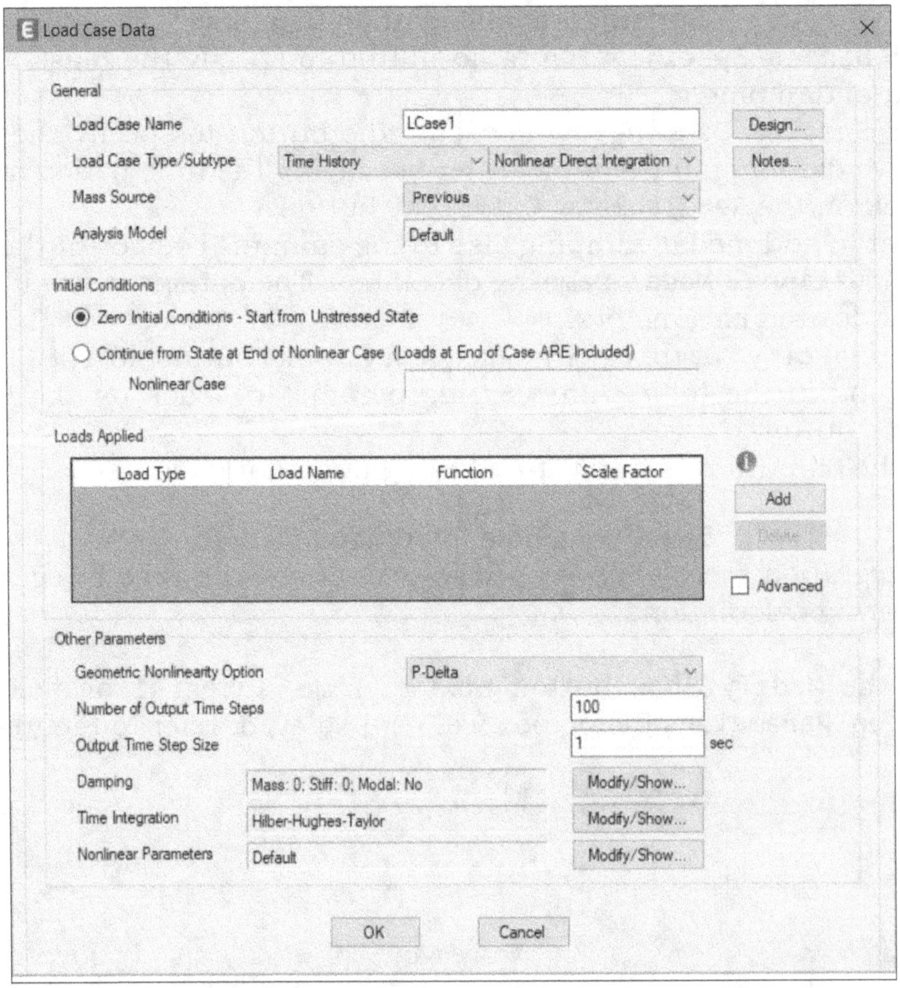

Figure-22. Load Case Data dialog box for Time History Nonlinear Direct Integration load case

- All the options in this dialog box have been discussed earlier. Set desired parameters and click on the **OK** button to create load case.

Creating Buckling Load Case

The Buckling load case is used when sudden failure of structure occurs due to buckling. The procedure to create this case is given next.

- Click on the **Add New Case** button from the **Load Cases** dialog box and select the **Buckling** option from the **Load Case Type** drop-down. The **Load Case Data** dialog box will be displayed as shown in Figure-23.
- Specify desired number of buckling modes to be found in the **Number of Buckling Modes** edit box.
- Specify desired value of convergence tolerance in the **Eigenvalue Convergence Tolerance** edit box.
- Set the other parameters in the dialog box as discussed earlier.
- After setting the parameters, click on the **OK** button.

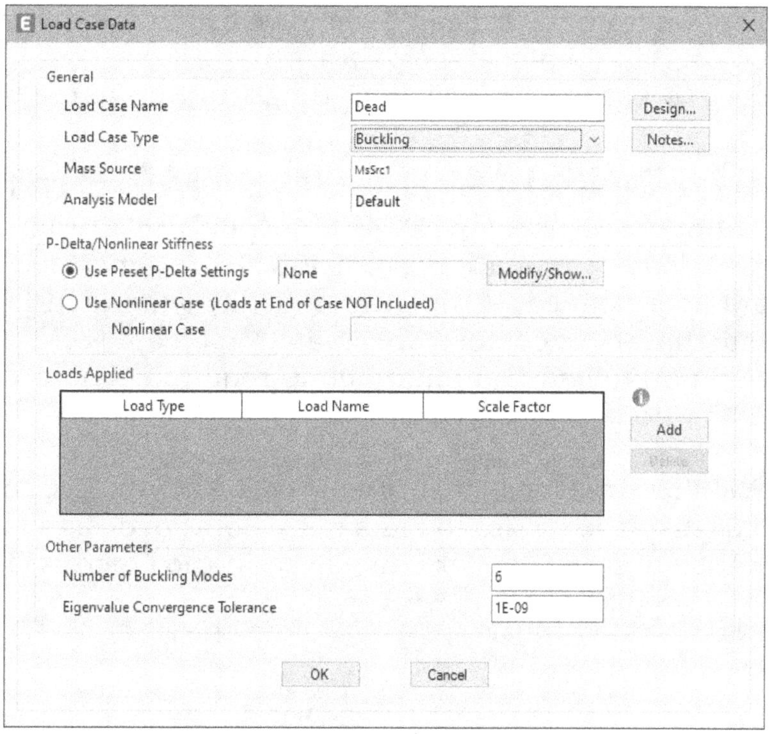

Figure-23. Load Case Data dialog box for Buckling load case

Creating Hyperstatic Load Case

A hyperstatic load case calculates the linear response of the structure with all supports removed and loaded only by the reactions from another linear static load case. This is typically used to calculate the secondary forces under prestress loading. The procedure to create this case is given next.

- Click on the **Add New Case** button from the **Load Cases** dialog box and select the **Hyperstatic** option from the **Load Case Type** drop-down. The **Load Case Data** dialog box will be displayed as shown in Figure-24.

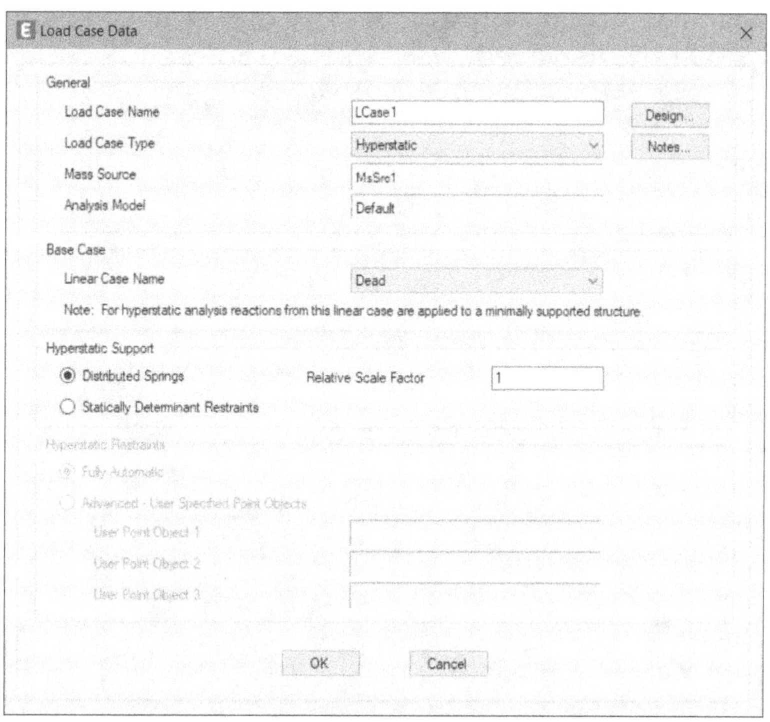

Figure-24. Load Case Data dialog box for Hyperstatic load case

- Select desired option from the **Linear Case Name** drop-down to set the load whose reaction will be used for analysis.
- After setting desired parameters, click on the **OK** button to create load case.

SETTING LOAD COMBINATIONS

The load combinations are used by ETABS to perform automatic designing of structure. In ETABS, load combinations, or combos, are generated automatically by the program or are user defined. If the automatically generated load combinations are acceptable (note that these combos are created/recreated after each design run), no definition of additional load combinations is required. The procedure to create and edit load combinations is given next.

- Click on the **Load Combinations** tool from the **Define** menu. The **Load Combinations** dialog box will be displayed; refer to Figure-25. Some of design load combinations created automatically by ETABS are displayed in the **Combinations** list box.

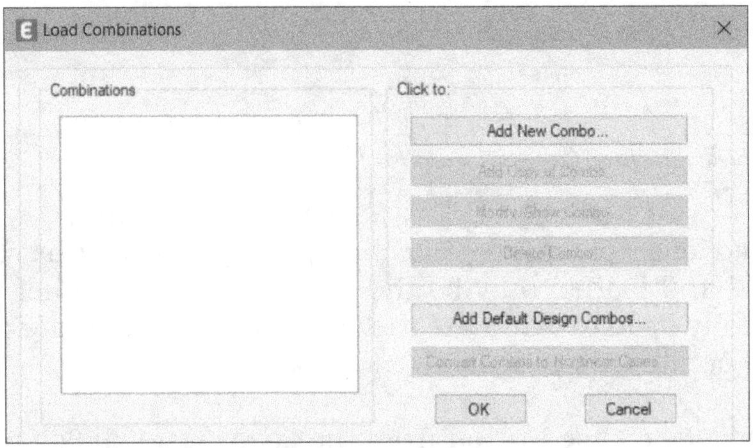

Figure-25. Load Combinations dialog box

Adding New Combination

- Click on the **Add New Combo** button from the dialog box. The **Load Combination Data** dialog box will be displayed; refer to Figure-26.

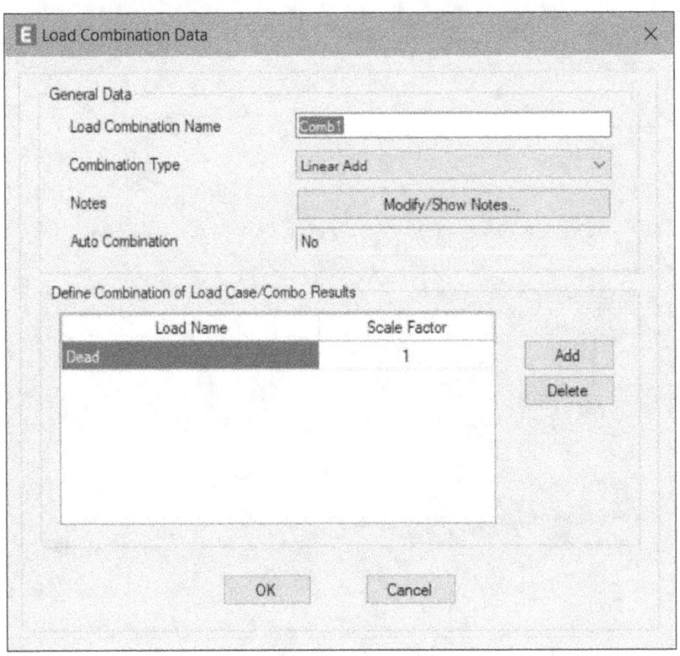

Figure-26. Load Combination Data dialog box

- Specify desired name in the **Load Combination Name** edit box.
- Select desired combination from the **Combination Type** drop-down. Select the **Linear Add** option to perform algebraic addition of loads. Select the **Envelope** option to set combo maximum, the maximum of all of the maximum values for each of the included analysis cases. Similarly, the combo minimum is the minimum of all of the minimum values for each of the included analysis cases. Select the **Absolute Add** option to set combo maximum, the sum of the larger absolute values for each of the included analysis cases. The combo minimum is the negative of the combo maximum. Select the **SRSS** option to set combo maximum, the square root of the sum of the squares of the larger absolute values for each of the included analysis cases. The combo minimum is the negative of the combo maximum. Select the **Range Add** option to set combination maximum, the sum of the positive maximum values from each of the contributing cases (a case with a negative maximum does not contribute), and the combined minimum, the sum of the negative minimum values from each of the contributing cases (a case with a positive minimum does not contribute). This combo type is useful for pattern or skip-type loading where all permutations of the contributing load case must be considered.
- Click on the **Modify/Show Notes** button to add or display notes to the load combination.
- Select desired load pattern from the **Load Name** column in the table for first load; refer to Figure-27. Specify desired factor value in the **Scale Factor** column.

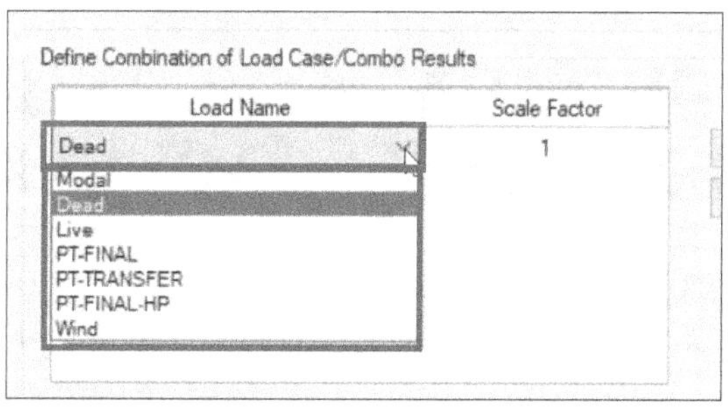

Figure-27. Selecting load

- Click on the **Add** button from the right of the table to add a new load and modify its parameters as discussed earlier.
- Set the other parameters as required and click on the **OK** button. The combination will be added in the **Combinations** list box.

Adding Default Design Combinations in the List

The procedure to add default design combinations is given next.

- Click on the **Add Default Design Combos** button from the dialog box to add default design combinations. The **Add Default Design Combinations** dialog box will be displayed; refer to Figure-28.
- Select desired check box to include load combinations from the different design codes. Like, select the **Steel Frame Design** check box to include default load combinations of steel frame design in the list.

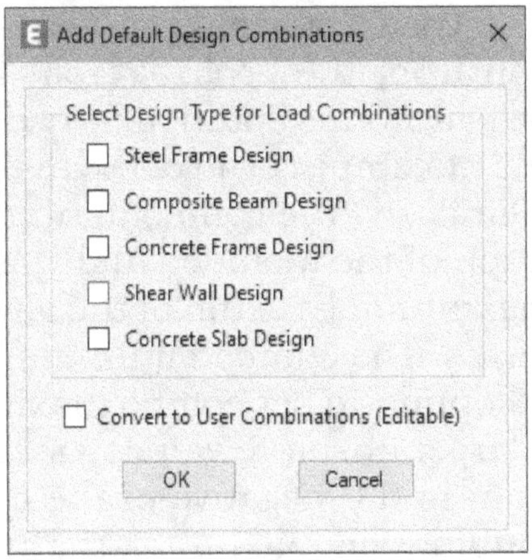

Figure-28. Add Default Design Combinations dialog box

- Select the **Convert to User Combinations (Editable)** check box to convert the default load combinations to user combinations so that they can be edited.
- After setting desired parameters, click on the **OK** button.

Converting Combination to Nonlinear Case

The **Convert Combos to Nonlinear Cases** button is used to convert selected load combination into a nonlinear case. Select desired combination and click on the **Convert Combos to Nonlinear Cases** button to set it as a no nlinear case.

CREATING AUTO CONSTRUCTION SEQUENCE LOAD CASE

An auto sequential construction case provides an automated method of creating static nonlinear staged construction cases that are specifically tailored to model construction sequence loading. Note that only one auto sequential construction case can be defined per model file. The procedure to create auto construction sequence load case is given next.

- Click on the **Auto Construction Sequence Case** tool from the **Define** menu. The **Auto Construction Sequence Load Case** dialog box will be displayed.
- Select the **Case is Active** check box to activate the case and options in this dialog box. The dialog box will be displayed as shown in Figure-29.
- Select desired option from the **Geometric Nonlinearity Option** drop-down to set the geometric non-linearity effect on all elements in the structure. Select the **None** option so that all equilibrium equations are considered in the undeformed configuration of the structure. Select the **P-Delta** option so that the equilibrium equations take into partial account the deformed configuration of the structure. Tensile forces tend to resist the rotation of elements and stiffen the structure, and compressive forces tend to enhance the rotation of elements and destabilize the structure. This may require a moderate amount of iteration. Select the **P-Delta plus Large Displacements** option so that all equilibrium equations are written in the deformed configuration of the structure. This may require a large amount of iteration.

- Specify desired value for number of stories to be combined in each construction sequence group in the **Combine this number of Stories in each Construction Sequence Group** edit box. Enter the value for the number of stories in each sequence. This means the structure will start with n stories in the first stage, n more stories will be added in the second stage for a total of 2n stories, and so on.
- Select the **Exclude this Group Until the Last Step** check box to exclude the group selected in the adjacent drop-down until last step analysis is performed.
- Select desired load pattern from the **Load Pattern Name** column and set the scale factor. Click on the **Add** button to add more loads.
- Select the **Replace Dead Type Load Cases with this Load Case in all Default Design Combinations** check box to set the combination created in this dialog box as default load combination.
- After setting desired parameters, click on the **OK** button.

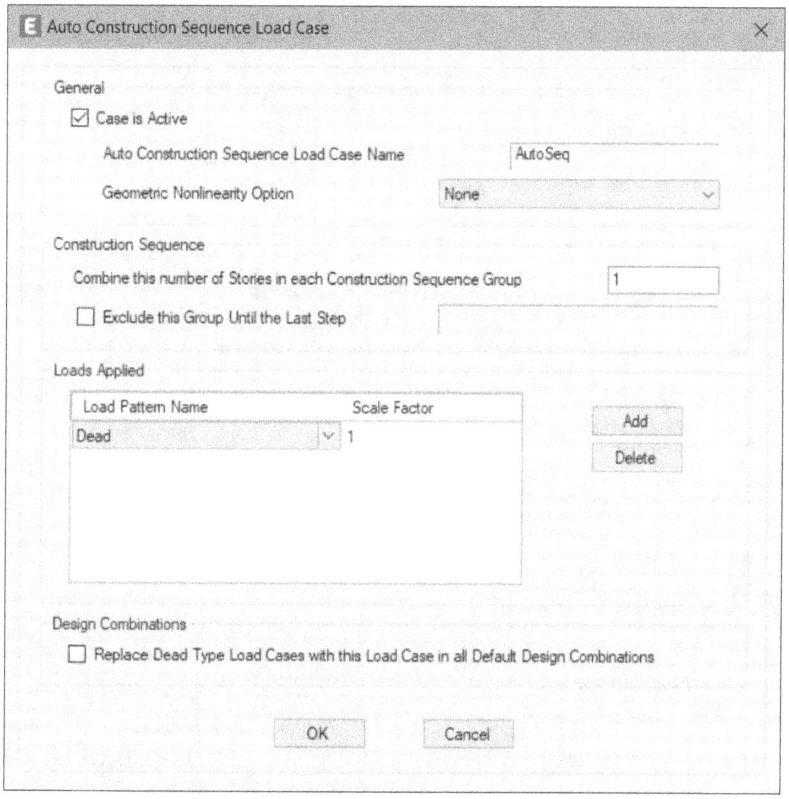

Figure-29. Auto Construction Sequence Load Case dialog box

CREATING WALKING VIBRATIONS

Walking vibrations are caused when people walk in the building. The procedure to create walking vibration load is given next.

- Click on the **Walking Vibrations** tool from the **Define** menu. The **Walking Vibrations** dialog box will be displayed; refer to Figure-30.
- Click on the **Add Walking Vibration** button from the dialog box. The **Walking Vibration Data** dialog box will be displayed; refer to Figure-31.
- Set desired name in the **Name** edit box.
- Select desired story from the **Story** drop-down where you want to apply walking vibrations.
- Set desired color for displaying walking vibration load case and specify desired notes.

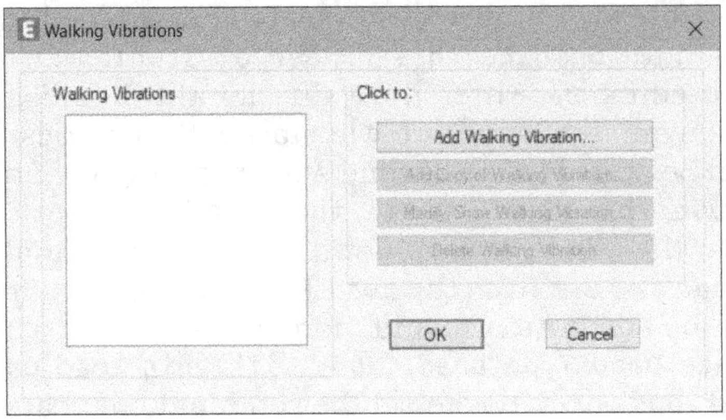

Figure-30. Walking Vibrations dialog box

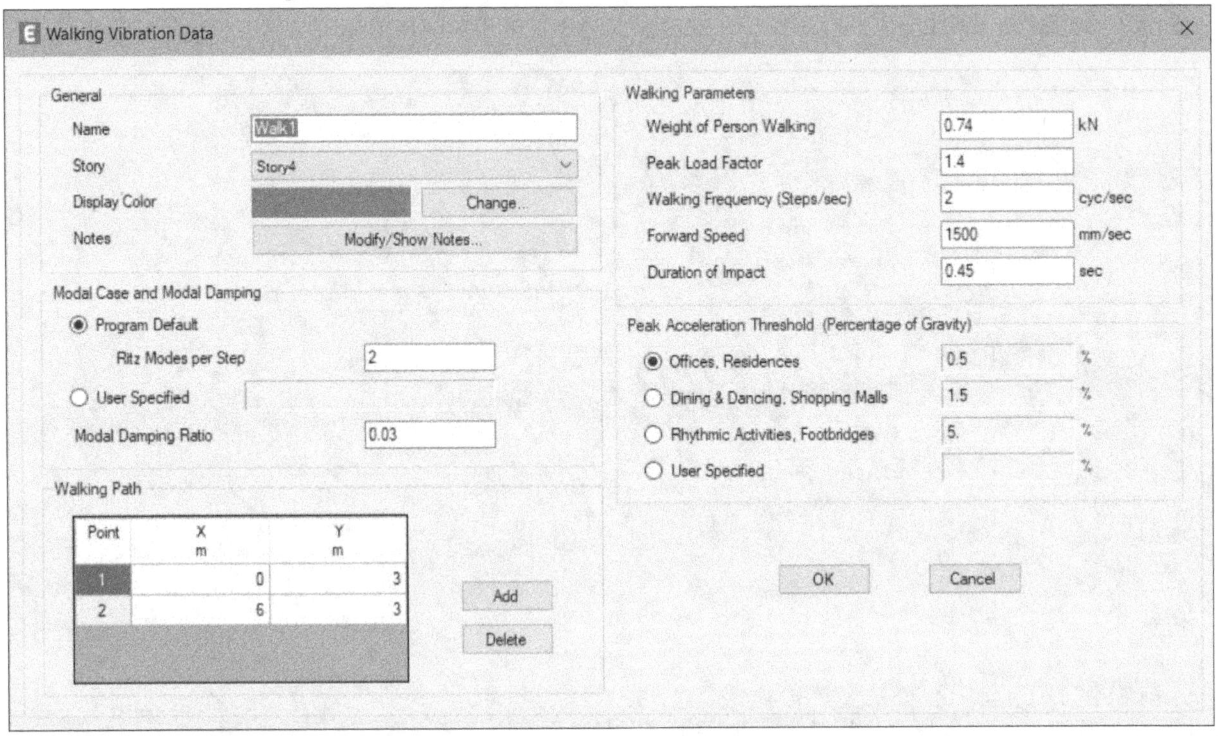

Figure-31. Walking Vibration Data dialog box

- Select the **Program Default** radio button so that program creates a modal case with the default P-delta using two Ritz modes for each step of the walking vibration case. This option uses the mass as defined in the Default Mass Source so make sure it includes vertical mass. Specify desired value for Ritz modes per step in **Ritz Modes per Step** edit box.
- Select the **User Specified** radio button to set user defined modal case and damping. Select desired option from the adjacent drop-down and specify value of modal damping ratio in the **Modal Damping Ratio** edit box.
- Double-click in desired field of table in **Walking Path** area of the dialog box to edit a walking path point location. Click on the **Add** button from the area to add a new walking path point.
- Specify the average weight of person walking through the building in the **Weight of Person Walking** edit box.
- Set desired value of peak load factor in the **Peak Load Factor** edit box to define the maximum load exerted due to walking.
- Specify the number of walking steps per second value in the **Walking Frequency (Steps/sec)** edit box.
- Specify desired value of moving speed in the **Forward Speed** edit box.

- Set the duration in seconds up to which there will be impact of foot steps on the floor in **Duration of Impact** edit box.
- Select desired radio button from the **Peak Acceleration Threshold (Percentage of Gravity)** area of the dialog box.
- After setting desired parameters, click on the **OK** button. A new walking vibration load will be added in the **Walking Vibrations** list box of **Walking Vibrations** dialog box. After setting desired number of walking vibrations, click on the **OK** button to apply loads.

CHECKING MODEL FOR ANALYSIS

The **Check Model** tool in the **Analyze** menu is used to check if all the parameters have been specified for performing analysis on the model. The procedure to use this tool is given next.

- Click on the **Check Model** tool from the **Analyze** menu. The **Check Model** dialog box will be displayed; refer to Figure-32.
- Select the check boxes from various Checks areas.
- Select the **Trim or Extend Frames and Move Joints to Fix Problems** check box to perform trimming and extending of frames for frame members that overlap, and move joints to fix problem.
- Select the **Joint Story Assignment** check box to assign joints to stories in the building automatically.

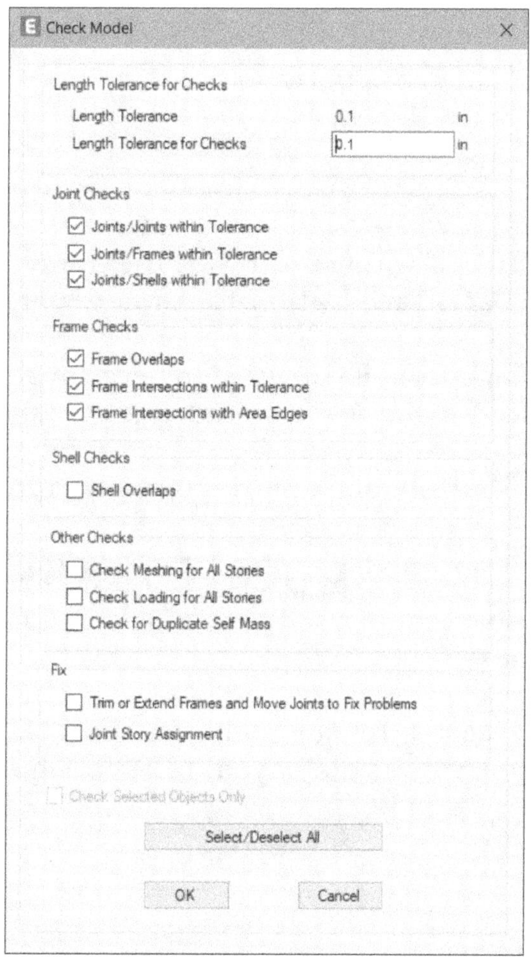

Figure-32. Check Model dialog box

- After setting desired parameters, click on the **OK** button. The **Warning** box will be displayed with warnings if there is any warning in the model otherwise the **Warning** box will be displayed as shown in Figure-33.

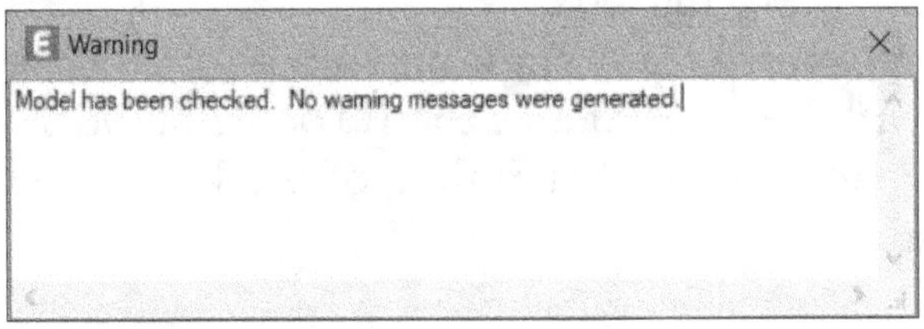

Figure-33. Warning box

SETTING DEGREE OF FREEDOM FOR ANALYSIS

The **Set Active Degrees of Freedom** tool is used to set the degree of freedom up to which the analysis will be performed on the building. The procedure to use this tool is given next.

- Click on the **Set Active Degrees of Freedom** tool from the **Analyze** menu. The **Active Degrees of Freedom** dialog box will be displayed; refer to Figure-34.

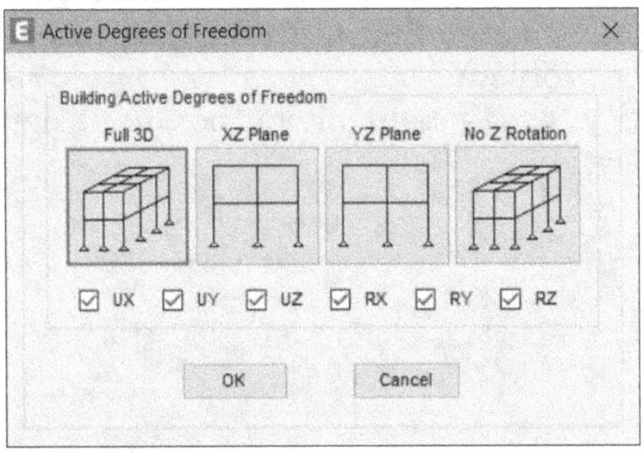

Figure-34. Active Degrees of Freedom dialog box

- Select the **Full 3D** button if you want to run the analysis in 3D.
- Select the **XZ Plane** or **YZ Plane** button to run a 2D analysis in XZ or YZ plane, respectively.
- Select the **No Z Rotation** button if you want to perform analysis in 3D environment where model cannot rotate about Z axis but can translate and rotate in all other directions.
- After selecting desired button, click on the **OK** button from the dialog box.

SETTING LOAD CASES TO RUN ANALYSIS

The **Set Load Cases To Run** tool is used to run selected load cases. The procedure to use this tool is given next.

- Click on the **Set Load Cases To Run** tool from the **Analyze** menu. The **Set Load Cases to Run** dialog box will be displayed; refer to Figure-35.

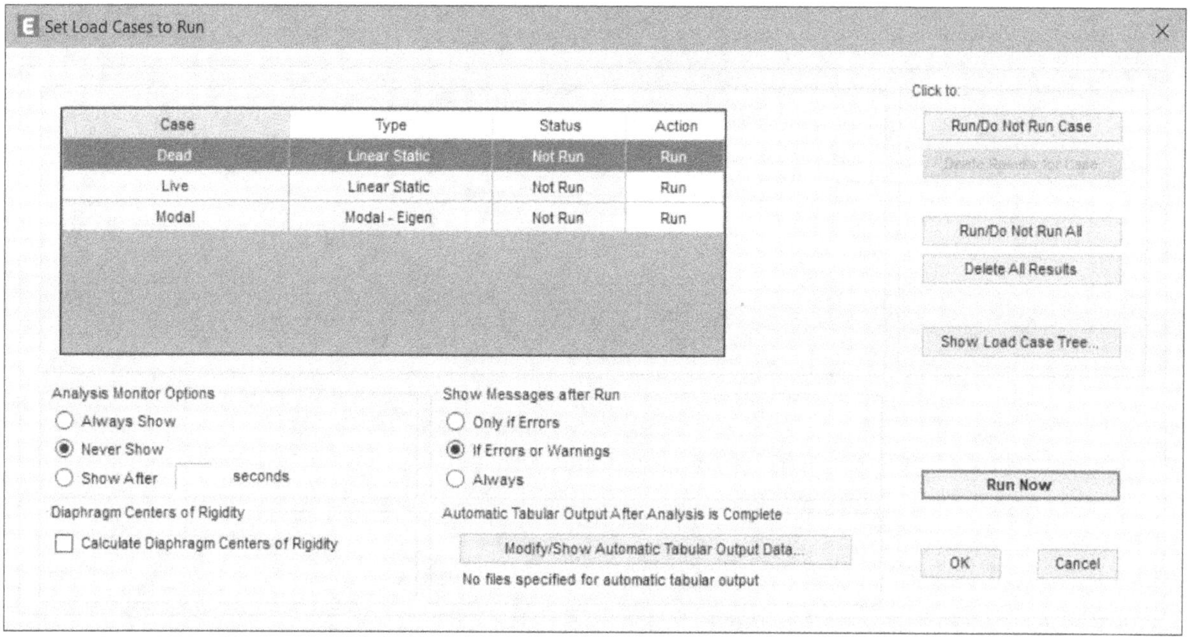

Figure-35. Set Load Cases to Run dialog box

- Select the case that you want to run or do not run from the table. Click on the **Run/Do Not Run Case** button to switch between running and not running the selected load case.
- Click on the **Run/Do Not Run All** button to switch between running and not running all the load cases.
- Select desired radio button from the **Analysis Monitor Options** area of the dialog box. Select the **Always Show** radio button to display the progress of analysis in the **Analysis** window. Select the **Show After** radio button and specify the time after which the **Analysis** window will be displayed in the adjacent edit box. Select the **Never Show** radio button to do not show the progress of analysis.
- Select the **Calculate Diaphragm Centers of Rigidity** check box to calculate the center of rigidity during the analysis. The original concept of center of rigidity dates back to manual rigidity analysis techniques associated with the lateral analysis of single-story shear wall buildings. Modern computer techniques do not require the explicit evaluation of the center of rigidity. However, the center of rigidity still needs to be evaluated because some building codes refer to it as a reference point to define design eccentricity requirements in multistory buildings.
- If you want to save data specified in this table in tabular form in a Microsoft Access or XML file then select the **Automatically save tables to Microsoft Access or XML after run completes** check box. Specify desired file name and location of file in the **Filename** edit box.
- Click on the **OK** button to set the load cases.

SETTING SAPFIRE SOLVER OPTIONS

The **Advanced SAPFire Options** tool is used to set the solver for analysis. The procedure to use this tool is given next.

- Click on the **Advanced SAPFire Options** tool from the **Analyze** menu. The **Advanced SapFire Options** dialog box will be displayed; refer to Figure-36.

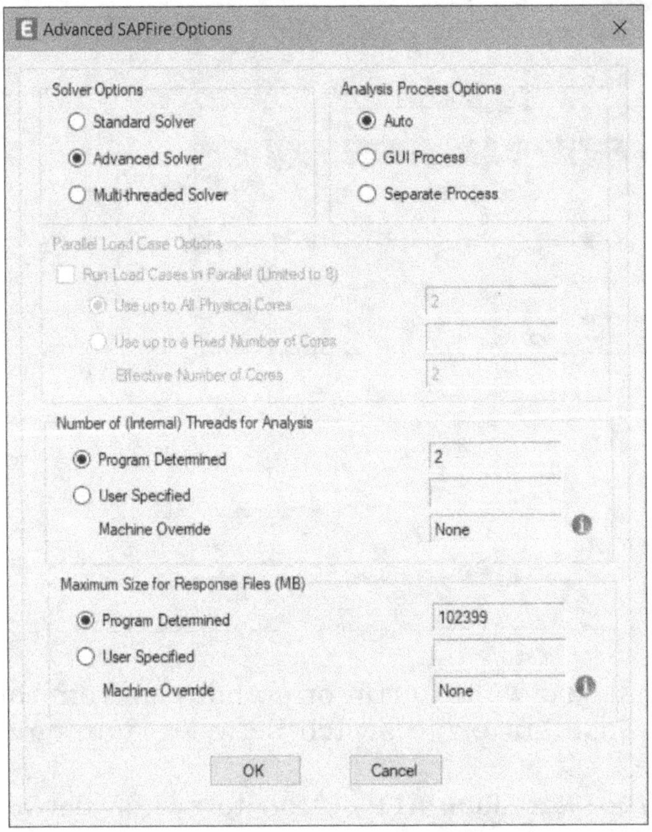

Figure-36. Advanced SapFire Options dialog box

- Select the **Standard Solver** radio button if you want to use only single core and check for model stability. This solver is used for smaller problems.
- Select the **Advanced Solver** radio button if you want to run analysis under default settings for medium to large problems. This option uses all cores of your CPUs.
- Select the **Multi-threaded Solver** radio button to solve medium to large problems. This option utilizes all CPU cores and full RAM during the analysis.
- Select desired options from the **Analysis Process Options** area. Description of these options is given next.

GUI process

- Best for small problems
- Analysis runs within the software, such as with SAFE.exe
- Benefit: less disk operations (I/O) are performed
- Drawback: the software itself consumes memory, leaving less available for analysis, which slows operations and prevents the ability to run larger models

Separate process

- Best for medium to large problems
- The analysis model is written to the disk and read by CSI.SAPFire.Driver.exe, then analysis is run within CSI.SAPFire.Driver.exe
- Benefit: the analysis engine has access to more memory, therefore larger problems can be solved and analysis runs faster
- Drawback: time is lost to write and read the analysis model

Auto

- Default setting
- The necessary memory is estimated, then compared to the physical ram available. If enough RAM is available, the analysis runs in GUI process. If not, it is shelled out to CSI.SAPFire.Driver.exe.

- If you have selected the **Separate Process** radio button from the **Analysis Process Options** area then you can set the parameters in **Parallel Load Case Options** area to define how many physical cores of processor can be used for analysis.
- After setting desired parameters in the dialog box and click on the **OK** button.

AUTOMATIC MESH SETTINGS FOR FLOORS

The **Automatic Mesh Settings for Floors** tool is used to set mesh size for floors in the building. The procedure to use this tool is given next.

- Click on the **Automatic Mesh Settings for Floors** tool from the **Analyze** menu. The **Automatic Mesh Options** dialog box will be displayed; refer to Figure-37.
- By default, the **General Mesh** radio button is selected and hence standard meshing of ETABS is used. The mesh generated will predominantly use 4 node quadrilateral elements, supplemented with triangular elements where needed.
- Select the **Rectangular Mesh** radio button to create rectangular mesh. Select the **Use Localized Meshing** check box if you do not want to create dense meshes in large object. On selecting this check box, mesh is created while passing through as many joints as possible while keeping mesh lines at least half the specified **Approximate Maximum Mesh Size** apart. Select the **Merge Joints Where Possible** check box to automatically merge joints which are overlapping or close to each other.
- Set desired mesh size in the **Approximate Maximum Mesh Size** edit box in feet.
- Set desired parameters and click on the **OK** button. The specified size of mesh will be applied.

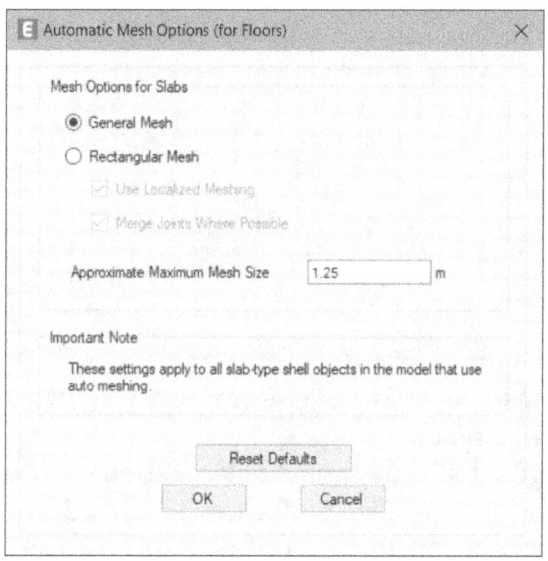

Figure-37. Automatic Mesh Options (for Floors) dialog box

SETTING MESH SIZE FOR WALLS

The **Automatic Rectangular Mesh Settings for Walls** tool is used to specify the size of mesh elements for walls. The procedure to use this tool is given next.

- Click on the **Automatic Rectangular Mesh Settings for Walls** tool from the **Analyze** menu. The **Automatic Rectangular Mesh Options (for Walls)** dialog box will be displayed; refer to Figure-38.
- Specify desired mesh size in the edit box and click on the **OK** button.

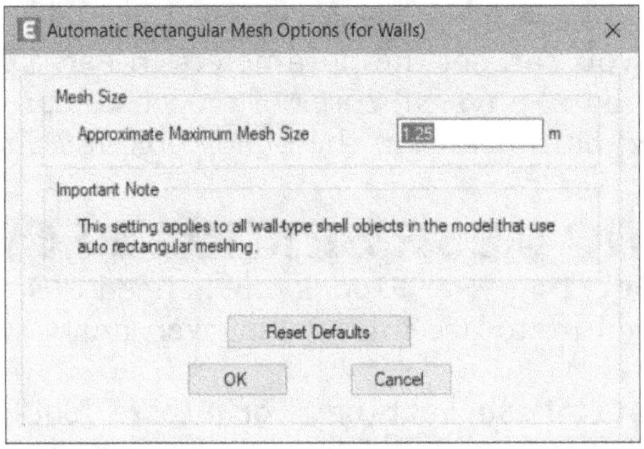

Figure-38. Automatic Rectangular Mesh Options (for Walls) dialog box

SETTING HINGES FOR ANALYSIS MODEL

The **Analysis Model for Nonlinear Hinges** tool is used to define how hinges will work in the analysis model. The procedure to use this tool is given next.

- Click on the **Analysis Model for Nonlinear Hinges** tool from the **Analyze** menu. The **Analysis Model for Nonlinear Hinges** dialog box will be displayed; refer to Figure-39.

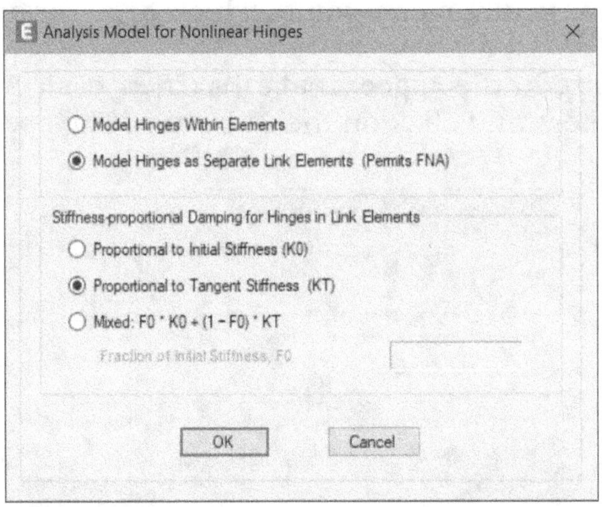

Figure-39. Analysis Model for Nonlinear Hinges dialog box

- Select the **Model Hinges Within Elements** radio button to make hinges integral part of frame and wall objects.
- Select the **Model Hinges as Separate Link Elements (Permits FNA)** radio button to model hinges with frames and walls using an additional link element. These hinges may be used in all nonlinear analyses, including the often faster and more stable Fast Nonlinear Analysis (FNA). On selecting this radio button, the radio buttons in the **Stiffness-proportional Damping for Hinges in Link Elements** area will become active.
- Select desired radio button to define damping for hinges.
- After setting desired parameters, click on the **OK** button.

CRACKING ANALYSIS OPTIONS

The **Cracking Analysis Options** tool is used to set options for performing cracking analysis. The procedure to use this tool is given next.

- Click on the **Cracking Analysis Options** tool from the **Analyze** menu. The **Reinforcement Options For Cracking Analysis** dialog box will be displayed; refer to Figure-40.
- Select desired option from the **Reinforcement Source** area to define source of reinforcement bars.
- Specify desired minimum reinforcing ratio for tension and compression in the respective edit boxes. Based on the specified value, the cracking analysis will be performed.

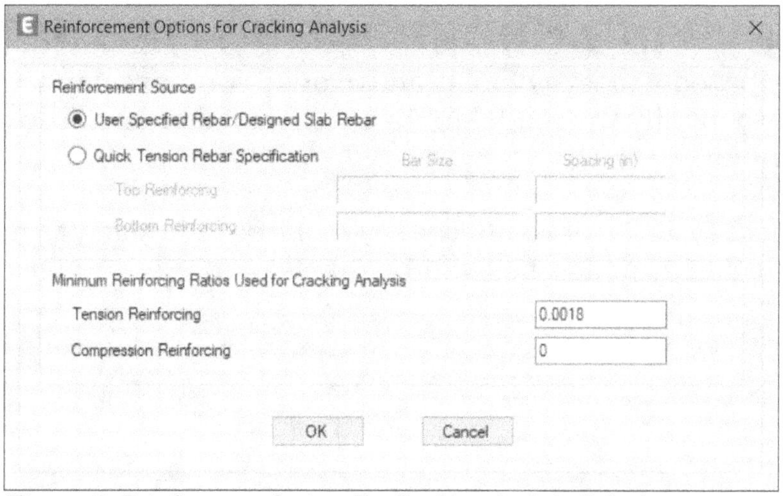

Figure-40. Reinforcement Options For Cracking Analysis dialog box

RUNNING THE ANALYSIS

The **Run Analysis** tool in **Analyze** menu is used to perform analysis. The procedure to perform analysis is given next.

- After setting all the parameters, click on the **Run Analysis** tool from the **Analyze** menu. The analysis will be performed and result will be displayed in selected viewport; refer to Figure-41.

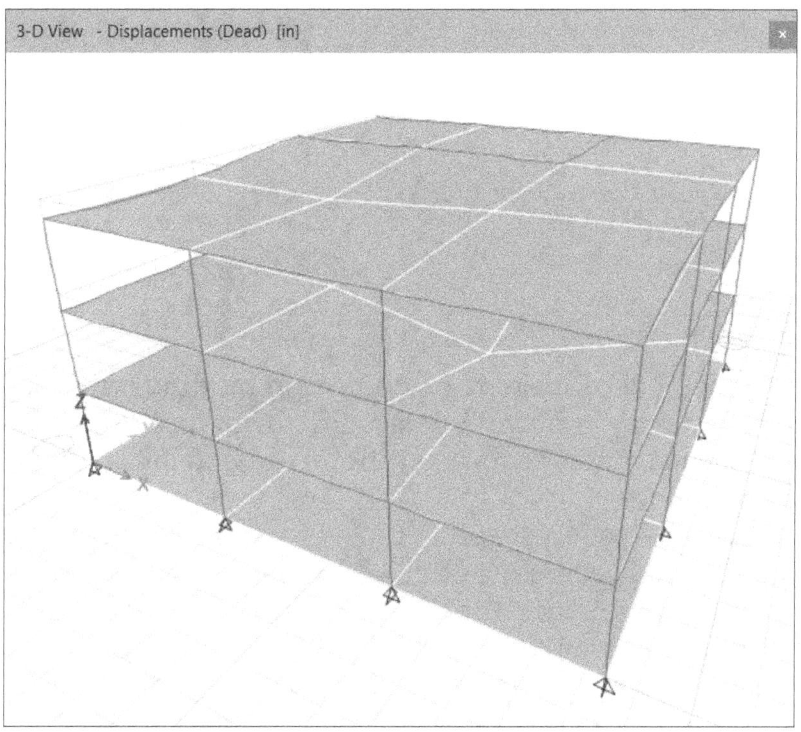

Figure-41. Result of analysis in viewport

APPLYING MODEL ALIVE

The **Model Alive** tool in the **Analyze** menu is a toggle button used to perform analysis automatically when the model is changed. The procedure is given next.

- Click on the **Model Alive** tool from the **Analyze** menu to toggle it on. The analysis mode will become active. Change the model as required and you will see the results accordingly.

MODIFYING UNDEFORMED GEOMETRY

The **Modify Undeformed Geometry** tool is active once you have performed the analysis. This tool is used to modify the model in its undeformed shape. The procedure to use this tool is given next.

- Click on the **Modify Undeformed Geometry** tool from the **Analyze** menu. The **Modify Undeformed Geometry** dialog box will be displayed; refer to Figure-42.

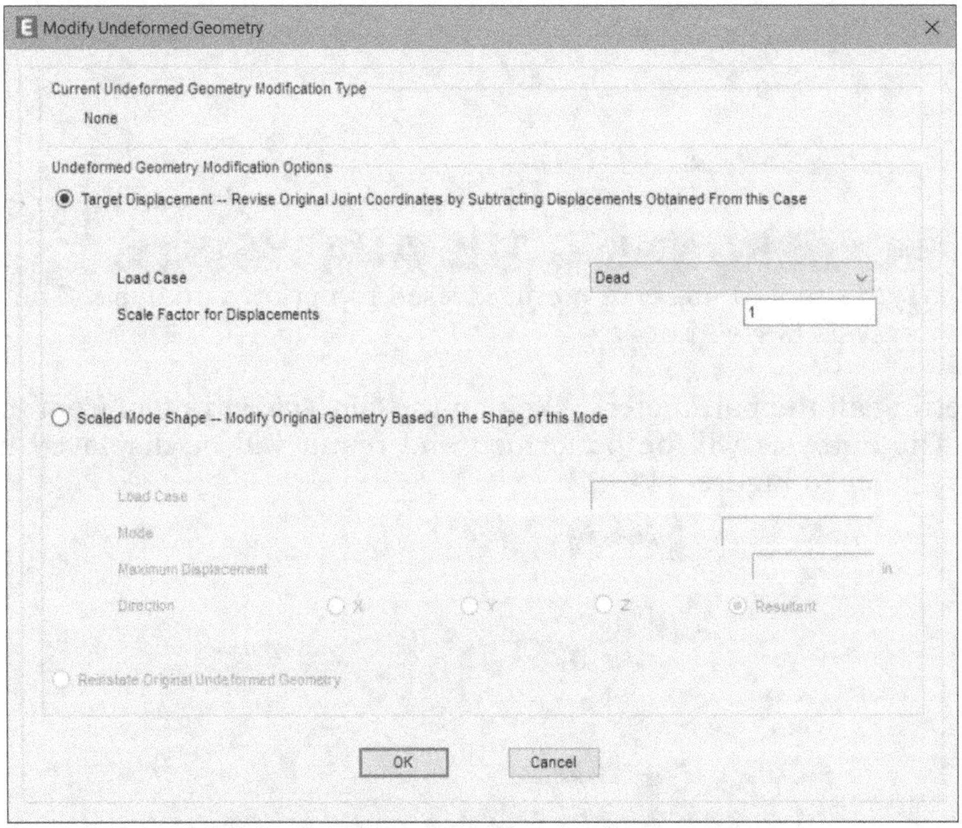

Figure-42. Modify Undeformed Geometry dialog box

- Selecting the **Target Displacement** radio button modifies the original geometry of the structure in order to achieve a desired deformed shape. When this option is selected, use the **Load Case** drop-down list to select the load case for which the geometry is to be modified.
- Selecting the **Scaled Mode Shape** radio button modifies the geometry of the structure based on a specified modal or buckling load case mode shape. This option is useful for modifying the input geometry to consider things such as imperfections. When this option is selected, use the **Load Case** drop-down list to select desired load case and the **Mode** drop-down list to select the associated

mode. The maximum displacement value represents the maximum change to be applied to the geometry, in the specified direction. The mode shape is scaled to the specified maximum displacement in the specified direction and used to modify the geometry of the structure. In order to modify the geometry based on another mode shape or to a different maximum displacement, it is necessary to first reinstate the original geometry.

- After setting desired parameters, click on the **OK** button.

CHECKING ANALYSIS LOG

The **Last Analysis Run Log** tool is used to check the details of analysis recently performed. To check the log, click on the **Last Analysis Run Log** tool from the **Analyze** menu. The **Analysis Complete** window will be displayed; refer to Figure-43.

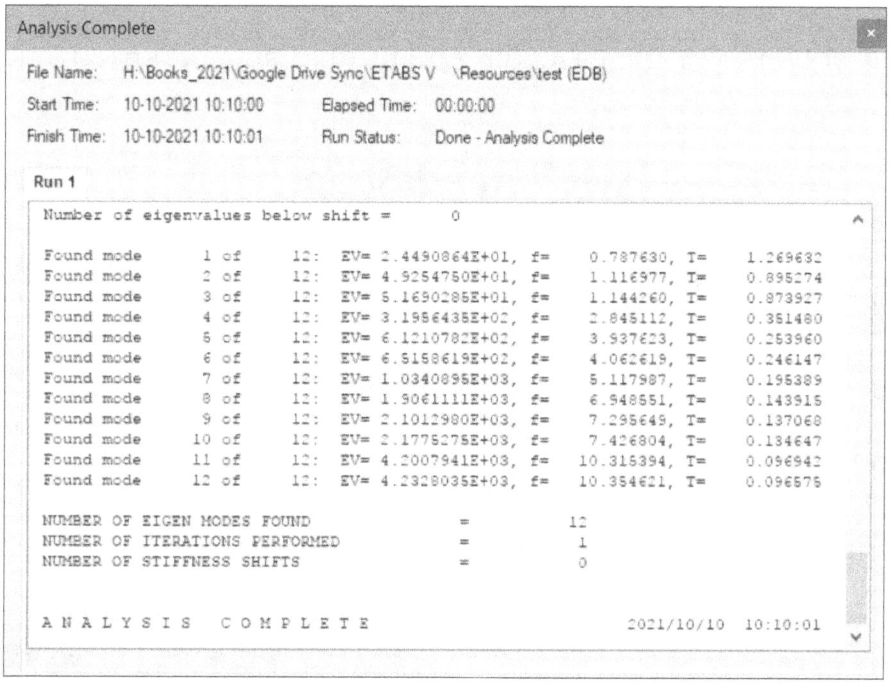

Figure-43. Analysis Complete dialog box

Similarly, you can use the **Show Analysis Messages** tool in **Analyze** menu to check warning messages of the solver after performing analysis.

UNLOCKING MODEL

Once the analysis is performed, the model gets locked for any changes. To perform changes, click on the **Unlock Model** button from the **Analyze** menu.

DISPLAY OPTIONS

The tools in the **Display** menu are used to display various properties of the model.

Undeformed Shape

The **Undeformed Shape** tool is used to view the model in undeformed shape in selected viewport. To use this tool, select desired viewport and click on the **Undeformed Shape** tool from the **Analyze** menu.

Load Assigns

The options in the **Load Assigns** cascading menu of **Display** menu are used to display various load assigns on the model in selected viewport; refer to Figure-44.

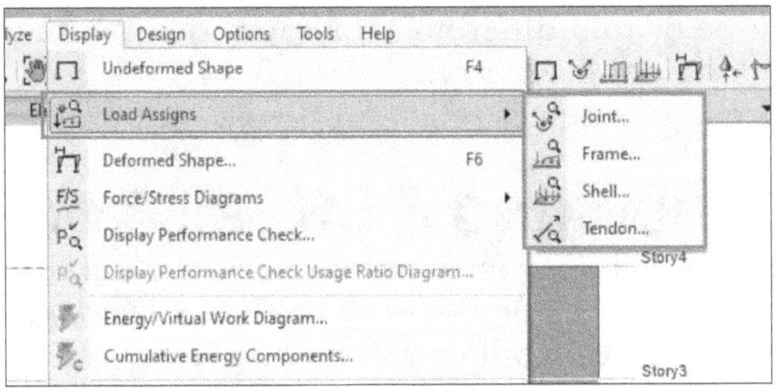

Figure-44. Load Assigns cascading menu

Select desired option from the cascading menu to display respective load assign in the viewport.

Deformed Shape

The **Deformed Shape** tool in the **Display** menu is used to display model in selected viewport as deformed. To use this tool, select the viewport and click on the **Deformed Shape** tool. The model will be displayed in deformed shape.

Force/Stress Diagrams

The tools in the **Force/Stress Diagrams** cascading menu are used to display forces, reactions, and stress on the model; refer to Figure-45. The tools in this menu are discussed next.

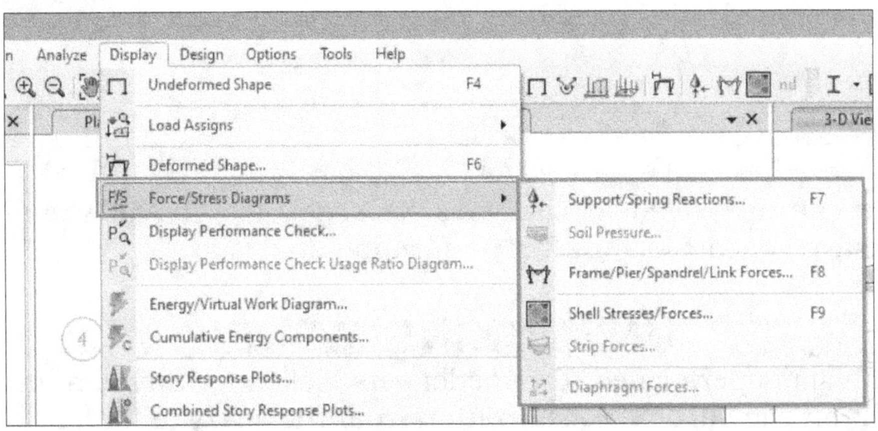

Figure-45. Force Stress Diagrams cascading menu

Support/Spring Reactions

The **Support/Spring Reactions** tool in the **Force/Stress Diagrams** cascading menu is used to display reaction forces of support/spring in the model. The procedure to use this tool is given next.

- Select desired viewport in which you want to display reaction forces. Click on the **Support/Spring Reactions** tool in **Force/Stress Diagrams** cascading menu of the **Display** menu. The **Reactions** dialog box will be displayed; refer to Figure-46.

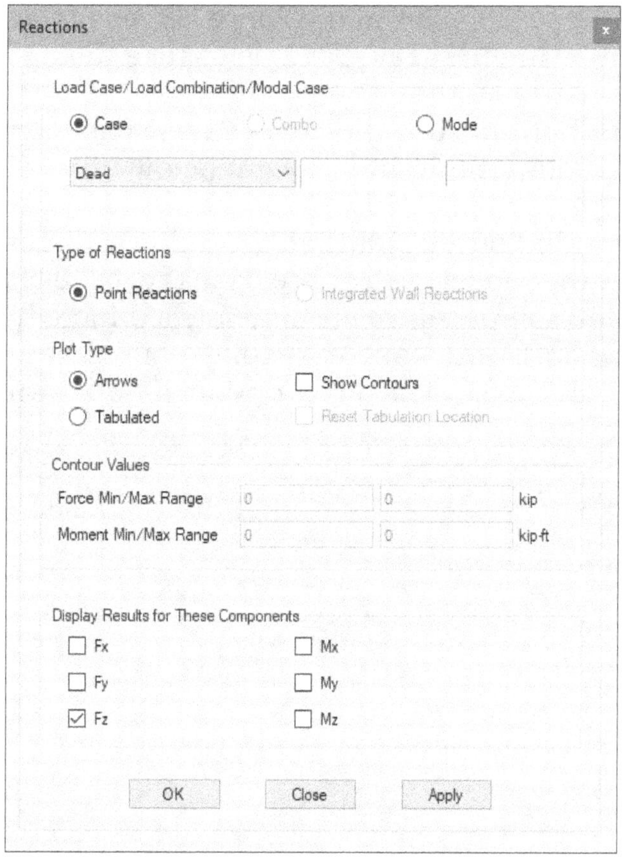

Figure-46. Reactions dialog box

- Select the **Case** radio button to check reaction force of selected load case. Select the **Combo** radio button to check the reaction force of selected load combination. Select the **Mode** radio button to check the reaction force of selected modal case. After selecting desired radio button, set the options in drop-downs of **Load Case/ Load Combination/Modal Case** area.
- Select **Point Reactions** radio button from the **Type of Reactions** area to display reactions at support or joints. Select **Integrated Wall Reactions** radio button to display integrated reactions along wall elements.
- Select desired radio button from the **Plot Type** area to define how reaction forces will be displayed in the viewport. Select the **Arrows** radio button to display reaction forces in the form of arrows. Select the **Tabulated** radio button to display reaction forces in the form of tabular data. Select the **Reset Tabulation Location** check box to reset the locations earlier set in the viewport.
- Click in the **Force Min/Max Range** edit box from the **Contour Values** area and specify desired value to define the minimum and maximum force range.
- Click in the **Moment Min/Max Range** edit box and specify desired value to define the minimum and maximum moment range.
- Select the check boxes for the components that you want to display in the viewport.
- After setting desired parameters, click on the **OK** button to display reaction forces.

Displaying Soil Pressure

The **Soil Pressure** tool in the **Force/Stress Diagrams** cascading menu is used to display soil pressure on the model. Note that you must have assigned an area spring properties to shell elements like slabs and had performed the analysis to display soil pressure in force/stress diagram. The procedure is given next.

- Click on the **Soil Pressure** tool from the **Force/Stress Diagrams** cascading menu of the **Display** menu. The **Display Soil Pressure** dialog box will be displayed; refer to Figure-47.
- Select desired radio button from the **Load Case/Load Combination/Modal Case** area and set the options in the drop-downs below it check soil pressure due to respective loads.
- Select desired option in **Contour Option** drop-down to define how soil pressure will be displayed in the model. Select the **Display on Undeformed Shape** option if you want to display soil pressure on undeformed model. Select the **Display on Deformed Shape** option if you want to display soil pressure in deformed shape. Select the **Display in Extruded Form** option if you want to display extruded contours of soil pressure.

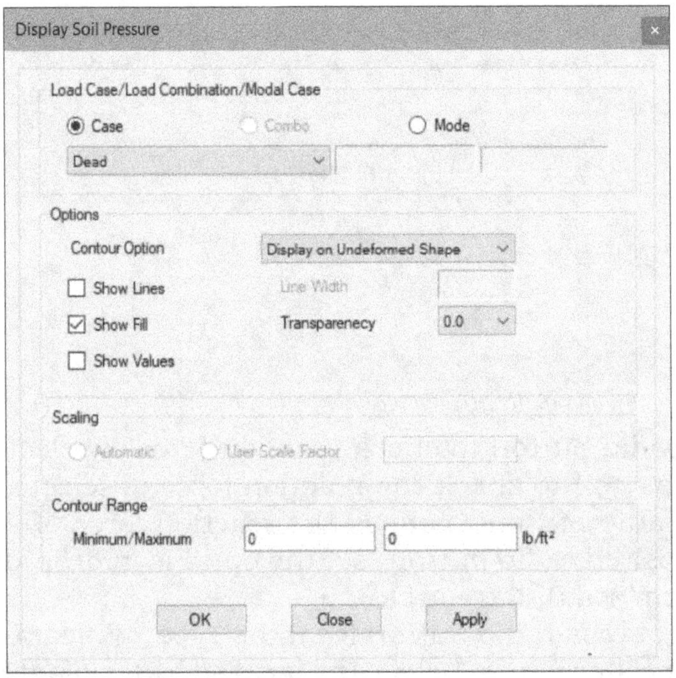

Figure-47. Display Soil Pressure dialog box

- Set desired parameters and click on the **OK** button. The soil pressure plot will be displayed; refer to Figure-48.

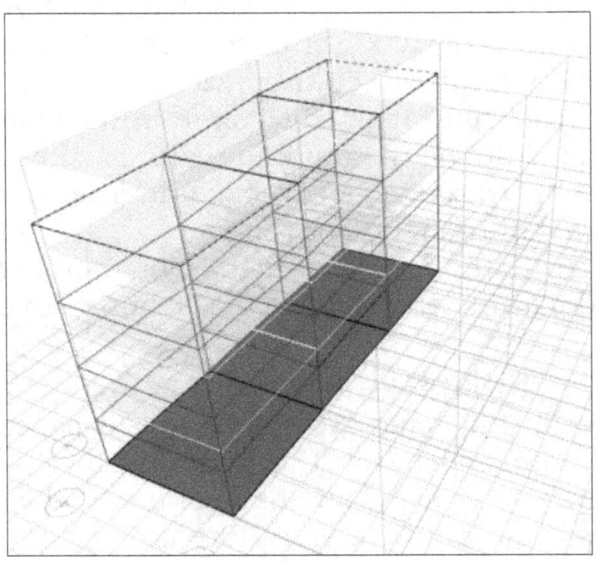

Figure-48. Soil pressure plot

Frame/Pier/Spandrel/Link Forces

The **Frame/Pier/Spandrel/Link Forces** tool is used to display component of selected force or moment in the selected viewport. The procedure is given next.

- Click on the **Frame/Pier/Spandrel/Link Forces** tool from the **Force/Stress Diagrams** cascading menu of the **Display** menu. The **Member Force Diagram for Frames/Piers/Spandrels/Links** dialog box will be displayed; refer to Figure-49.
- Select desired option from the **Component** area to display respective force/moment component in the model.
- Select the objects to be included in the diagram to display forces/moments from the **Include** area.
- Set the other parameters as discussed earlier and click on the **OK** button. The forces/moments will be displayed in the selected viewport.

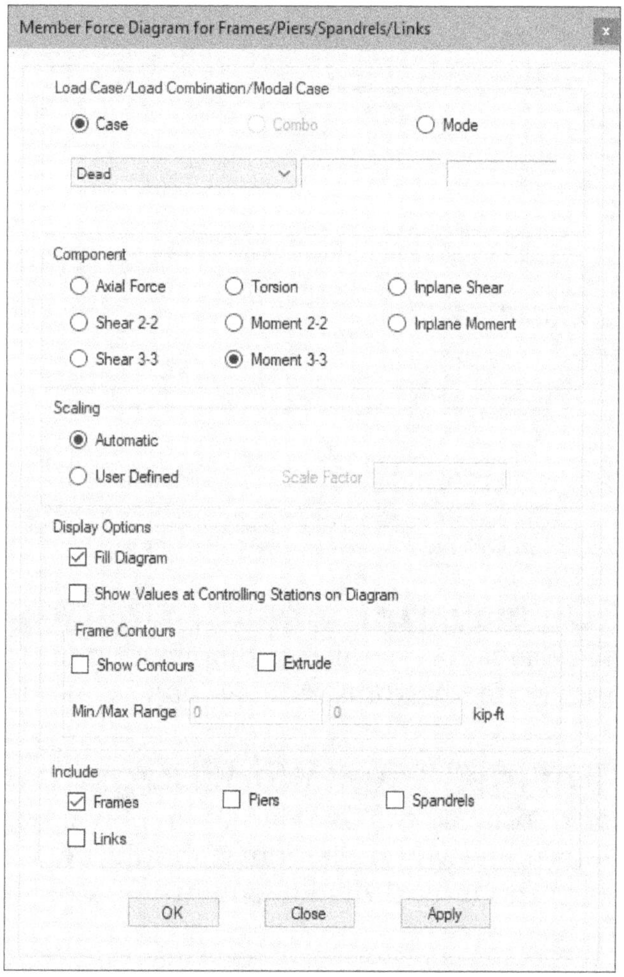

Figure-49. Member Force Diagram for Frames/Piers/Spandrels/Links dialog box

Shell Stresses/Forces

The **Shell Stresses/Forces** tool is used to display the stresses/forces exerted on the shell members. The procedure to use this tool is given next.

- Click on the **Shell Stresses/Forces** tool from the **Force/Stress Diagrams** cascading menu of the **Display** menu. The **Shell Forces/Stresses** dialog box will be displayed; refer to Figure-50.
- Set the parameters as discussed earlier and click on the **OK** button.

Similarly, you can use the **Strip Forces** and **Diaphragm Forces** tools.

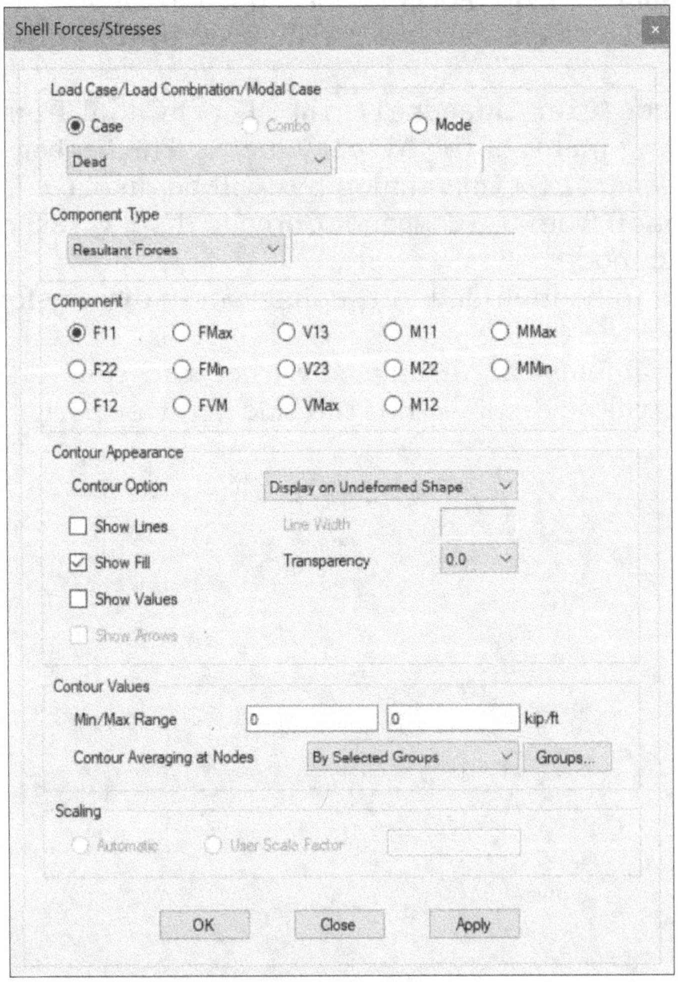

Figure-50. Shell Forces/Stresses dialog box

Display Performance Check

The **Display Performance Check** tool is used to display demand/capacity (D/C) ratios for hinges after a nonlinear time history analysis has been run.

Energy/Virtual Work Diagram

The **Energy/Virtual Work Diagram** tool is used to display energy diagrams that can be used as an aid to determine which elements should be stiffened to achieve the most efficient control over the lateral displacements of a structure. The procedure to use this tool is given next.

- Click on the **Energy/Virtual Work Diagram** tool from the **Display** menu. The **Energy/Virtual Work Diagram** dialog box will be displayed; refer to Figure-51.
- Select desired load cases and load steps from the **Forces** and **Displacement** areas.
- Click on the **OK** button. The viewport will be displayed with forces and displacements annotated.

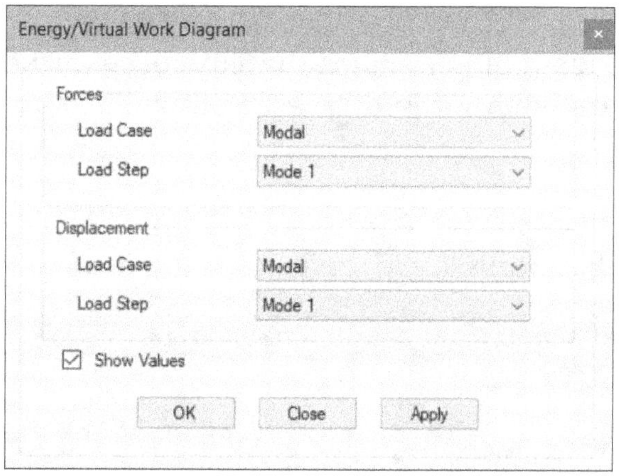

Figure-51. Energy/Virtual Work Diagram dialog box

Cumulative Energy Components

The **Cumulative Energy Components** tool plots all energy components in a cumulative manner after a modal time history analysis has been run. Click on the **Cumulative Energy Components** tool from the **Display** menu. The **Cumulative Energy Components** dialog box will be displayed; refer to Figure-52.

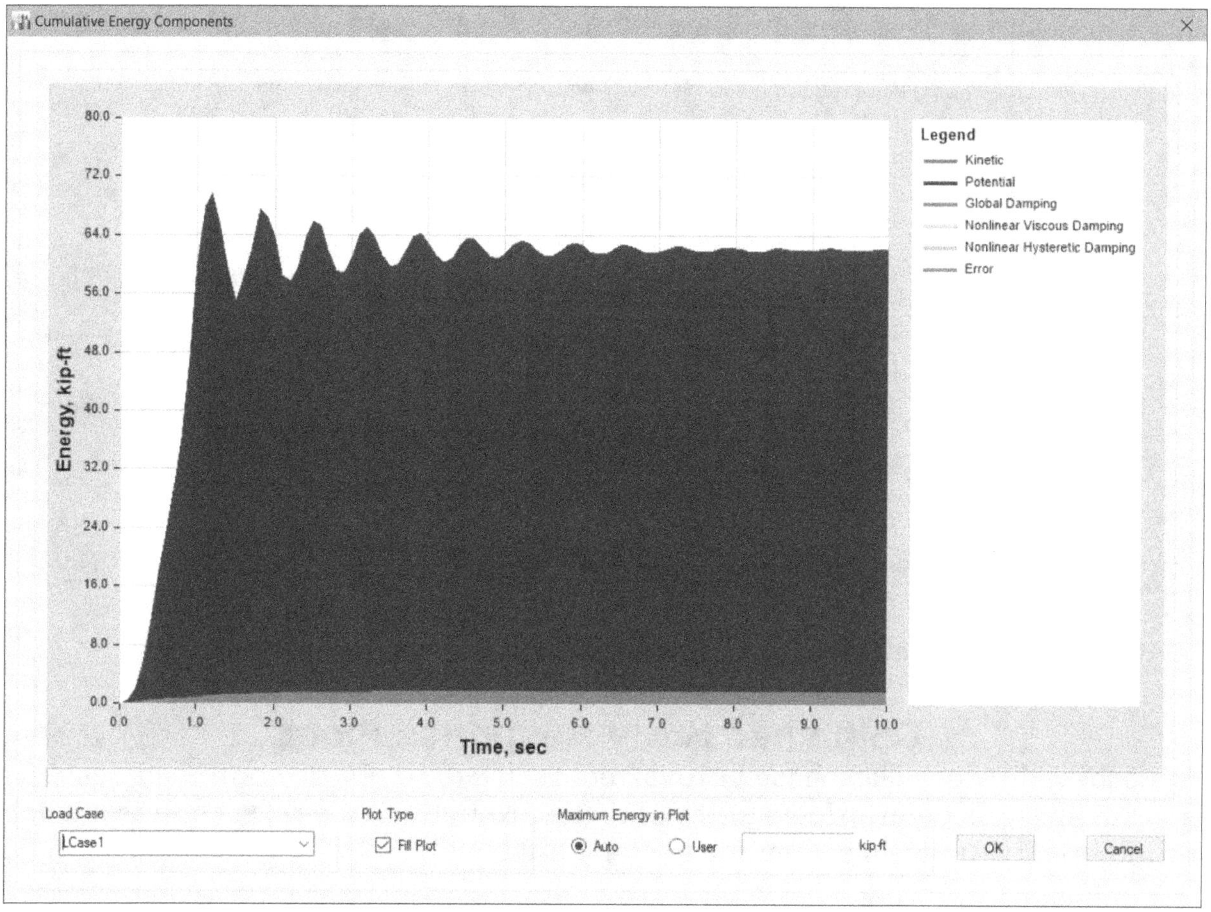

Figure-52. Cumulative Energy Components dialog box

Select desired load case from the **Load Case** drop-down. Set the other options as required and click on the **OK** button.

Story Response Plots

The **Story Response Plots** tool is used to display force and displacement responses for specified stories as a new tab in a display window. The procedure to use this tool is given next.

- Click on the **Story Response Plots** tool from the **Display** menu. The **Story Response** window will be displayed; refer to Figure-53.

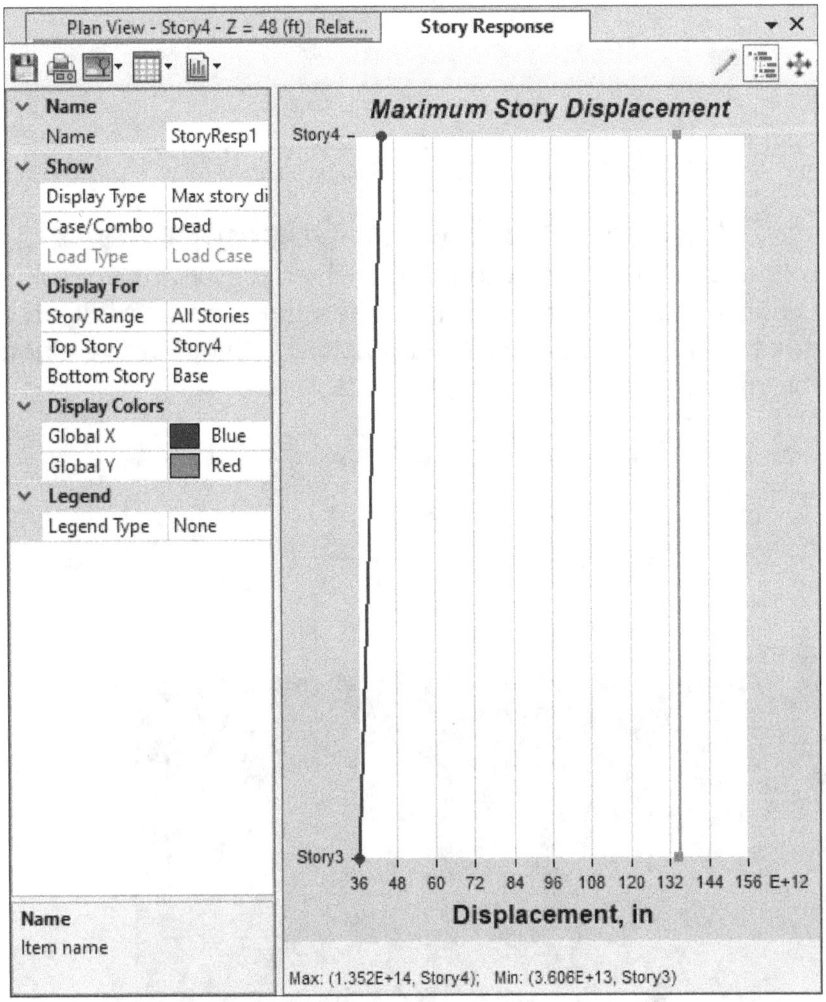

Figure-53. Story Response window

- Set desired parameters in the table at the left of the graph.

Combined Story Response Plots

The **Combined Story Response Plots** tool is used to check combined results of response analysis on multiple stories of model. Note that this plot result is available for time history analysis or nonlinear static analysis.

- Click on the **Combined Story Response Plots** tool from the **Display** menu after running the time history or nonlinear static analysis. The **Combined Story Response** window will be displayed; refer to Figure-54.
- Set desired value in the **Load Step** spinner to check the results.
- Click on the **OK** button to exit the window.

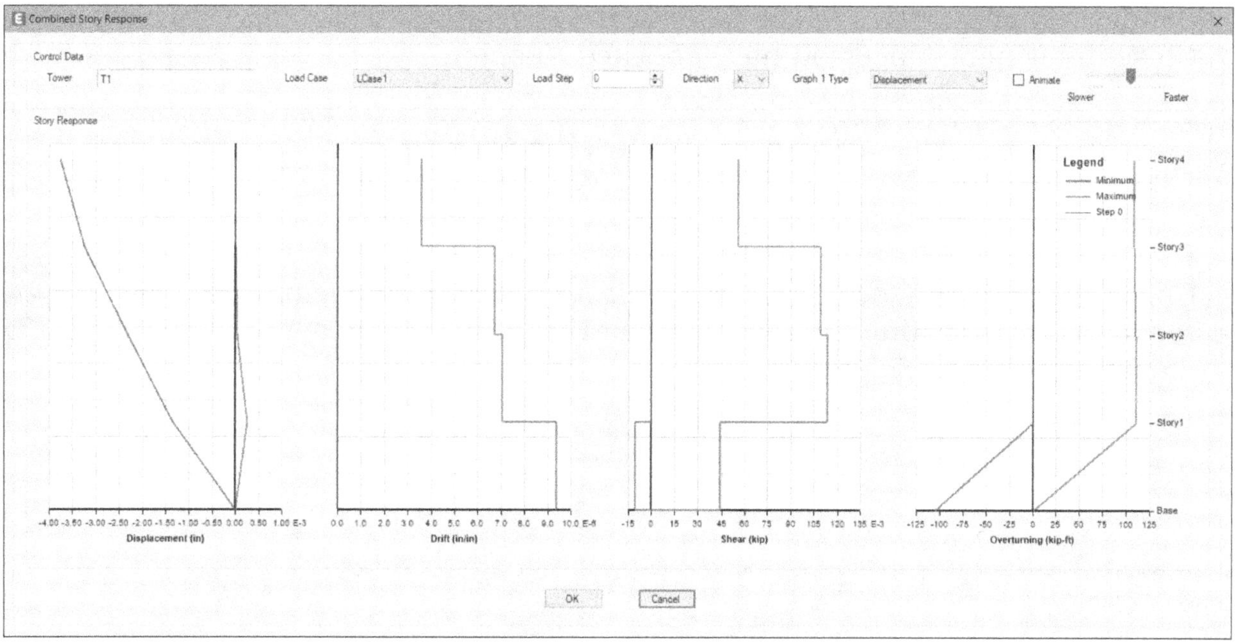

Figure-54. Combined Story Response window

Plot Functions

The **Plot Functions** tool in the **Display** menu is used to plot time history graph of the building. The procedure to use this tool is given next.

- After performing the time history analysis, click on the **Plot Functions** tool from the **Display** menu. The **Time History Plot** window will be displayed; refer to Figure-55.

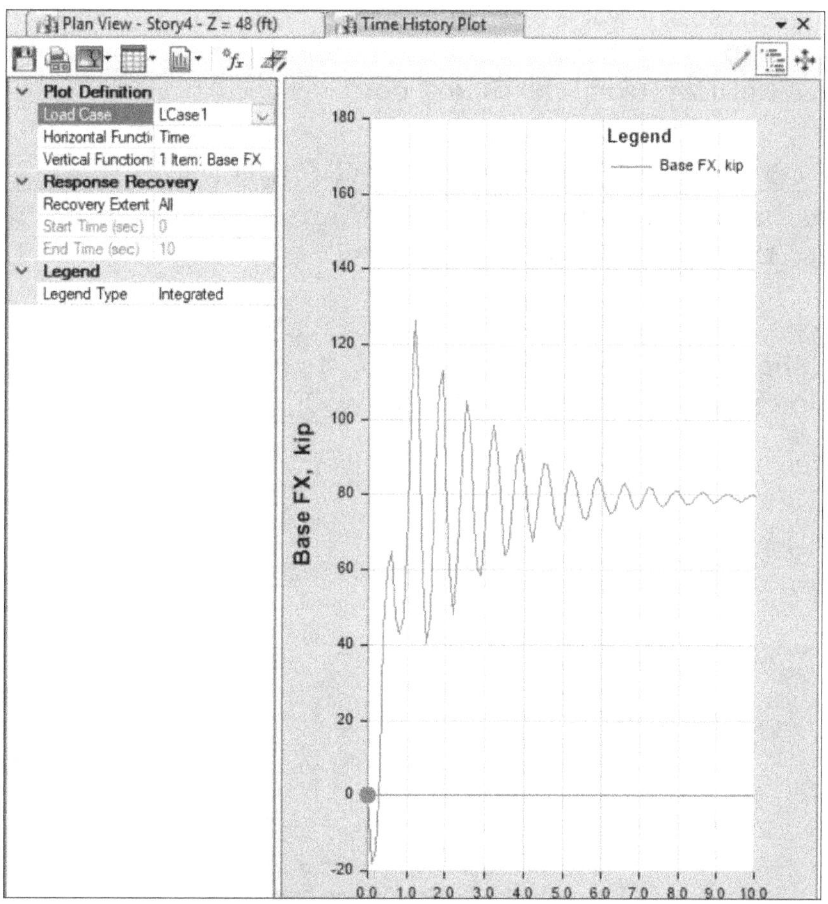

Figure-55. Time History Plot window

- Set desired parameters in the table at the left of the window and check the plot.
- Close the window once you have checked the plot.

Saving Named Display

The **Save Named Display** tool is used to save current result view. The procedure to use this tool is given next.

- Click on the **Save Named Display** tool from the **Display** menu after orienting the model and generating desired results. The **Named Displays** dialog box will be displayed; refer to Figure-56.

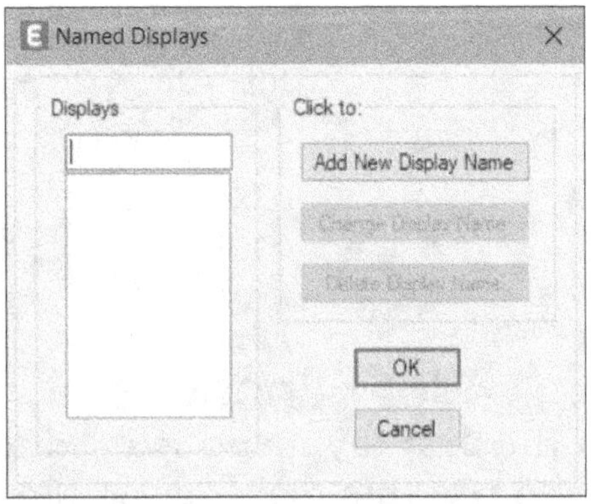

Figure-56. Named Displays dialog box

- Type desired name for the current view in the **Displays** edit box and click on the **Add New Display Name** button. The new display will be created and added in the list.
- Use the **Change Display Name** button to modify name of the selected display view.
- Use the **Delete Display Name** button to delete selected display view.
- Click on the **OK** button from the dialog box.

Showing Named Display

The **Show Named Display** tool is used to show an earlier saved named display. The procedure to use this tool given next.

- After saving named display, you can anytime show this saved display. To do so, click on the **Show Named Display** tool from the **Display** menu. The **Select Named Display** dialog box will be displayed; refer to Figure-57.

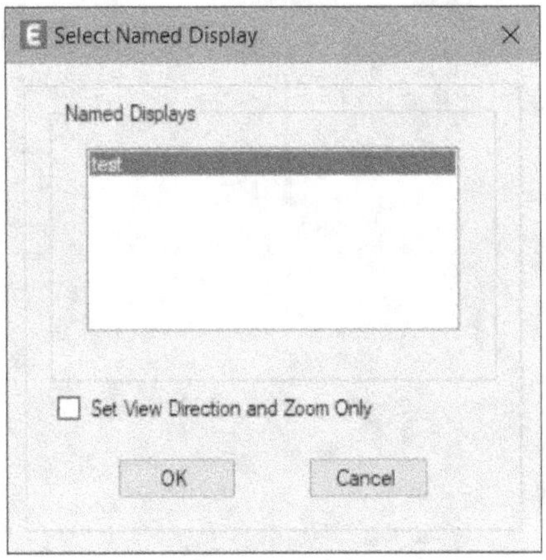

Figure-57. Select Named Display dialog box

- Select desired named display and click on the **OK** button from the dialog box. The selected view will be set in the current viewport.

Showing Data Tables

The **Show Tables** tool is used to display various data tables containing parameters related to model elements. The procedure to use this tool is given next.

- Click on the **Show Tables** tool from the **Display** menu. The **Choose Tables for Display** dialog box will be displayed; refer to Figure-58.

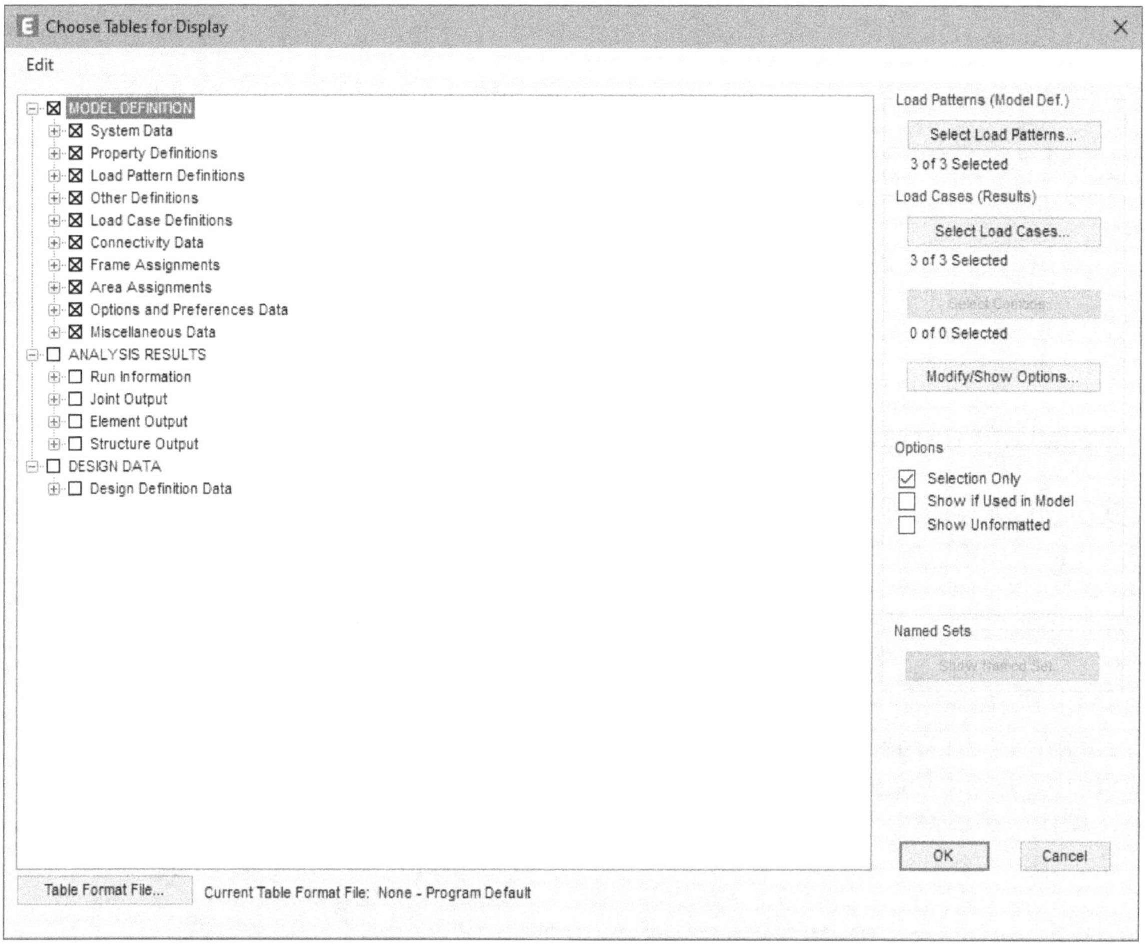

Figure-58. Choose Tables for Display dialog box

- Select check boxes for desired tables to be displayed from the list box and click on the **OK** button from the dialog box. The application window to show tables will be displayed; refer to Figure-59.
- Click on the **Done** button from the dialog box to exit.

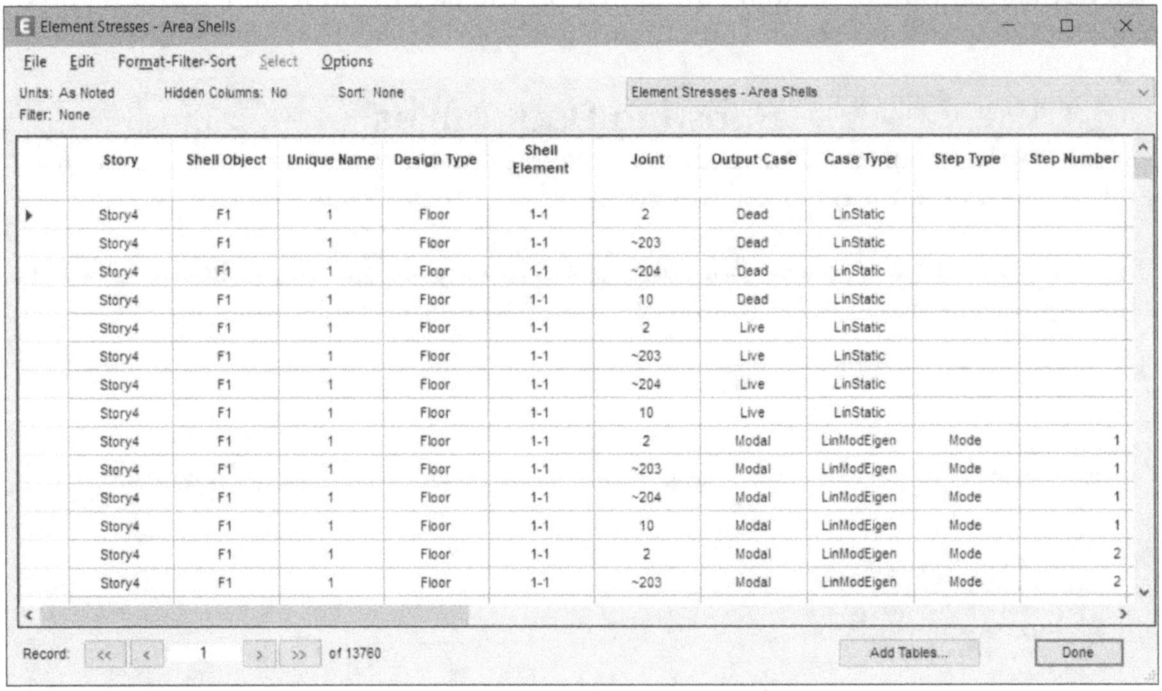

Element Stresses - Area Shells

File Edit Format-Filter-Sort Select Options

Units: As Noted Hidden Columns: No Sort: None

Filter: None

Element Stresses - Area Shells

Story	Shell Object	Unique Name	Design Type	Shell Element	Joint	Output Case	Case Type	Step Type	Step Number
Story4	F1	1	Floor	1-1	2	Dead	LinStatic		
Story4	F1	1	Floor	1-1	~203	Dead	LinStatic		
Story4	F1	1	Floor	1-1	~204	Dead	LinStatic		
Story4	F1	1	Floor	1-1	10	Dead	LinStatic		
Story4	F1	1	Floor	1-1	2	Live	LinStatic		
Story4	F1	1	Floor	1-1	~203	Live	LinStatic		
Story4	F1	1	Floor	1-1	~204	Live	LinStatic		
Story4	F1	1	Floor	1-1	10	Live	LinStatic		
Story4	F1	1	Floor	1-1	2	Modal	LinModEigen	Mode	1
Story4	F1	1	Floor	1-1	~203	Modal	LinModEigen	Mode	1
Story4	F1	1	Floor	1-1	~204	Modal	LinModEigen	Mode	1
Story4	F1	1	Floor	1-1	10	Modal	LinModEigen	Mode	1
Story4	F1	1	Floor	1-1	2	Modal	LinModEigen	Mode	2
Story4	F1	1	Floor	1-1	~203	Modal	LinModEigen	Mode	2

Record: << < 1 > >> of 13760 Add Tables... Done

Figure-59. Table results

You can create the other plots in the same way.

Once you have performed the analysis then its time to perform design study so that size and parameters can be updated as per the analysis results. The tools to perform design study are discussed in next chapter.

SELF-ASSESSMENT

Q1. What is the first step for applying load in the ETABS model?
A. Creating Load Cases
B. Defining Load Patterns
C. Running Analysis
D. Creating Mesh

Q2. Which menu contains the Load Pattern tool?
A. Analyze menu
B. Assign menu
C. Define menu
D. Edit menu

Q3. Which option is selected from the Type drop-down to apply wind load?
A. Dead
B. Live
C. Wind
D. Modal

Q4. What is the purpose of the Self-weight Multiplier edit box?
A. To define mesh size
B. To define weight added to the current load
C. To define damping ratio
D. To define stiffness

Q5. Which button is clicked to add a new load pattern?
A. Modify Load
B. Add New Load
C. Add Case
D. Load Pattern

Q6. Which button is used to modify lateral loading of wind or seismic load?
A. Modify Load
B. Modify/Show
C. Modify Lateral Load
D. Add New Load

Q7. Which radio button specifies wind load individually to each diaphragm?
A. Exposure from Frame and Shell Objects
B. Exposure from Extents of Diaphragms
C. User Defined Exposure
D. Program Determined Exposure

Q8. What does the Windward direction define?
A. Direction where wind stops
B. Direction where wind is reflected
C. Direction from where wind is coming
D. Direction of earthquake force

Q9. Which exposure type is used for urban and suburban areas?
A. Type A
B. Type B
C. Type C
D. Type D

Q10. What is the purpose of the Gust Factor?
A. Defines mesh density
B. Defines soil pressure
C. Examines structure against peak wind speed effects
D. Defines modal damping

Q11. Which tool is used to define how load patterns are applied and analyzed?
A. Load Combination tool
B. Load Pattern tool
C. Load Cases tool
D. Mesh tool

Q12. Which Load Case Type is used for fixed loads without dynamic effects?
A. Response Spectrum
B. Linear Static
C. Time History
D. Hyperstatic

Q13. Which load case type is used for time-varying loads?
A. Buckling
B. Linear Static
C. Time History
D. Modal

Q14. Which option is recommended for most tall buildings to consider P-Delta effects?
A. No P-Delta
B. Use Preset P-Delta Settings
C. Ignore Gravity Load
D. Use Modal Analysis

Q15. Which P-Delta method captures local buckling effects better?
A. Non-iterative - Based on Mass
B. Iterative - Based on Loads
C. Static Mass Method
D. Modal Method

Q16. Which load case is used to find natural frequencies of the model?
A. Buckling
B. Hyperstatic
C. Modal
D. Time History

Q17. Which vectors are generally recommended for dynamic analysis?
A. Eigen vectors
B. Ritz vectors
C. Shell vectors
D. Buckling vectors

Q18. Which response spectrum modal combination method accounts for modal damping?
A. SRSS
B. Absolute
C. CQC
D. Range Add

Q19. Which Time History Motion Type is used when loads repeat indefinitely?
A. Transient
B. Periodic
C. Dynamic
D. Static

Q20. What is the purpose of the Buckling Load Case?
A. To check vibration response
B. To check sudden failure due to buckling
C. To calculate walking load
D. To define seismic loads

Q21. Which combination type performs algebraic addition of loads?
A. Envelope
B. SRSS
C. Linear Add
D. Range Add

Q22. Which tool checks whether all parameters are specified for analysis?
A. Load Cases
B. Check Model
C. Auto Mesh
D. Walking Vibrations

Q23. Which button is selected to run full 3D analysis?
A. XZ Plane
B. YZ Plane
C. No Z Rotation
D. Full 3D

Q24. Which solver option uses all CPU cores and full RAM during analysis?
A. Standard Solver
B. Advanced Solver
C. Multi-threaded Solver
D. GUI Process Solver

Q25. Which mesh option predominantly uses 4-node quadrilateral elements?
A. Rectangular Mesh
B. General Mesh
C. Localized Mesh
D. Manual Mesh

Q26. Which tool is used to set the solver for analysis in ETABS?
A. Load Cases Tool
B. Advanced SAPFire Options Tool
C. Mesh Settings Tool
D. Display Tool

Q27. Which solver uses only a single core and checks model stability?
A. Multi-threaded Solver
B. Advanced Solver
C. Standard Solver
D. Auto Solver

Q28. Which solver utilizes all CPU cores and full RAM during analysis?
A. Standard Solver
B. Multi-threaded Solver
C. GUI Solver
D. Manual Solver

Q29. Which Analysis Process Option is best for small problems?
A. Separate Process
B. Auto
C. GUI Process
D. Parallel Process

Q30. What is the main drawback of GUI Process analysis?
A. Requires internet connection
B. Consumes software memory and slows large models
C. Cannot analyze frames
D. Cannot use RAM

Q31. Which Analysis Process Option runs analysis through CSI.SAPFire.Driver.exe?
A. GUI Process
B. Auto
C. Separate Process
D. Quick Process

Q32. Which mesh option is selected by default for floors?
A. Rectangular Mesh
B. General Mesh
C. Hexagonal Mesh
D. Circular Mesh

Q33. Which mesh type predominantly uses 4-node quadrilateral elements?
A. General Mesh
B. Rectangular Mesh
C. Triangular Mesh
D. Linear Mesh

Q34. Which check box avoids creating dense meshes in large objects?
A. Merge Joints Where Possible
B. Use Localized Meshing
C. Display Mesh
D. Refine Mesh

Q35. Which tool is used to specify mesh size for walls?
A. Automatic Mesh Settings for Floors
B. Wall Thickness Tool
C. Automatic Rectangular Mesh Settings for Walls
D. Shell Stress Tool

Q36. Which radio button models hinges as additional link elements?
A. Model Hinges Within Elements
B. Model Hinges as Separate Link Elements
C. Merge Hinges
D. Nonlinear Hinges Off

Q37. Which nonlinear analysis method is permitted when hinges are modeled as separate link elements?
A. SRSS
B. Ritz Analysis
C. Fast Nonlinear Analysis (FNA)
D. Buckling Analysis

Q38. Which tool is used to set options for cracking analysis?
A. Cracking Analysis Options
B. Load Pattern Tool
C. Story Response Tool
D. Mesh Tool

Q39. Which menu contains the Run Analysis tool?
A. Display menu
B. Edit menu
C. Analyze menu
D. Assign menu

Q40. What is the function of the Model Alive tool?
A. Deletes analysis results
B. Automatically performs analysis when model changes
C. Locks the model
D. Creates load cases

Q41. Which tool modifies the structure geometry in undeformed shape?
A. Modify Undeformed Geometry
B. Deformed Shape
C. Plot Functions
D. Shell Forces

Q42. Which option modifies geometry based on modal or buckling mode shape?
A. Target Displacement
B. Scaled Mode Shape
C. Full Geometry
D. Static Shape

Q43. Which tool displays details of the most recent analysis run?
A. Show Tables
B. Last Analysis Run Log
C. Plot Functions
D. Force Diagram

Q44. What happens to the model after analysis is performed?
A. It gets deleted
B. It becomes hidden
C. It gets locked for changes
D. It resets automatically

Q45. Which Display menu tool shows the model in undeformed shape?
A. Deformed Shape
B. Load Assigns
C. Undeformed Shape
D. Shell Forces

Q46. Which tool displays the model in deformed condition?
A. Deformed Shape
B. Story Plot
C. Mesh Display
D. Load Assigns

Q47. Which tool displays support or spring reaction forces?
A. Shell Stresses
B. Support/Spring Reactions
C. Plot Functions
D. Walking Vibrations

Q48. Which Plot Type displays reaction forces in tabular form?
A. Arrows
B. Contours
C. Tabulated
D. Extruded

Q49. Which tool is used to display soil pressure on the model?
A. Soil Pressure Tool
B. Shell Forces Tool
C. Story Plot Tool
D. Modal Tool

Q50. Which option displays soil pressure in extruded contour form?
A. Display on Undeformed Shape
B. Display on Deformed Shape
C. Display in Extruded Form
D. Display in Wireframe Form

Chapter 6

Running Design Study

Topics Covered

The major topics covered in this chapter are:

- ***Introduction***
- ***Steel Frame Design***
- ***Concrete Frame Design***
- ***Composite Beam/Column Design***
- ***Overwrite Frame Design Procedure***
- ***Shear Wall Design and Concrete Slab Design***
- ***Live Load Reduction Factors***
- ***Setting Lateral Displacement Targets***
- ***Setting Time Period Targets***

INTRODUCTION

In previous chapters, you have learned to draw frame members and create properties of different frame members. You have also performed analysis on the model for specified load conditions. In this chapter, you will learn to design the structural members based on specified loads.

STEEL FRAME DESIGN

The parameters related to steel frame design are available in the **Steel Frame Design** cascading menu of the **Design** menu; refer to Figure-1. Various options in this menu are discussed next.

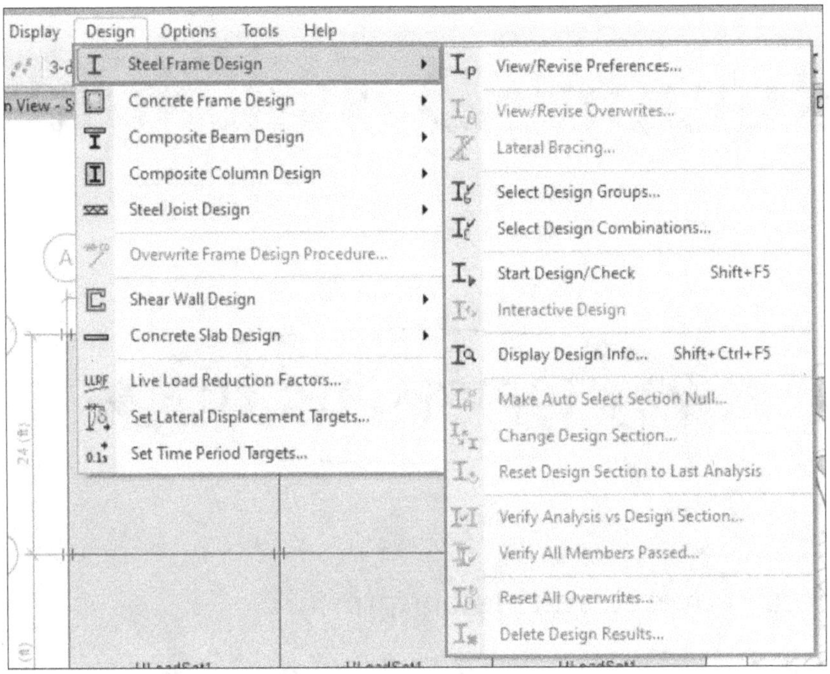

Figure-1. Steel Frame Design cascading menu

Steel Frame Preferences

Steel Frame Preferences are the set of rules to be abide by all steel frame members in the project. The procedure to set these rules is given next.

- Click on the **View/Revise Preferences** tool from the **Design** menu. The **Steel Frame Design Preferences** dialog box will be displayed; refer to Figure-2.
- Select desired design code from the **Design Code** field in the table. There are various standards as per the construction industry.
- Select desired option from the **Multi-Response Case Design** field. The **Step-by-Step-All** is used when analysis is to be performed step by step for Time History first, then Multi-step static, and then Nonlinear static in the end. Select the **Envelopes-All** option if you want to run the analysis all along and compile a final result. Select the **Last Step** option if you want to considers the last values of Time History, Multi-step static, and Nonlinear static analysis for generating design. You can use the other options in the same way.

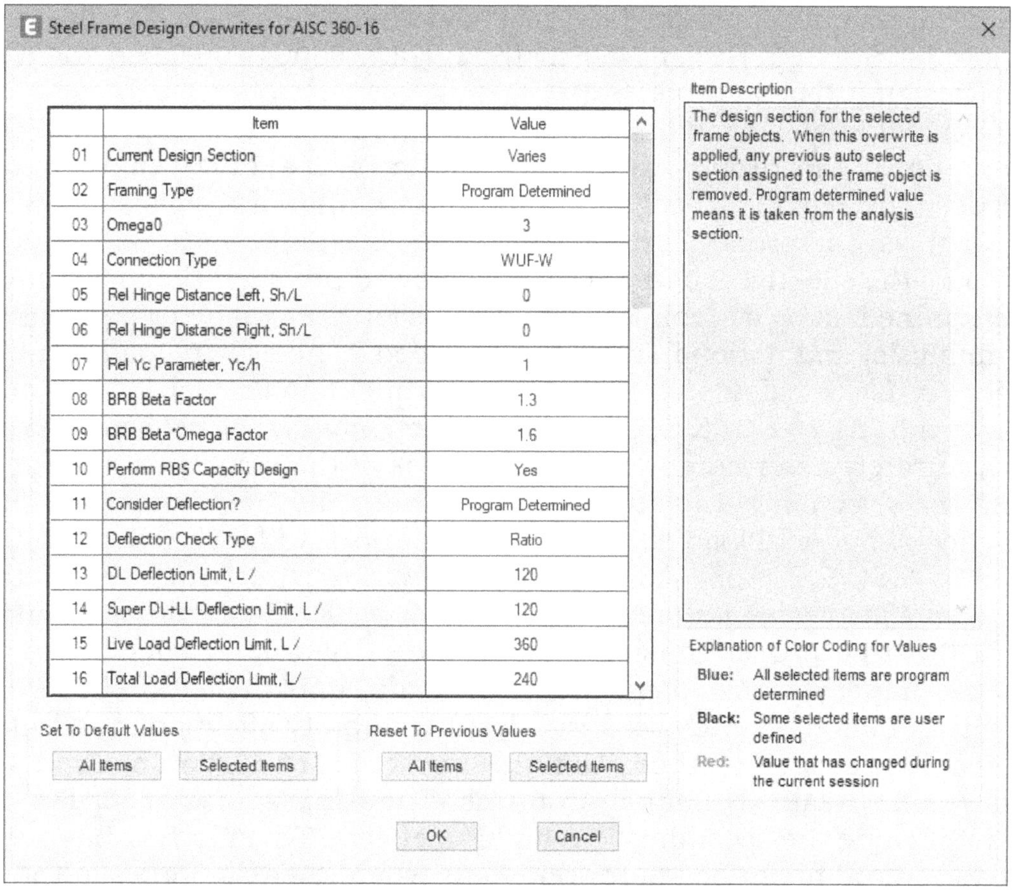

Figure-2. Steel Frame Design Preferences for AISC 360-10 dialog box

- Select desired option from the **Framing Type** drop-down. Various parameters in the drop-down and their respective references are given in the next table.

Framing Type	References
SMF (Special Moment Frame)	AISC SEISMIC 9
IMF (Intermediate Moment Frame)	AISC SEISMIC 10
OMF (Ordinary Moment Frame)	AISC SEISMIC 11
STMF (Special Truss Moment Frame)	AISC SEISMIC 12
SCBF (Special Concentrically Braced Frame)	AISC SEISMIC 13
OCBF (Ordinary Concentrically Braced Frame)	AISC SEISMIC 14
EBF (Eccentrically Braced Frame)	AISC SEISMIC 15
BRBF (Buckling Restrained Braced Frame)	AISC SEISMIC 16
SPSW (Special Plate Shear Wall)	AISC SEISMIC 17

- Select desired option from the **Seismic Design Category** drop-down. These options are related to seismic design.
- Set desired value of importance factor in the **Importance Factor** edit box.
- Similarly, you can set desired value of redundancy factor (Design System Rho), Design Spectral Acceleration Parameter (Design System Sds), Response Modification Factor (Design System R), System Over strength Factor (Design System Omega0), and Deflection Amplification Factor (Design System Cd) in their respective edit boxes.

- Select desired option from the **Design Provision** drop-down. The first method is the Allowable Strength Design (**ASD**) method. The second is the Load and Resistance Factor Design (**LRFD**) method.
- Select desired analysis method from the **Analysis Method** drop-down. There are three options in this drop-down; **Direct Analysis**, **Effective Length**, and **Limited 1st Order**. The **Effective Length** method is a traditional analysis method used to solve small second order effects of load on the building. In the AISC 360-05/IBC 2006 code, the effective length method is allowed provided the member demands are determined using a second-order analysis (either explicit or by amplified first-order analysis) and notional loads are included in all gravity load combinations. K-factors must be calculated to account for buckling (except for braced frames, or where $\Delta 2 / \Delta 1 < 1.0$, $K = 1.0$). The **Direct Analysis Method** is expected to more accurately determine the internal forces of the structure, provided care is used in the selection of the appropriate methods used to determine the second-order effects, notional load effects and appropriate stiffness reduction factors as defined in AISC 2.2, App. 7.3(3). Additionally, the Direct Analysis Method does not use an effective length factor other than $k = 1.0$. The rational behind the use of $k = 1.0$ is that proper consideration of the second-order effects (P-Δ and P-δ), geometric imperfections (using notional loads) and inelastic effects (applying stiffness reductions) better accounts for the stability effects of a structure than the **Effective Length** method. Use the **Limited 1st Order** method when the analysis is for relatively simple problems. The **Limited First Order Analysis** does not include the secondary P-Δ and P-δ effects. This method has very limited applicability and might be appropriate only when the axial forces in the columns are very small compared to their Euler buckling capacities.
- Select desired method from the **Second Order Method** field in the table. These methods are only useful when you have selected **Direct Analysis** or **Effective Length** option from the **Analysis Method** field.
- Select desired option from the **Stiffness Reduction Method** drop-down in the table. There are three options in this drop-down; **Tau-b Variable**, **Tau-b Fixed**, and **No Modification**. Select the **Tau-b Variable** or **Tau-b Fixed** option if you want the apply reduction factor.
- Select the **Yes** option from the **Add Notional load cases into seismic combos** drop-down if you want to add the notional loads to seismic load desired from standards.
- Similarly, set the other parameters in the table and click on the **OK** button.

View/Revise Overwrite for Steel Frame Sections

The **View/Revise Overwrite** tool in the **Steel Frame Design** cascading menu is used to overwrite the default preferences for selected steel frame section. The procedure to use this tool is given next.

- Select desired steel section from the model and click on the **View/Revise Overwrite** tool from the **Steel Frame Design** cascading menu of the **Design** menu. The **Steel Frame Design Overwrites** dialog box will be displayed; refer to Figure-3.

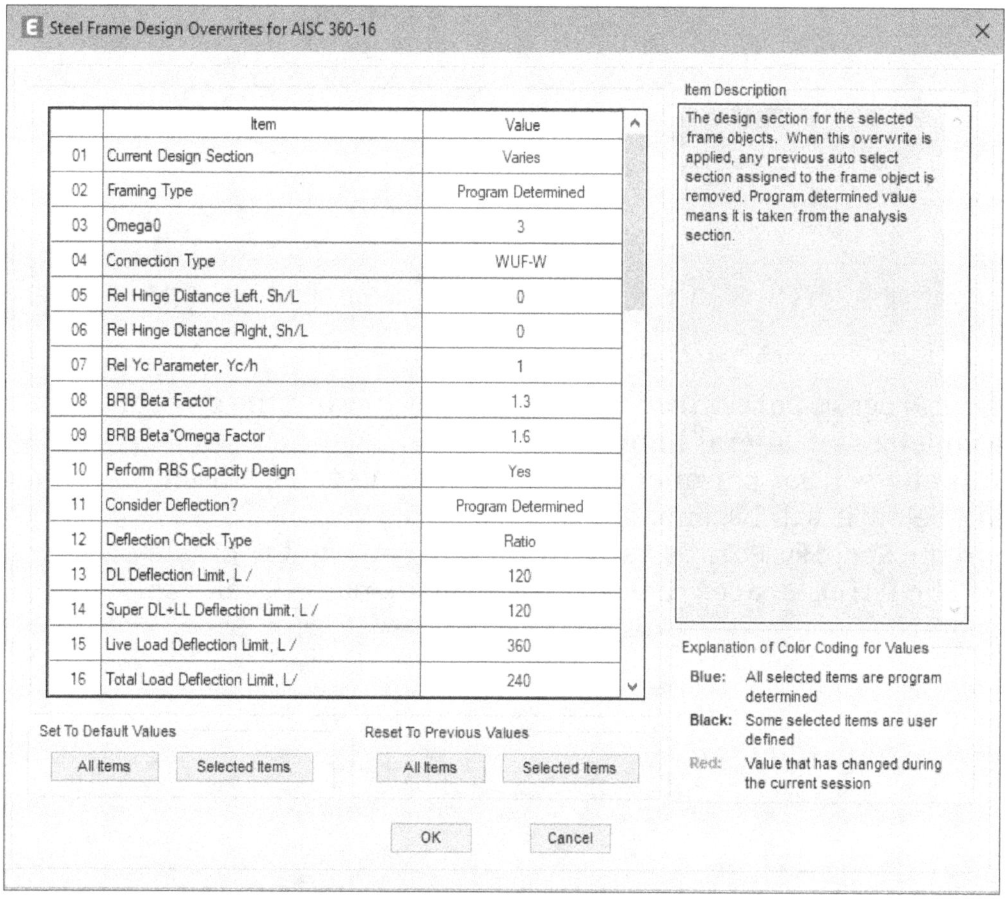

Figure-3. Steel Frame Design Overwrites dialog box

- Set desired parameters in the table and click on the **OK** button.

Setting Lateral Brace Parameters

Lateral bracing is the term we use to refer to any pieces on a bridge that help keep the sides (trusses) from twisting. It also helps keep the top chords of the bridge from bending or deforming in or out. In Figure-4, the lateral bracing is marked red. The procedure to set lateral bracing is given next.

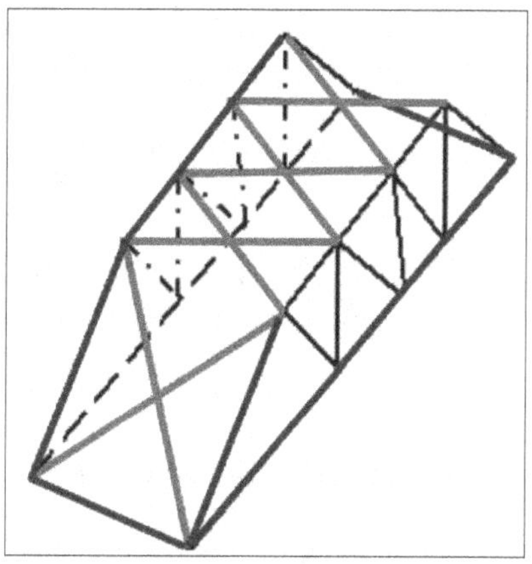

Figure-4. Lateral bracing

- Click on the **Lateral Bracing** tool from the **Steel Frame Design** cascading menu of the **Design** menu. The **Lateral Bracing** dialog box will be displayed; refer to Figure-5.

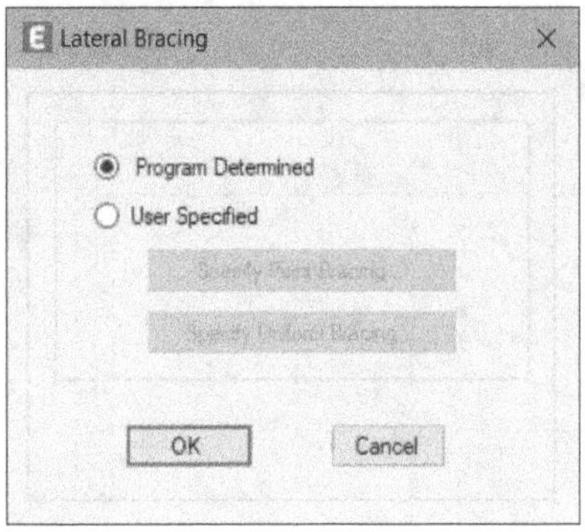

Figure-5. Lateral Bracing dialog box

- Select the **Program Determined** radio button if you want the system to automatically set parameters for lateral bracing.
- To manually set the parameters, click on the **User Specified** radio button. The buttons below it will become active.
- Click on the **Specify Point Bracing** button if you want to create bracing at desired points. The **Point Braces** dialog box will be displayed; refer to Figure-6.

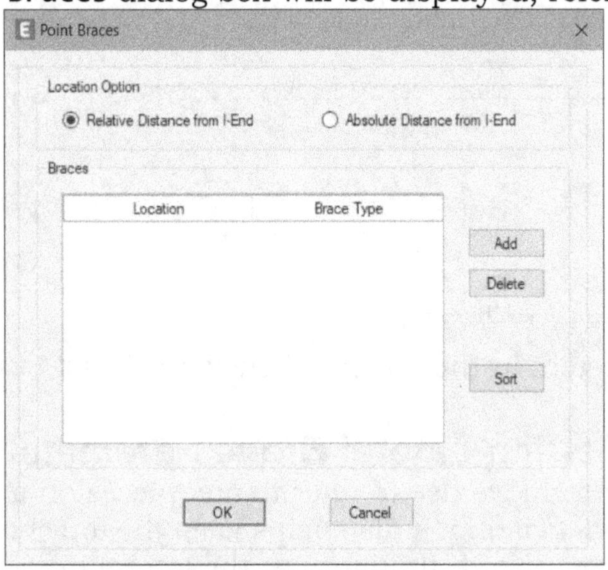

Figure-6. Point Braces dialog box

- Select the **Relative Distance from I-End** radio button if you want to specify the distance in ratio of the total length. Select the **Absolute Distance from I-End** radio button if you want to specify the absolute value of distance from the end point of selected beam.
- Click on the **Add** button from the dialog box. A new location of brace will be added in the table.
- Specify desired value of location in the **Location** field and set the brace type.
- Click on the **OK** button to apply braces.
- Similarly, you can create uniform bracing by using the **Specify Uniform Bracing** button.
- After creating the braces, click on the **OK** button from the dialog box.

Selecting Design Group

The **Select Design Group** tool is used to select a design group earlier created by using the **Group Definitions** tool in the **Define** menu for select components of structure. The procedure is given next.

- Click on the **Select Design Groups** tool from the **Steel Frame Design** cascading menu of the **Design** menu. The **Steel Frame** dialog box will be displayed as shown in Figure-7.

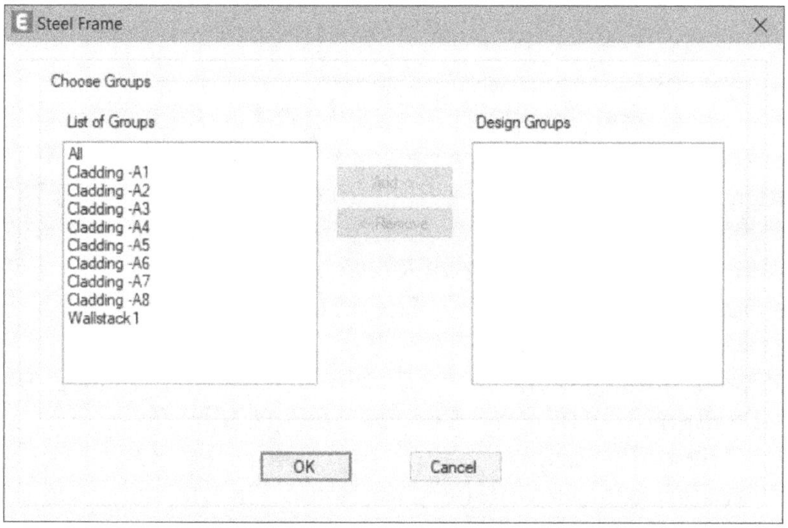

Figure-7. Steel Frame dialog box

- Double-click on the design groups that you want to be included in the current project and click on the **OK** button. The selected groups will be available for assignment to various structural components.

Selecting Design Combinations

The **Select Design Combinations** tool is used to select the design load combinations applicable in current project. The procedure to use this tool is given next.

- Click on the **Select Design Combinations** tool from the **Steel Frame Design** cascading menu of the **Design** menu. The **Design Load Combinations Selection** dialog box will be displayed; refer to Figure-8.

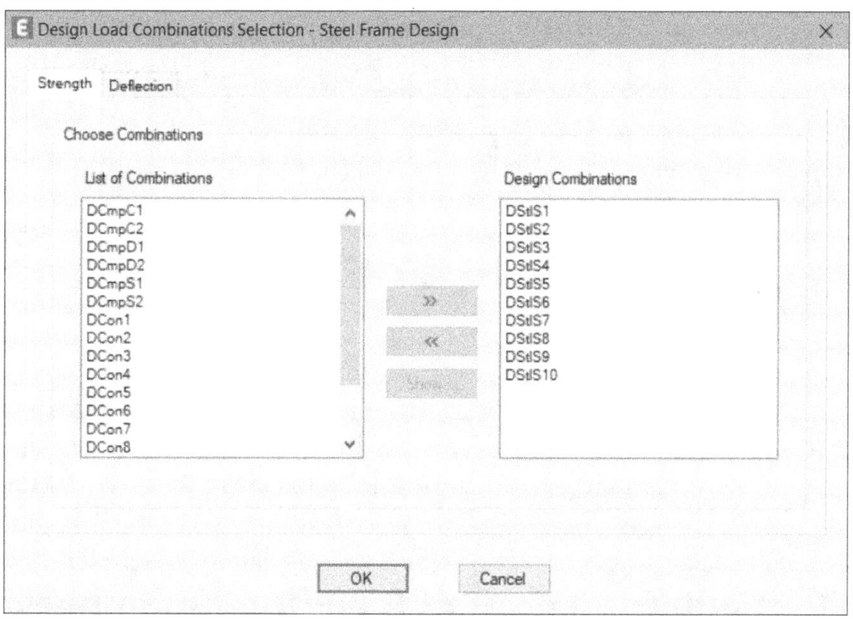

Figure-8. Design Load Combinations Selection dialog box

- There are two load groups in the dialog box viz. **Strength** and **Deflection**. Select the **Strength** tab in the dialog box if you want to use strength based design load combinations in your project. Select the **Deflection** tab if you want use the deflection based design load combinations in your project.

- Double-click on the design load combination from the **List of Combinations** list box in the dialog box to add the selected combination in the project.
- Click on the **OK** button from the dialog box to apply the changes.

Performing Design Check

The **Start Design/Check** tool is used to perform design checks on the structure based on specified load combinations and selected design groups. This tool is available when you have performed analysis by using the **Run Analysis** tool in the **Analyze** menu. You need to check the topics of Load assignment and performing analysis in next chapters before using this tool. The procedure to use this tool is given next.

- Click on the **Start Design/Check** tool from the **Steel Frame Design** cascading menu in the **Design** menu. The **Steel Design/Check** information box will be displayed while the calculations are being performed. Once the analysis is done, the results of design check will be displayed in the selected viewport; refer to Figure-9.

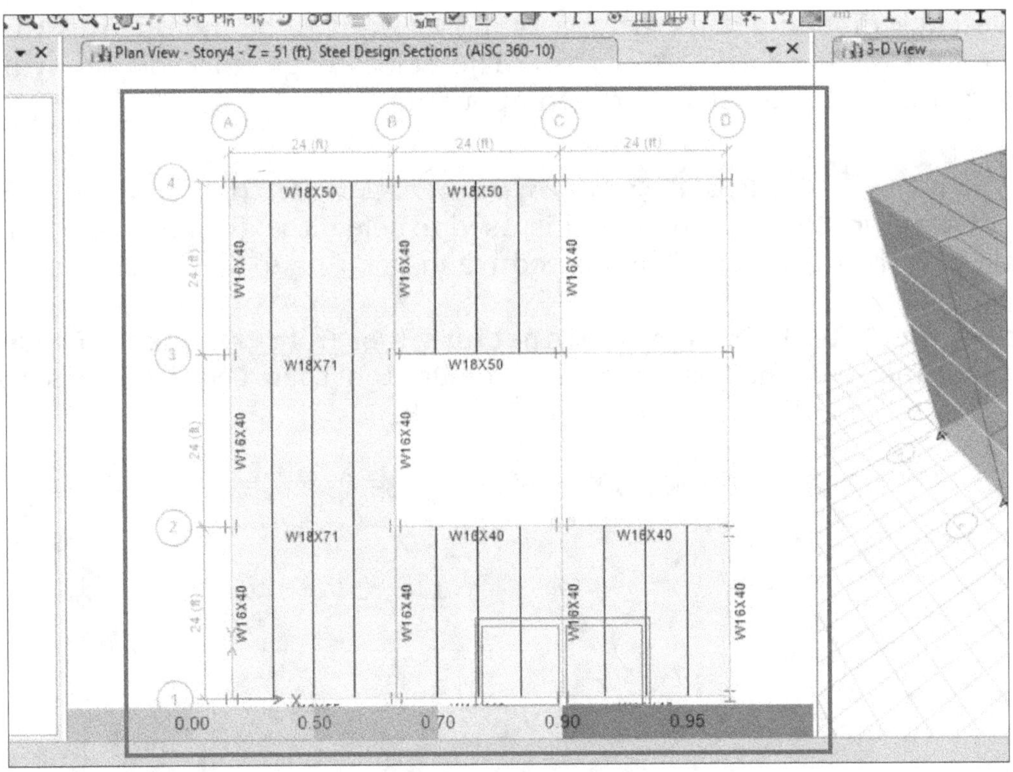

Figure-9. Results of design check

In the design results, each member of structure is displayed with its dimensional name and colors are assigned to them based on their load conditions. Like the structural members in red color are showing high deflection as compared to structural members in cyan color.

Stress Information of Structural members

Once you have performed the design check, it is required to check the details of stress induced on high deflecting components. To do so, follow the steps given next.

- To check this information, click on desired structural member from the design check result and right-click on it. The **Steel Stress Check Information** dialog box will be displayed; refer to Figure-10.

The details of each column in the table are as follows:

COMBO ID: The name of the design load combination considered.

STATION LOC: The location of the station considered, measured from the i-end of the frame element.

RATIO: The total PMM stress ratio for the element. When stress ratios are reported for this item, they are followed by (T) or (C). The (T) indicates that the axial component of the stress ratio is tension. The (C) indicates that the axial component of the stress ratio is compression. Note that typically the interaction formulas are different depending on whether the axial stress is tension or compression.
AXL: The axial component of the PMM stress ratio.
B-MAJ: The bending component of the PMM stress ratio for bending about the major axis.

B-MIN: The bending component of the PMM stress ratio for bending about the minor axis.

MAJ SHR RATIO: The shear stress ratio for shear acting in the major direction of the frame element.

MIN SHR RATIO: The shear stress ratio for shear acting in the minor direction of the frame element.

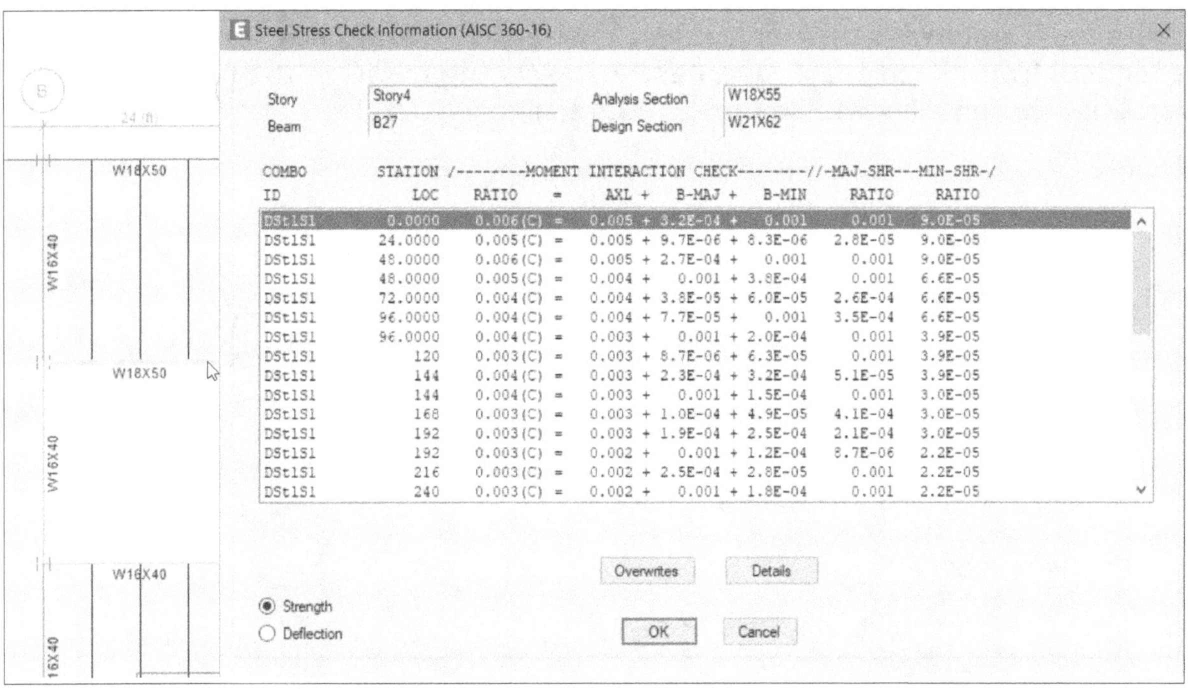

Figure-10. Stress information for selected member

- To check further details of stress on selected structure member, select it from the table, and click on the **Details** button. Complete summary of the selected member will be displayed; refer to Figure-11.

- Click on the **Envelope** tab to check the strength under envelope type load combinations.
- Close the dialog box and click on the **OK** button from the **Steel Stress Check Information** dialog box to exit.

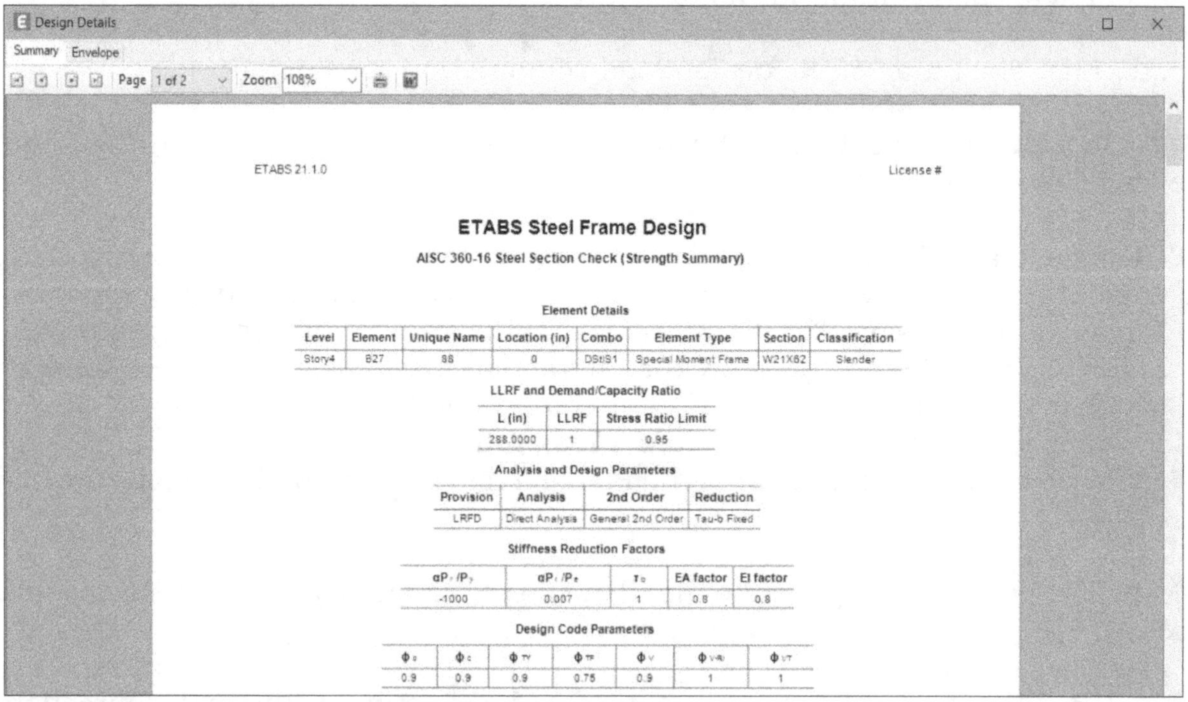

Figure-11. Strength summary of selected structural member

Interactive Design View

While designing the structure, you can check the interactive view of the frame members in shades according to their deformation under specified standard load conditions. To activate this tool, select desired view and click on the **Interactive Design** tool from the **Steel Frame Design** cascading menu in the **Design** menu. The interactive design view will be displayed; refer to Figure-12.

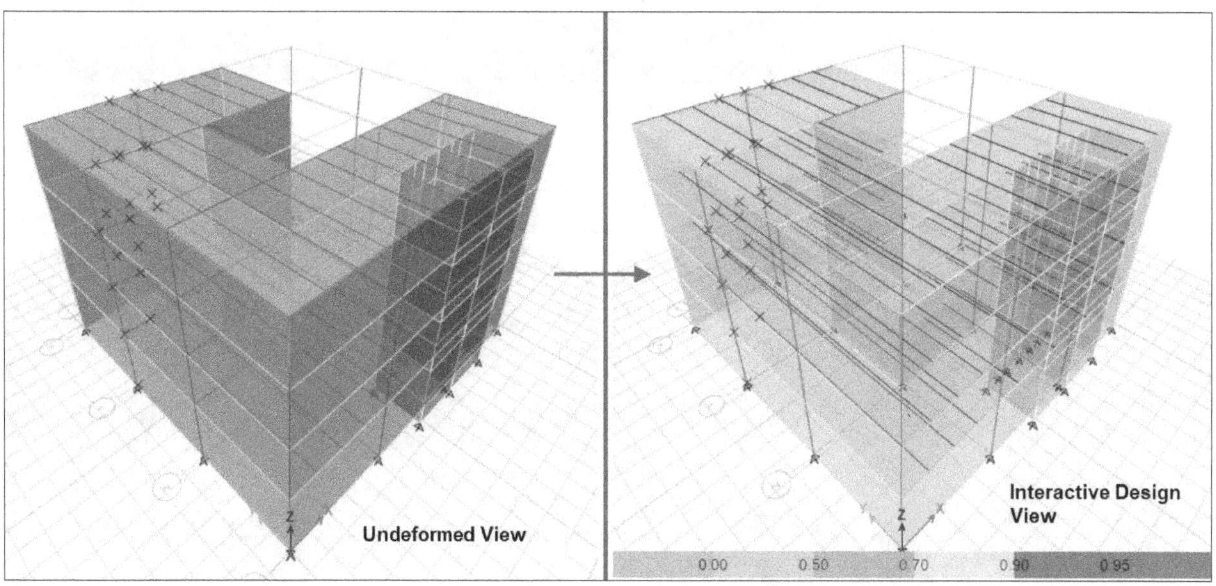

Figure-12. Interactive design view

Display Design Info

The **Display Design Info** tool is used to check different design parameters of the selected structure member. The procedure to use this tool is given next.

- Select desired view in which you want to display the design parameters and click on the **Display Design Info** tool from the **Steel Frame Design** cascading menu of the **Design** menu. The **Display Steel Frame Design Results** dialog box will be displayed; refer to Figure-13.

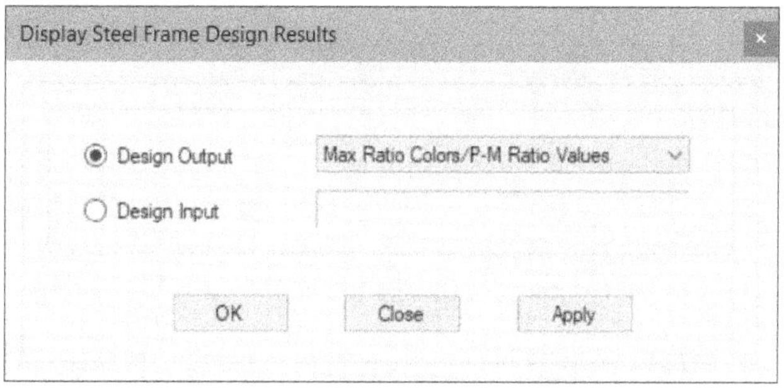

Figure-13. Display Steel Frame Design Results dialog box

- Select desired radio button to display design output parameters or design input parameters.
- Click in the drop-down and select the parameter you want to display in the view.
- Click on the **OK** button to apply the modifications.

Removing Auto Select Sections

The **Make Auto Select Section Null** tool to clear the list of auto select sections after performing analysis. The procedure to use this tool is given next.

- Click on the **Make Auto Select Section Null** tool from the **Steel Frame Design** cascading menu of the **Design** menu. The **ETABS** information box will be displayed; refer to Figure-14.

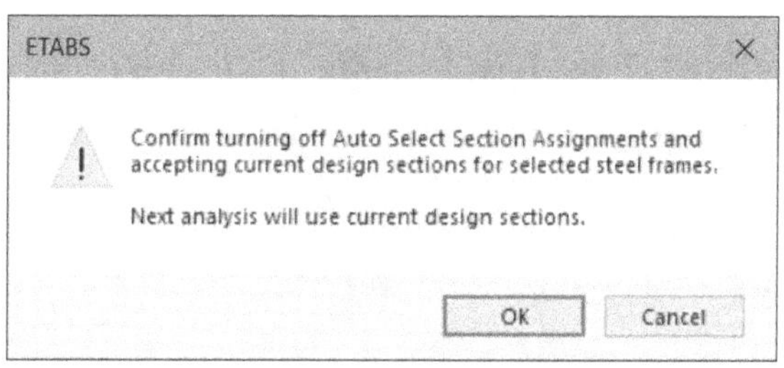

Figure-14. ETABS information box

- Click on the **OK** button from the box to disable auto selection of sections and assign the present section sizes to various members of structure.

Changing Steel Frame Sections

The **Change Design Section** tool is used to change the size of selected sections of the frame. This tool is active only when a frame member is selected. The procedure to use this tool is given next.

- Select desired frame member and click on the **Change Design Section** tool from the **Steel Frame Design** cascading menu of the **Design** menu. The **Select Sections** dialog box will be displayed; refer to Figure-15.

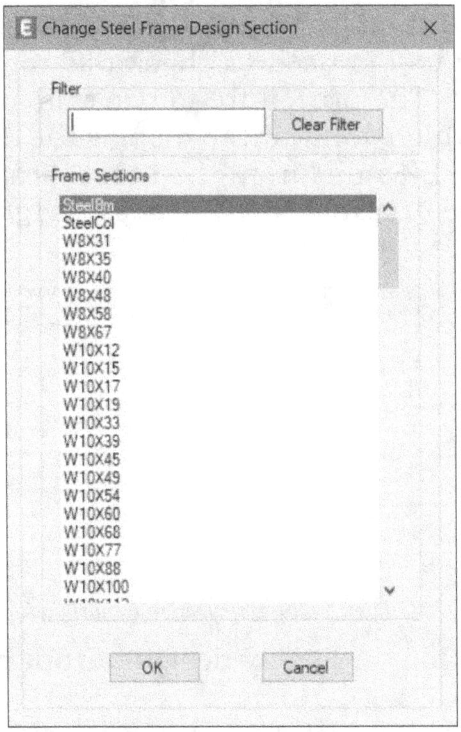

Figure-15. Select Sections dialog box

- Select desired section and click on the **OK** button from the dialog box to apply the changes.

Resetting the Design Sections to Previous Analysis Values

The **Reset Design Section to Last Analysis** tool is used to reset the sections of frame to previous values generated by last analysis. To do so, click on the **Reset Design Section to Last Analysis** tool from the **Steel Frame Design** cascading menu of the **Design** menu. The values will be reset automatically.

Verifying Analysis V/s Design Section

The **Verify Analysis vs Design Section** tool is used to check whether the assigned section values are same as generated by analysis. If there is any section different from generated after analysis then an information box will be displayed. The procedure to use this tool is given next.

- Click on the **Verify Analysis vs Design Section** tool from the **Steel Frame Design** cascading menu of the **Design** menu. If there is no difference between the values of section in model and the values generated by analysis then the information box will be displayed as shown in Figure-16. If there is a difference between the values then the information box will be displayed as shown in Figure-17. Click on the **Yes** button from the dialog box to select all the differing sections.

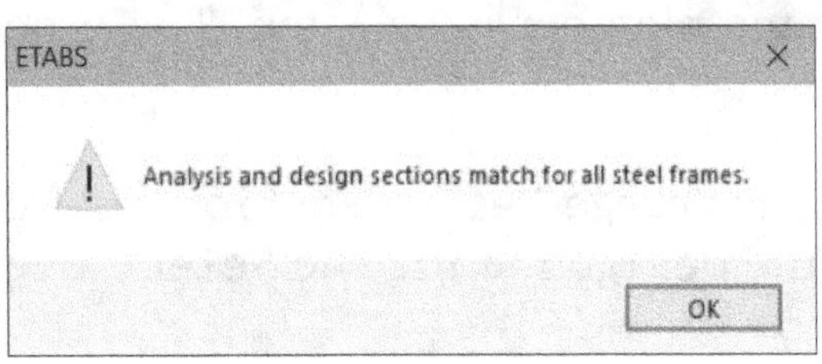

Figure-16. Information box for matching steel sections

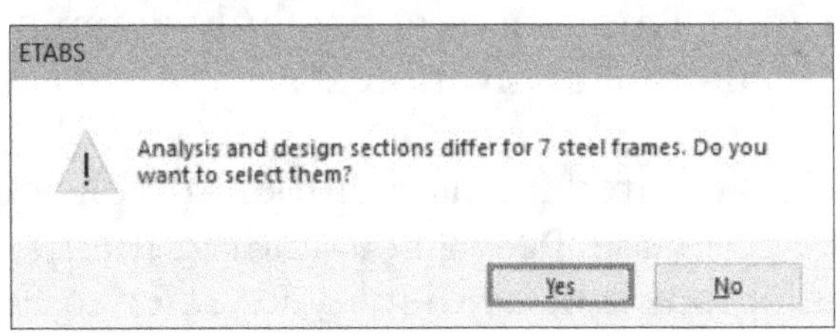

Figure-17. Information box with differing steel sections

- Now, you can perform the required operations on the selected frame members like changing them to design sections.

Verifying All Members according to Analysis Parameters

The **Verify All Members Pass** tool is used to check if all the members of frame have passed the analysis or not. The procedure to use this tool is given next.

- Click on the **Verify All Members Pass** tool from the **Steel Frame Design** cascading menu in the **Design** menu. The information box will be displayed as shown in Figure-18 if there is any member on which the analysis has not been performed. Click on the **Yes** button to select the left members. Perform the analysis again to set parameters as per the analysis. If all the members have passed the analysis then the related information box will be displayed. Click on the **OK** button to exit. Note that you need to change the section of frame members to bigger if they do not pass the analysis.

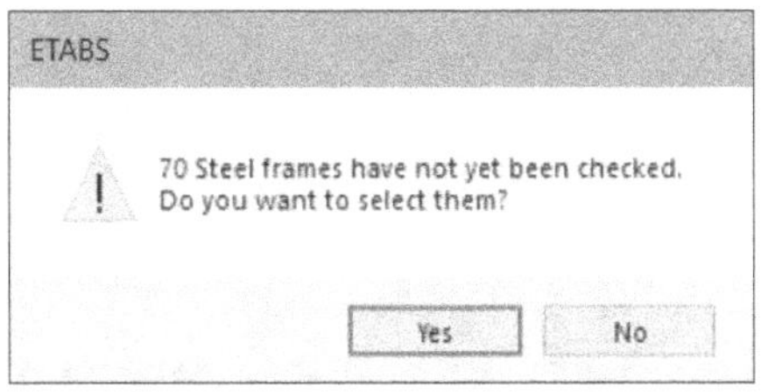

Figure-18. Information box for frame members not checked

Resetting All Overwrites

The **Reset All Overwrites** tool is used to reset all the parameters of steel frame to design default values. To reset overwrites, click on the **Reset All Overwrites** tool from the **Steel Frame Design** cascading menu of the **Design** menu. The dialog box for resetting parameters will be displayed as shown in Figure-19. Click on the **OK** button from the dialog box to reset steel frame members.

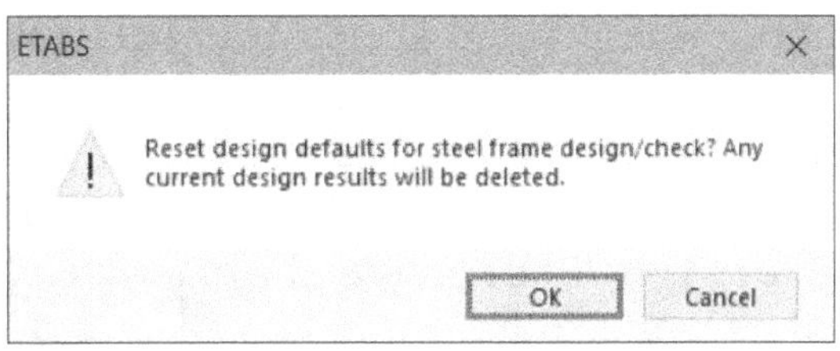

Figure-19. Dialog box for resetting steel frame parameters

Deleting Design Results

The **Delete Design Results** tool is used to delete the results of previous design check performed.

CONCRETE FRAME DESIGN

The tools in the **Concrete Frame Design** cascading menu are used to design concrete frames based on analysis and design check; refer to Figure-20. The tools in this cascading menu work in the same way as discussed for Steel Frame Design. Note that the tools in this menu will be active only when you have assigned concrete beams/ columns to the frame and you have performed the analysis.

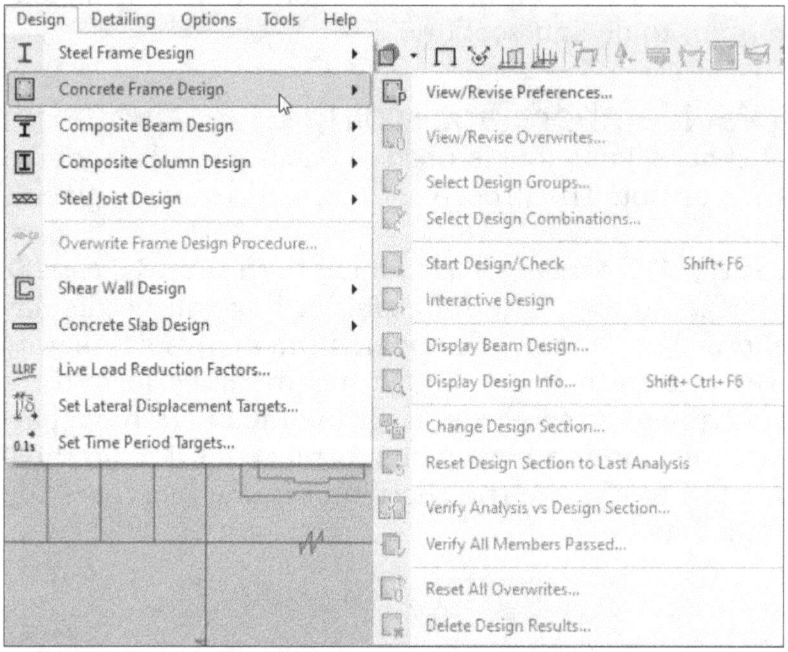

Figure-20. Concrete Frame Design cascading menu

COMPOSITE BEAM DESIGN, COMPOSITE COLUMN DESIGN, AND STEEL JOIST DESIGN

The tools in the **Composite Beam Design**, **Composite Column Design**, and **Steel Joist Design** cascading menus are same as discussed for Steel Frame Design.

OVERWRITE FRAME DESIGN PROCEDURE

The **Overwrite Frame Design Procedure** tool is used to overwrite the design postprocessor for selected frames. Using this tool, you can set composite design postprocessor for selected steel frame members or you can set steel postprocessor for selected composite beam. The procedure to use this tool is given next.

- Select the frame member for which you want to change the design postprocessor and click on the **Overwrite Frame Design Procedure** tool from the **Design** menu. The **Overwrite Frame Design Procedure** dialog box will be displayed; refer to Figure-21.

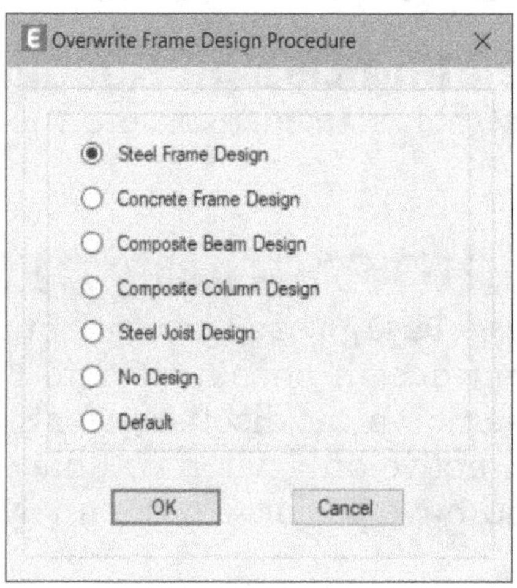

Figure-21. Overwrite Frame Design Procedure dialog box

- Select desired radio button from the dialog box and click on the **OK** button. The changes will be applied to selected members.

Note that there is a limitation to overwriting frame design procedures and depending on the frame section assignment made to the frame object, several of the options may not be available. The conditions for this tool are given next.

1. A concrete frame element can be switched between the Concrete Frame Design and the No Design procedures. Assign a concrete frame element the No Design (Null) design procedure if you do not want it designed by the Concrete Frame Design postprocessor.
2. A steel frame element can be switched between the Steel Frame Design, Composite Beam Design (if it qualifies), and the No Design design procedures. In this form a steel frame element qualifies for the Composite Beam Design procedure if it meets all of the following criteria.
   - The frame type is Beam, that is, the frame object is horizontal.
   - The frame element is oriented with its positive local 2-axis in the same direction as the positive global Z-axis (vertical upward).
   - The frame element has I-section or channel section properties.
   - Assign a steel frame element the No Design procedure if you do not want it designed by either the Steel Frame Design or the Composite Beam Design postprocessor.
3. A composite beam element can be switched between the Steel Frame Design and the No Design (see above).
4. A composite column element can be switched to No Design.
5. A steel joist element can be switched to No Design. Only beams that are assigned joist sections can be designed as steel joists. The joist sections can be selected from the built-in program joist section database, or they can be user defined. The user-defined sections can be specified using the Define menu > Frame Sections command and the Steel Joist option.
6. No Design option. This option is available for all frame objects. It means that the frame object is not designed by any design postprocessor. These frame objects are given the Null design procedure.
7. Default option. This option is available for all frame objects. It means that the frame object is to be given the default design procedure (postprocessor).

SHEAR WALL DESIGN AND CONCRETE SLAB DESIGN

The tools in the **Shear Wall Design** cascading menu are used to design load bearing walls; refer to Figure-22. The tools in the **Concrete Slab Design** cascading menu are used to design concrete slabs; refer to Figure-23. The tools in these cascading menus work in the same way as the tools in **Steel Frame Design** cascading menu.

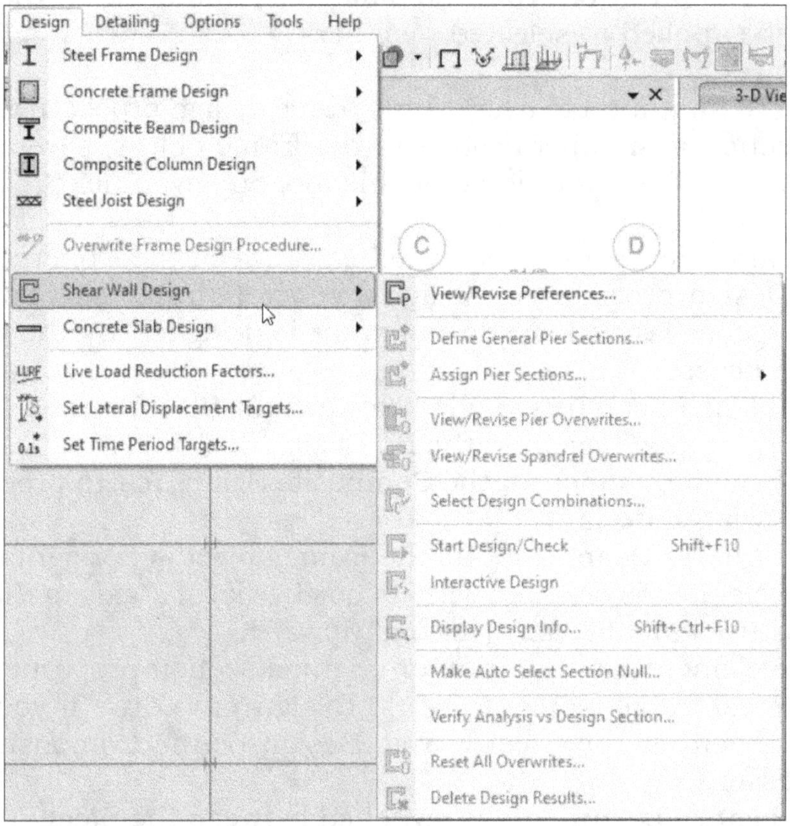

Figure-22. Shear Wall Design cascading menu

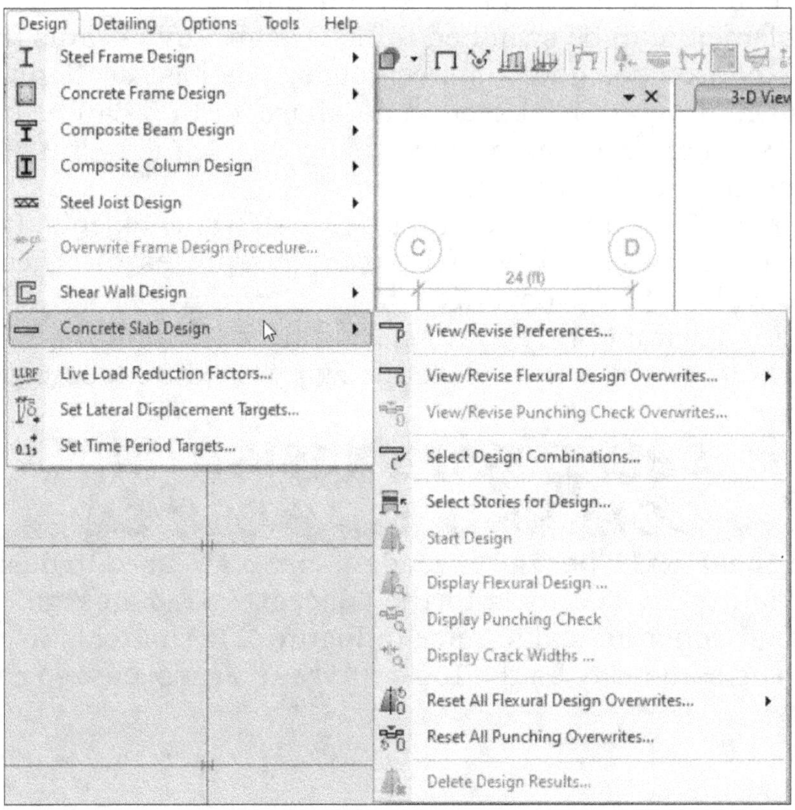

Figure-23. Concrete Slab Design cascading menu

LIVE LOAD REDUCTION FACTORS

The **Live Load Reduction Factors** tool is used to set preferences for reduction in live load during the course of analysis. Note that to use this tool, you must have specified a live reducible load in load patterns (the process to assign reducible live load has been discussed in previous chapter). The procedure to use this tool is given next.

- Click on the **Live Load Reduction Factors** tool from the **Design** menu. The **Live Load Reduction Factor** dialog box will be displayed; refer to Figure-24.

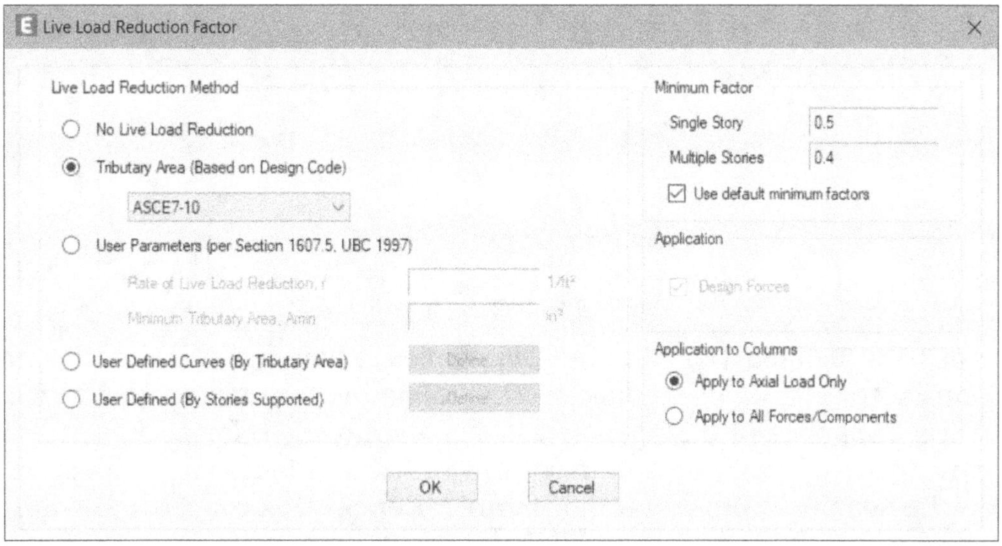

Figure-24. Live Load Reduction Factor dialog box

- Select the **No Live Load Reduction** button if you want no reduction in the live loads.
- Select the **Tributary Area (Based on Design Code)** radio button if you want to set the reduction of load based on the design code selected in the next drop-down.
- Select the **User Parameters (per Section 1607.5, UBC 1997)** radio button if you want to specify the rate of reduction and minimum tributary area manually in the edit boxes below this radio button. Tributary area is a loaded area that contributes to the load on the structural members supporting that area. It can also be called the load periphery.
- Select the **User Defined Curves (By Tributary Area)** radio button to define the curves for load reduction. Click on the **Define** button, the **Live Load Reduction Factor Curves** dialog box will be displayed; refer to Figure-25. Specify desired curve data in the **Trib. Area, ft2**, and **Red. Factor** edit boxes, and then click on the **Add** button on the right of the table to add it to the curve. Repeat this procedure to create the required curve. Click on the **Add Curve** button to add the next curve, if required. Preview of the curve will be displayed in the graph on the right in the dialog box. Click on the **OK** button from the dialog box to apply the changes.

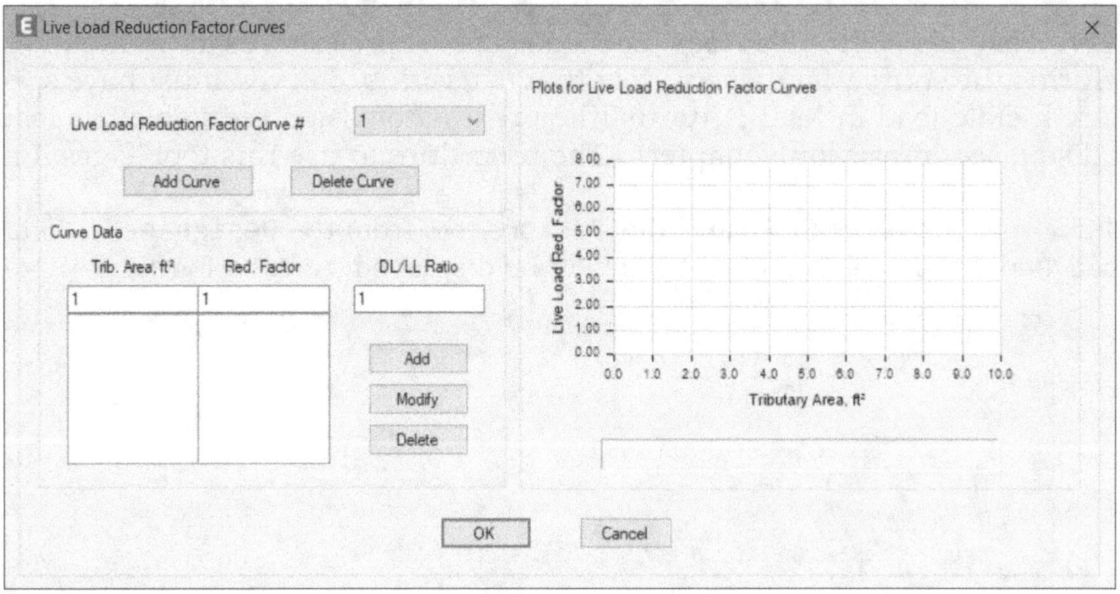

Figure-25. Live Load Reduction Factor Curves dialog box

- Select the **User Defined (By Stories Supported)** radio button if you want to define reduction factor story-wise. Click on the **Define** button, the **Live Load Reduction By Stories Supported** dialog box will be displayed; refer to Figure-26. Specify desired number of stories supported and reduction factor in respective edit box. Click on the **Add** button to add it in the table. Repeat this procedure to add required number of parameters and then click on the **OK** button.

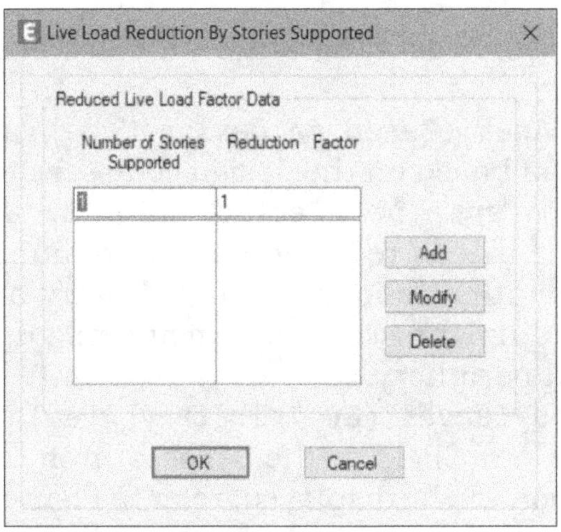

Figure-26. Live Load Reduction By Stories Supported dialog box

- Specify desired parameters in the **Minimum Factor** area of the dialog box if you have selected **Tributary Area (Based on Design Code)** or **User Parameters** radio button in the **Live Load Reduction Factor** dialog box.
- If you want to apply the load reduction only to axial forces then select the **Apply to Axial Load only** radio button otherwise select the **Apply to All Forces/ Components** radio button.
- Click on the **OK** button from the dialog box to apply the changes.

SETTING LATERAL DISPLACEMENT TARGETS

The **Set Lateral Displacement Targets** tool is used to define the target structural members which would increase in size after application of load. By default, ETABS predicts which members should be increased in size to control the displacements based on the energy per unit volume in the members. The members with more energy per unit volume are increased in size a larger percentage than those with smaller energies per unit volume. Some members with small energy per unit volume may be decreased in size if they are still acceptable for strength considerations. The procedure to use this tool is give next.

- Click on the **Set Lateral Displacement Targets** tool from the **Design** menu. The **Lateral Displacement Targets** dialog box will be displayed; refer to Figure-27.

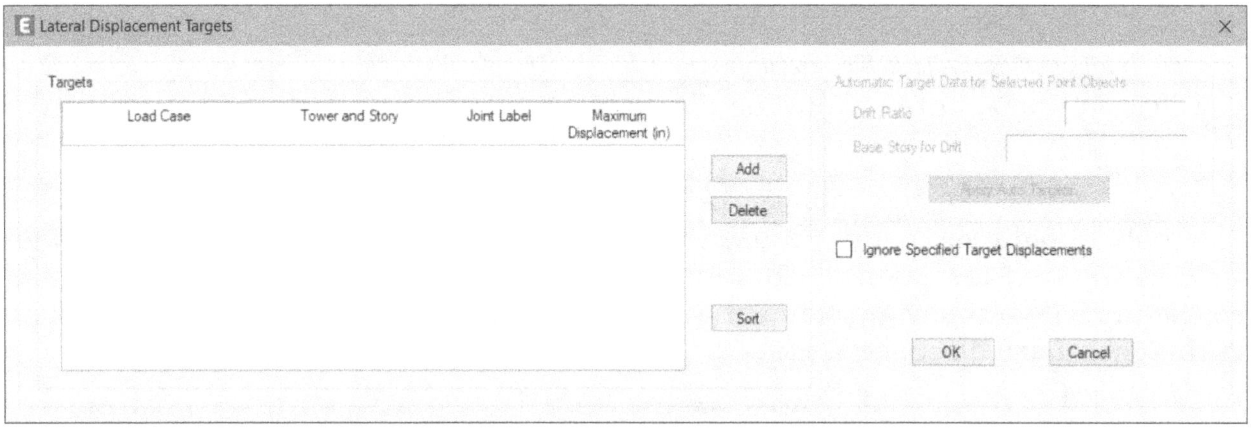

Figure-27. Lateral Displacement Targets dialog box

- Click on the **Add** button to add a new target. A new load case at default joint label will be added in the table. Set desired parameters in the table. Repeat this step to specify as many targets as you want.
- Select the **Ignore Specified Target Displacements** check box if you want the earlier specified displacement targets by design iterations to be ignored.
- The options in the **Automatic Target Data for Selected Point Objects** area are active only when you have selected joints before selecting this tool. Specify desired value for drift ratio in the **Drift Ratio** edit box. Select desired story level from the **Base Story for Drift** drop-down.
- Select the **Apply Auto Parts** button if you want to add auto calculated targets to selected points.
- After setting desired parameters, click on the **OK** button.

SETTING TIME PERIOD TARGETS

The **Set Time Period Targets** tool is used to set the mode and time period of load. On the basis of time period, the design iteration will modify the structure members. The procedure to use this tool is given next.

- Click on the **Set Time Period Targets** tool from the **Design** menu. The **Time Period Targets** dialog box will be displayed; refer to Figure-28.
- Click on the **Add** button from the dialog box. A new load case will be added. Specify the parameters as required. Repeat this step to get as many load cases as required.
- Select the **Ignore Specified Time Period Targets** check box if you want to ignore the automatically specified targets during design iteration.

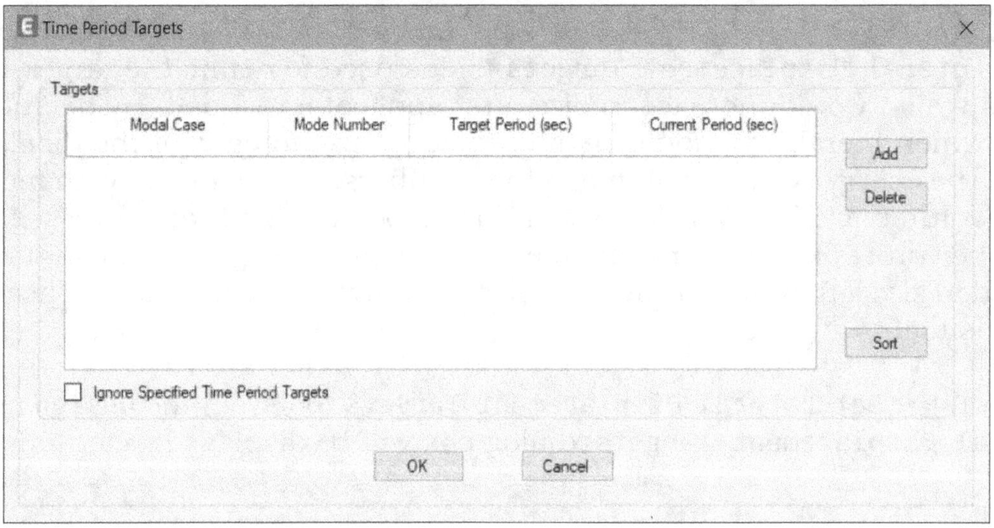

Figure-28. Time Period Targets dialog box

- Click on the **OK** button from the dialog box to apply the changes.

OPTIONS

The tools in **Options** menu are used to manage overall preferences of the software for creating and displaying model in graphics area; refer to Figure-29. Depending on the model, you may get some of the tools inactive in this menu. Various options in this menu are discussed next.

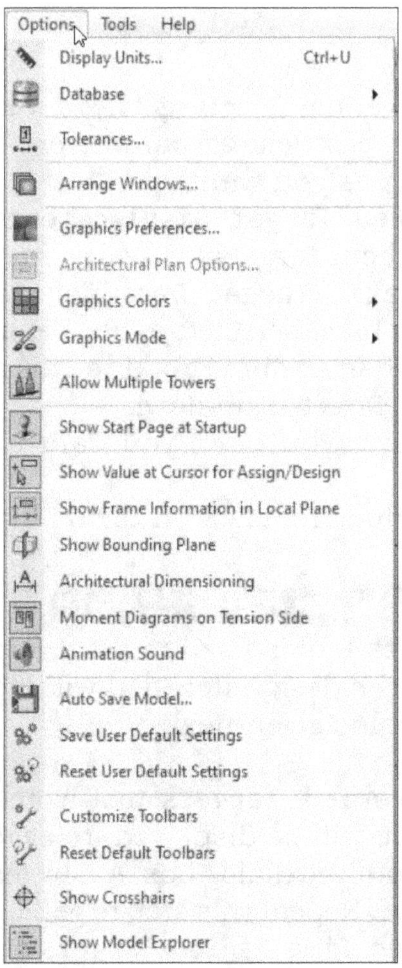

Figure-29. Options menu

Setting Display Units

The **Display Units** tool is used to set units for various parameters of the model like length, area, force, moment and so on. On clicking this tool, the **Display Units** dialog box is displayed; refer to Figure-30. The edit-able fields of table will be displayed in white color and inactive fields will be displayed in gray color. Click in the field you want to edit, the unit drop-down will be displayed. Click on the drop-down button and select desired unit available. Select the **Expanded Form** check box to display all the parameters in expanded dialog box. Select the **Use Architectural Dimensioning** check box to display fractions of unit as 1/2, 1/4, 1/8, 1/16, and so on. After setting desired parameters, click on the **OK** button from the dialog box.

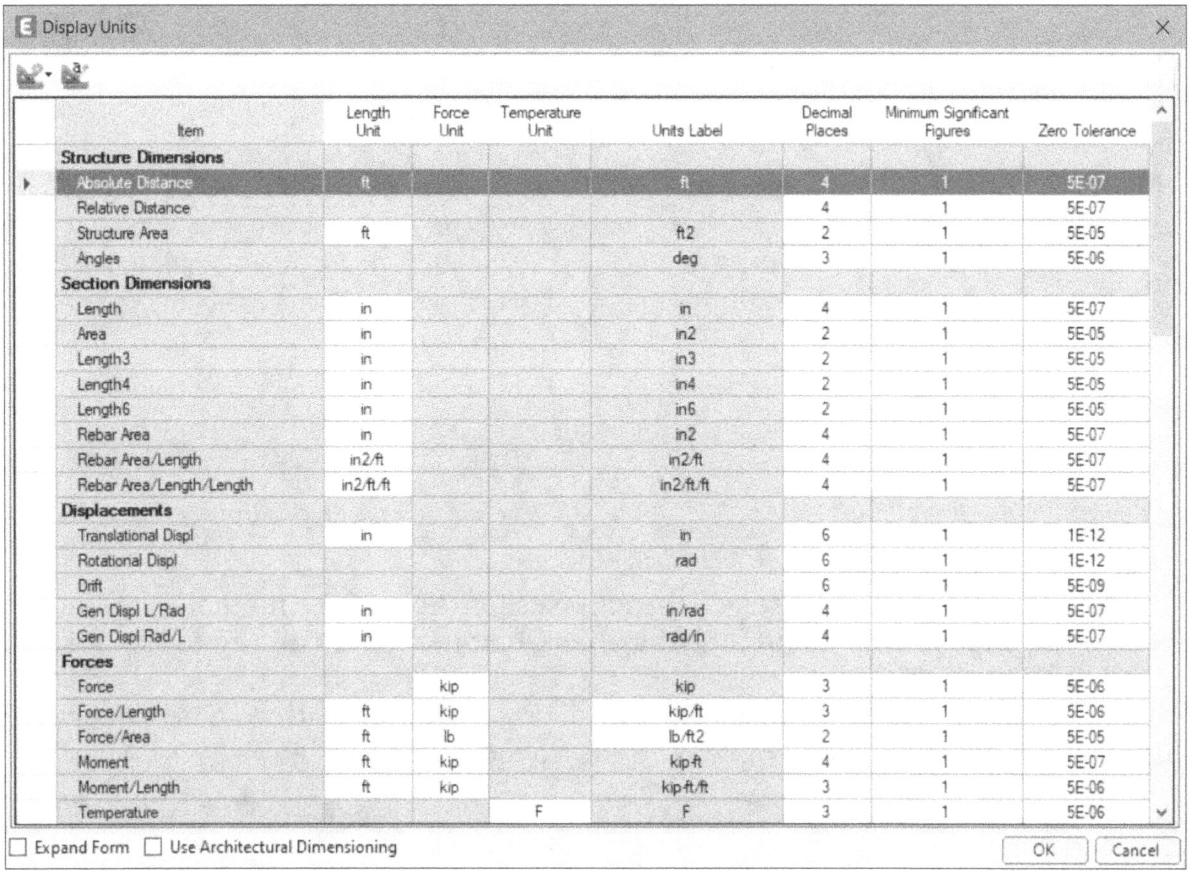

Item	Length Unit	Force Unit	Temperature Unit	Units Label	Decimal Places	Minimum Significant Figures	Zero Tolerance
Structure Dimensions							
Absolute Distance	ft			ft	4	1	5E-07
Relative Distance					4	1	5E-07
Structure Area	ft			ft2	2	1	5E-05
Angles				deg	3	1	5E-06
Section Dimensions							
Length	in			in	4	1	5E-07
Area	in			in2	2	1	5E-05
Length3	in			in3	2	1	5E-05
Length4	in			in4	2	1	5E-05
Length6	in			in6	2	1	5E-05
Rebar Area	in			in2	4	1	5E-07
Rebar Area/Length	in2/ft			in2/ft	4	1	5E-07
Rebar Area/Length/Length	in2/ft/ft			in2/ft/ft	4	1	5E-07
Displacements							
Translational Displ	in			in	6	1	1E-12
Rotational Displ				rad	6	1	1E-12
Drift					6	1	5E-09
Gen Displ L/Rad	in			in/rad	4	1	5E-07
Gen Displ Rad/L	in			rad/in	4	1	5E-07
Forces							
Force		kip		kip	3	1	5E-06
Force/Length	ft	kip		kip/ft	3	1	5E-06
Force/Area	ft	lb		lb/ft2	2	1	5E-05
Moment	ft	kip		kip-ft	4	1	5E-07
Moment/Length	ft	kip		kip-ft/ft	3	1	5E-06
Temperature			F	F	3	1	5E-06

Figure-30. Display Units dialog box

Database Management

The tools to manage database of various objects in ETABS model are available in the **Edit** menu and **Database** cascading menu of **Options** menu. These tools are discussed next.

Interactive Database Editing

The **Interactive Database** tool in **Edit** menu is used to modify selected tables of database interactively. These tables hold the data of all the objects created in the model. The procedure to use this tool is given next.

- Click on the **Interactive Database** tool from the **Edit** menu. The **Choose Tables for Interactive Editing** dialog box will be displayed; refer to Figure-31.

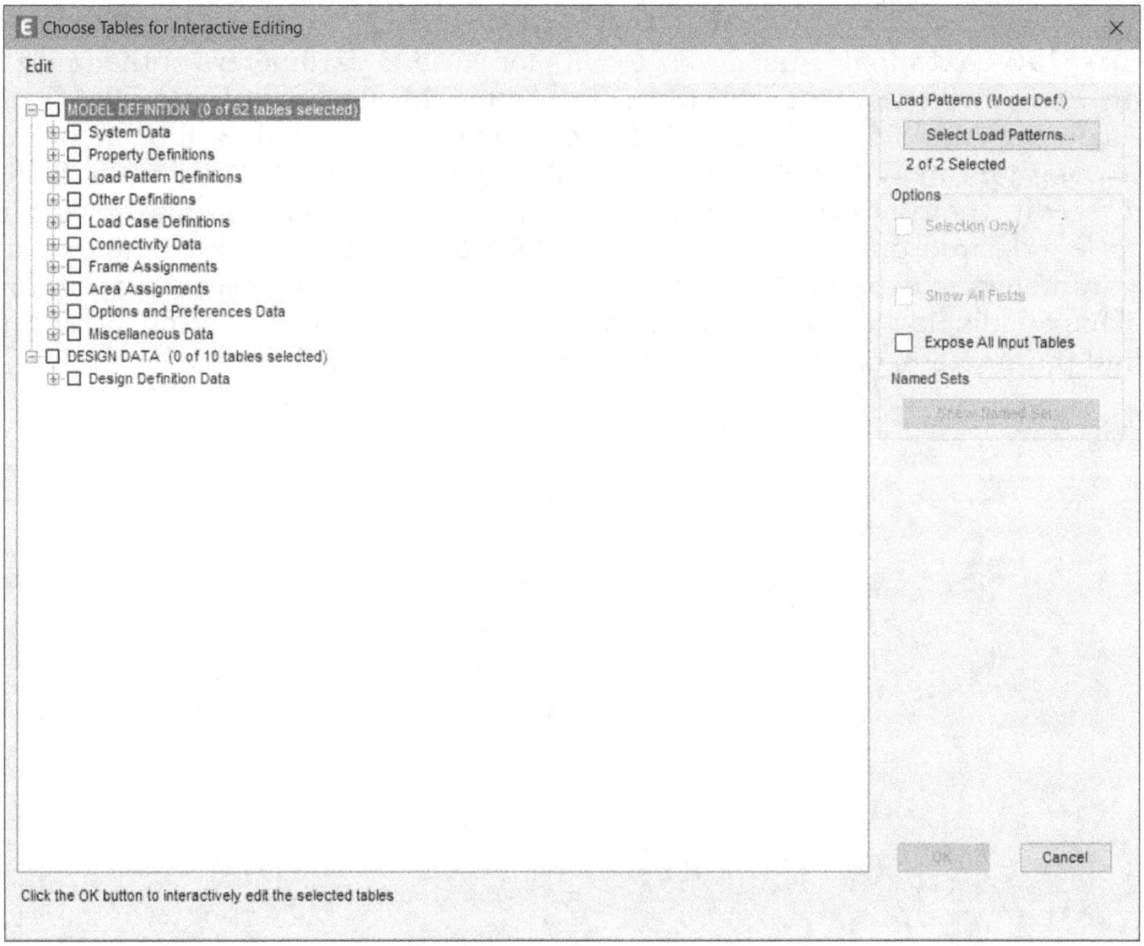

Figure-31. Choose Tables for Interactive Editing dialog box

- Expand the categories and select check boxes for tables to be modified interactively.
- Select the **Expose All Input Tables** check box to display all the tables for which data can be specified.
- After selecting tables to be modified, click on the **OK** button. The **Interactive Database Editing** dialog box will be displayed; refer to Figure-32.

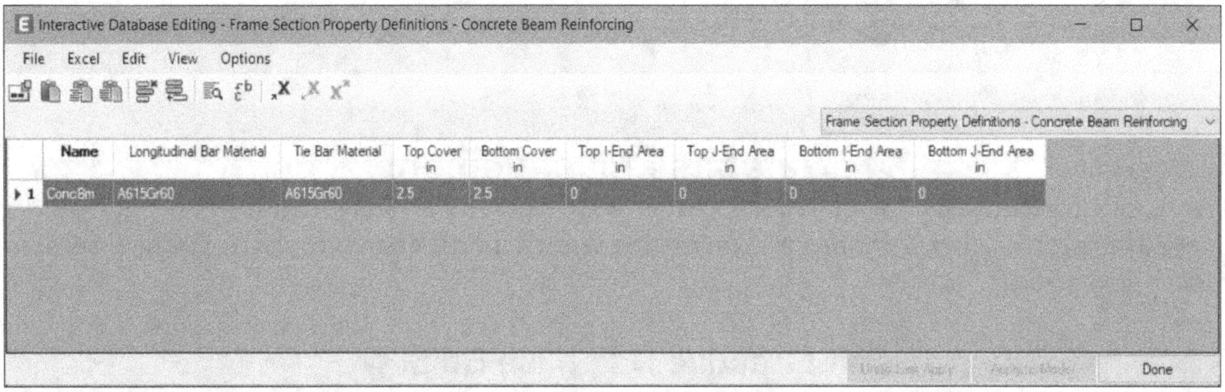

Figure-32. Interactive Database Editing window

- Triple-click in the field you want to edit in the table and specify desired value.
- Select the table to be modified from the drop-down at the top-right corner in the window; refer to Figure-33.

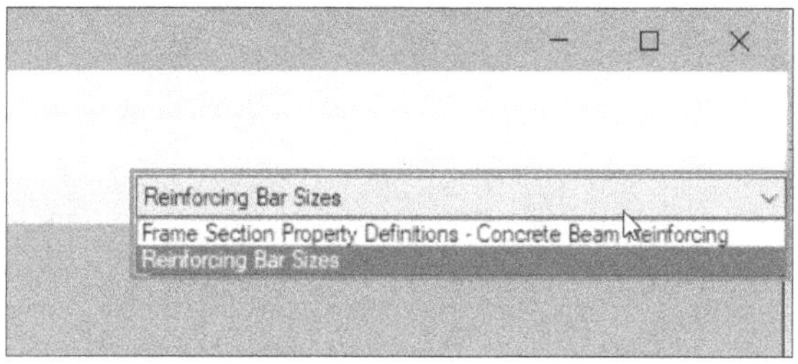

Figure-33. Drop-down for selecting database table

- Using the tools in **Toolbar** of window, you can copy, paste, append, delete, and perform other operations in rows of the table.
- To edit the table in Microsoft Excel, click on the **To Excel** tool from **Toolbar**. The file will be displayed in Microsoft Excel. Modify the table in Excel and click on the **Retrieve Table from Excel** tool from the **Excel** menu in the window. The values will be modified accordingly in the table.
- After performing desired changes in database, click on the **Done** button from the **File** menu. The warning message will be displayed whether you want to apply changes in model or not based on changes in the table. Click on the **Yes** button to modify model.

Setting Database Table Format File

The **Set Current Format File Source** tool is used to select file for modifying table format. The procedure to use this tool is given next.

- Click on the **Set Current Format File Source** tool from the **Database** cascading menu of the **Options** menu. The **Set Database Table Format File** dialog box will be displayed; refer to Figure-34.

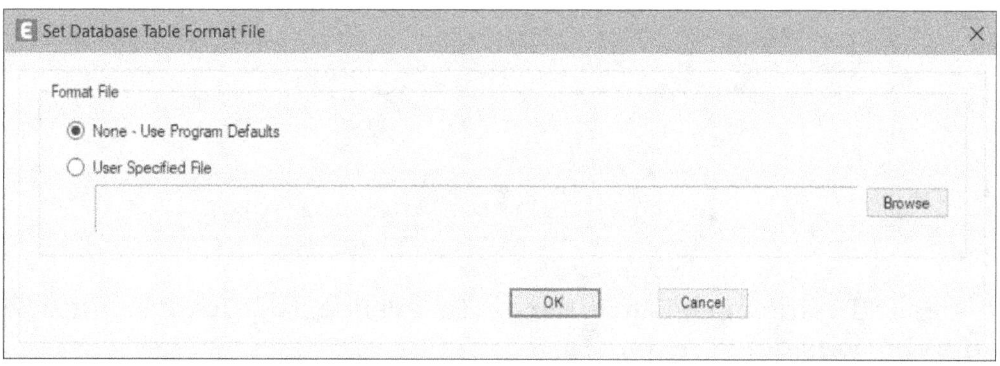

Figure-34. Set Database Table Format File dialog box

- Select the **None** radio button to use default formats specified in software preferences.
- Select the **User Specified File** radio button to select file for defining formats of tables using selected format file. On selecting the radio button, **Set Current Table XML Format File** dialog box will be displayed; refer to Figure-35.

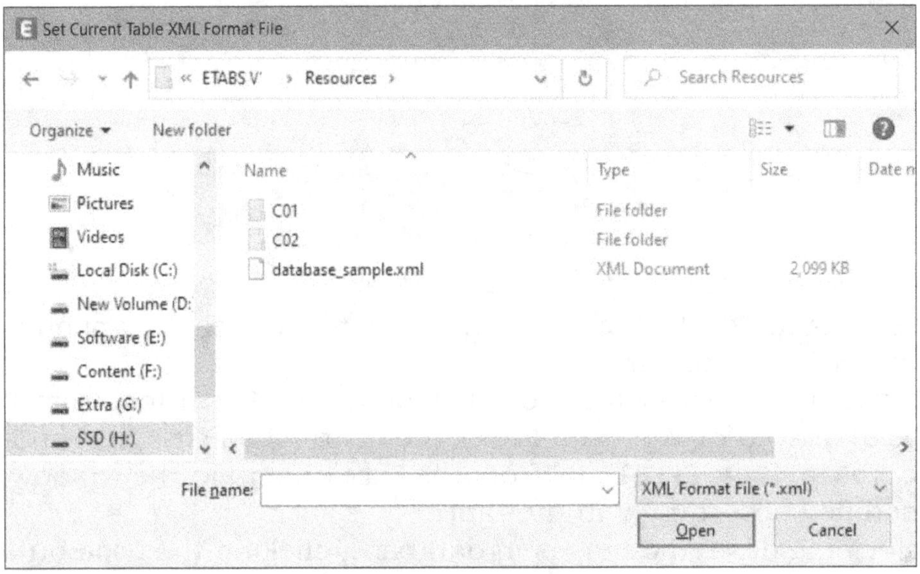

Figure-35. Set Current Table XML Format File dialog box

- Select desired file and click on the **Open** button to use selected format file.

Similarly, you can use other tools in the **Database** cascading menu.

Setting Tolerances

The **Tolerances** tool in **Options** menu is used to define the gap which will be considered as one continuous object. If the gap between two joining objects is less than specified tolerance then two objects will be considered as joint. On clicking this tool, the **Set Tolerances** dialog box will be displayed; refer to Figure-36.

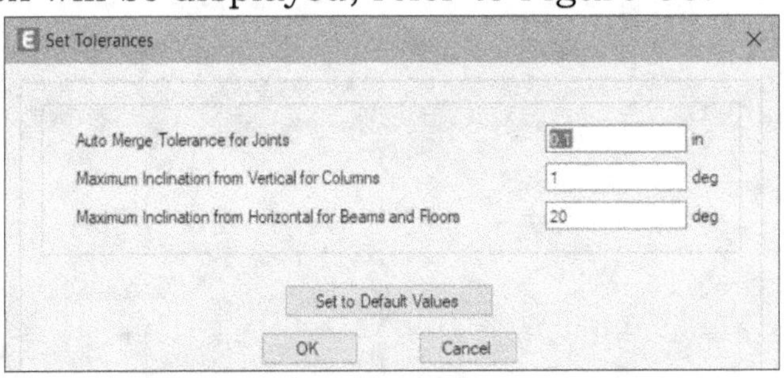

Figure-36. Set Tolerances dialog box

- Specify desired value in the **Auto Merge Tolerance for Joints** edit box to define the gap to be considered as joined.
- The value specified in **Maximum Inclination from Vertical for Columns** edit box is used to define the angle value up to which inclined vertical objects will be considered as columns.
- The value specified in **Maximum Inclination from Horizontal for Beams and Floors** edit box is used to define the angle value up to which inclined horizontal objects will be considered as beams and floors.
- Click on the **Set to Default Values** button to set program default values in the edit boxes.
- Click on the **OK** button to set the tolerances.

Arranging Windows

The **Arrange Windows** tool in **Options** menu is used to automatically arrange viewports side by side in the graphics area. This tool is useful when viewports are overlapping each other to arrange them properly.

Graphics Preferences

The **Graphics Preferences** tool is used to modify fonts and quality of objects displaying in the graphics area. The procedure to use this tool is given next.

- Click on the **Graphics Preferences** tool from the **Options** menu. The **Graphics Preferences** dialog box will be displayed; refer to Figure-37.

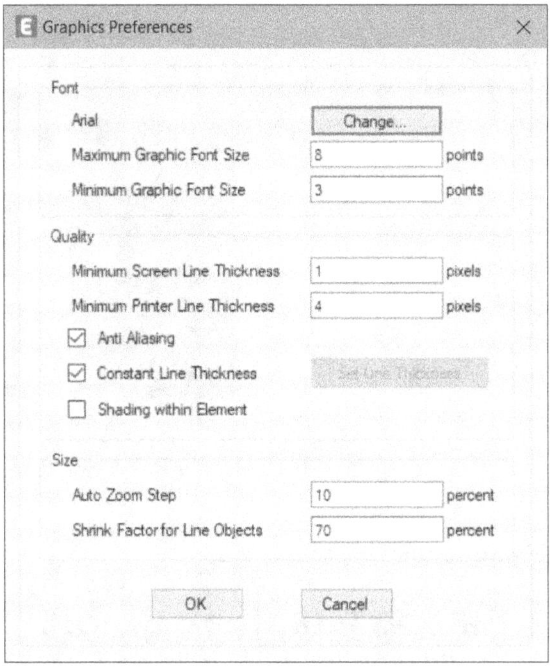

Figure-37. Graphics Preferences dialog box

- Click on the **Change** button from the dialog box to change the font for model. The **Font** dialog box will be displayed. Set desired font, style, and size for model and then click on the **OK** button.
- Specify desired values in the **Maximum Graphic Font Size** and **Minimum Graphic Font Size** edit boxes to define range within which the font size will be rendered in graphics area depending on model size.
- Specify desired values in the **Minimum Screen Line Thickness** and **Minimum Printer Line Thickness** edit boxes to define minimum thickness of lines of model in graphics area and printed copy.
- Select the **Anti Aliasing** check box to eliminate jaggies in the model display at the cost of some processing power of computer.
- Select the **Constant Line Thickness** check box to keep the thickness of lines uniform.
- Select the **Shading within Elements** check box to add shading effect on objects. Make sure to select **Object Fill** check box in **Set View Options** dialog box to use this option.
- Specify the zoom step in percentage in the **Auto Zoom Step** edit box to define the amount of reduction/increase in size of object when using zoom function.
- Similarly, specify desired value in the **Shrink Factor for Line Objects** edit box to define the percentage by which object will shrink when using the **Object Shrink Toggle** button.
- After setting desired parameters, click on the **OK** button.

Architectural Plan Options

The **Architectural Plan Options** tool is used to manage display of various layers of architectural objects imported in the model. On clicking this tool, the **Architectural Plan Options** dialog box will be displayed; refer to Figure-38. Set desired color and visibility of layers in the table and click on the **OK** button.

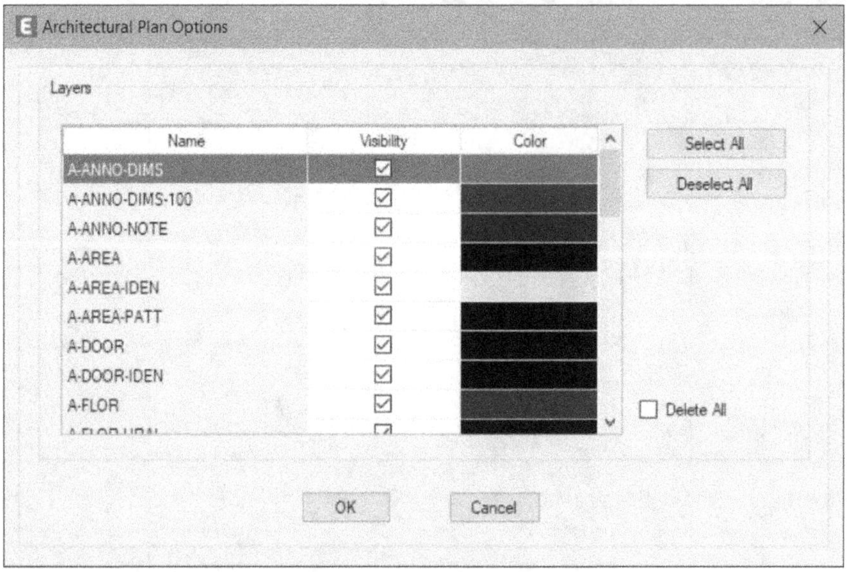

Figure-38. Architectural Plan Options dialog box

Setting Graphics Colors

The options in the **Graphics Colors** cascading menu are used to modify colors for various objects in the graphics area. Select the **Display** option from the cascading menu to set colors for columns, beams, walls, and so on. The **Set Display Colors** dialog box will be displayed; refer to Figure-39. Click in the color field for desired object to change the color. Similarly, set the other parameters and click on the **OK** button.

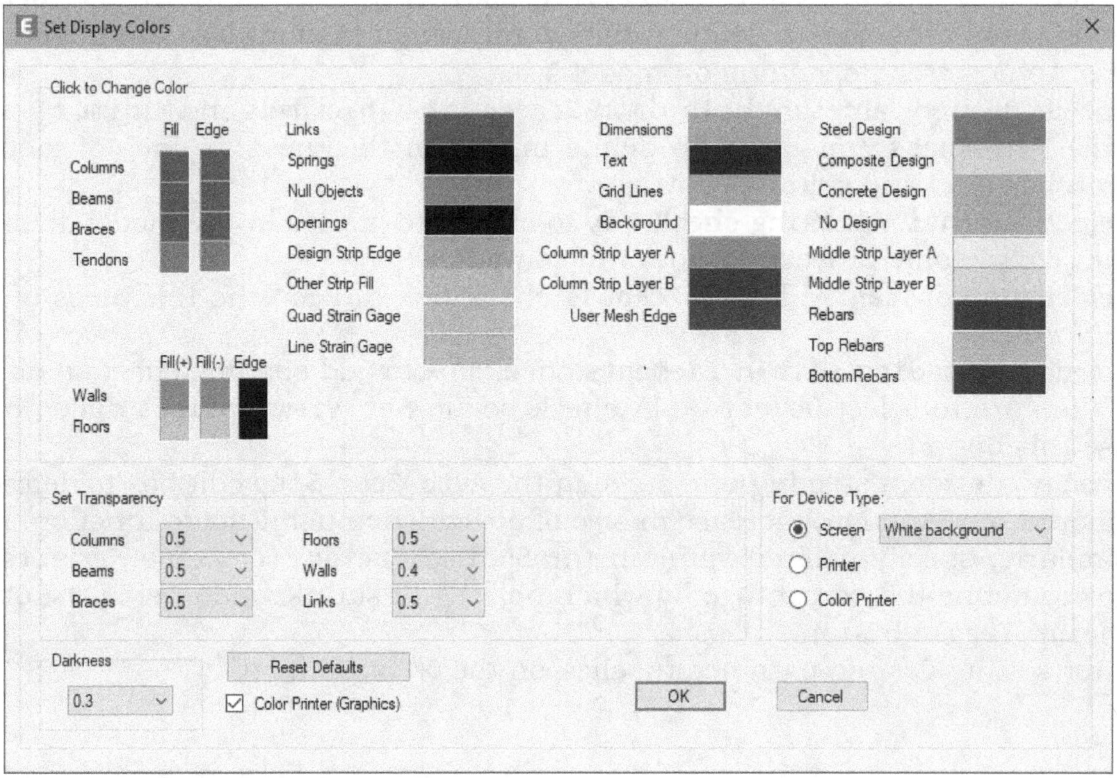

Figure-39. Set Display Colors dialog box

Similarly, you can use **Output** option from the **Graphics Colors** cascading menu of the **Options** menu to define output colors.

Setting Graphics Mode

The options in the **Graphics Mode** cascading menu are used to set mode to be used for rendering objects in the application. Select the DirectX option to use software rendering.

Allowing Multiple Towers

Select the **Allow Multiple Towers** toggle button from the **Options** menu to create and manage multiple towers; refer to Figure-40.

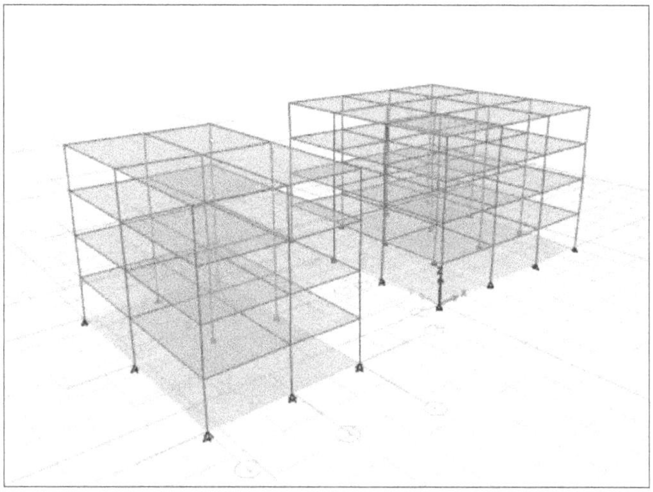

Figure-40. Multiple models

Creating Tower Property

To work with multiple towers in ETABS, you need to create multiple tower properties so that different objects can be assigned to their respective towers. The procedure to create tower properties is given next.

- After activating **Allow Multiple Towers** toggle button, click on the **Edit Towers, Stories and Grid Systems** tool from the **Edit** menu. The **Edit Towers, Stories and Grid Systems** dialog box will be displayed; refer to Figure-41.

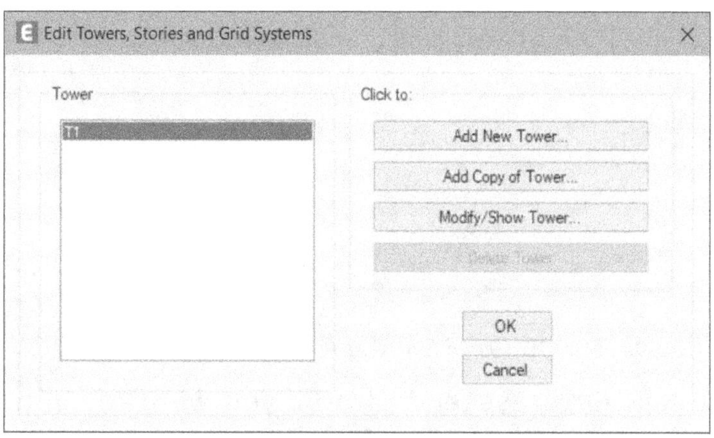

Figure-41. Edit Towers, Stories and Grid Systems dialog box

- Click on the **Add New Tower** button from the dialog box to create a new tower property. The **Quick Story Definition for New Tower** dialog box will be displayed; refer to Figure-42.

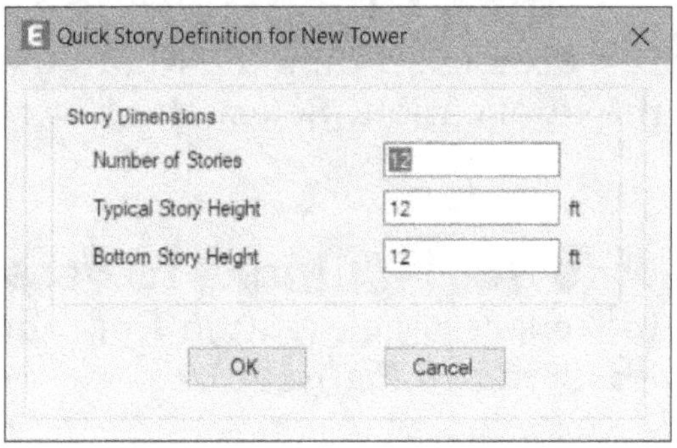

Figure-42. Quick Story Definition for New Tower dialog box

- Specify desired number of stories and other data in the edit boxes and then click on the **OK** button. The **Tower, Story and Grid System Data** dialog box will be displayed; refer to Figure-43.
- Specify the name of new tower in **Tower Name** edit box.

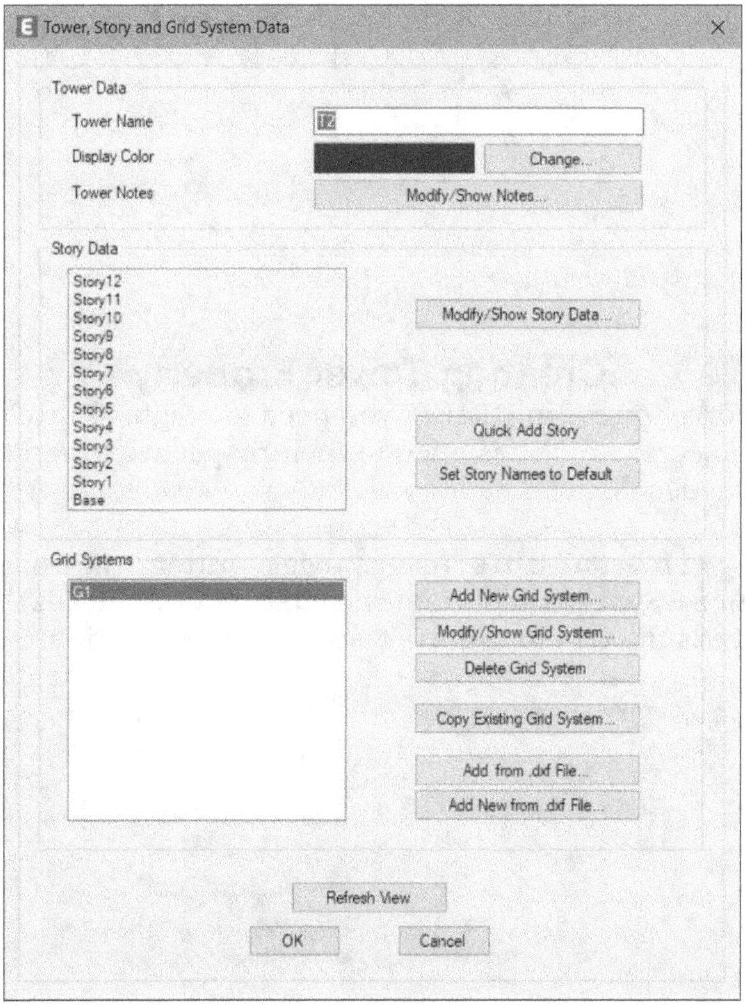

Figure-43. Tower Story and Grid System Data dialog box

- Click on the **Add New Grid System** button if you want to change the layout of the building. The **Add New Grid System Quick** dialog box will be displayed; refer to Figure-44.

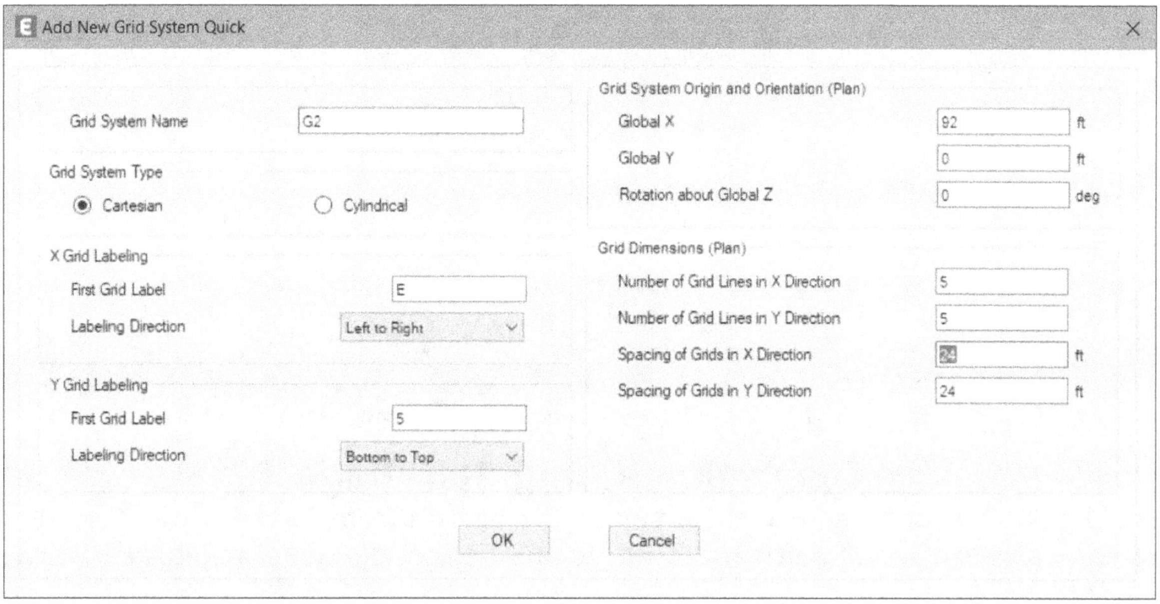

Figure-44. Add New Grid System Quick dialog box

- Specify the distance for new grid system from origin in **Global X** and **Global Y** edit boxes along X and Y axes of origin, respectively.
- Set the other parameters as discussed earlier and click on the **OK** button. The **Tower, Story and Grid System Data** dialog box will be displayed again. Set the parameters as discussed earlier and click on the **OK** button to create the tower property. The **Edit Towers, Stories and Grid Systems** dialog box will be displayed.
- Click on the **OK** button to exit the dialog box. The grid system of new tower will be displayed; refer to Figure-45.

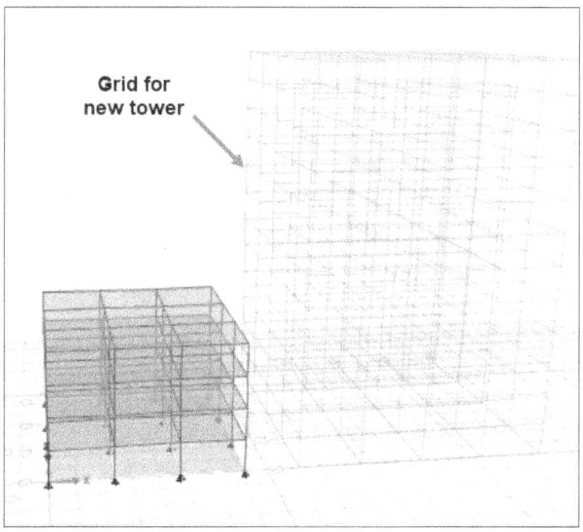

Figure-45. Grid for new tower

- Draw the objects like beams, columns, and so on as discussed earlier on new grid lines to add them to new tower.

Assigning Objects to Towers

You can assign selected objects to a tower if you are working with multiple towers. To do so, select the objects to be assigned to a tower and click on the **Assign Objects to Tower** tool from the **Assign** menu. The **Assign Objects to Tower** dialog box will be displayed; refer to Figure-46. Select desired tower from the list and click on the **OK** button to assign objects to selected tower.

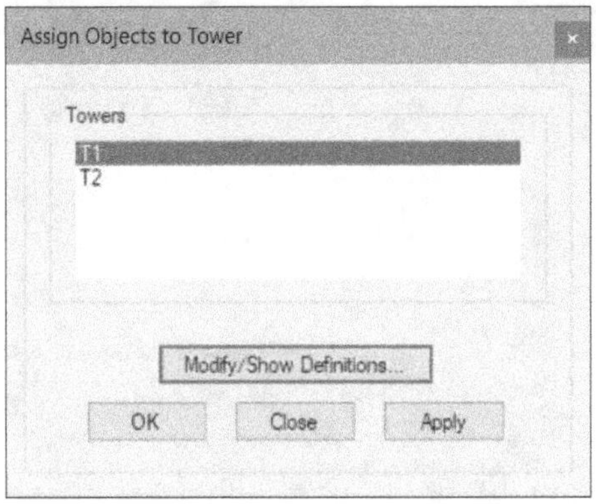

Figure-46. Assign Objects to Tower dialog box

Displaying/Hiding Start Page

Select the **Show Start Page at Startup** toggle button from **Options** menu to display start page at the start of software. To hide start page, de-select the toggle button.

Showing Value at Cursor for Assign/Design

Select the **Show Value at Cursor for Assign/Design** toggle button from the **Options** menu to display/hide assigned property of objects in model when you hover cursor on them; refer to Figure-47.

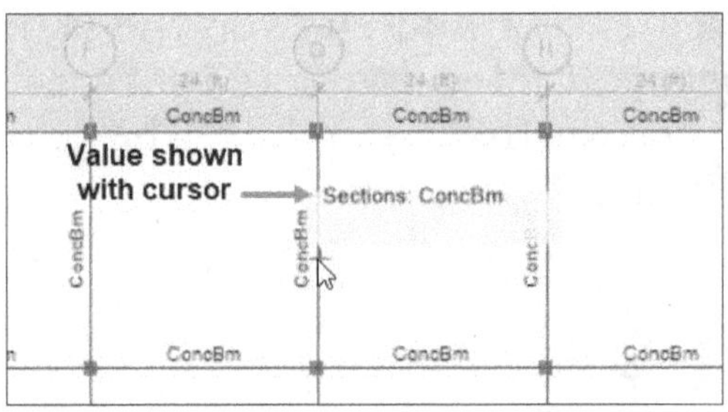

Figure-47. Value shown at cursor

Showing Bounding Plane

Select the **Show Bounding Plane** toggle button from the **Options** menu to display/ hide bounding plane; refer to Figure-48.

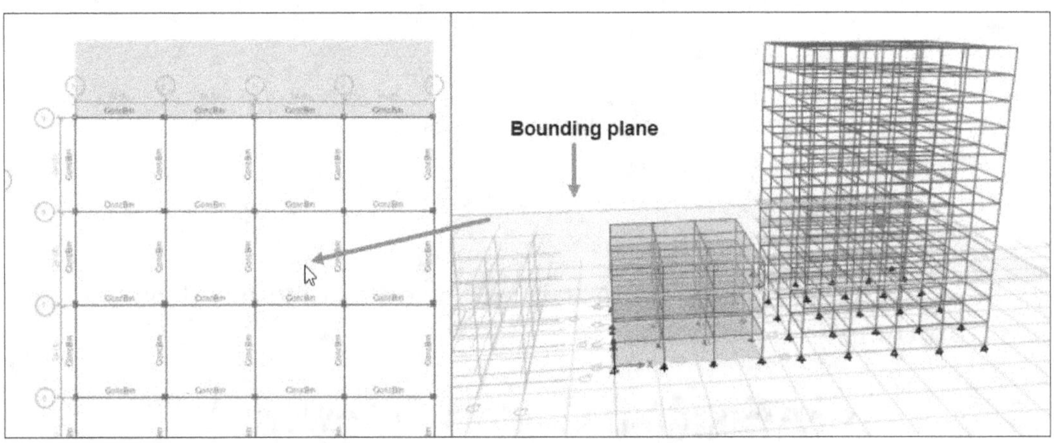

Figure-48. Bounding plane preview

Displaying Architectural Dimensioning

Select the **Architectural Dimensioning** toggle button from the **Options** menu to display architectural dimensioning in graphics area.

Moment Diagrams on Tension Side

Select the **Moment Diagrams on Tension Side** toggle button from the **Options** menu to display bending moment diagrams in results at lower side of horizontal axis (-ve Y).

Playing/Stopping Animation Sound

By default, animation sound is played when you run result animation. To stop playing sound during animation, deselect the **Animation Sound** toggle button from the **Options** menu.

Defining Auto Save Time for Model

Click on the **Auto Save Model** option from the **Options** menu to set time duration after which model will automatically saved. The **Model Auto Save Options** dialog box will be displayed; refer to Figure-49. Select the **Auto Save Model File at Specified Interval** check box to specify time in minutes after which model will be saved automatically and click on the **OK** button.

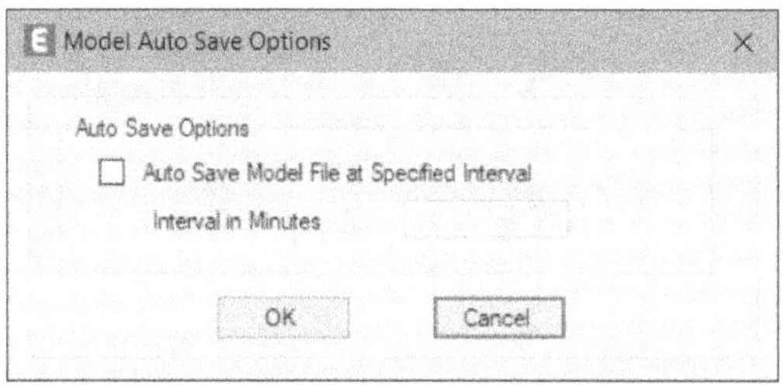

Figure-49. Model Auto Save Options dialog box

Saving and Resetting User Default Settings

Click on the **Save User Default Settings** option from the **Options** menu, the user settings defined in the software will be saved as default settings. Click on the **Reset User Default Settings** option from the menu to reset user settings for software defaults.

Customizing Toolbars

The **Customize Toolbars** tool is used to add or remove tools from the toolbars. On clicking this tool, the **Customize** dialog box will be displayed; refer to Figure-50. Select check boxes for the toolbars that you want to display in the interface. Click on the **Keyboard** tab and specify shortcut key for selected tool in the dialog box; refer to Figure-51.

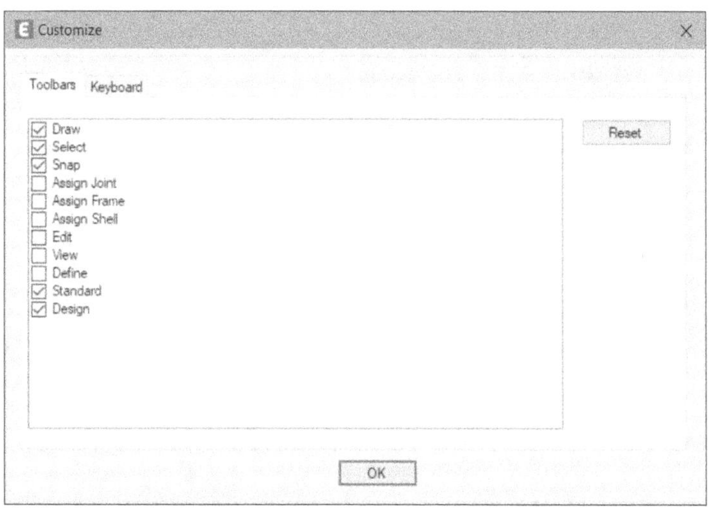

Figure-50. Customize dialog box

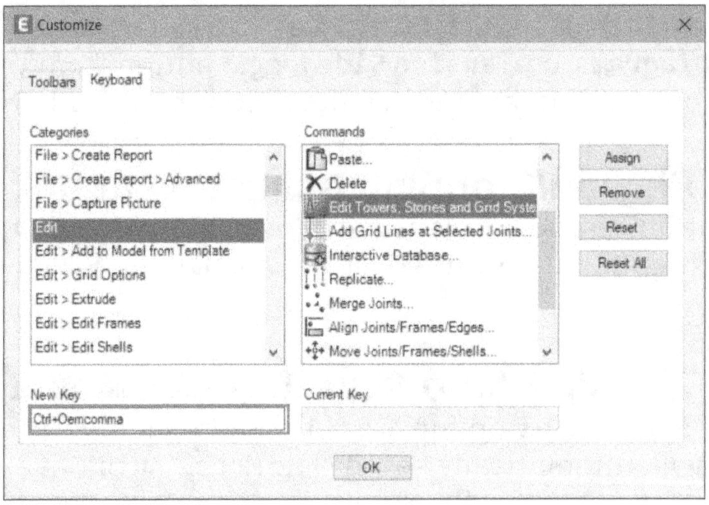

Figure-51. Specifying shortcut key

After setting desired parameters, click on the **OK** button from the dialog box. If you want to reset toolbar to software default then click on the **Reset Default Toolbars** tool from the **Options** menu.

Showing/hiding Crosshairs

The **Show Crosshairs** toggle button is used to display/hide crosshair when you hover cursor on a snap point.

Showing/hiding Model Explorer

The **Show Model Explorer** toggle button is used to display/hide Model Explorer at the left in the application window.

SELF-ASSESSMENT

Q1. Which menu contains the Steel Frame Design options in ETABS?
A. File menu
B. Design menu
C. Display menu
D. Options menu

Q2. What are Steel Frame Preferences used for?
A. Drawing walls
B. Creating loads
C. Setting rules for all steel frame members
D. Printing tables

Q3. Which tool is used to open the Steel Frame Design Preferences dialog box?
A. View/Revise Preferences
B. Start Design/Check
C. Display Design Info
D. Verify All Members Pass

Q4. Which option is used to perform Time History, Multi-step Static, and Nonlinear Static analysis step by step?
A. Last Step
B. Step-by-Step-All
C. Envelopes-All
D. Direct Analysis

Q5. Which Multi-Response Case Design option compiles a final result after complete analysis?
A. Step-by-Step-All
B. Last Step
C. Envelopes-All
D. Limited 1st Order

Q6. Which framing type refers to Special Moment Frame?
A. IMF
B. OMF
C. SMF
D. EBF

Q7. Which design provision method stands for Load and Resistance Factor Design?
A. ASD
B. LRFD
C. STMF
D. BRBF

Q8. Which analysis method does NOT include secondary P-Δ and P-δ effects?
A. Direct Analysis
B. Effective Length
C. Limited 1st Order
D. Nonlinear Static

Q9. Which analysis method uses K = 1.0 instead of effective length factors?
A. Direct Analysis
B. Effective Length
C. Limited 1st Order
D. ASD Method

Q10. Which stiffness reduction option applies no reduction factor?
A. Tau-b Variable
B. Tau-b Fixed
C. No Modification
D. Direct Method

Q11. Which tool is used to overwrite default preferences for selected steel frame sections?
A. Change Design Section
B. View/Revise Overwrite
C. Display Design Info
D. Reset All Overwrites

Q12. What is the purpose of lateral bracing in steel structures?
A. To reduce dead load
B. To prevent trusses from twisting
C. To create load combinations
D. To change material properties

Q13. Which radio button automatically sets parameters for lateral bracing?
A. User Specified
B. Program Determined
C. Absolute Distance
D. Relative Distance

Q14. Which option allows specifying brace distance as a ratio of total length?
A. Absolute Distance from I-End
B. User Specified
C. Relative Distance from I-End
D. Point Braces

Q15. Which tool is used to select design load combinations for the project?
A. Select Design Groups
B. Select Design Combinations
C. Start Design/Check
D. Display Design Info

Q16. Which tabs are available in the Design Load Combinations Selection dialog box?
A. Load and Force
B. Graph and Table
C. Strength and Deflection
D. Model and Design

Q17. Which tool is used to perform design checks on the structure?
A. Start Design/Check
B. Verify All Members Pass
C. Interactive Design
D. Change Design Section

Q18. In design check results, which color generally indicates high deflection?
A. Cyan
B. Green
C. Blue
D. Red

Q19. What does the RATIO column represent in Steel Stress Check Information?
A. Load factor
B. Total PMM stress ratio
C. Story drift
D. Grid spacing

Q20. What does (T) beside the stress ratio indicate?
A. Torsion
B. Temperature
C. Tension
D. Thickness

Q21. Which component represents bending about the major axis?
A. B-MAJ
B. B-MIN
C. AXL
D. MIN SHR RATIO

Q22. Which tool displays frame members in shades according to deformation?
A. Display Design Info
B. Interactive Design
C. Verify Analysis vs Design Section
D. Show Tables

Q23. Which tool is used to clear the list of auto select sections?
A. Make Auto Select Section Null
B. Reset Design Section to Last Analysis
C. Change Design Section
D. Reset All Overwrites

Q24. Which tool is used to change the size of selected frame sections?
A. Verify All Members Pass
B. Change Design Section
C. Display Design Info
D. Start Design/Check

Q25. Which tool checks whether assigned section values match analysis-generated values?
A. Verify Analysis vs Design Section
B. Reset All Overwrites
C. Select Design Groups
D. Live Load Reduction Factors

Q26. Which tool is used to automatically arrange overlapping viewports in ETABS?
A. graphics preferences
B. arrange windows
C. architectural plan options
D. graphics colors

Q27. The Arrange Windows tool is available in which menu?
A. analyze
B. display
C. options
D. assign

Q28. Which tool is used to modify fonts and display quality of objects in graphics area?
A. graphics preferences
B. graphics mode
C. display units
D. model explorer

Q29. Which button is clicked to change the font for the model in Graphics Preferences?
A. apply
B. modify
C. change
D. format

Q30. Which option is used to remove jaggies in model display?
A. shading within elements
B. anti aliasing
C. constant line thickness
D. object fill

Q31. Which check box keeps the thickness of lines uniform?
A. anti aliasing
B. object fill
C. constant line thickness
D. shrink factor

Q32. Which option adds shading effect on objects in graphics area?
A. shading within elements
B. anti aliasing
C. directx
D. architectural dimensioning

Q33. The Auto Zoom Step edit box defines:
A. font size
B. line thickness
C. zoom percentage
D. object color

Q34. Which option defines the percentage by which an object shrinks?
A. auto zoom step
B. shrink factor for line objects
C. minimum line thickness
D. anti aliasing

Q35. Which tool is used to manage visibility of imported architectural layers?
A. graphics mode
B. architectural plan options
C. arrange windows
D. model explorer

Q36. Which cascading menu is used to modify colors of objects in graphics area?
A. graphics preferences
B. graphics colors
C. graphics mode
D. display units

Q37. Which option in Graphics Colors is used to define colors for beams and columns?
A. display
B. output
C. render
D. format

Q38. Which option in Graphics Colors is used to define output colors?
A. render
B. output
C. display
D. export

Q39. Which option in Graphics Mode is selected to use software rendering?
A. opengl
B. vulkan
C. directx
D. renderx

Q40. Which toggle button is used to create and manage multiple towers?
A. allow multiple towers
B. show start page
C. show bounding plane
D. animation sound

Q41. Which tool is used to create a new tower property?
A. add new section
B. edit towers, stories and grid systems
C. graphics preferences
D. assign objects to tower

Q42. Which button is clicked to create a new tower in ETABS?
A. add new story
B. add new tower
C. add grid system
D. create structure

Q43. Which dialog box appears after clicking Add New Tower?
A. graphics preferences
B. tower data
C. quick story definition for new tower
D. assign tower

Q44. In which edit box is the name of a new tower specified?
A. tower id
B. tower name
C. structure name
D. story name

Q45. Which button is used to create a new grid layout for tower property?
A. add new grid system
B. add story grid
C. grid layout
D. create axis

Q46. Global X and Global Y edit boxes are used to specify:
A. load values
B. font size
C. grid system distance from origin
D. tower height

Q47. Which tool is used to assign selected objects to a specific tower?
A. assign objects to tower
B. tower assign mode
C. object grouping
D. assign grid

Q48. Which toggle button displays the start page when ETABS starts?
A. show model explorer
B. show start page at startup
C. show bounding plane
D. show crosshairs

Q49. Which toggle button displays assigned properties when hovering the cursor on objects?
A. show value at cursor for assign/design
B. show crosshairs
C. show model explorer
D. show bounding plane

Q50. Which toggle button is used to display or hide the bounding plane?
A. show start page
B. show bounding plane
C. animation sound
D. architectural dimensioning

Chapter 7

Detailing

Topics Covered

The major topics covered in this chapter are:

- ***Introduction***
- ***Bar Selection and Curtailment Rules***
- ***Clearing Detailing***
- ***Concrete Detailing Checks***
- ***Showing/Editing Detailing***
- ***Steel Detailing***
- ***Drawing Management***
- ***Generating/Showing Beam Groups***
- ***Checking Log Information***

INTRODUCTION

The tools in the **Detailing** menu are used to create and detail the model; refer to Figure-1. You can use these tools to define size of rebars in various structural elements of building and document them. Various tools in this menu are discussed next.

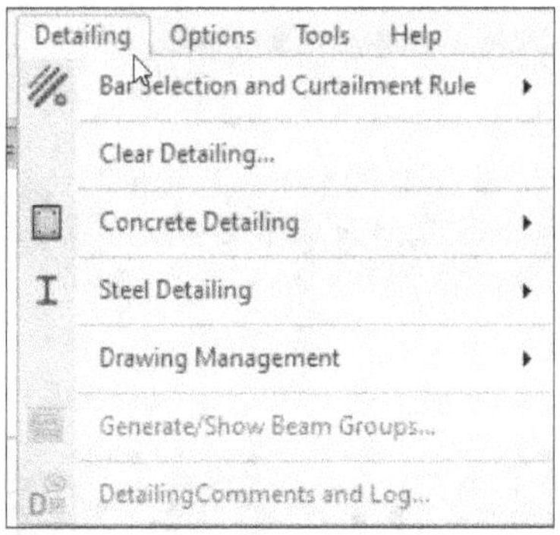

Figure-1. Detailing menu

BAR SELECTION AND CURTAILMENT RULE

The tools in **Bar Selection and Curtailment Rule** cascading menu are used to define rebar types and sizes to be used in the construction of structure; refer to Figure-2. Also, you can define the curtailment rules to optimize size of rebars in concrete. The tools in this cascading menu are discussed next.

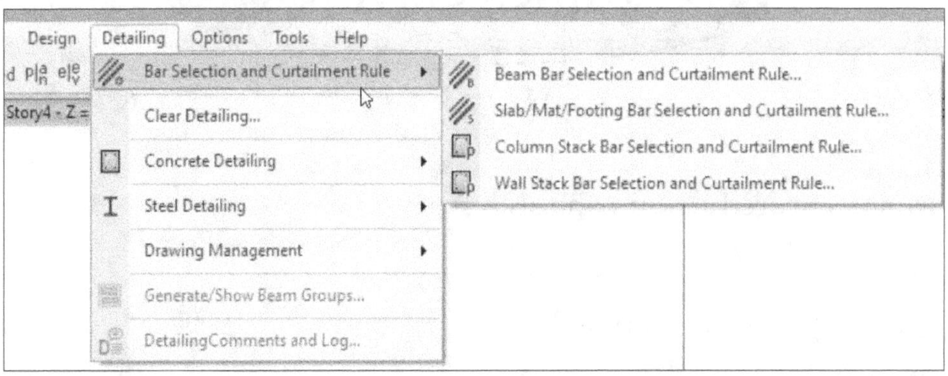

Figure-2. Bar Selection and Curtailment Rule cascading menu

Beam Bar Selection and Curtailment Rule

The **Beam bar selection and curtailment rule** tool is used to display the Bar Selection and Curtailment rule form for beams. You can use this tool to define how reinforced concrete beams should be detailed and displayed in the detailing drawings. The procedure to use this tool is discussed next.

- Click on the **Beam Bar Selection and Curtailment Rule** tool from the **Bar Selection and Curtailment Rule** cascading menu of the **Detailing** menu. The **Concrete Beam: Bar Selection and Curtailment Rule** dialog box will be displayed; refer to Figure-3.

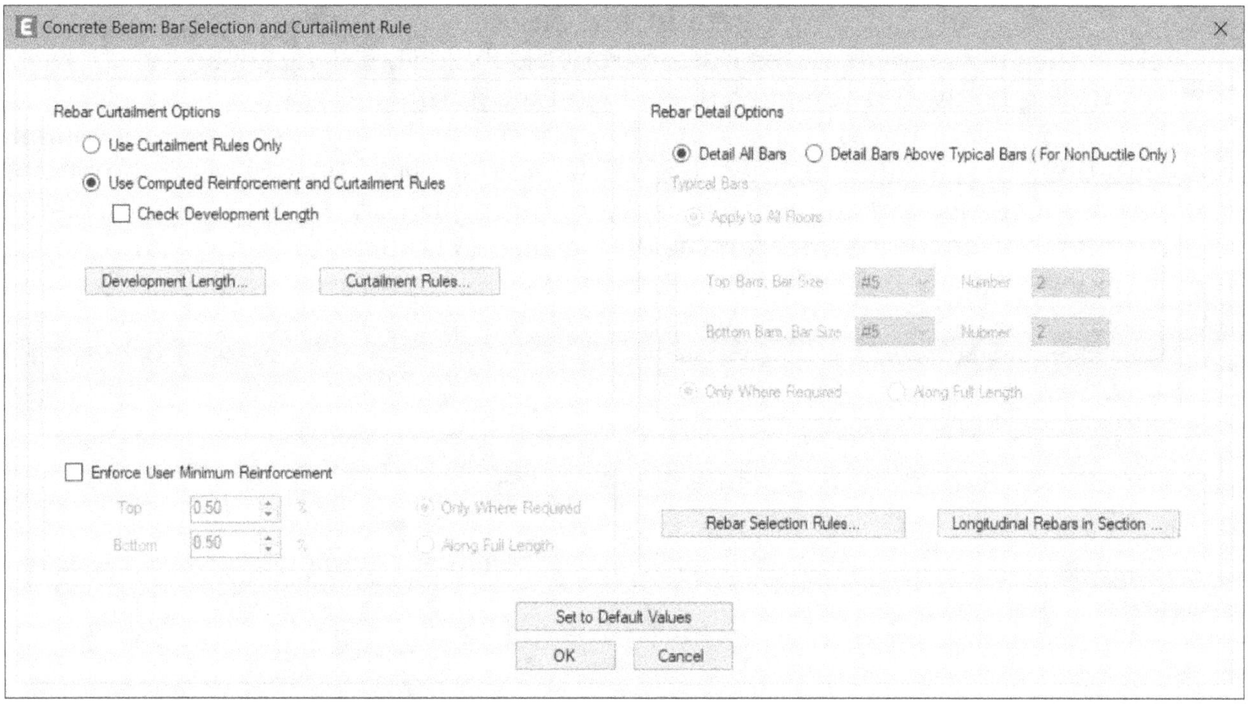

Figure-3. Concrete Beam Bar Selection and Curtailment Rule dialog box

- Select the **Use Curtailment Rules Only** radio button to define rules for curtailment of rebars in concrete. The size and position of reinforcement will not be controlled by rules but rather by parameters specified.
- Select the **Use Computed Reinforcement and Curtailment Rules** to define curtailment rules along with rules for reinforcement. Select the **Check Development Length** check box to calculate development length of rebar and add it to cut off location when computing reinforcement.

Development Length

- Click on the **Development Length** button from the **Rebar Curtailment Options** area of the dialog box. The **Bar Development Rules** dialog box will be displayed; refer to Figure-4. Select desired option from the **Method** drop-down to define method to be used for development calculation. Select desired option from the **Bar Coating Type** drop-down to define whether bars will be uncoated or epoxy coating when performing calculation. Similarly, you can set the type of concrete by weight in the **Concrete Type** drop-down in the dialog box. After setting desired parameters, click on the **OK** button from the dialog box.

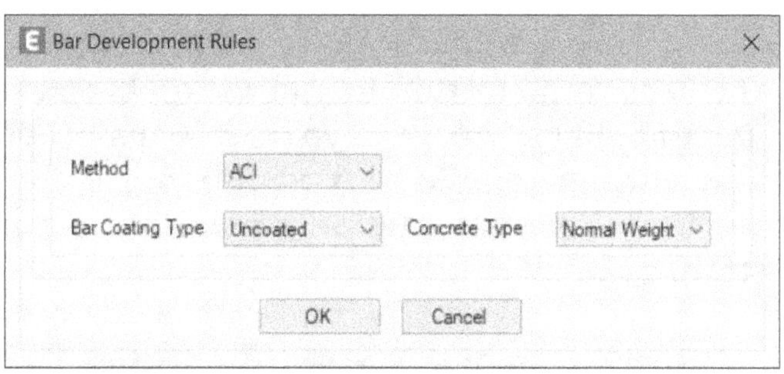

Figure-4. Bar Development Rules dialog box

Curtailment Rules

- Click on the **Curtailment Rules** button from the dialog box. The **Rebar Curtailment Rules** dialog box will be displayed; refer to Figure-5.
- Select the **Seismic Detailing** radio button from the dialog box to include seismic calculations in curtailment rules. Select the **Non Seismic Detailing** radio button if you want to use curtailment rules for detailing which do not include seismic calculations. Set the parameters in dialog box to control the shape and size of curtailment and then click on the **OK** button.

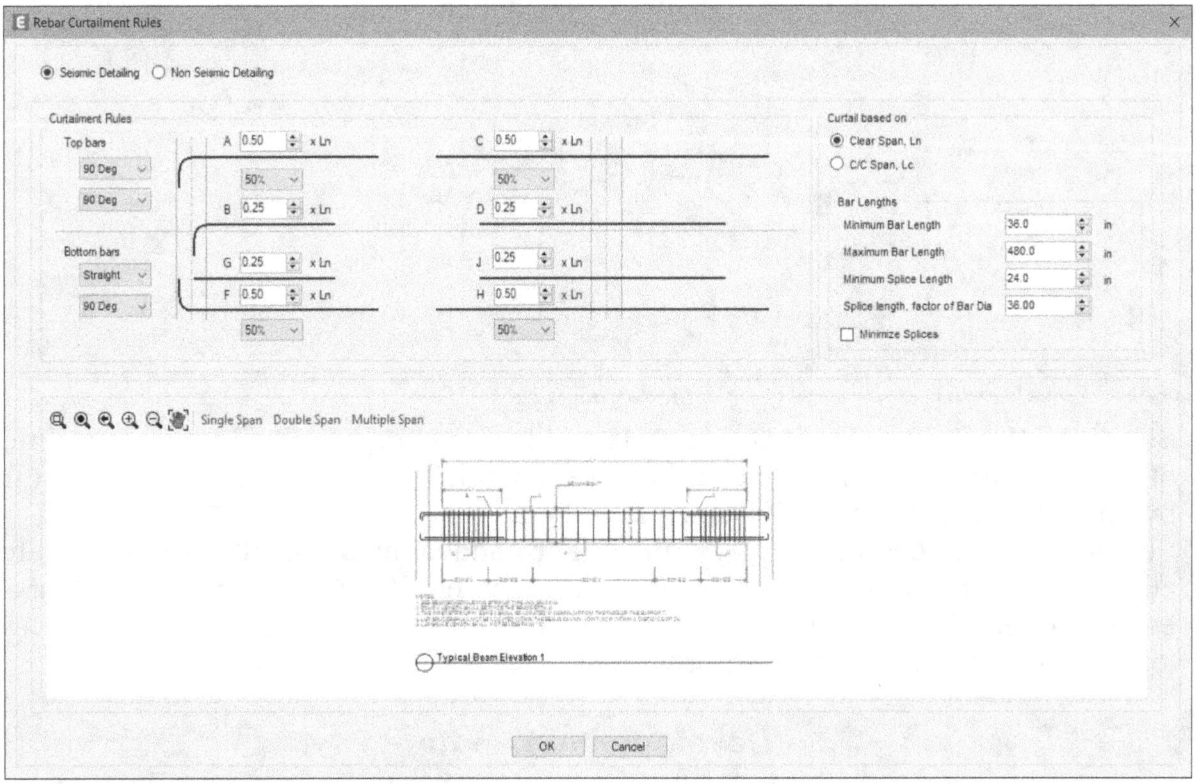

Figure-5. Rebar Curtailment Rules dialog box

- Select the **Enforce User Minimum Reinforcement** check box to define minimum amount of enforcement to be created in the rebars in detailing. The options below the check box will become active.
- Specify desired value in the **Top** and **Bottom** edit boxes to define minimum amount of reinforcement. Select the **Only Where Required** radio button to apply minimum reinforcement only when the reinforcement is lower than minimum. Select the **Along Full Length** radio button to keep minimum reinforcement at full length of rebar.
- Select the **Detail All Bars** radio button to include all bars of structure in detailing report. Select the **Detail Bars Above Typical Bars (For NonDuctile Only)** radio button to define settings related to typical bar size above which bars will be included in detailing. The options to define typical bar settings will be displayed below the radio button.

Setting Rebar Selection Rules

- Click on the **Rebar Selection Rules** button from the dialog box. The **Rebar Selection Rules** dialog box will be displayed; refer to Figure-6.

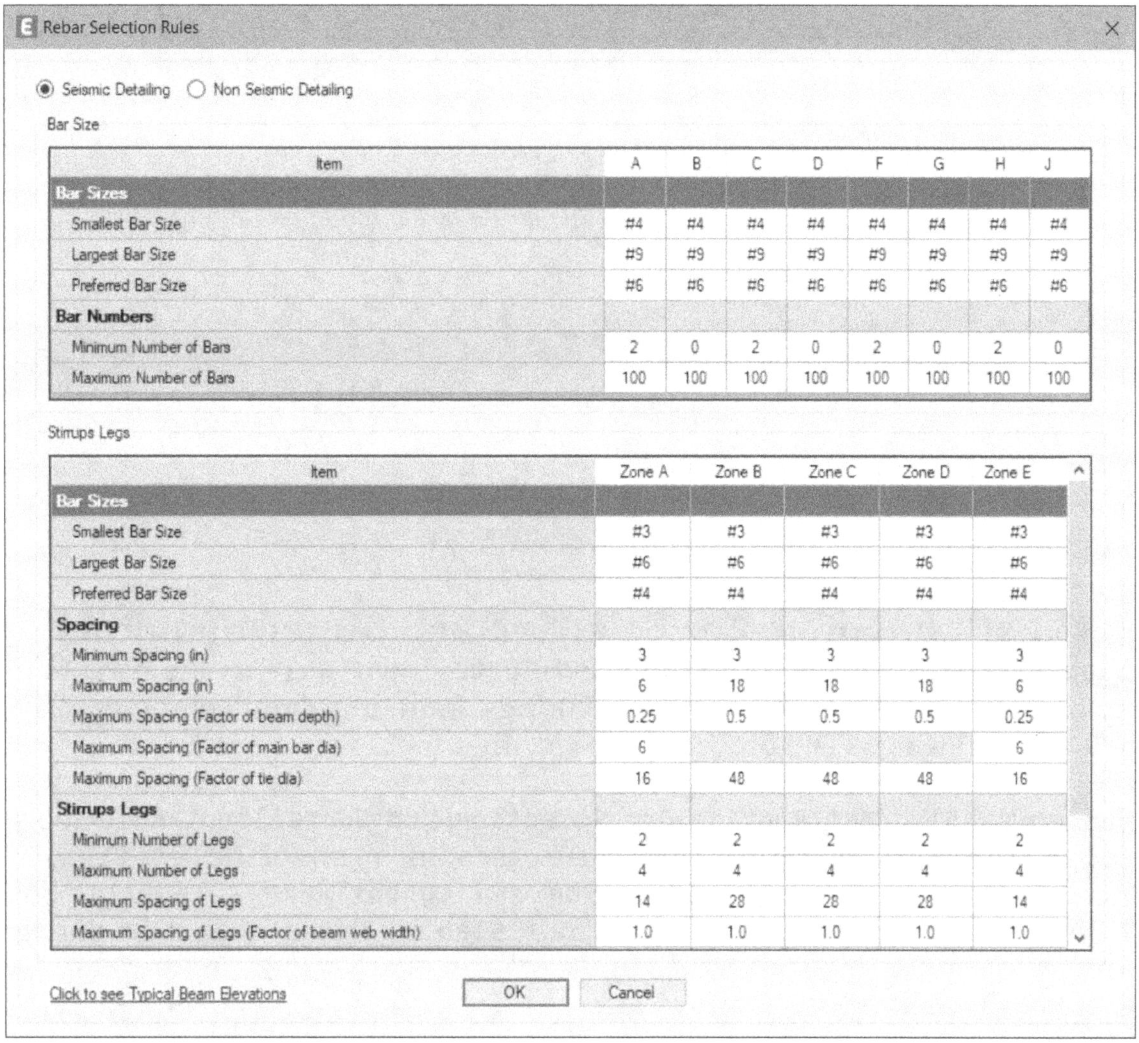

Item	A	B	C	D	F	G	H	J
Bar Sizes								
Smallest Bar Size	#4	#4	#4	#4	#4	#4	#4	#4
Largest Bar Size	#9	#9	#9	#9	#9	#9	#9	#9
Preferred Bar Size	#6	#6	#6	#6	#6	#6	#6	#6
Bar Numbers								
Minimum Number of Bars	2	0	2	0	2	0	2	0
Maximum Number of Bars	100	100	100	100	100	100	100	100

Item	Zone A	Zone B	Zone C	Zone D	Zone E
Bar Sizes					
Smallest Bar Size	#3	#3	#3	#3	#3
Largest Bar Size	#6	#6	#6	#6	#6
Preferred Bar Size	#4	#4	#4	#4	#4
Spacing					
Minimum Spacing (in)	3	3	3	3	3
Maximum Spacing (in)	6	18	18	18	6
Maximum Spacing (Factor of beam depth)	0.25	0.5	0.5	0.5	0.25
Maximum Spacing (Factor of main bar dia)	6				6
Maximum Spacing (Factor of tie dia)	16	48	48	48	16
Stirrups Legs					
Minimum Number of Legs	2	2	2	2	2
Maximum Number of Legs	4	4	4	4	4
Maximum Spacing of Legs	14	28	28	28	14
Maximum Spacing of Legs (Factor of beam web width)	1.0	1.0	1.0	1.0	1.0

Figure-6. Rebar Selection Rules dialog box

- Set desired size values for various types of rebars in the table and click on **OK** button from the dialog box.

Defining Longitudinal Rebar Parameters

- Click on the **Longitudinal Rebars in Section** button from the dialog box. The **Longitudinal Rebars in Section** dialog box will be displayed; refer to Figure-7.
- Specify desired parameters in the dialog box to define spacing between bars, number of bars, number of layers in rebar, and other related objects.
- Click on the **OK** button from the dialog box to apply the parameters.
- After setting desired parameters in the **Concrete Beam: Bar Selection and Curtailment Rule** dialog box, click on the **OK** button to apply parameters and exit the dialog box.

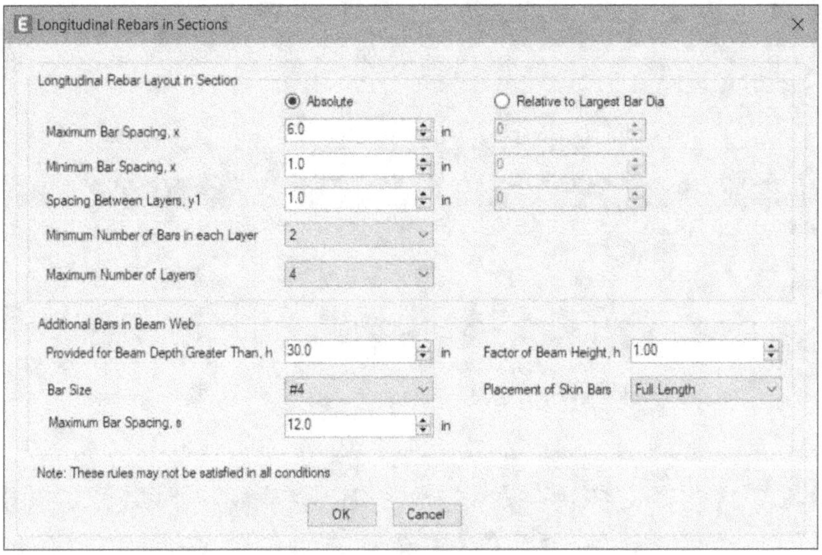

Figure-7. Longitudinal Rebars in Sections dialog box

Slab/Mat/footing Bar Selection and Curtailment Rule

The **Slab/Mat/Footing Bar Selection and Curtailment Rule** tool is used to define how reinforced concrete slabs, mats, or footings should be detailed and displayed in the detailing drawings. The procedure to use this tool is discussed next.

- Click on the **Slab/Mat/Footing Bar Selection and Curtailment Rule** tool from the **Bar Selection and Curtailment Rule** cascading menu of the **Detailing** menu. The **Slab/Mat/Footing: Bar Selection and Curtailment Rule** dialog box will be displayed; refer to Figure-8. By default, **Slab** tab is selected in the dialog box.

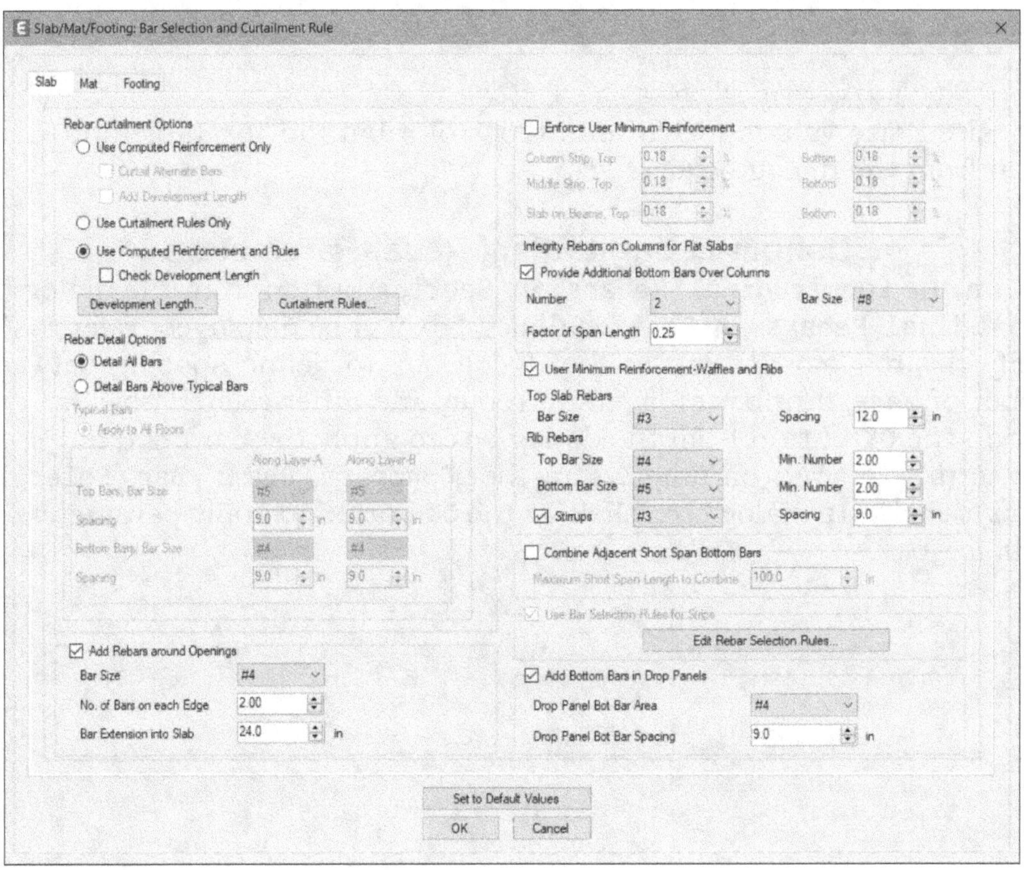

Figure-8. Slab Bar Selection and Curtailment Rule

- Click on the **Use Computed Reinforcement Only** radio button to create rebars based on computed reinforcement only. On selecting this radio button, check boxes below the radio button will become active. Select the **Curtail Alternate Bars** check box to modify only alternative bars in reinforcement based on curtailment parameters generated by analysis. Select the **Add Development Length** check box to add development length at the end of bars.
- Select the **Add Rebars around Openings** check box to define how rebars will be added at opening where slab foundation meets mat reinforcement.
- Select the **Enforce User Minimum Reinforcement** check box to define the minimum amount of reinforcement to be applied in slab structure and specify related parameters.
- Similarly, specify the other parameters in the dialog box to define reinforcement for Slab.
- Click on the **Mat** tab from the **Slab/Mat/Footing: Bar Selection and Curtailment Rule** dialog box to define reinforcement for mat structure in building; refer to Figure-9. The options similar to **Slab** tab will be displayed which have been discussed earlier.

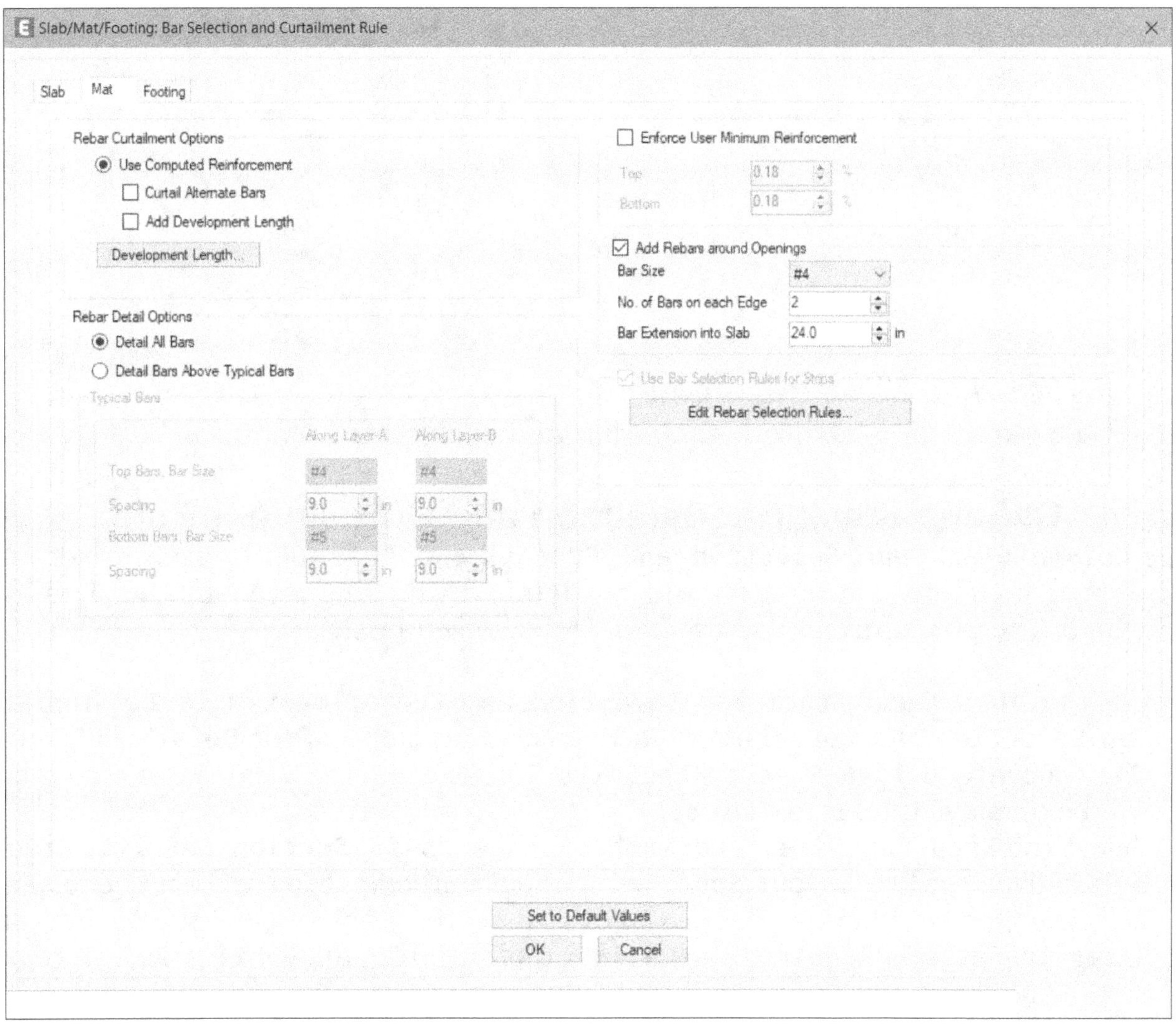

Figure-9. Mat Bar Selection and Curtailment Rule

- Similarly, you can define reinforcement parameters for footing structure using the **Footing** tab of **Slab/Mat/Footing: Bar Selection and Curtailment Rule** dialog box; refer to Figure-10.

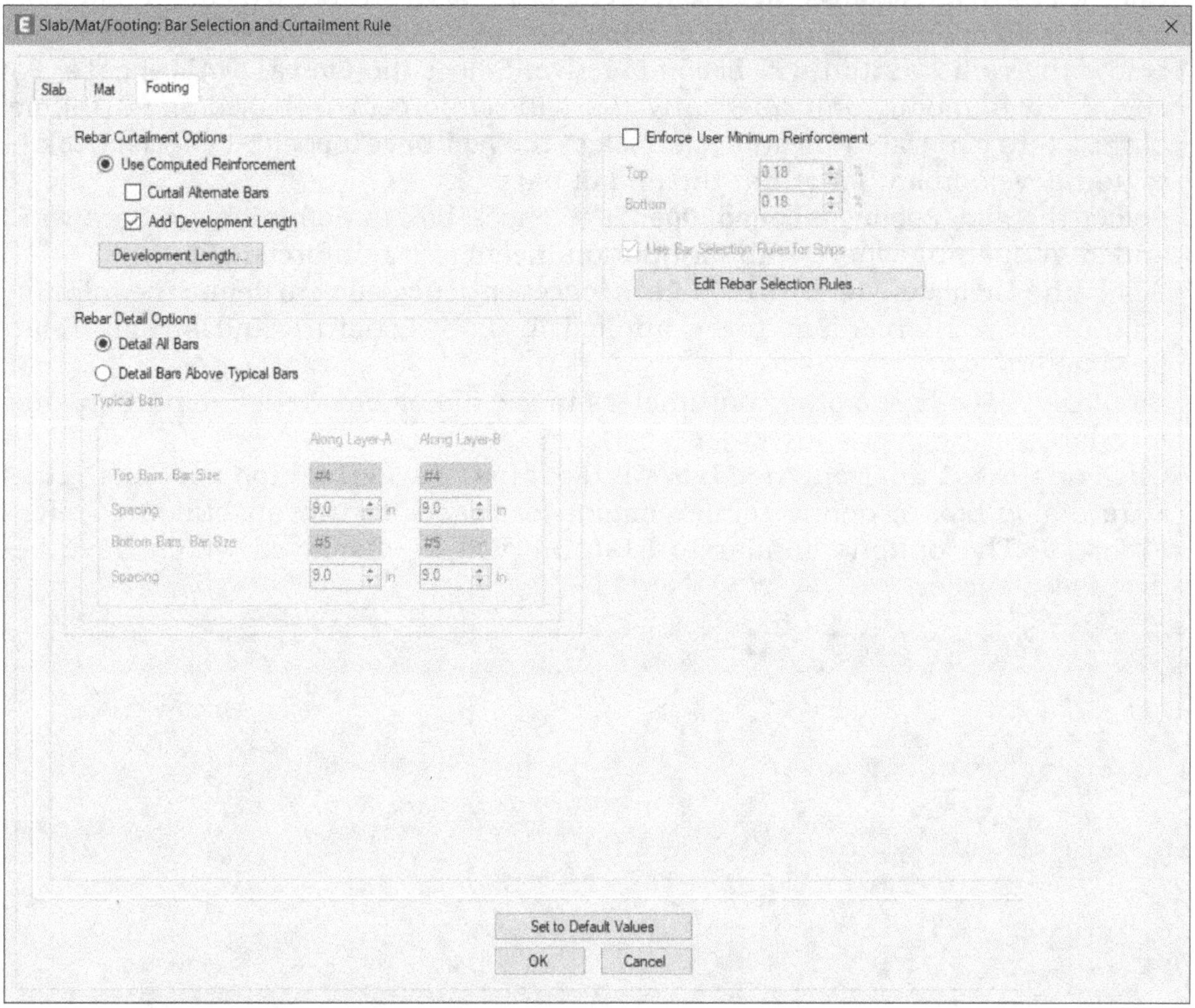

Figure-10. Footing Bar Selection and Curtailment Rule

- Set the parameters as discussed earlier and click on the **OK** button.

Column Stack Bar Selection and Curtailment Rule

The **Column Stack Bar Selection and Curtailment Rule** tool is used to set how reinforced concrete columns should be detailed and displayed in the detailing drawings. The procedure to use this tool is discussed next.

- Click on the **Column Stack Bar Selection and Curtailment Rule** tool from the **Bar Selection and Curtailment Rule** cascading menu of the **Detailing** menu. The **Concrete Column Stack: Bar Selection and Curtailment Rule** dialog box will be displayed; refer to Figure-11.
- Select the **Keep Bar Sizes and Numbers same as in Section Definition** radio button from the **Use Layout Specified in Section Definition** section to use same bar sizes and numbers as defined in design section table. Select the **Select Bar Sizes to Satisfy Required As/Ratio** radio button to allow free selection of bar sizes while keeping the design ratio of elements same.
- Select the **Automatically Select Bar Layout** radio button to allow ETABS to automatically select bar sizes/numbers based on design calculations. Select the **Prefer Larger Bar Number** radio button to prioritize increase in bar numbers in detailing. Select the **Prefer Larger Bar Size** radio button to prioritize increase in bar size in detailing based on design calculations.

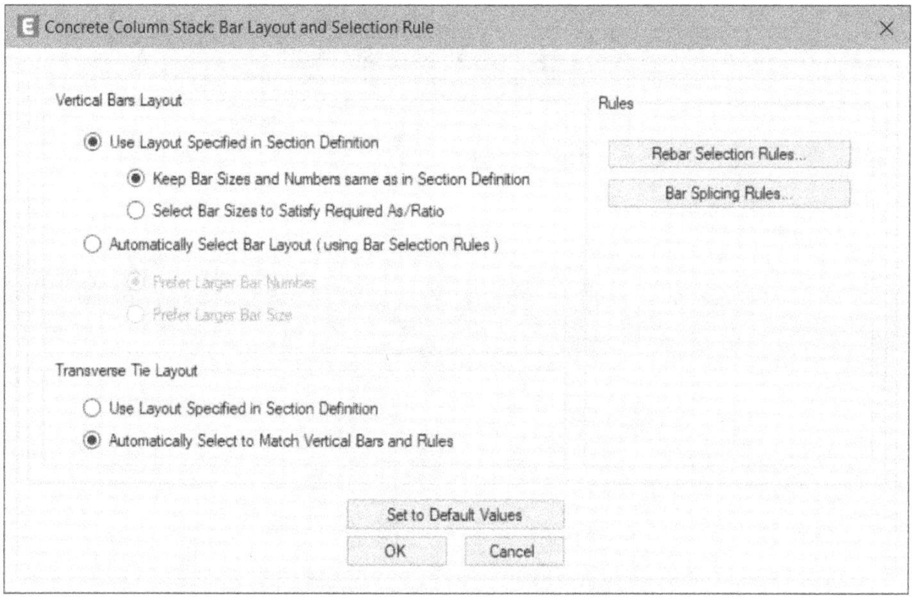

Figure-11. Concrete Column Stack: Bar Layout and Selection Rule

- The options in **Transverse Tie Layout** section of dialog box are used to define how continuous reinforcements are created in detailing drawings.
- Click on the **Rebar Selection Rules** button from the **Rules** section in dialog box. The **Rebar Selection Rules** dialog box will be displayed; refer to Figure-12. Set desired sizes of bars for reinforcement in the table as discussed earlier.
- Click on the **Confinement** button from the bottom in the dialog box. The **Confinement: Non Seismic Detailing** dialog box will be displayed if the **Non Seismic Detailing** radio button is selected; refer to Figure-13. The **Confinement: Seismic Detailing** dialog box will be displayed; refer to Figure-14. Set desired parameters in the dialog box to define confinement zone for cutting extra portion of rebars. After setting desired parameters, click on the **OK** button from the dialog box.

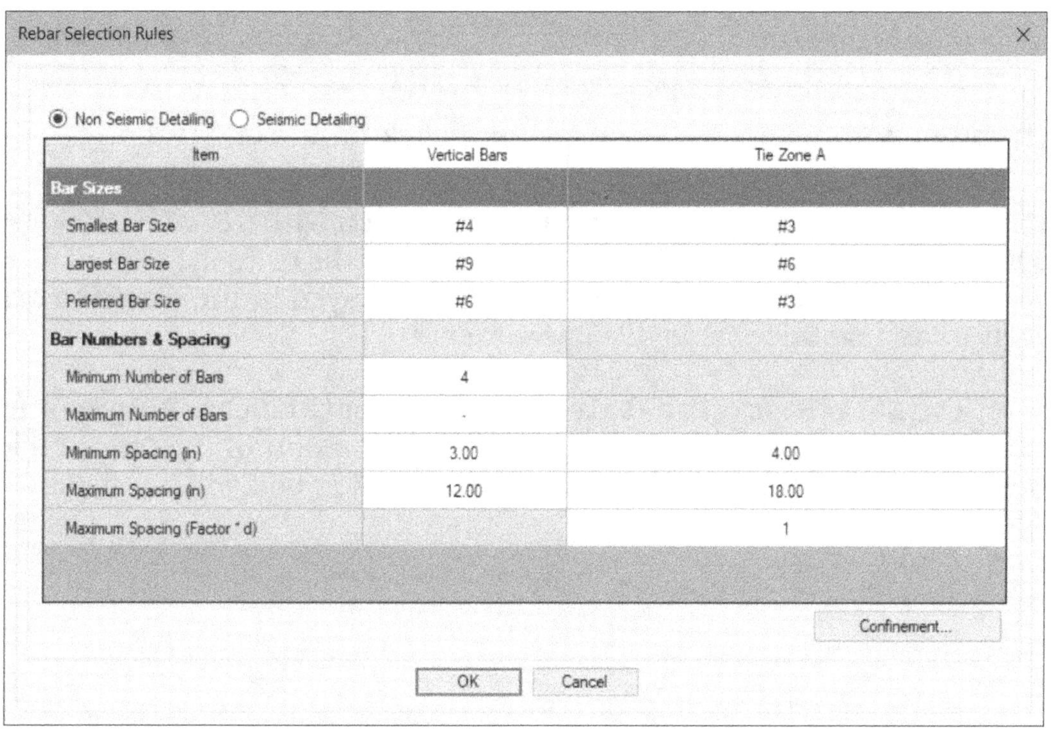

Figure-12. Rebar Selection Rules dialog box

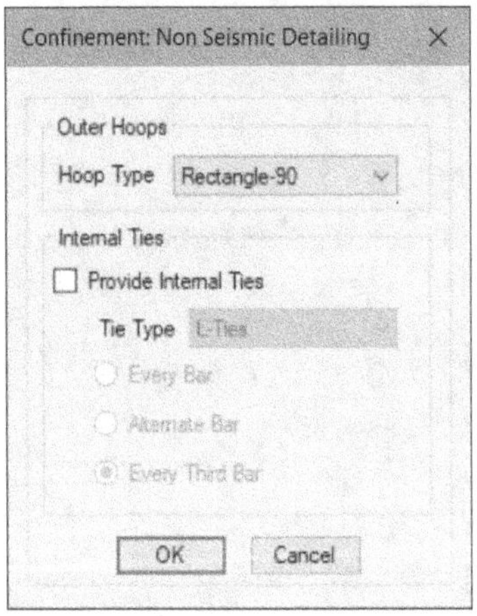

Figure-13. Confinement Non Seismic Detailing dialog box

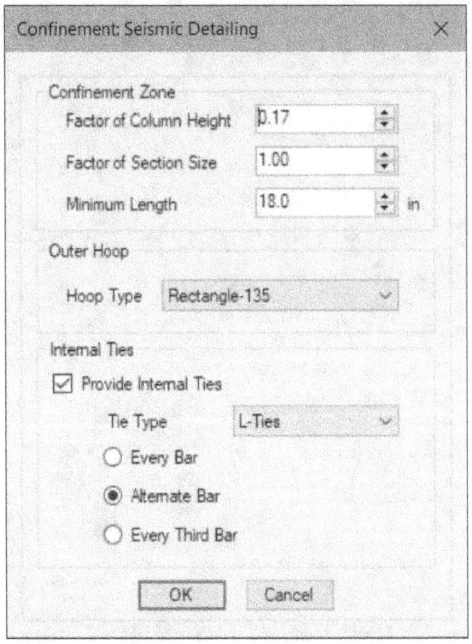

Figure-14. Confinement Seismic Detailing dialog box

- Click on the **Bar Splicing Rules** button from the **Rules** section in the dialog box. The **Bar Splices** dialog box will be displayed; refer to Figure-15. Set desired parameters in the dialog box to define method of splicing bars in reinforcement.

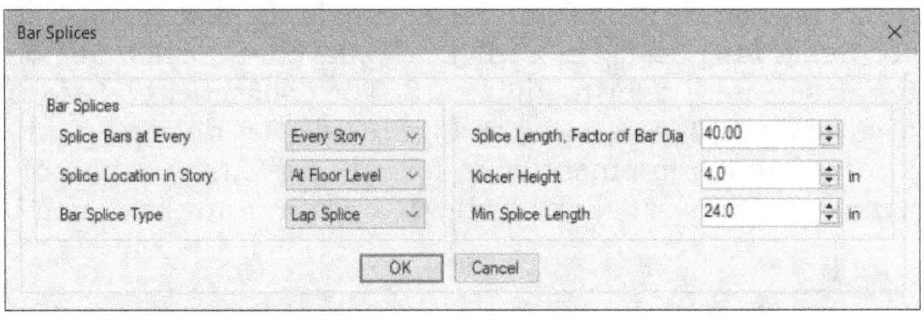

Figure-15. Bar splices dialog box

- Set the other parameters in dialog box and click on the **OK** button.

Wall Stack Bar Selection and Curtailment Rule

The **Wall Stack Bar Selection and Curtailment Rule** tool is used to set how reinforced concrete wall should be detailed and displayed in the detailing drawings. The procedure to use this tool is discussed next.

- Click on the **Wall Stack Bar Selection and Curtailment Rule** tool from the **Bar Selection and Curtailment Rule** cascading menu of the **Detailing** menu. The **Concrete Wall Stack: Detailing Rules** dialog box will be displayed; refer to Figure-16.

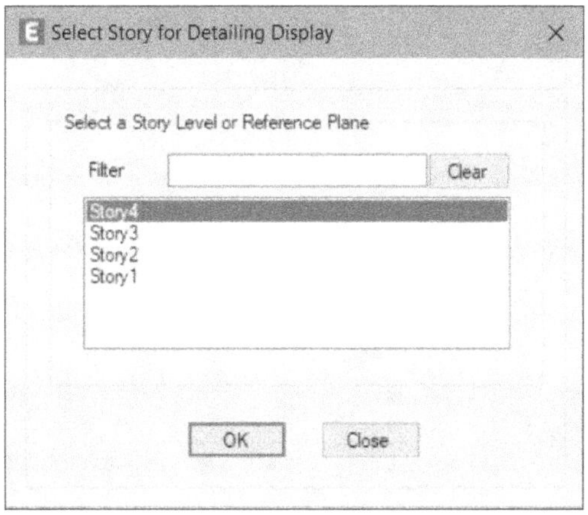

Figure-20. Select Story for Detailing Display dialog box

- Select desired story from the list to create story detail drawings and click on the **OK** button from the dialog box. The **Beam Detailing Editor** application window will be displayed; refer to Figure-21.

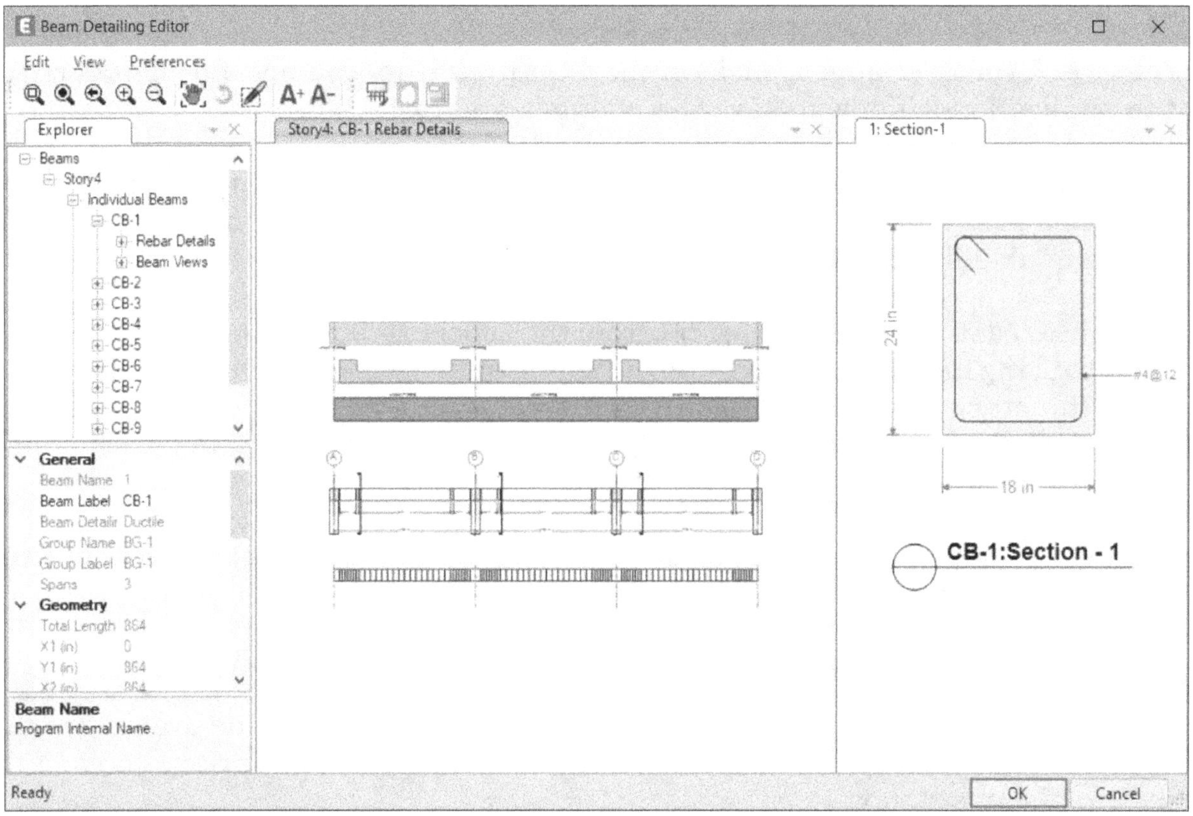

Figure-21. Beam Detailing Editor window

- Select desired beam element from top left area in **Explorer** and set desired parameters in the table below to modify beam element size; refer to Figure-22.

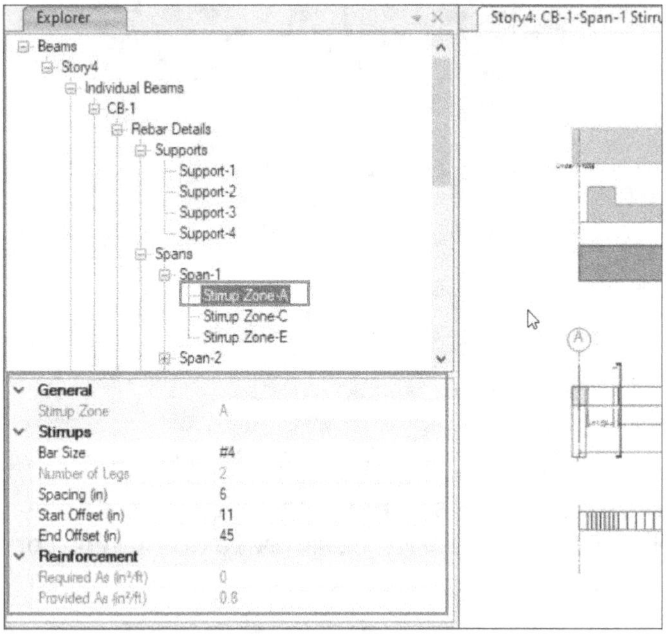

Figure-22. Modifying beam elements

- After performing changes, click on the **OK** button from the bottom right corner of the application window to save the changes and exit.

- Similarly, you can use the **Run/Update Column Stack Detailing/Check** tool from the **Concrete Detailing** cascading menu of the **Detailing** menu to edit and generate detailing data of column in reinforcement.

Use the **Show/Edit Beam Detailing** tool from the **Concrete Detailing** cascading menu of the **Detailing** menu to display beam detailing data of selected story. Similarly, you can use the other Show/Edit tools from the cascading menu to display/edit other structural elements. Use the Clear tools from the cascading menu to clear detailing parameters applied earlier.

If you are working with steel structure design then you can use the tools of **Steel Detailing** cascading menu as discussed earlier.

DRAWING MANAGEMENT

The tools in **Drawing Management** cascading menu are used to define drawing related parameters and generate the drawing data; refer to Figure-23. Various tools of this menu are discussed next.

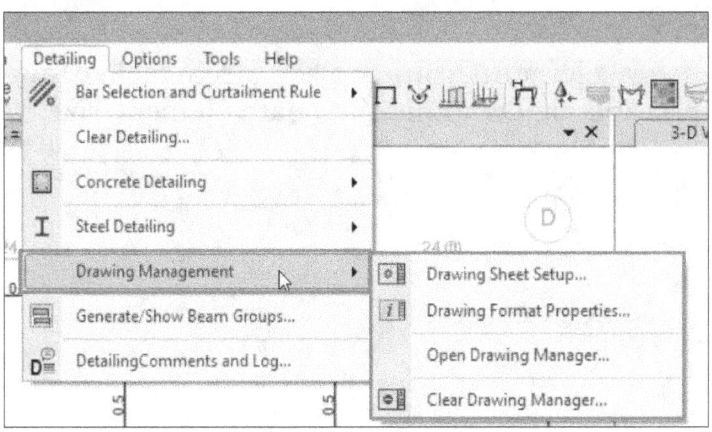

Figure-23. Drawing Management cascading menu

Drawing Sheet Setup

The **Drawing Sheet Setup** tool is used to define settings for drawing units and sizes, specify margins, specify text, and so on. The procedure to use this tool is discussed next.

- Click on the **Drawing Sheet Setup** tool from the **Drawing Management** cascading menu of the **Detailing** menu. The **Drawing Sheet Setup** dialog box will be displayed; refer to Figure-24.

Figure-24. Drawing Sheet Setup dialog box

- Specify desired unit system for drawing sheet from the **Sheet Type** drop-down in **Drawing Units and Size** area of the dialog box.
- Select the size of drawing sheet from the **Sheet Size** drop-down. The width and height of sheet will be displayed in the **Sheet Width** and **Sheet Height** edit boxes, respectively.
- Select the **Custom** option from the **Sheet Size** drop-down to specify user defined width and height of sheet.
- Select desired option from the **Plans** drop-down in the **View Placement Options** area to define how plans will be placed in drawing sheet when auto placement of views is applied. Select the **Scale to Fit** option to change size of views so that all of them fit in the drawing sheet. Select the **Autowrap** option from the drop-down automatically organize views to fit them in drawing sheet. Select the **None** option from the **Plans** drop-down if you do not want to apply other automatic placement strategies.
- Similarly, you can set options in the **Tables** and **Elevations** drop-downs.
- Select the **Beam Elevation on New Line** check box to place beam elevations using new line as alignment reference.

- Specify desired parameters in the edit boxes of the **Text, Symbol, and Gap Sizes** area to define text height, size, thickness, and so on.
- Similarly, you can specify other parameters in the dialog box. After setting parameters, click on the **OK** button from the dialog box.

Drawing Format Properties

The **Drawing Format Properties** tool is used to modify the formatting preferences such as thickness of lines, style of lines, colors, and so on. The procedure to use this tool is discussed next.

- Click on the **Drawing Format Properties** tool from the **Drawing Management** cascading menu of the **Detailing** menu. The **Drawing Format Properties** dialog box will be displayed; refer to Figure-25.

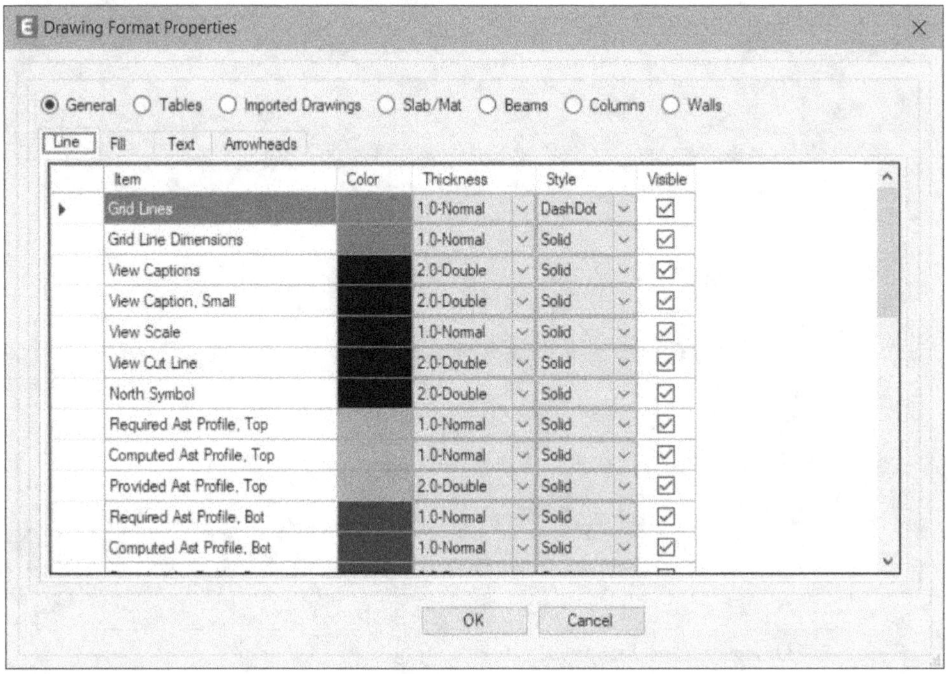

Figure-25. Drawing Format Properties dialog box

- Select desired radio button from the top area in the dialog box to define which drawing element is to be modified in the drawing.
- Select the **Line** tab from the dialog box to define line color, thickness, style, and visibility using options in respective columns. Similarly, you can set parameters for **Fill**, **Text**, and **Arrowheads**.
- After setting desired parameters, click on the **OK** button from the dialog box.

Drawing Manager

The **Open Drawing Manager** tool is used to define which detailing will be included and updated in the drawing. The procedure to use this tool is given next.

- Click on the **Open Drawing Manager** tool from the **Drawing Management** cascading menu of the **Detailing** menu. The **Drawing Management Component Selection** dialog box will be displayed; refer to Figure-26.

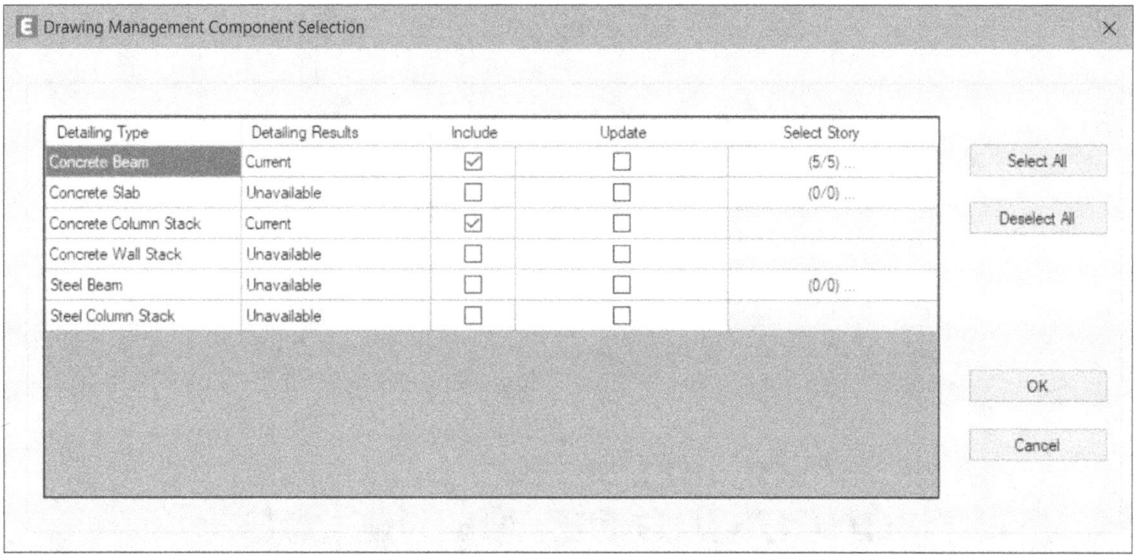

Figure-26. Drawing Management Component Selection dialog box

- Select check boxes from **Include** column for detailing elements to be included in the drawing.
- Select check boxes from the **Update** column to update detailing elements in the drawing.
- After selecting desired check boxes, click on the **OK** button from the dialog box. The **Drawing Manager** will be displayed. You can check drawings of various elements in the **Drawing Manager**. After checking drawings, click on the **OK** button from the Manager to exit.

Clearing Drawing Manager

The **Clear Drawing Manager** tool is used to clear all the drawings earlier generated by **Drawing Manager**.

GENERATING/SHOWING BEAM GROUPS

The **Generate/Show Beam Groups** tool is used to modify and add beam groups. The procedure to use this tool is given next.

- Click on the **Generate/Show Beam Groups** tool from the **Detailing** menu. The **Define Groups** dialog box will be displayed; refer to Figure-27.
- Click in the color field of desired group to change color.
- If you want to delete a group then select it from the table and click on the **Delete** button from the dialog box.
- If you want to add a new beam group then click on the **Add** button from the dialog box. The new group will be added in the dialog box.
- Click on the name of group twice with a small pause to change its name.
- After setting desired parameters, click on the **OK** button from the dialog box.

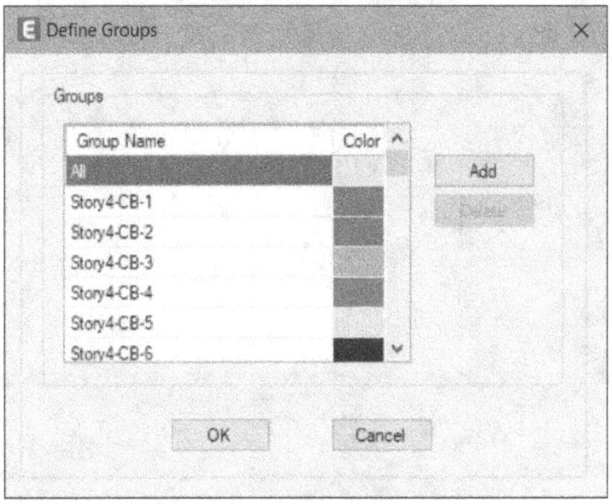

Figure-27. Define Groups dialog box

CHECKING DETAILING LOG

The **DetailingComments and Log** tool is used to check log of beam detailing. On clicking this tool, the **Log** information box will be displayed; refer to Figure-28. Close the information box by using **Close** button (X) at top right corner.

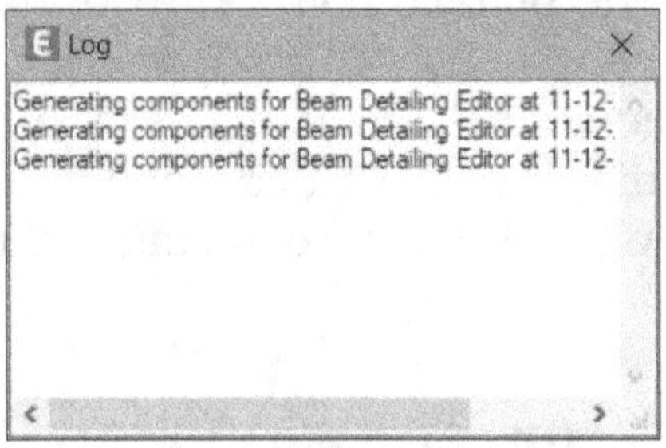

Figure-28. Log dialog box

SELF-ASSESSMENT

Q1. Which cascading menu is used to define rebar types and sizes in ETABS?
A. Concrete detailing
B. Bar selection and curtailment rule
C. Drawing management
D. Graphics preferences

Q2. The Beam Bar Selection and Curtailment Rule tool is used for:
A. Steel beam design
B. Slab meshing
C. Detailing reinforced concrete beams
D. Drawing management

Q3. Which radio button is selected to define only curtailment rules?
A. Use computed reinforcement only
B. Use curtailment rules only
C. Enforce minimum reinforcement
D. Detail all bars

Q4. Which option defines both reinforcement and curtailment rules?
A. Use computed reinforcement and curtailment rules
B. Use curtailment rules only
C. Add development length
D. Seismic detailing

Q5. Which check box calculates development length of rebar?
A. Detail all bars
B. Add rebars around openings
C. Check development length
D. Apply auto parts

Q6. Which button opens the Bar Development Rules dialog box?
A. Confinement
B. Development length
C. Curtailment rules
D. Rebar selection rules

Q7. The Method drop-down in Bar Development Rules is used to define:
A. Concrete color
B. Development calculation method
C. Drawing scale
D. Beam size

Q8. Which drop-down defines whether bars are epoxy coated or uncoated?
A. Concrete type
B. Method
C. Bar coating type
D. Reinforcement type

Q9. Which button opens the Rebar Curtailment Rules dialog box?
A. Confinement
B. Curtailment rules
C. Development length
D. Splicing rules

Q10. Which radio button includes seismic calculations in curtailment rules?
A. Seismic detailing
B. Non seismic detailing
C. Detail all bars
D. Computed reinforcement

Q11. Which check box is used to define minimum reinforcement in rebars?
A. Detail all bars
B. Enforce user minimum reinforcement
C. Add development length
D. Curtail alternate bars

Q12. Which radio button applies minimum reinforcement along the entire rebar length?
A. Only where required
B. Seismic detailing
C. Along full length
D. Detail all bars

Q13. Which radio button includes all bars in the detailing report?
A. Detail all bars
B. Along full length
C. Seismic detailing
D. User minimum reinforcement

Q14. Which button opens the Rebar Selection Rules dialog box?
A. Confinement
B. Bar splicing rules
C. Rebar selection rules
D. Development length

Q15. The Longitudinal Rebars in Section dialog box is used to define:
A. Drawing scale
B. Line thickness
C. Bar spacing and layers
D. Story height

Q16. Which tool is used to define detailing rules for slabs, mats, and footings?
A. Wall stack bar selection tool
B. Slab/mat/footing bar selection and curtailment rule
C. Beam detailing generation
D. Drawing sheet setup

Q17. Which tab is selected by default in the Slab/Mat/Footing dialog box?
A. Footing
B. Mat
C. Slab
D. Column

Q18. Which radio button creates rebars based only on computed reinforcement?
A. Use computed reinforcement only
B. Use curtailment rules only
C. Detail all bars
D. Seismic detailing

Q19. Which check box modifies only alternate bars in reinforcement?
A. Add development length
B. Curtail alternate bars
C. Detail all bars
D. Enforce user minimum reinforcement

Q20. Which check box adds development length at the end of bars?
A. Add development length
B. Curtail alternate bars
C. Detail all bars
D. Add rebars around openings

Q21. Which check box adds rebars around slab openings?
A. Detail all bars
B. Add rebars around openings
C. Enforce user minimum reinforcement
D. Curtail alternate bars

Q22. Which tool is used to define detailing rules for concrete columns?
A. Wall stack bar selection tool
B. Column stack bar selection and curtailment rule
C. Beam detailing generation
D. Drawing manager

Q23. Which radio button keeps bar sizes same as section definition?
A. Automatically select bar layout
B. Keep bar sizes and numbers same as in section definition
C. Prefer larger bar size
D. Select bar sizes to satisfy required ratio

Q24. Which option allows free selection of bar sizes while maintaining design ratio?
A. Keep bar sizes same as section definition
B. Select bar sizes to satisfy required As/Ratio
C. Prefer larger bar number
D. Seismic detailing

Q25. Which radio button allows ETABS to automatically select bar sizes and numbers?
A. Prefer larger bar size
B. Automatically select bar layout
C. Detail all bars
D. Add development length

Q26. Which radio button prioritizes increase in bar numbers?
A. Prefer larger bar number
B. Prefer larger bar size
C. Detail all bars
D. Seismic detailing

Q27. Which radio button prioritizes increase in bar size?
A. Automatically select bar layout
B. Prefer larger bar size
C. Prefer larger bar number
D. Use computed reinforcement only

Q28. Which button opens the Confinement dialog box?
A. Development length
B. Confinement
C. Spandrel rebar rules
D. Drawing setup

Q29. Which button opens the Bar Splices dialog box?
A. Confinement
B. Bar splicing rules
C. Rebar selection rules
D. Curtailment rules

Q30. The Wall Stack Bar Selection and Curtailment Rule tool is used for:
A. Slab detailing
B. Steel detailing
C. Wall detailing
D. Footing analysis

Q31. Which button modifies spandrel rebar rules?
A. Confinement
B. Spandrel rebar rules
C. Drawing manager
D. Bar splicing rules

Q32. Which tool is used to remove previous detailing?
A. Clear detailing
B. Drawing manager
C. Graphics preferences
D. Generate groups

Q33. Which button selects all detailing types for clearing?
A. Add all
B. Select all
C. Clear all
D. Apply all

Q34. Which tool generates beam detailing documents?
A. Run/update beam detailing/check
B. Clear detailing
C. Drawing manager
D. Generate beam groups

Q35. Which dialog box appears after running Beam Detailing Generation?
A. Drawing sheet setup
B. Select story for detailing display
C. Graphics preferences
D. Define groups

Q36. Which application window is displayed for beam detailing editing?
A. Beam detailing editor
B. Drawing manager
C. Model explorer
D. Table editor

Q37. Which tool displays beam detailing data of selected story?
A. Clear detailing
B. Show/edit beam detailing
C. Generate beam groups
D. Detailing log

Q38. Which cascading menu is used to define drawing related parameters?
A. Concrete detailing
B. Drawing management
C. Graphics colors
D. Bar selection rules

Q39. Which tool defines settings for drawing units and sheet sizes?
A. Drawing manager
B. Drawing sheet setup
C. Drawing format properties
D. Detailing log

Q40. Which drop-down is used to select drawing sheet size?
A. Sheet type
B. Plans
C. Sheet size
D. Tables

Q41. Which option changes view size to fit all views in drawing sheet?
A. None
B. Autowrap
C. Scale to fit
D. Update

Q42. Which option automatically organizes views in drawing sheet?
A. Autowrap
B. Scale to fit
C. None
D. Placement fit

Q43. Which check box places beam elevations on a new line?
A. Beam elevation on new line
B. Auto placement
C. Object fill
D. Detail all bars

Q44. Which tool modifies line thickness, style, and colors in drawings?
A. Drawing manager
B. Drawing format properties
C. Beam detailing
D. Detailing log

Q45. Which tab in Drawing Format Properties defines line color and style?
A. Fill
B. Text
C. Line
D. Arrowheads

Q46. Which tool is used to define included detailing in drawings?
A. Open drawing manager
B. Drawing sheet setup
C. Clear detailing
D. Generate beam groups

Q47. Which column in Drawing Manager is used to include detailing elements?
A. Update
B. Include
C. Detail
D. Select

Q48. Which tool clears drawings generated by Drawing Manager?
A. Clear detailing
B. Clear drawing manager
C. Drawing sheet setup
D. Beam detailing generation

Q49. Which tool is used to add and modify beam groups?
A. Detailing comments and log
B. Generate/show beam groups
C. Drawing manager
D. Clear drawing manager

Q50. Which tool is used to check the beam detailing log?
A. Generate beam groups
B. Detailing comments and log
C. Beam detailing generation
D. Drawing sheet setup

Chapter 8

Project

Topics Covered

The major topics covered in this chapter are:

- ***Introduction***
- ***Project 1***
- ***Project 2***

PROJECT 1

In this chapter, we will work on a real world project of structural designing. We will design structure of 5 story building with ground floor used as parking. The material used for structure is concrete.

Starting Project

- Start ETABS from the **Start** menu and click on the **New Model** button. The **Model Initialization** dialog box will be displayed. Select the **Use Build-in Settings With:** radio button.
- Select **U.S. Customary** option from the **Display Units** drop-down. Select the **United States** option from the **Region for Default Materials** drop-down. Select the **AISC14** option from the **Steel Section Database** drop-down. Select the **AISC 360-10** option from the **Steel Design Code** drop-down. Select the **ACI 318-14** option from the **Concrete Design Code** drop-down.
- After setting desired parameters, click on the **OK** button. The **New Model Quick Templates** dialog box will be displayed; refer to Figure-1.

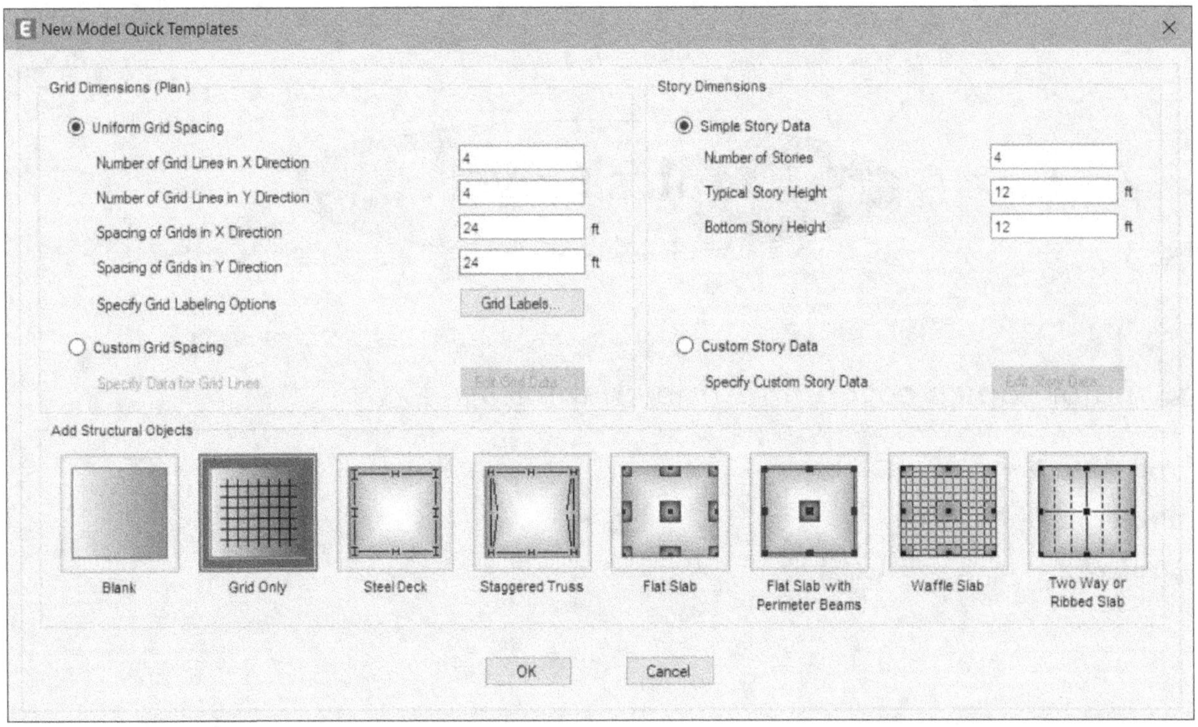

Figure-1. New Model Quick Templates dialog box

- Set the number of stories as **5** in the **Number of Stories** edit box. Set typical story height and bottom story height as **12** in respective edit boxes.
- Keep the other parameters as default and make sure the **Grid Only** button is selected.
- Click on the **OK** button. The grid of building will be displayed; refer to Figure-2.

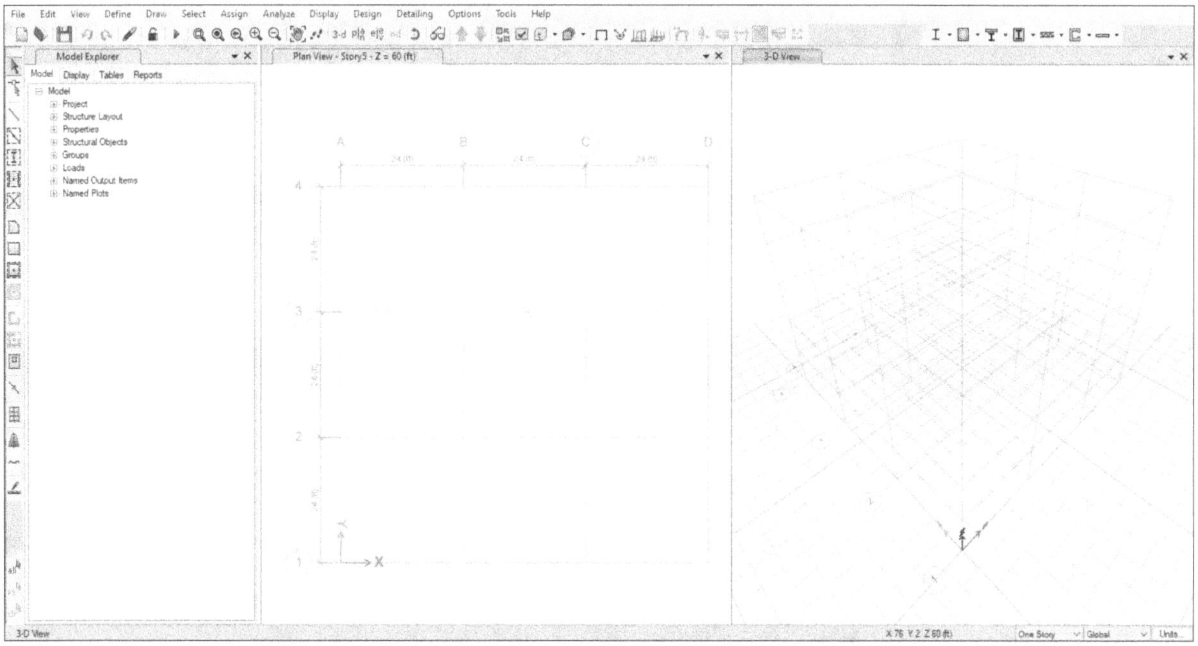

Figure-2. Grid created for project

Adding Material

- Click on the **Material Properties** tool from the **Define** menu. The **Define Materials** dialog box will be displayed; refer to Figure-3.

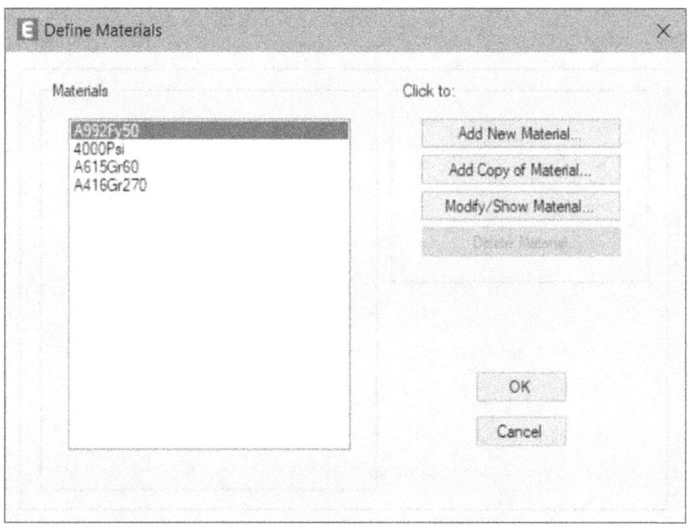

Figure-3. Define Materials dialog box

- Click on the **Add New Material** button from the dialog box. The **Add New Material Property** dialog box will be displayed; refer to Figure-4.

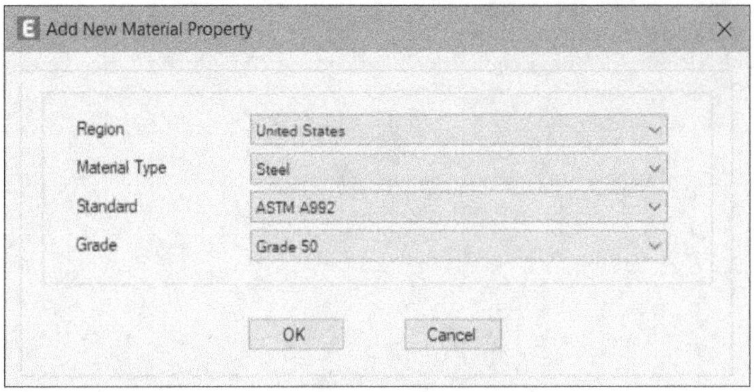

Figure-4. Add New Material Property dialog box

- Select the **United States** option from **Region** drop-down, **Concrete** option from **Material Type** drop-down, **Customary** option from the **Standard** drop-down, and **f'c 4000 psi** option from the **Grade** drop-down.
- After setting the parameters, click on the **OK** button. The **Material Property Data** dialog box will be displayed; refer to Figure-5.

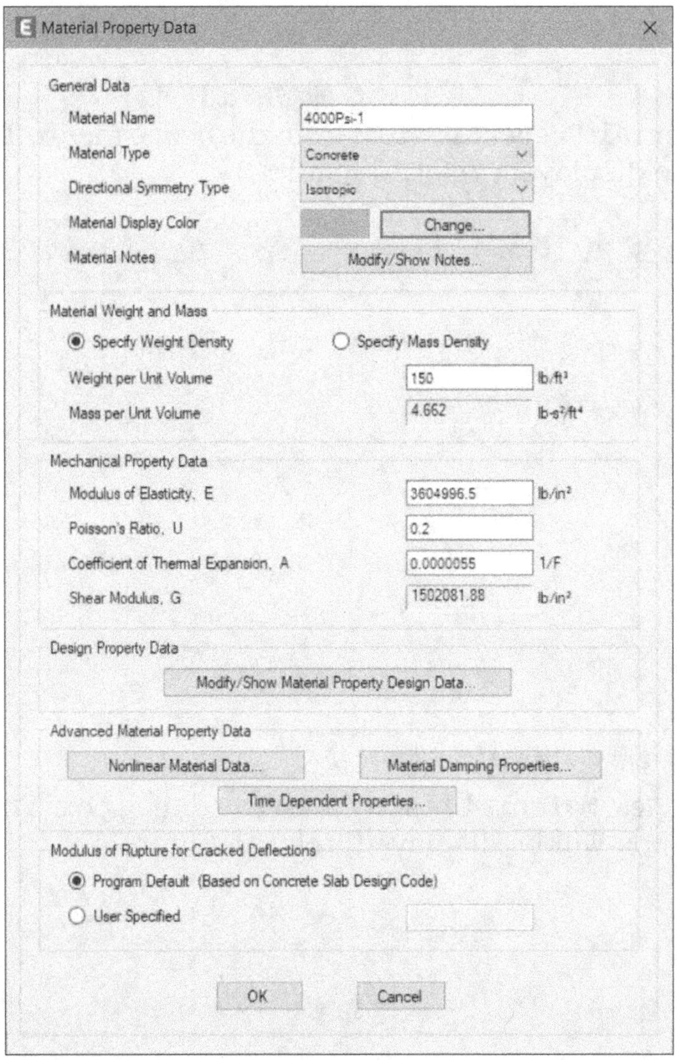

Figure-5. Material Property Data dialog box

- Set the parameters as shown in Figure-5 and click on the **OK** button. The material will be created and displayed in the **Define Materials** dialog box.
- Click on the **OK** button from the **Define Materials** dialog box.

Creating Frame Section

- Click on the **Frame Sections** tool from the **Section Properties** cascading menu of the **Define** menu. The **Frame Properties** dialog box will be displayed.
- Click on the **Add New Property** button from the dialog box. The **Frame Property Shape Type** dialog box will be displayed; refer to Figure-6.

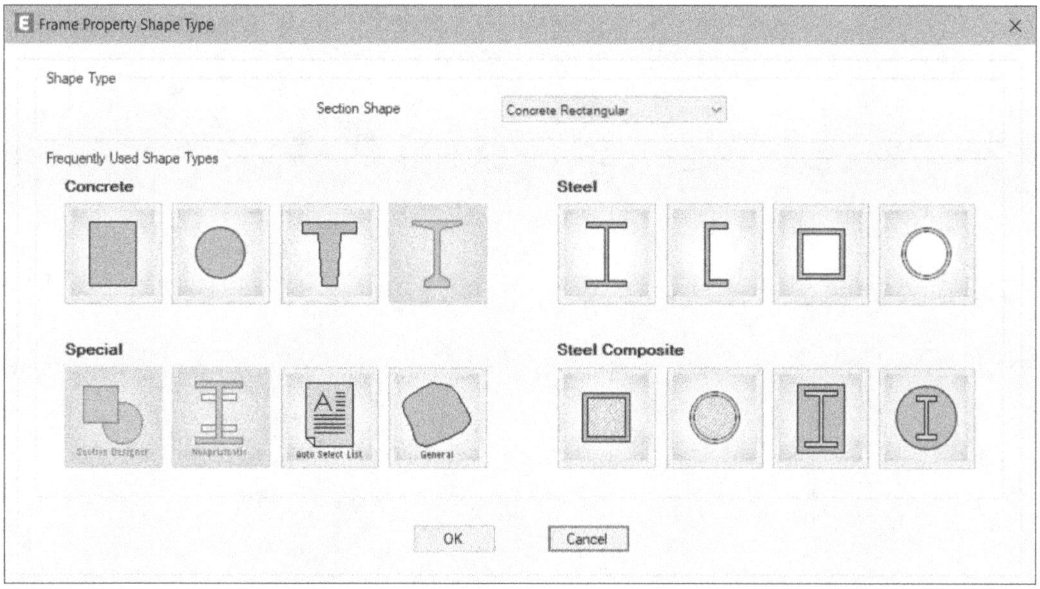

Figure-6. Frame Property Shape Type dialog box

- Select the **Concrete Rectangular** option from the **Section Shape** drop-down at the top in the dialog box and click on the **OK** button. The **Frame Section Property Data** dialog box will be displayed; refer to Figure-7.

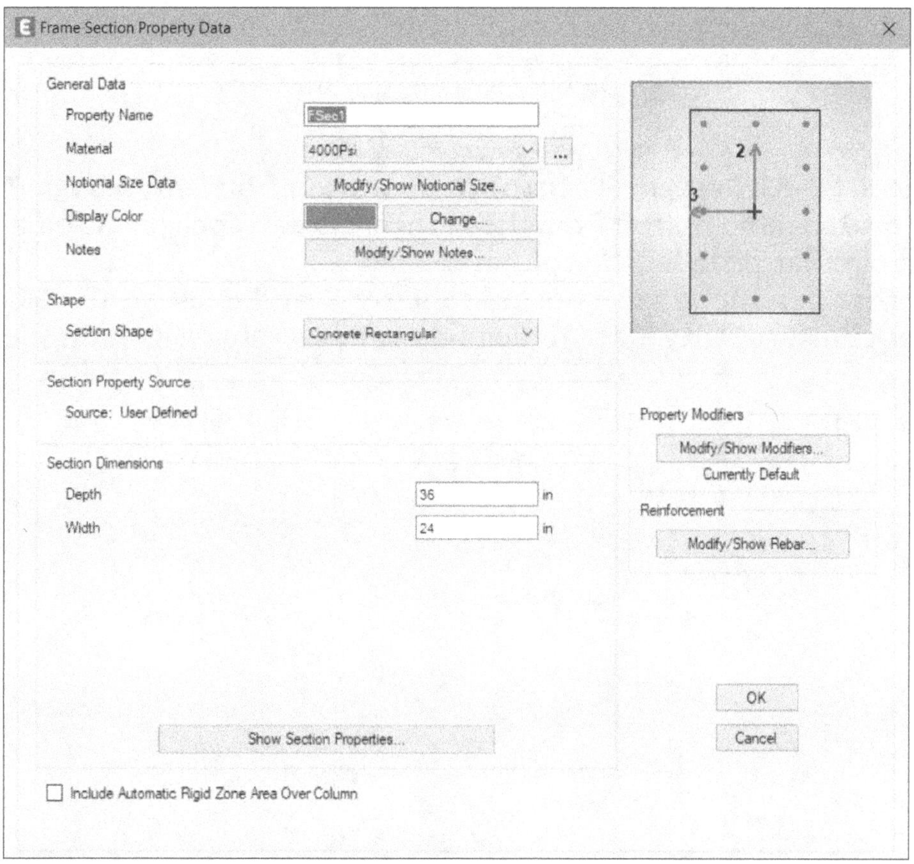

Figure-7. Frame Section Property Data dialog box

- Set the name **Conc Column** for frame property in the **Property Name** edit box.
- Select the newly created material **4000Psi** from the **Material** drop-down.
- Click on the **Modify/Show Rebar** button in the **Reinforcement** area of the dialog box to define how rebars are placed in the concrete columns. The **Frame Section Property Reinforcement Data** dialog box will be displayed; refer to Figure-8.

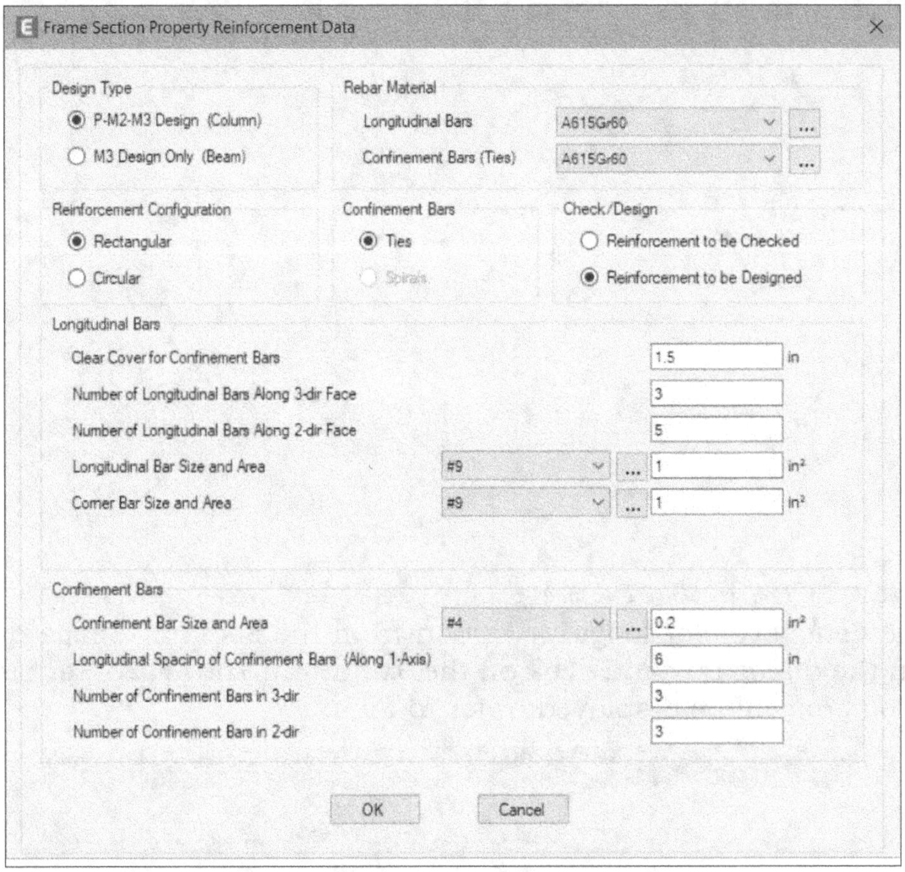

Figure-8. Frame Section Property Reinforcement Data dialog box

- Select the **P-M2-M3 Design (Column)** radio button from the **Design Type** area.
- Select **Circular** radio button from the **Reinforcement Configuration** area to create rebars in circular array in the columns.
- Set the other parameters as shown in Figure-9 and click on the **OK** button. The **Frame Section Property Data** dialog box will be displayed again.

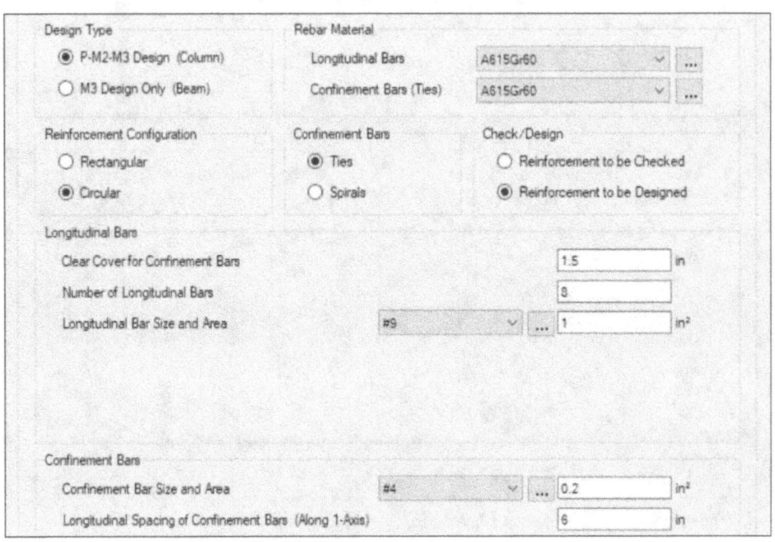

Figure-9. Circular reinforcement data

- Set the depth of column as **36** inch and width as **24** inch in respective edit boxes of the dialog box. Click on the **OK** button to create the property.
- Create the **Conc Beam** property in the same way with reinforcement options set as shown in Figure-10. Make sure the size is 36 x 24 inches.

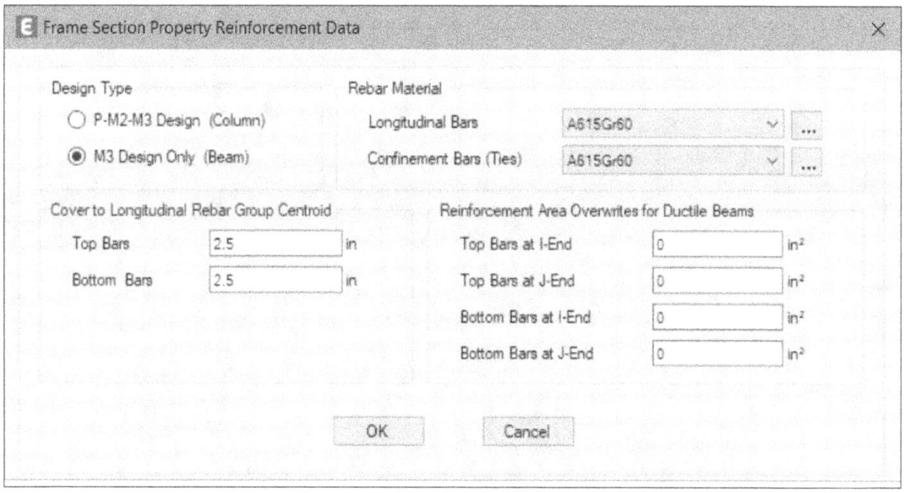

Figure-10. Frame Section Property Reinforcement Data dialog box for beam

- After creating frame section properties, click on the **OK** button.

Defining Slab Section

- Click on the **Slab Sections** tool from the **Section Properties** cascading menu of the **Define** menu. The **Slab Properties** dialog box will be displayed; refer to Figure-11.

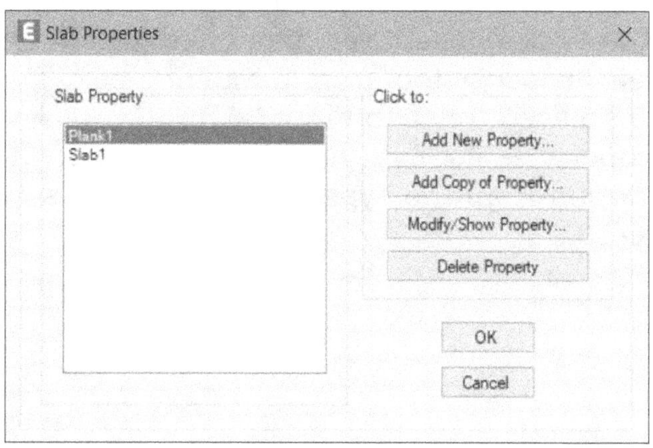

Figure-11. Slab Properties dialog box

- Click on the **Add New Property** button. The **Slab Property Data** dialog box will be displayed; refer to Figure-12.

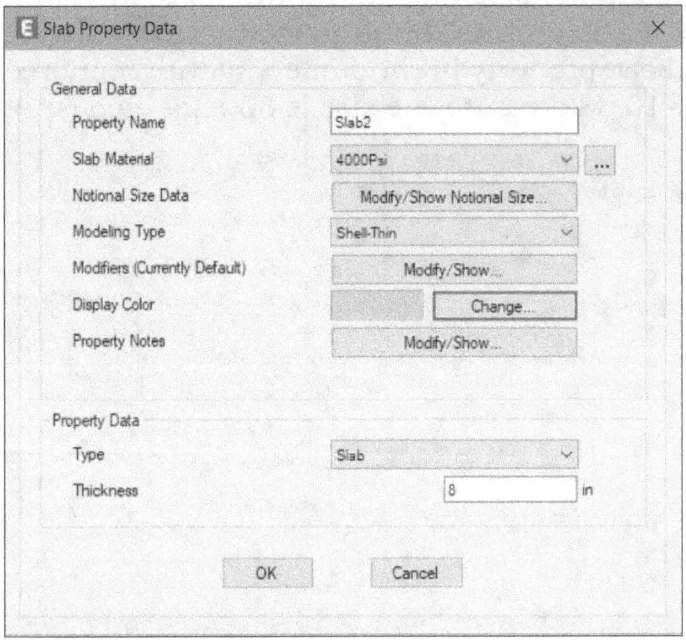

Figure-12. Slab Property Data dialog box

- Specify the name of slab as **Floor slab** in the **Property Name** edit box.
- Set the thickness of slab as **9** inch and other parameters as in Figure-12.
- After setting the parameters, click on the **OK** button. A new property will be added in the slab properties. Click on the **OK** button from the **Slab Properties** dialog box.

Defining Deck Section

- Click on the **Deck Sections** tool from the **Section Properties** cascading menu of the **Define** menu. The **Deck Properties** dialog box will be displayed.
- Click on the **Add New Property** tool from the dialog box. The **Deck Property Data** dialog box will be displayed; refer to Figure-13.
- Set the name of property as **Roof** in the **Property Name** edit box.
- Select the **Filled** option from the **Type** drop-down and set the other parameters as shown in Figure-13.

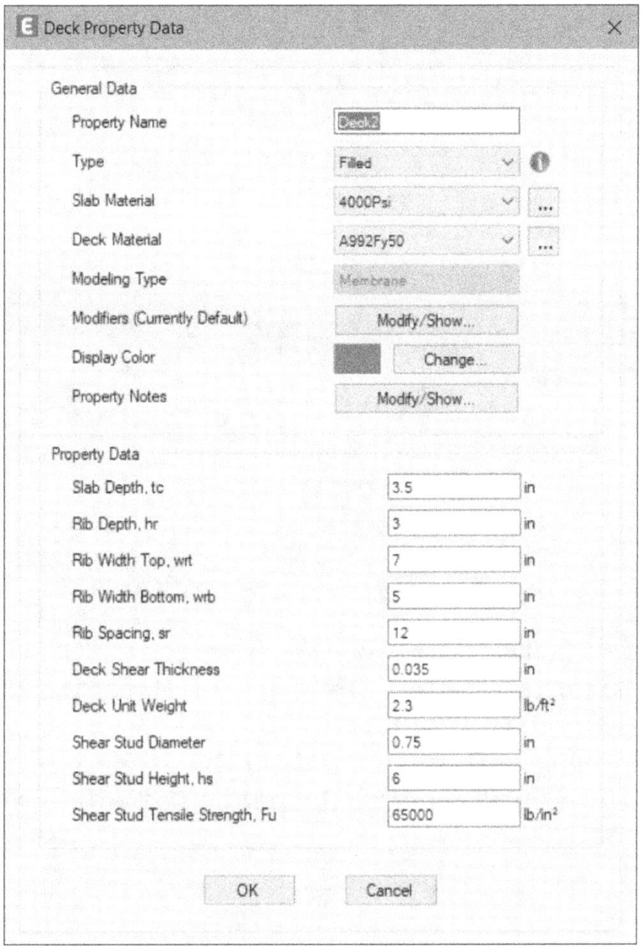

Figure-13. Deck Property Data dialog box

- After setting desired parameters, click on the **OK** button. The property will be added in the **Deck Properties** dialog box. Click on the **OK** button.

Defining Wall Section

- Click on the **Wall Sections** tool from the **Section Properties** cascading menu of the **Define** menu. The **Wall Properties** dialog box will be displayed.
- Click on the **Add New Property** button. The **Wall Property Data** dialog box will be displayed; refer to Figure-14.
- Set the name of property as **Wall section** in the **Property Name** edit box.

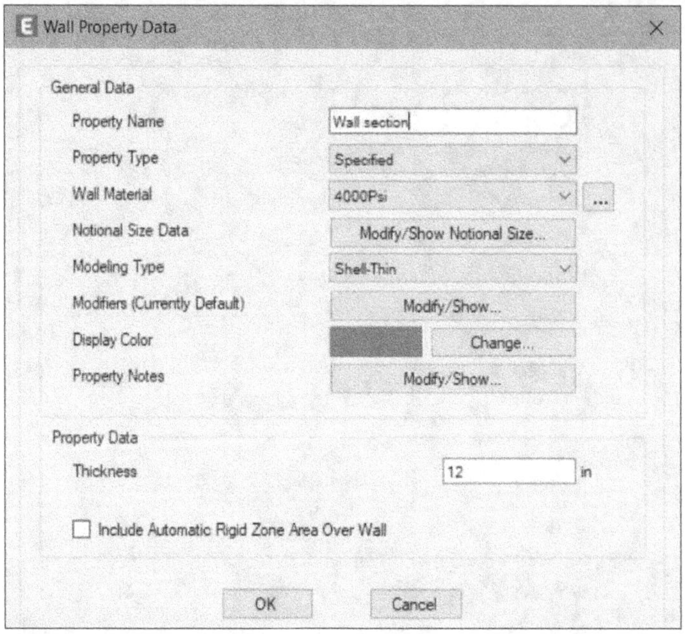

Figure-14. Wall Property Data dialog box

- Click on the **Add Material** button next to **Wall Material** drop-down. The **Define Materials** dialog box will be displayed.
- Click on the **Add New Material** button and set the properties as shown in Figure-15. Click on the **OK** button. The **Material Property Data** dialog box will be displayed.

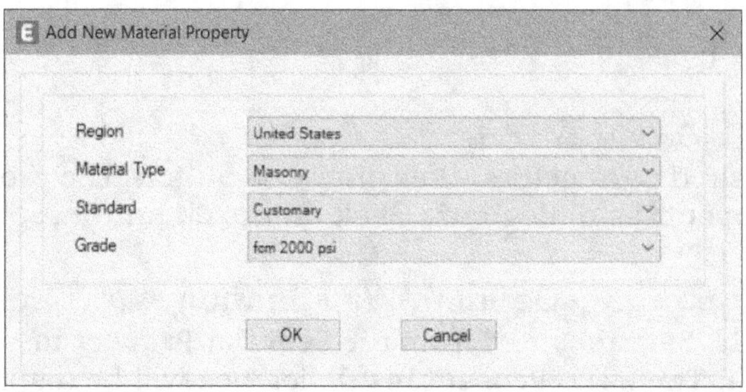

Figure-15. Add New Material Property dialog box

- Specify the name of material as **Wall Material** in the **Material Name** edit box and set the parameters as shown in Figure-16.
- Click on the **OK** button from the dialog box. The **Define Materials** dialog box will be displayed again. Click on the **OK** button. The **Wall Property Data** dialog box will be displayed again.
- Select **Wall Material** option from the **Wall Material** drop-down.
- Specify the thickness of wall as **9** inch in the **Thickness** edit box.
- Click on the **OK** button from the **Wall Property Data** dialog box to create wall section.

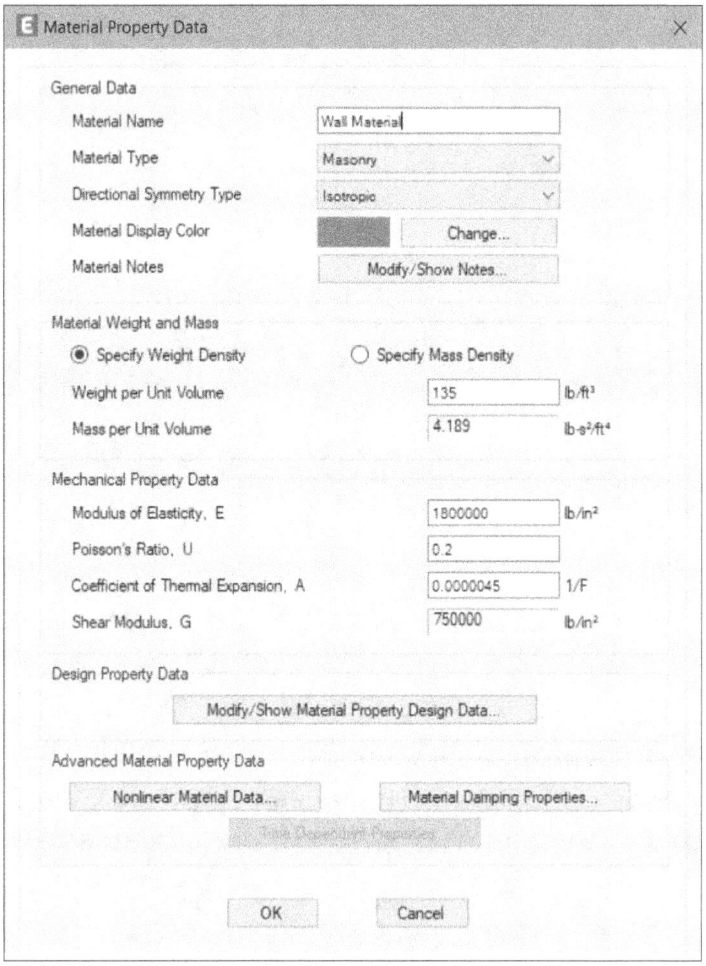

Figure-16. Material Property Data dialog box

- The **Wall Properties** dialog box will be displayed again. Click on the **OK** button from the dialog box to exit.

Creating Structure

- Select the **All Stories** option from the **Stories** drop-down at the bottom right corner of the window.
- Click on the **Quick Draw Beams/Columns (Plan, Elev, 3D)** tool from the **Draw Beam/Column/Brace Objects** cascading menu of the **Draw** menu. The **Properties of Object** tab will be displayed below **Model Explorer**.
- Select the **Conc Column** option from the **Property** drop-down and select all the vertical lines of top floor from the 3D View. The concrete columns will be created; refer to Figure-17.

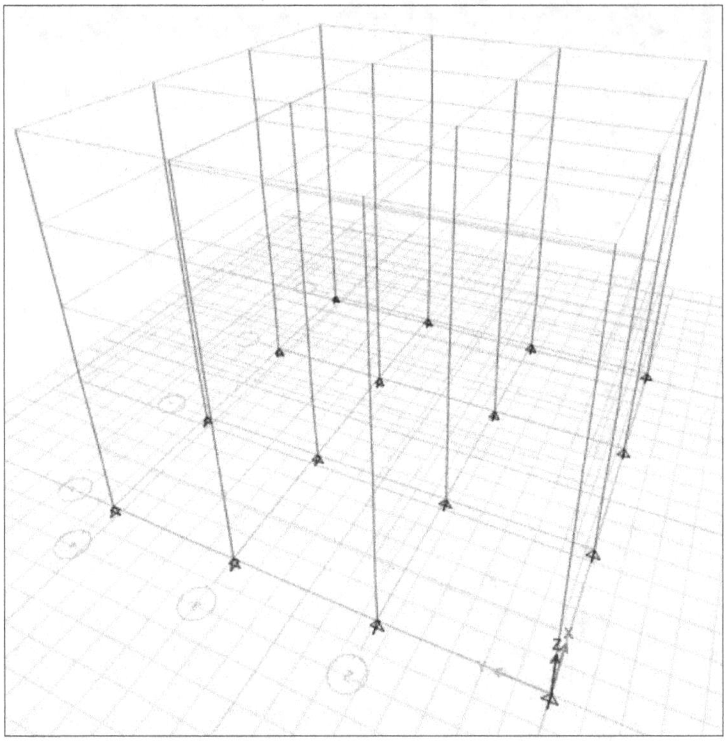

Figure-17. Concrete columns created

- Select the **Conc Beam** option from the **Property** drop-down and select the lines along X and Y axes in the Plan view. The concrete beams will be created; refer to Figure-18.

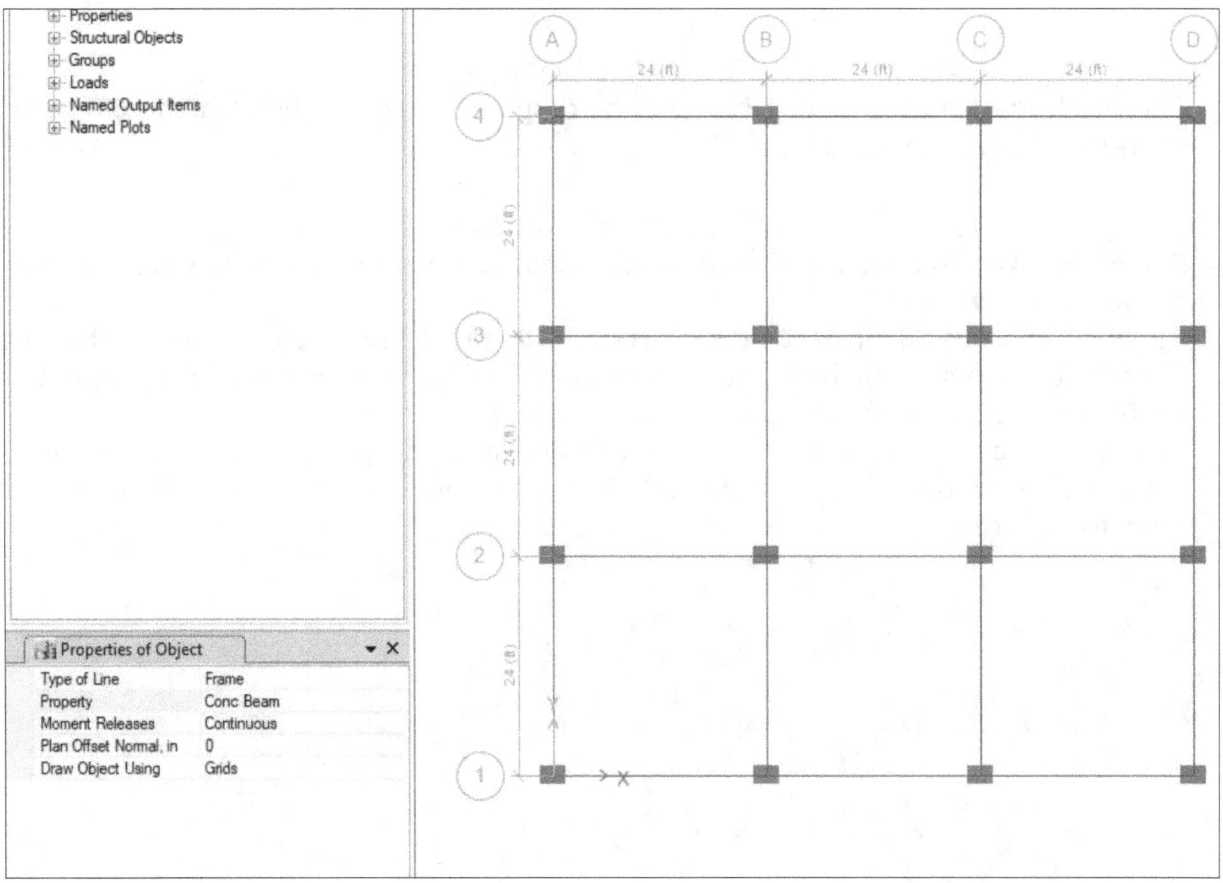

Figure-18. Concrete beams created

- Click on the **Quick Draw Floor/Wall (Plan, Elev)** tool from the **Draw Floor/ Wall Objects** cascading menu of the **Draw** menu. The **Properties of Object** tab will be displayed and you will be asked to specify location.
- Select the **Floor slab** option from the **Property** drop-down and click in the rectangular boxes in elevation views. The model will be displayed as shown in Figure-19.

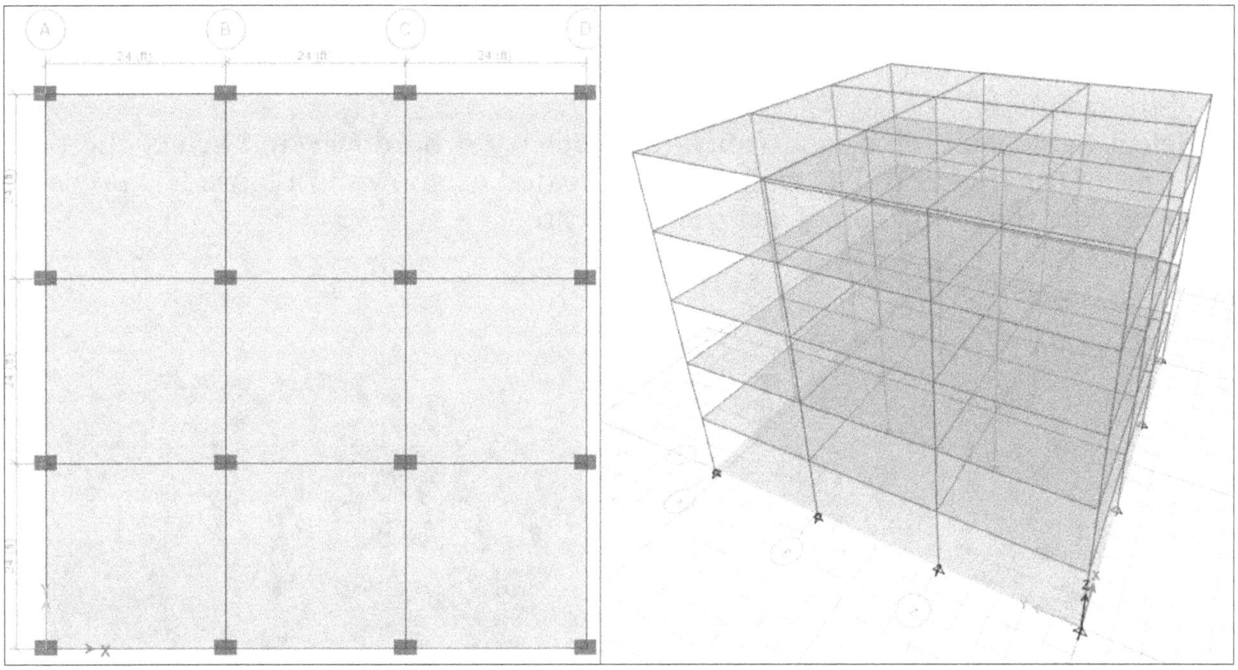

Figure-19. Floors created

- Select the **One Story** option from the **Stories** drop-down and select the floor slabs of top floor created recently from the Plan view; refer to Figure-20.

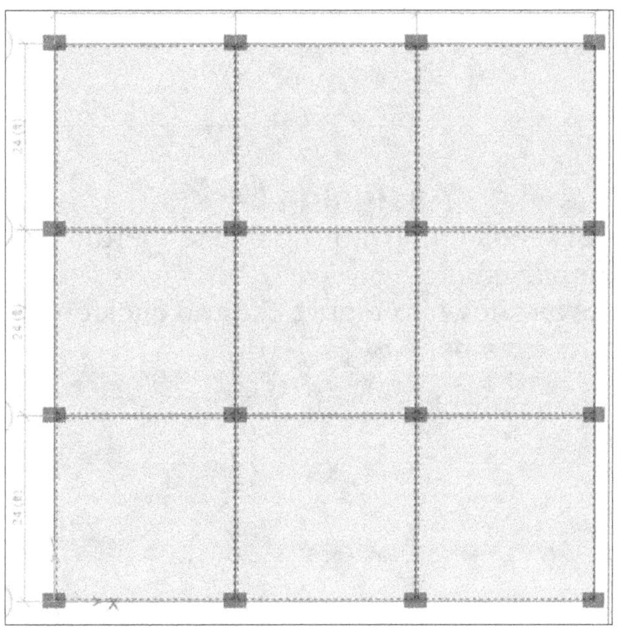

Figure-20. Floor slabs selected for deletion

- Press the **DELETE** button from the keyboard to delete floor slabs created at the top.

- Click on the **Quick Draw Floor/Wall (Plan, Elev)** tool from the **Draw Floor/ Wall Objects** cascading menu of the **Draw** menu again. Select the **Roof** property from the **Property** drop-down.
- Click in the rectangular boxes of plan view which you deleted earlier. The roof slabs will be created.
- Click on the **Quick Draw Walls (Plan)** tool from the **Draw Floor/Wall Objects** cascading menu of the **Draw** menu. You will be asked to specify location for creating wall.
- Select **All Stories** option from the **Stories** drop-down at the bottom right corner of the window.
- Select **Wall section** option from the **Property** drop-down and select the outer edges of the building. The walls will be created as shown in Figure-21. Press **ESC** to exit the tool. This completes the structure.

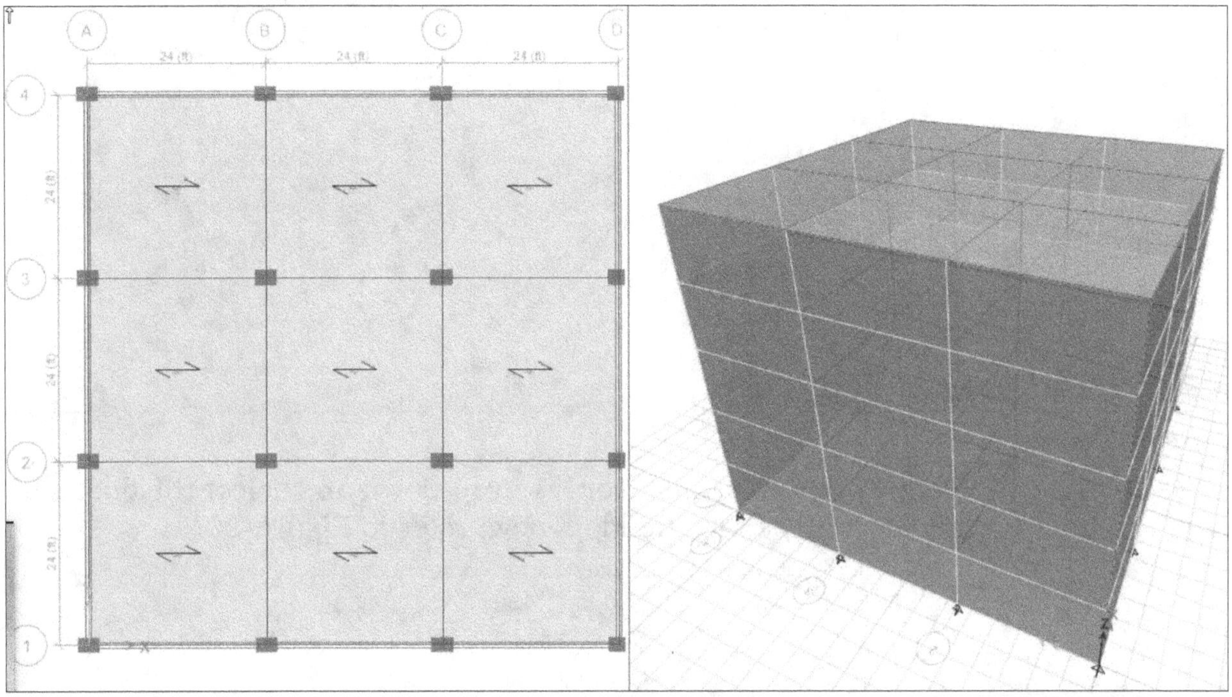

Figure-21. Walls created in project

Assigning Loads

- Click on the **Load Patterns** tool from the **Define** menu. The **Define Load Patterns** dialog box will be displayed.
- Enter the parameters as shown in Figure-22 and click on the **Add New Load** button. Wind load will be added as per ASCE 7-10.

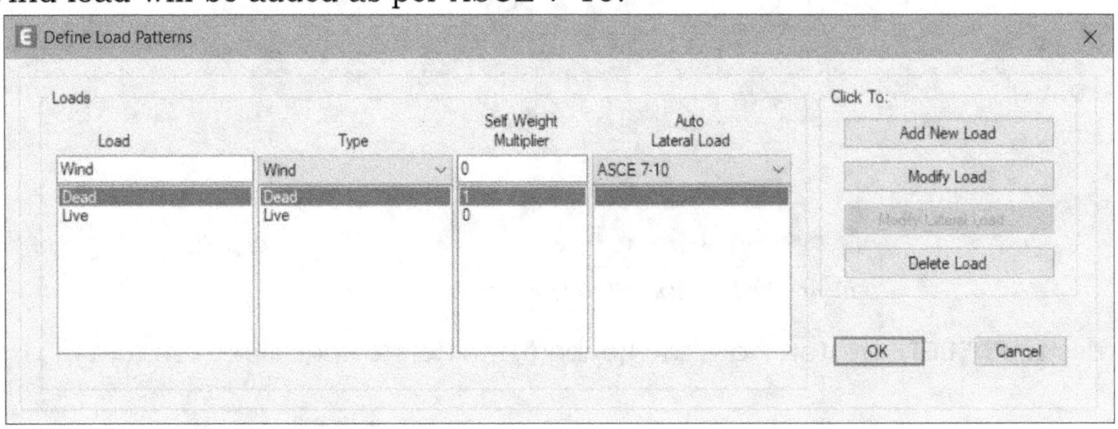

Figure-22. Wind load pattern created

- Click on the **OK** button from the dialog box to create the load pattern.
- Click on the **Uniform** tool from the **Shell Loads** cascading menu of the **Assign** menu. The **Shell Load Assignment - Uniform** dialog box will be displayed.
- Select the **Live** option from the **Load Pattern Name** drop-down. Specify the load value as **25** in **Load** edit box.
- Select **One Story** option from the **Stories** drop-down. Select the top roof shells; refer to Figure-23.

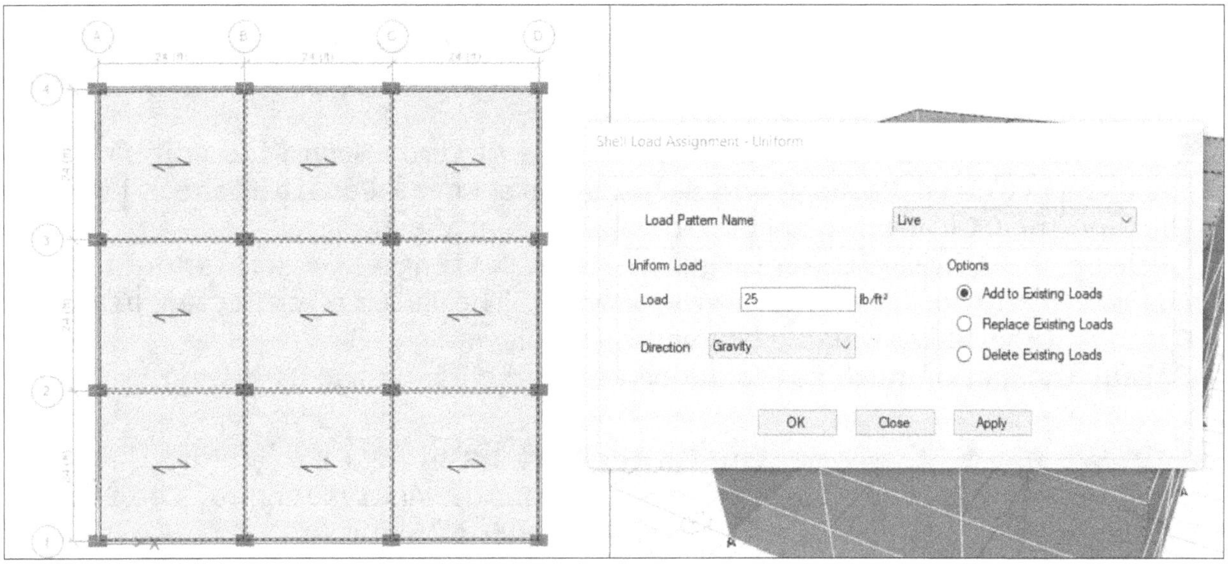

Figure-23. Assigning load on roof

- Make sure you select **Add to Existing Loads** radio button and then click on the **OK** button.
- Click on the **Floor slab** object in the **Slab Sections** node of the **Model** tab in the **Model Explorer** and right-click on it. A shortcut menu will be displayed; refer to Figure-24.

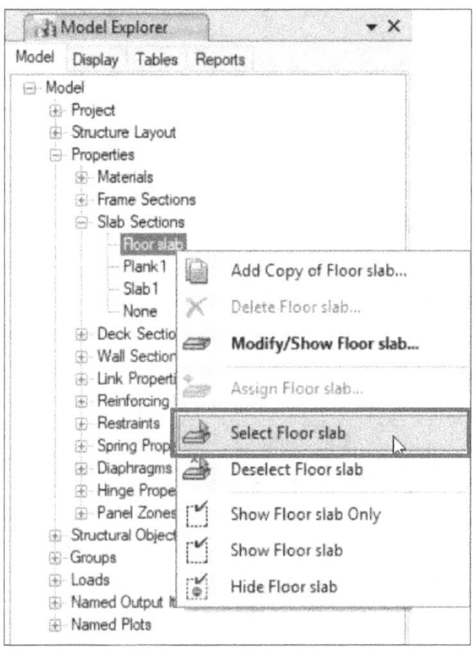

Figure-24. Selecting floor slabs

- Click on the **Select Floor slab** option from the shortcut menu. All the floor slabs will be selected.

- Click on the **Uniform** tool from the **Shell Loads** cascading menu of the **Assign** menu. The **Shell Load Assignment-Uniform** dialog box will be displayed.
- Specify the load value as **100** lb/ft^2 and select the **Add to Existing Loads** radio button.
- Make sure the **Live** option is selected in the **Load Pattern Name** edit box and click on the **OK** button.

Setting Mesh Size

- Click on the **Automatic Mesh Settings for Floors** tool from the **Analyze** menu to define mesh size for floors. The **Automatic Mesh Options (for Floors)** dialog box will be displayed.
- Set the mesh size as **4** feet in the **Approximate Maximum Mesh Size** edit box. Select the **Use Localized Meshing** and **Merge Joints where Possible** check boxes.
- Click on the **OK** button to specify the size of floor mesh.
- Click on the **Automatic Rectangular Mesh Settings for Walls** tool from the **Analyze** menu to define mesh size of walls. The **Automatic Rectangular Mesh Options (for Walls)** dialog box will be displayed.
- Specify the size of mesh in the dialog box and click on the **OK** button.

Running Analysis

- Click on the **Set Load Cases to Run** tool from the **Analyze** menu. The **Set Load Cases to Run** dialog box will be displayed; refer to Figure-25.

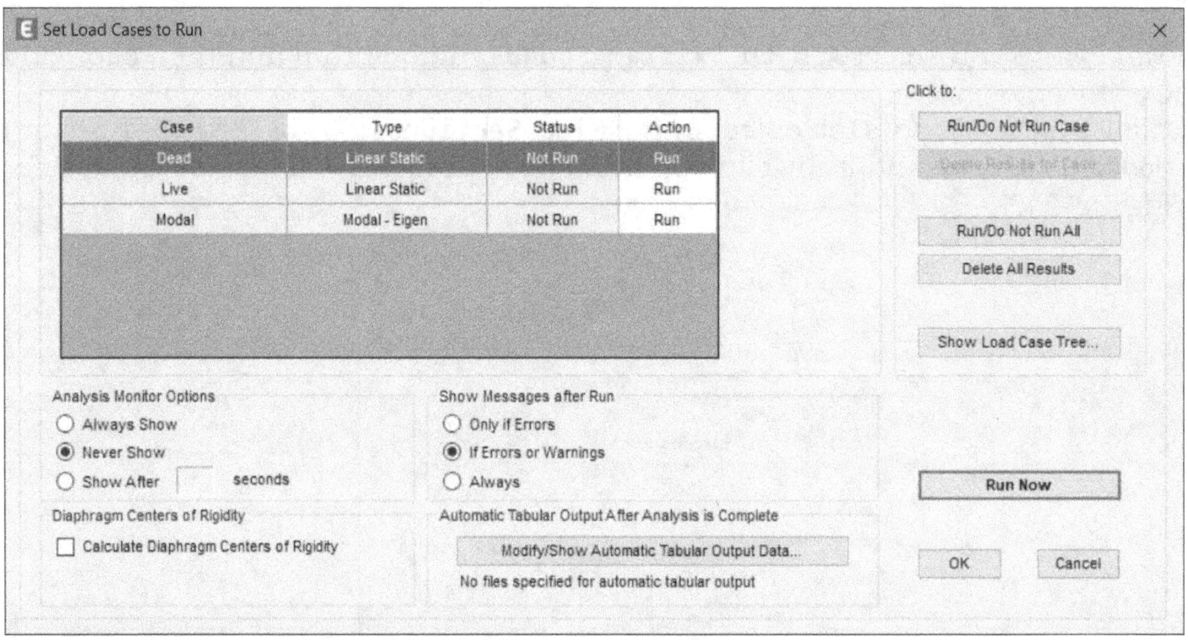

Figure-25. Set Load Cases to Run dialog box

- Make sure all the cases are selected and select the **Always Show** radio button from the **Analysis Monitor Options** area.
- Click on the **Run Now** button. Once the analysis is performed, the deformed model will be displayed in 3D View; refer to Figure-26.

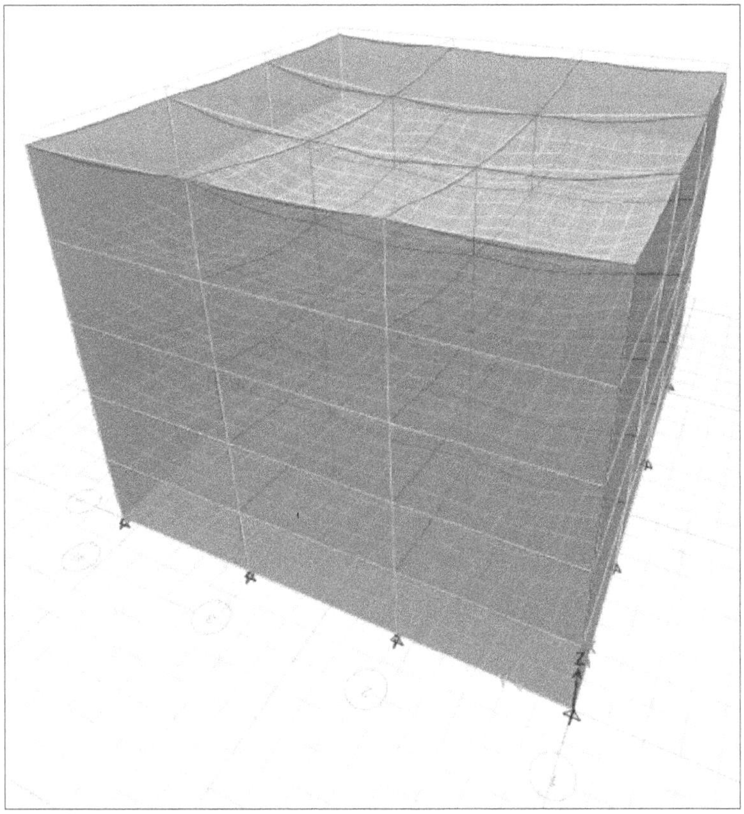

Figure-26. Deformed model after analysis

Performing Design Check

- Click on the **View/Revise Preferences** tool from the **Concrete Frame Design** cascading menu of the **Design** menu. The **Concrete Frame Design Preferences** dialog box will be displayed.
- Select the **ACI 318-14** option from the **Design Code** field of the **Concrete Frame Design** dialog box and click on the **OK** button.
- Click on the **Start Design/Check** tool from the **Concrete Frame Design** cascading menu of **Design** menu. The size of different reinforcements of concrete beams will be displayed in the plan view; refer to Figure-27.

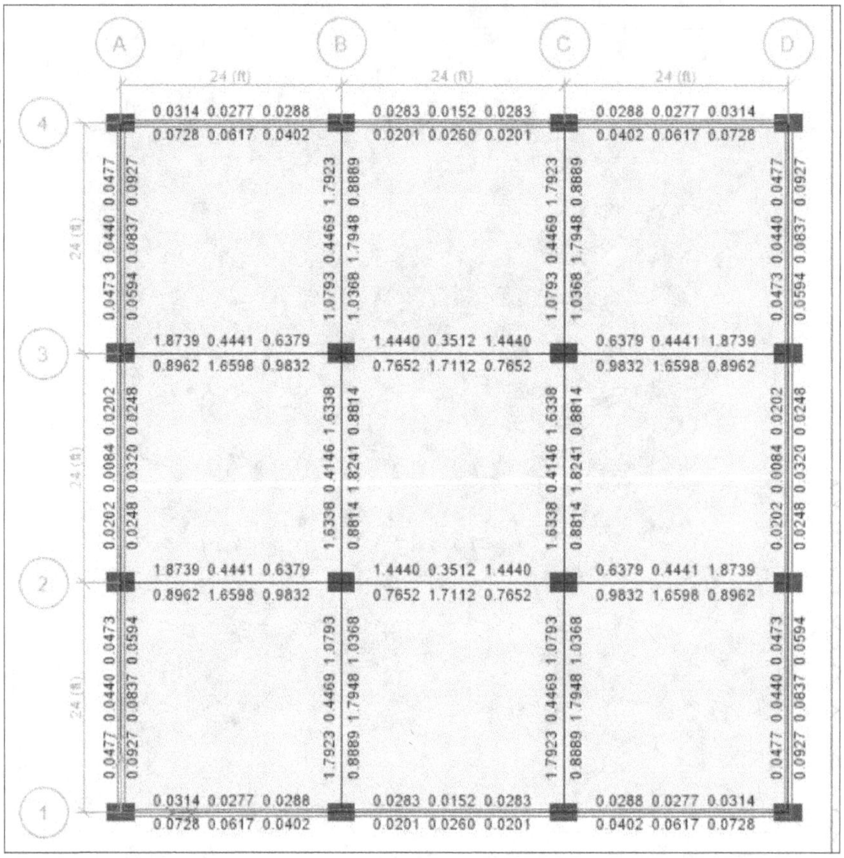

Figure-27. Reinforcement of concrete beam

- Click on the **Verify All Members Passed** tool from the **Concrete Frame Design** cascading menu of the **Design** menu to make sure all the concrete members have passed design check. The message box will be displayed as shown in Figure-28. Click on the **OK** button. Modify the concrete frame elements that did not pass the test and then rerun the test.

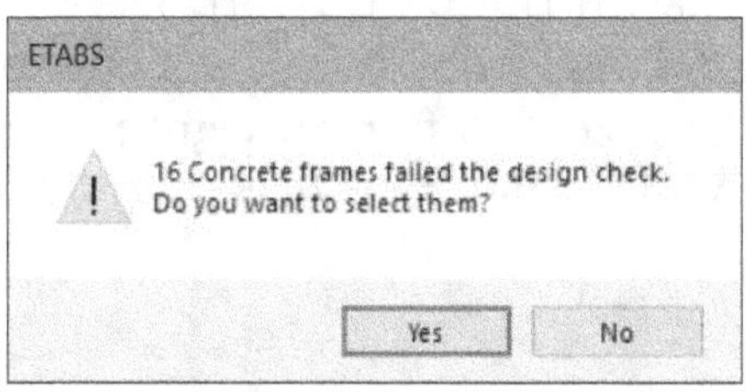

Figure-28. ETABS message box

Creating Project Report

- Click on the **Show Project Report** tool from the **Create Report** cascading menu of the **File** menu. The complete report of project will be created and displayed in the **Report Viewer** tab.
- You can export this report to word document by using the **Export to Word Document** button and you can print this report by using the **Print Report** button in the **Report Viewer** tab.
- You can also print the model in graphics by selecting desired viewport from drawing area and clicking on the **Print Graphics** tool from the **File** menu or by pressing **CTRL+P** key from the keyboard.

Save the file and close the project by using the **Close** button.

DESIGNING TALL BUILDINGS

Designing tall building involves effect of various loads which could be ignored for one story or two story buildings like wind load, and so on.

Lateral Stability Systems in Tall Buildings

The key feature of the tall building from a structural engineer's point of view is its lateral stability structural system. Various structural systems are:

Moment resisting frame (MRF) system

Moment resisting frame (MRF) system (also be called as rigid frame) is a system where the beams and columns are rigidly connected to provide lateral resistance. Bracings and shear walls (primarily core walls) are used widely as the lateral resisting system in tall buildings, they are also widely used in different type of low-rise buildings. For tall buildings, purely rely on the core wall system to resist lateral loading is not sufficient, which will result in a very thick wall thickness. Therefore, it is quite common that firm horizontal members (such as outriggers) are used to connect the main core to the exterior columns.

Tube structures

Tube structures are one of the major lateral resisting systems in tall buildings. A building is designed to act like a hollow cylinder, cantilevered perpendicular to the ground in order to resist lateral loads. There are different types of tube structures, such as framed tubes, braced tubes, tube-in-tube and bundle tube.

Diagrid system

Diagrid system is another lateral stability system. With their structural efficiency as a adapted version of the bracing systems, diagrid structures have been emerging as a new aesthetic trend for tall buildings in this era. The triangular geometry of diagrid structures can effectively prevent structural failure as a result of both horizontal and vertical loads resulted by the externally imposed lateral loads and gravity load.

Super frame (or mega frame) structure System

Super frame (or mega frame) structures use mega columns and mega girders to work together as the primary lateral resisting structural systems. Figure-29 shows the structure systems and their respective story limits.

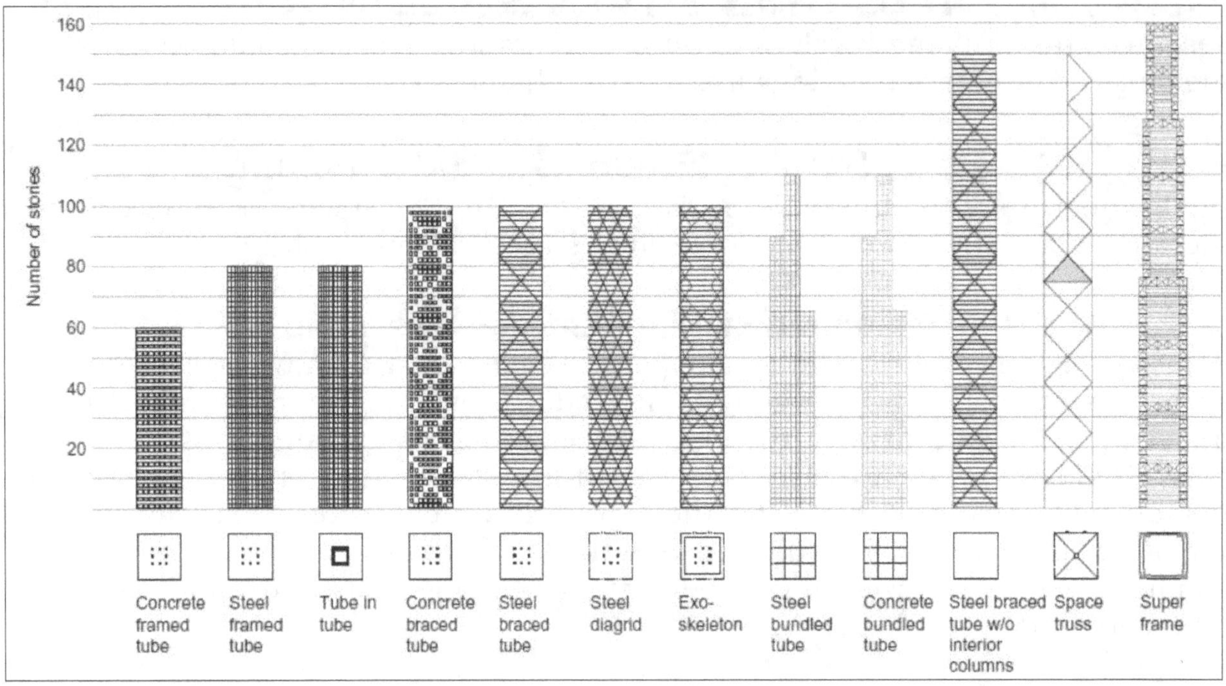

Figure-29. Structure systems and story limits

Note that generally more than one type of lateral stability systems is employed in the building for providing lateral strength. Apart from lateral stability systems, floor system also plays an important role in resisting the lateral loadings. Floor diaphragm should be designed to transmit the lateral load from cladding to the lateral load-resisting systems at each level. Horizontal trusses or deep beams are also used where the floor diaphragm is interrupted by penetrations or openings. It is important that designers allow for the concentrations of horizontal stress within the floor diaphragm around openings and penetrations through the provision of diagonal corner steel within the flooring.

General Loads on Tall Buildings

Various loads applicable on tall buildings are generally specified in the ASCE/SEI 2010. Various loads applicable are discussed next.

Dead Loads

The dead loads are loads that permanently stay in a building, including self-weights of various structural members such as beams, columns, and core walls.

Live Loads

The live loads are loads that vary magnitude and location. An example of live loads can be people, furniture, and equipment. Depending on type of building (residential or commercial), you can select loads from design codes.

Snow Loads

The snow load is a type of live loads. But depending on geography and unique features of building, this load is considered as separate category. In design code such as ASCE/SEI 7-10, snow loads are determined from a zone map reporting 50-year recurrence interval. A designer can check it for the determination of the value of snow load.

Wind Loads

Effects of wind depend on density and flow of air, angle of incidence, shape and stiffness of the structure, and roughness of surface. Its load value also depends on roof geometry, wind exposure, location, and its importance. For buildings with height less than 200m, EN1991-1-4 code is applicable and for building height more than 200m, a wind tunnel test is needed.

Earthquake Loads

The earthquake load is a type of dynamic load where magnitude and direction of load keeps changing. The response of the building depends on the mass and stiffness of the structure. If the building is very flexible with large masses, induced motion will cause small accelerations of the building and large relative displacement. For earthquake analysis, three types of structural analysis methods are used for tall buildings; Response Spectrum Analysis, Pushover Analysis, and Time History Analysis.

Response Spectrum Analysis

The Response Spectrum Analysis is a type of linear dynamic analysis which defines the response (such as acceleration, velocity, or displacement) spectrum by enveloping and smoothing the spectra corresponding to different earthquake time histories.

Pushover Analysis

The Pushover analysis is a type of non linear analysis in which gravity load and monotonically increasing horizontal loads (until target displacement is obtained) are applicable on the structure.

Time History Analysis

The Time History analysis can be linear or nonlinear dynamic analysis. It is used to analyze the building under a ground acceleration time history with the frequency content matches the design spectrum.

One other way to design building earthquake resistant is by using Ductile structure.

Key Points for Tall building Design

Various important points to be considered while designing tall buildings are discussed next.

- Design ductile structure of building so that in case of earthquake, the structure can dissipate extra energy in elastic deformation. You can use reducing beam section for building design.
- You should not use soft story-strong column weak beam design. Building is classified as soft story if the stiffness value is less than 80% of average of three stories in the building. Soft story buildings are vulnerable to collapse in a moderate-to-severe intensity earthquake is a phenomenon known as soft story collapse, as it will sometime results in the collapse of the entire building.
- There should be vertical and horizontal regularity as much as possible so that there is no abrupt change in the overall strength/stiffness of structure causing soft-story collapse.
- If the structure is in an area with high frequency of earthquakes then you can use special active and passive devices to reduce excitation. Various devices for this purpose are base isolation, shock absorber, damping systems, and so on.

- The nearby tall buildings can also affect the flow of wind and can cause venturi effect so you need to consider dynamic pressure and motion of wind for analysis.
- The building structure should also be checked by progressive collapse analysis when certain columns are removed from the structure. You can use linear static, linear dynamic, nonlinear static, and nonlinear dynamic studies for this purpose.
- Fire safety is an important aspect of structure design to ensure minimum property and life damage in case of fire.
- Cladding can be used to prevent the transmission of sound, provide thermal insulation, create an external facade, and prevent the spread of fire.

PROJECT 2

In this project, we will design structure of a tall building with more than 30 stories. We will analyze this building for vertical loads using Linear Static analysis and then we will perform seismic and wind analysis which are dynamic in nature. The plan view and 3D view of model are shown in Figure-30.

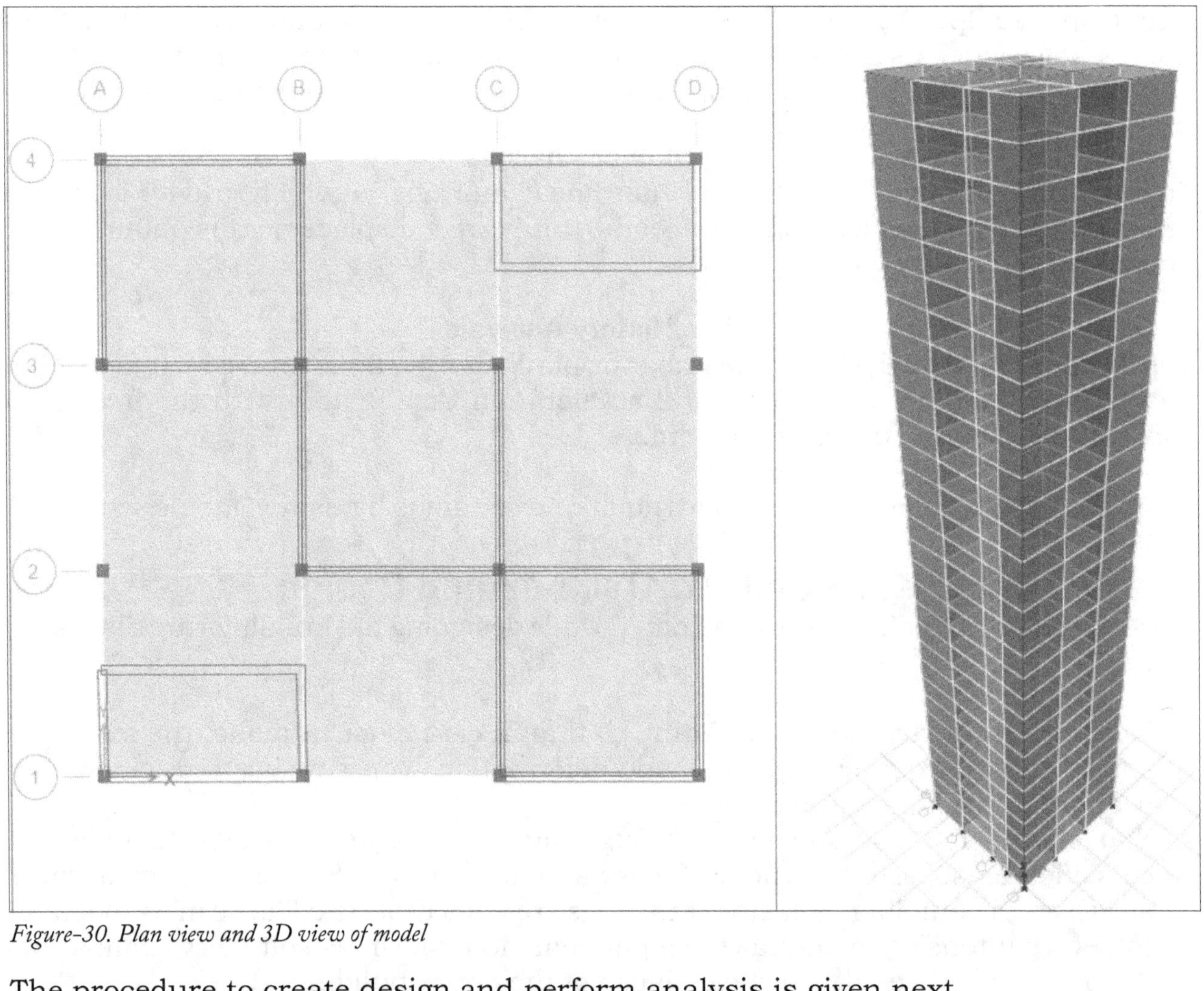

Figure-30. Plan view and 3D view of model

The procedure to create design and perform analysis is given next.

Creating Structure

- Start ETABS and click on the **New Model** tool from the **File** menu. The **Model Initialization** dialog box will be displayed.

- Select the **Use Built-in Settings With** radio button and set the parameters as shown in Figure-31.

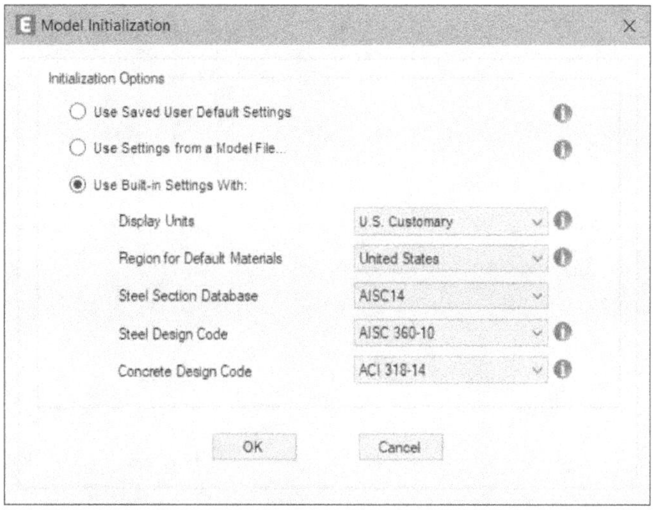

Figure-31. Parameters for project

- Click on the **OK** button from the dialog box to apply settings. The **New Model Quick Templates** dialog box will be displayed. Set the parameters as shown in Figure-32 to define parameters for 30 stories.

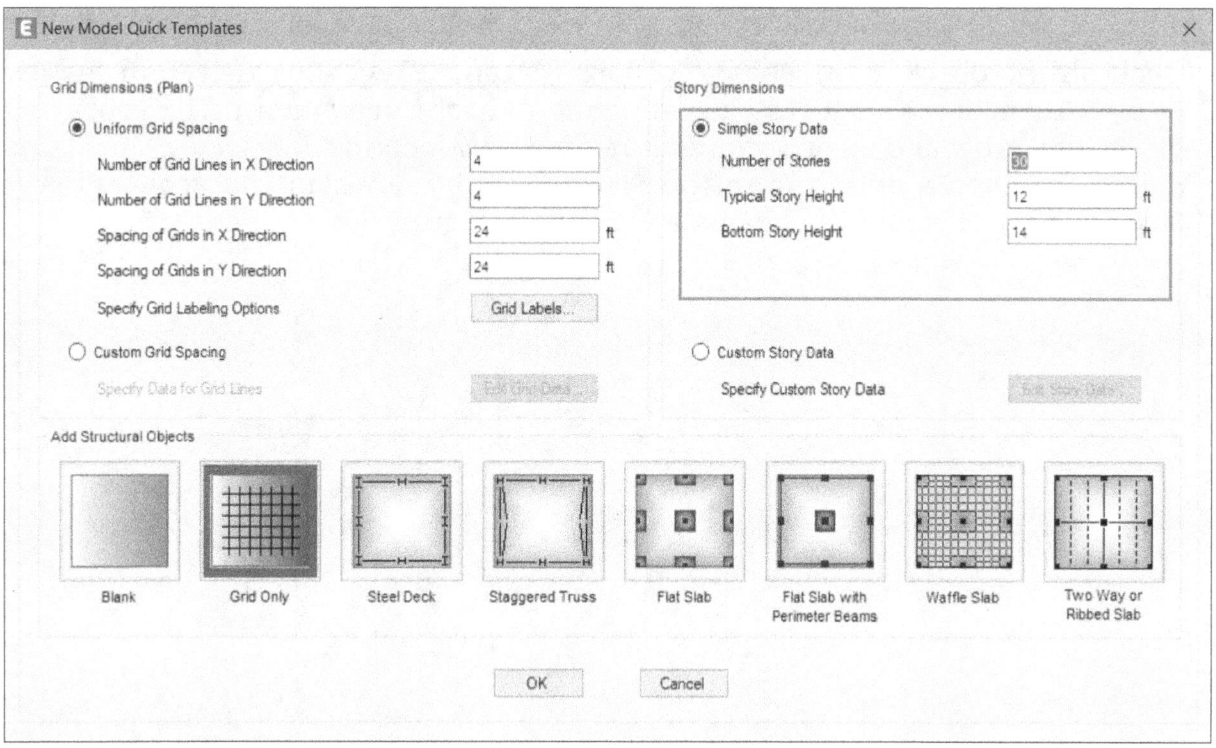

Figure-32. Building story parameters

- Click on the **OK** button from the dialog box. The grids of model will be displayed in plan and 3D view in different viewports.
- Click on the **Quick Draw Columns (Plan, 3D)** tool from the **Draw Beam/Column/ Brace Objects** cascading menu in the **Draw** menu. The **Properties of Object** tab will be displayed below **Model Explorer**.
- Select the **ConcCol** option from the **Property** drop-down in the **Properties of Object** tab.

- Select the **All Stories** option from the **Story** drop-down and create columns at each grid point; refer to Figure-33.

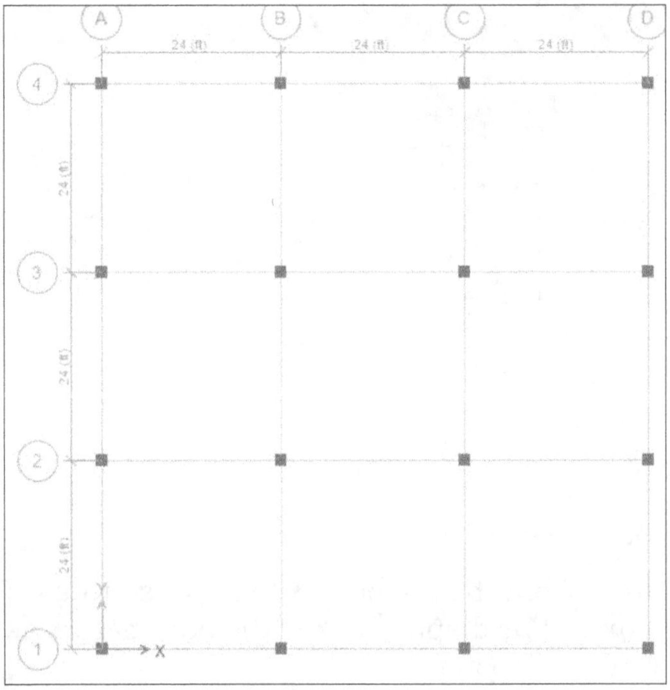

Figure-33. Columns created on grid points

- Click on the **Quick Draw Beams/Columns (Plan, Elev, 3D)** tool from the **Draw Beam/Column/Brace Objects** cascading menu in the **Draw** menu. The **Properties of Object** tab will be displayed with related parameters.
- Select the **ConcBm** option from the **Property** drop-down in the **Properties of Object** tab and create the beams as shown in Figure-34.

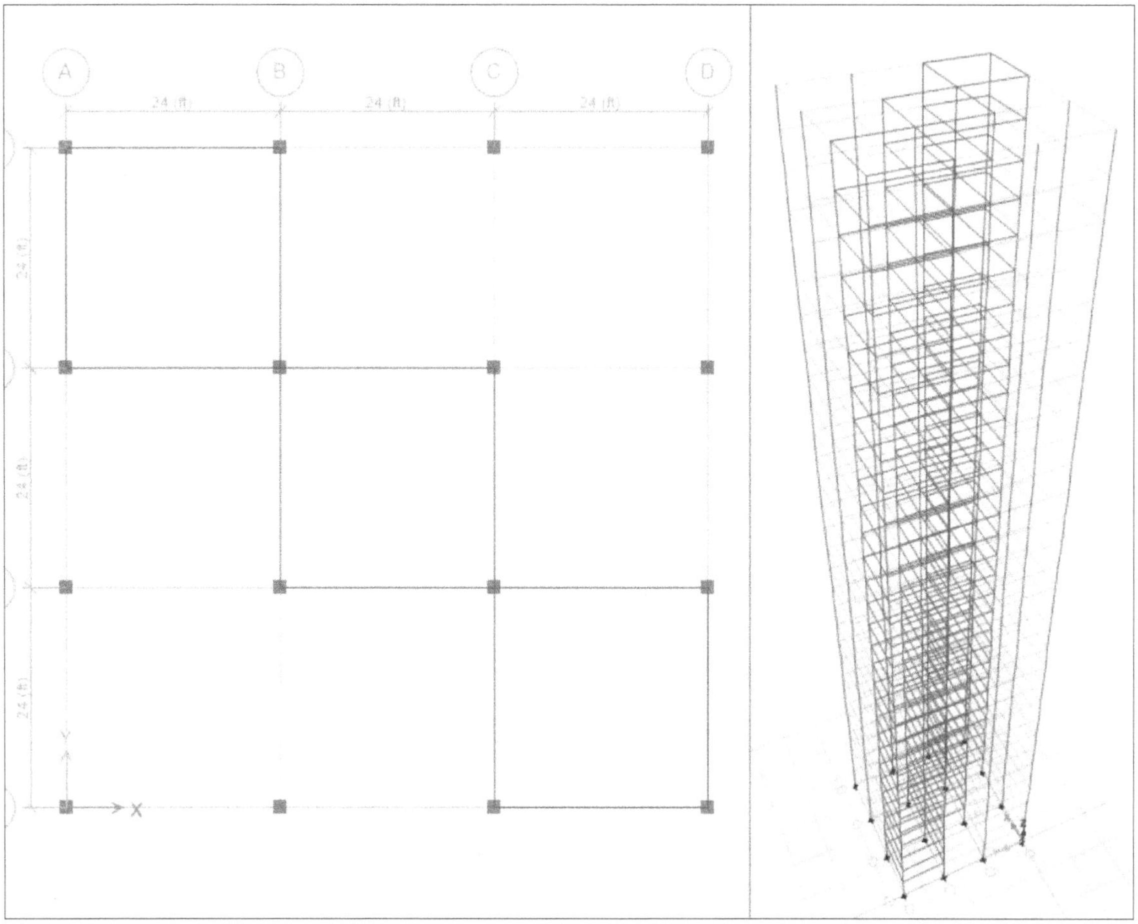

Figure-34. Beams created

- Now, we will define thickness of slab as 10 inch for creating floors in the building. Click on the **Slab Sections** tool from the **Section Properties** cascading menu of the **Define** menu. The **Slab Properties** dialog box will be displayed.
- Select the **Slab1** property from the **Slab Property** list box and click on the **Modify/ Show Property** button. The **Slab Property Data** dialog box will be displayed.
- Specify the value **10** in **Thickness** edit box of the **Property Data** area of the dialog box; refer to Figure-35 and click on the **OK** button.

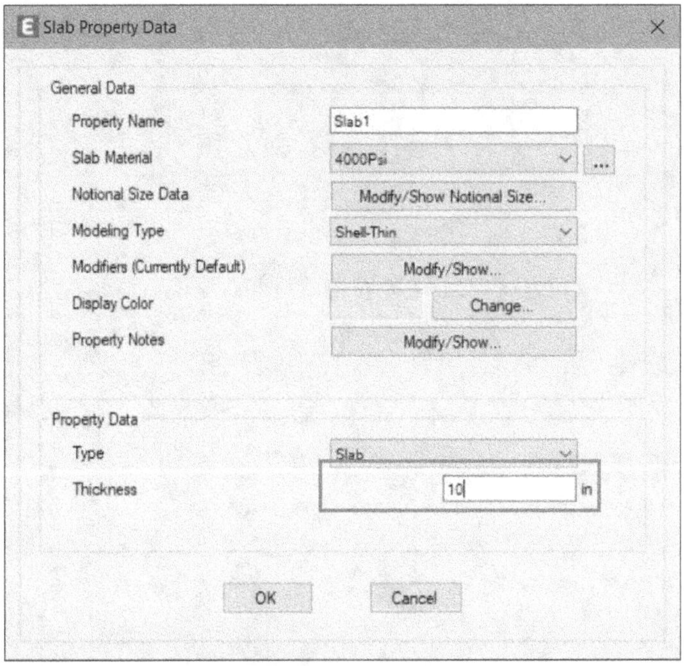

Figure-35. Thickness specified for slab

- Click on the **OK** button from the **Slab Properties** dialog box to exit.
- To draw the floor, click on the **Quick Draw Floor/Wall (Plan, Elev)** tool from the **Draw Floor/Wall Objects** cascading menu of the **Draw** menu and select the **Slab1** option from the **Property** drop-down in the **Properties of Object** tab.
- Draw the slab as shown in Figure-36.

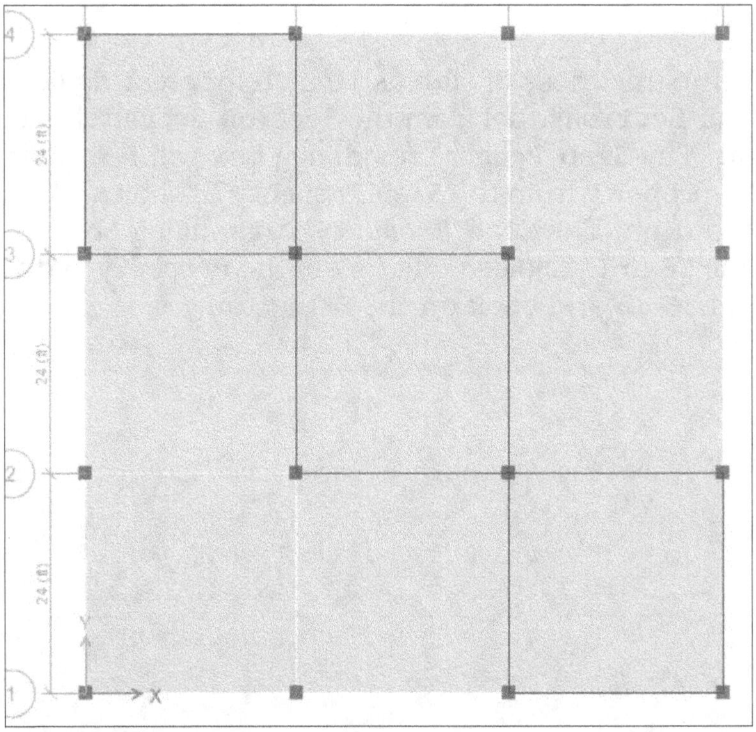

Figure-36. Drawing floor

- Click on the **Quick Draw Walls (Plan)** tool from the **Draw Floor/Wall Objects** cascading menu of the **Draw** menu. The **Properties of Object** tab will be displayed.
- Select the **Wall1** option from the **Property** drop-down and **Pier** from **Type of Area** drop-down.

- Select the grid lines where you have created beams earlier; refer to Figure-37.

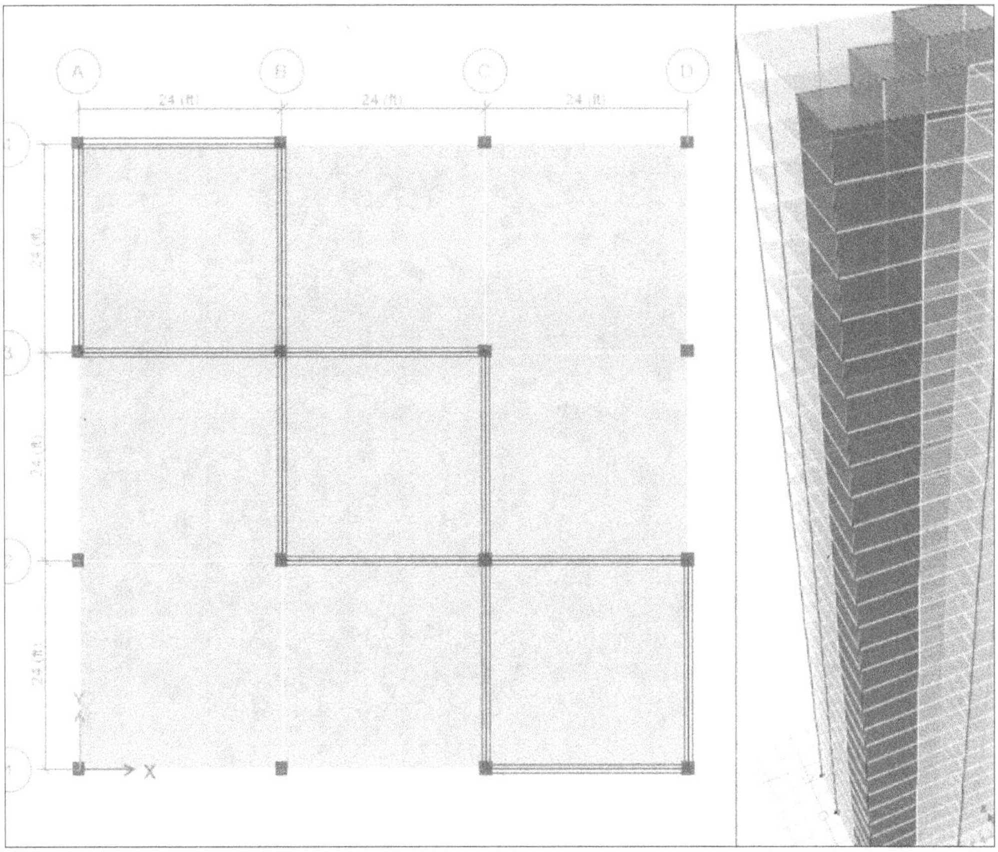

Figure-37. Walls created on grid lines

- Click on the **Draw Walls (Plan)** from the **Draw Floor/Wall Objects** cascading menu in the **Draw** menu. You will be asked to specify start point of wall. Make sure **Wall1** option is selected in the **Property** drop-down and **Pier** option is selected in the **Type of Area** drop-down.
- Create the walls as shown in Figure-38.

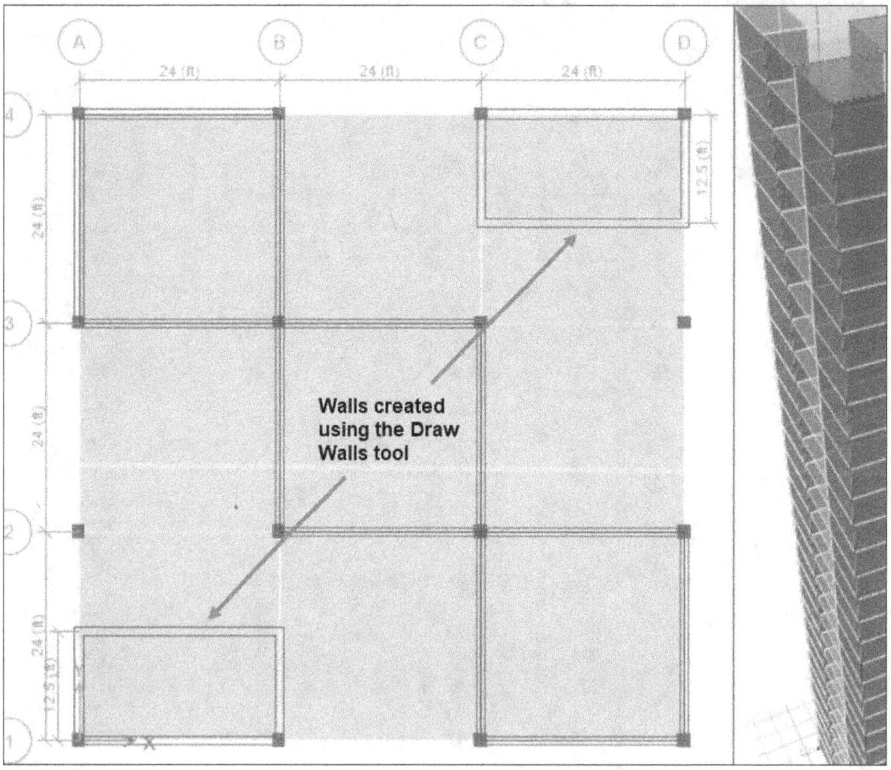

Figure-38. Walls created using Draw Walls tool

- Click on the **Auto Draw Cladding** tool from the **Draw** menu. The **Cladding Options** dialog box will be displayed; refer to Figure-39.

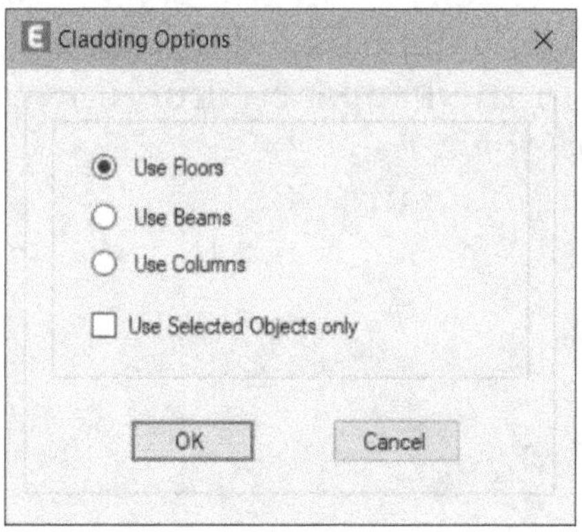

Figure-39. Cladding Options dialog box

- Select the **Use Floors** radio button from the dialog box and click on the **OK** button. The cladding will be created.

Defining Load Patterns, Load Cases, and Load Combinations

- Click on the **Load Patterns** tool from the **Define** menu. The **Define Load Patterns** dialog box will be displayed.
- Specify **Wind** in the **Load** edit box, **Wind** in the **Type** drop-down, **0** in **Self Weight Multiplier**, and **ASCE 7-05** in the **Auto Lateral Load** drop-down; refer to Figure-40. Click on the **Add New Load** button from the dialog box. The load will be added in the list.

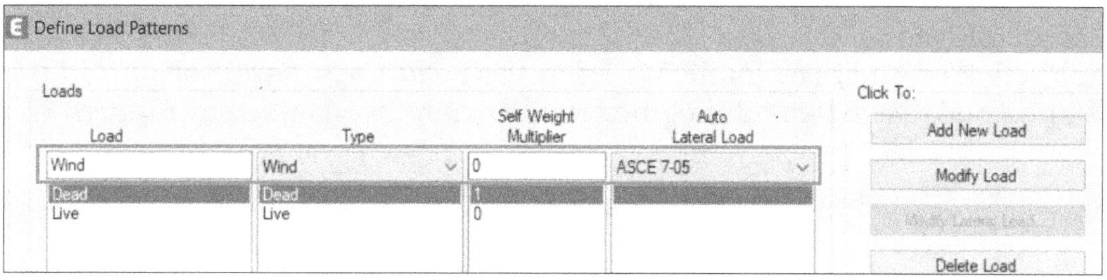

Figure-40. Specifying wind load parameters

- Similarly, create earthquake load with parameters **EQ** in **Load** edit box, **Seismic** in **Type** drop-down, **0** in **Self Weight Multiplier**, and **ASCE 7-10** in the **Auto Lateral Load** drop-down.
- Now, select newly created **Wind** load from the list and click on the **Modify Lateral Load** button. The **Wind Load Pattern** dialog box will be displayed.
- Specify the speed of wind as **100** in the **Wind Speed (mph)** edit box and **C** in the **Exposure Type** drop-down. Click on the **OK** button to modify lateral load parameters for wind load.
- Select the **EQ** load from the list and click on the **Modify Lateral Load** button. The **Seismic Loading** dialog box will be displayed.
- Select the **Ss and S1 from USGS Database - by Zip Code** radio button to specify seismic load parameters based on geological survey database and specify desired zip code (like 00501 for New York) in the **Site Zip Code** edit box in the **Seismic Coefficients** area of the dialog box.
- Select desired option from the **Site Class** drop-down to define soil property for foundation of building.
- Click on the **OK** button to apply changes to the load. The **Define Load Patterns** dialog box will be displayed again. Click on the **OK** button to apply the changes.
- Now, we will create shell uniform load set for applying load on floors. Click on the **Shell Uniform Load Sets** tool from the **Define** menu. The **Shell Uniform Load Sets** dialog box will be displayed.
- Click on the **Add New Load Set** button from the dialog box. The **Shell Uniform Load Set Data** dialog box will be displayed.
- Click on the **Add** button, add live load - **40** lb/ft^2 and dead load - **70** lb/ft^2; refer to Figure-41. After setting parameters, click on the **OK** button. Click on the **OK** button from the **Shell Uniform Load Sets** dialog box to exit.

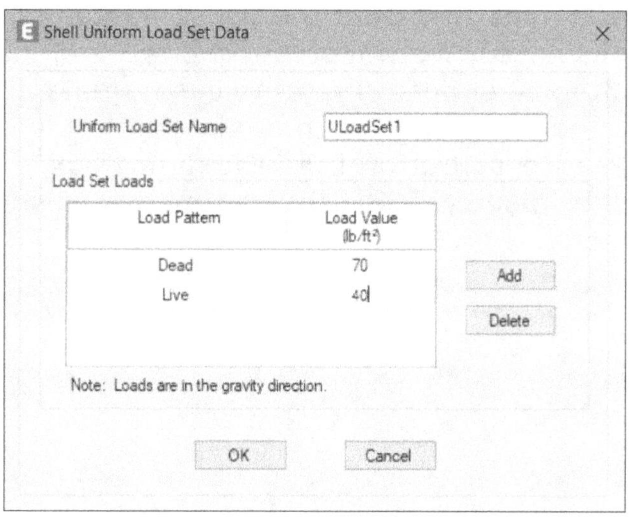

Figure-41. Load parameters for shell load

- Click on the **Load Cases** tool from the **Define** menu. The **Load Cases** dialog box will be displayed.
- Double-click on the **Wind** load case from the **Load Case Name** column in the dialog box. The **Load Case Data** dialog box will be displayed; refer to Figure-42.

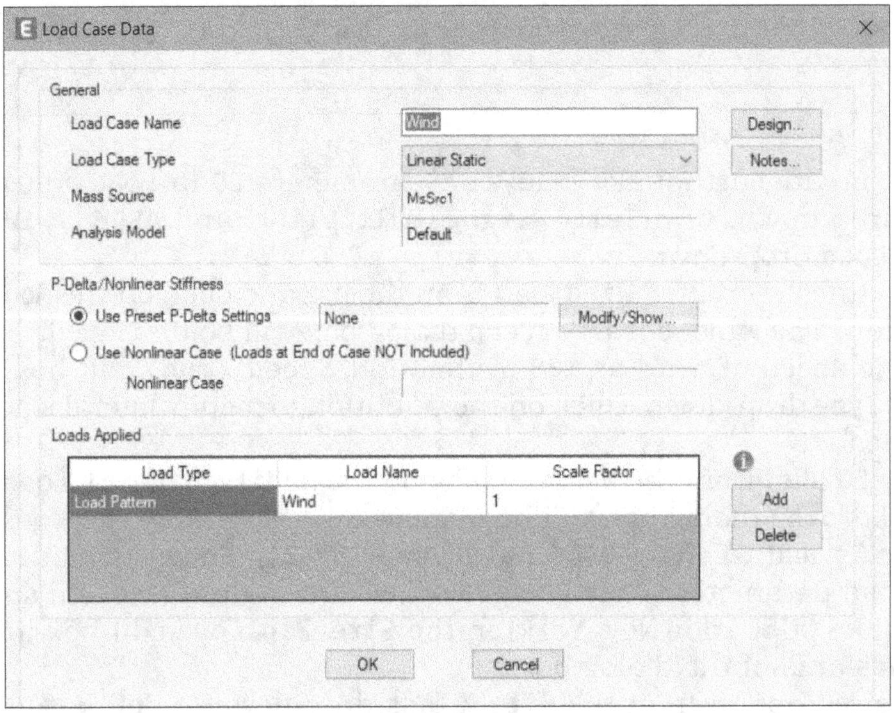

Figure-42. Load Case Data dialog box for wind load

- Select the **Nonlinear Static** option from the **Load Case Type** drop-down. Set desired parameters in the dialog box and click on the **OK** button.
- Double-click on **EQ** load case from the **Load Case Name** column. The **Load Case Data** dialog box will be displayed for earth quake load.
- Select the **Response Spectrum** option from the **Load Case Type** drop-down.
- Click on the **Add** button from the **Loads Applied** area. Acceleration type load will be added in the table. Set desired parameters and click on the **OK** button. The **Load Cases** dialog box will be displayed again.
- Click on the **OK** button from the **Load Cases** dialog box to exit.
- Now, we will create load combinations for apply loads on floors and walls. Click on the **Load Combinations** tool from the **Define** menu. The **Load Combinations** dialog box will be displayed. We will create two types of load combinations: one for floors which will include dead, live, and earth quake loads; and another for walls which will include wind and earth quake loads.
- Click on the **Add New Combo** button from the dialog box. The **Load Combination Data** dialog box will be displayed. Set the parameters as shown in Figure-43 and click on the **OK** button.

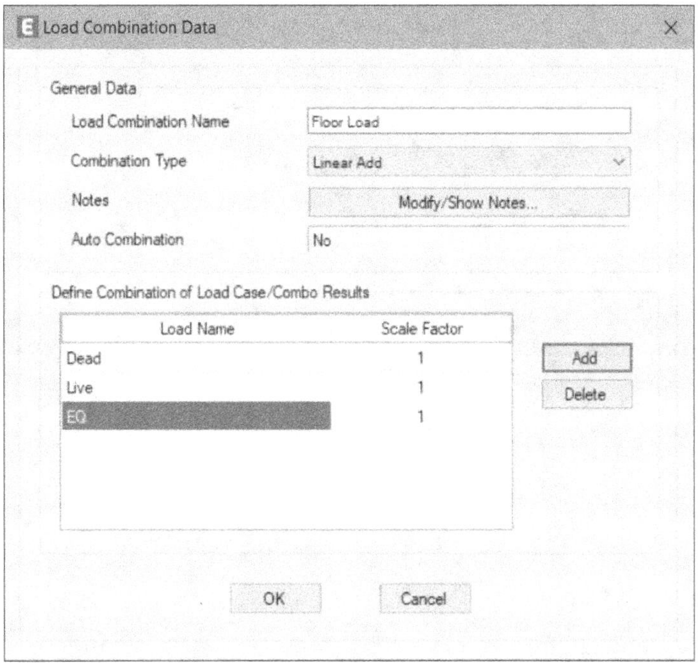

Figure-43. Load combination for floor

- Similarly, create Wall Load combination which includes EQ and Wind loads with 1 scale factor each; refer to Figure-44.

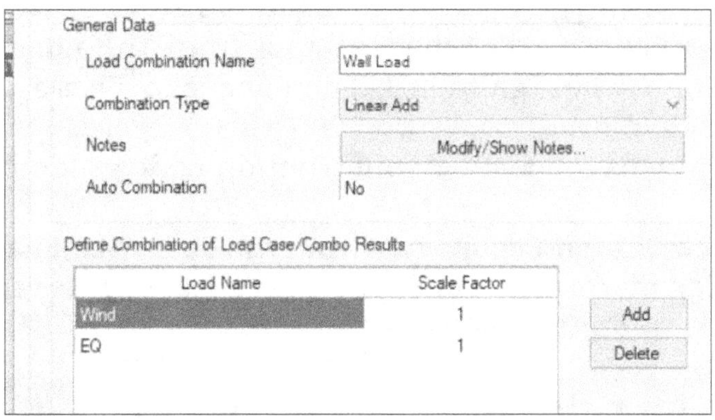

Figure-44. Parameters for wall load

- Click on the **OK** button from the **Load Combinations** dialog box to exit.

Assigning Loads and Running Analysis

- Select all the floors in the model and click on the **Uniform Load Sets** tool from the **Shell Loads** cascading menu of the **Assign** menu. The **Shell Assignment - Uniform Load Set** dialog box will be displayed.
- Select the earlier created load set which is **ULoadSet1** from the list and click on the **OK** button.
- Click on the **Set Load Cases To Run** tool from the **Analyze** menu. The **Set Load Cases to Run** dialog box will be displayed.
- Set all the load cases in the table to **Run** in **Action** column; refer to Figure-45.

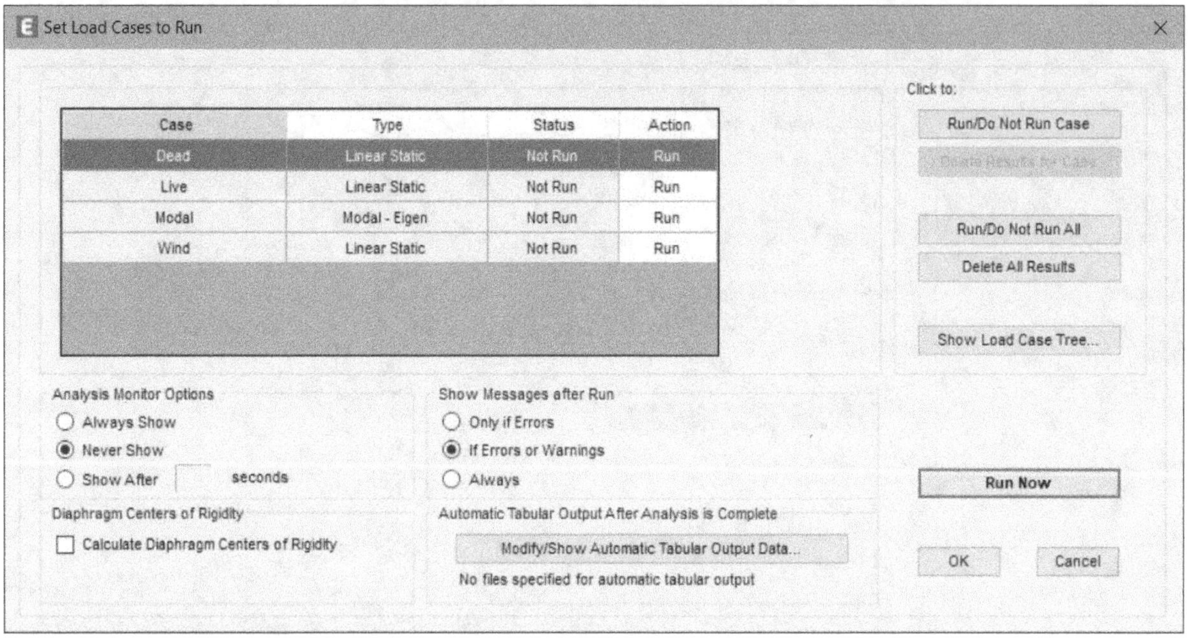

Figure-45. Setting load cases for analysis

- Click on the **Run Now** button to perform the analysis. Once the analysis is complete, the model will be locked.

Running Concrete Frame Design Study

- Click on the **View/Revise Preferences** tool from the **Concrete Frame Design** cascading menu of the **Design** menu. The **Concrete Frame Design Preferences** dialog box will be displayed.
- Set desired preferences as per the construction code guidelines and click on the **OK** button.
- Click on the **Select Design Groups** tool from the **Concrete Frame Design** cascading menu of the **Design** menu. The **Concrete Design Group Selection** dialog box will be displayed; refer to Figure-46.

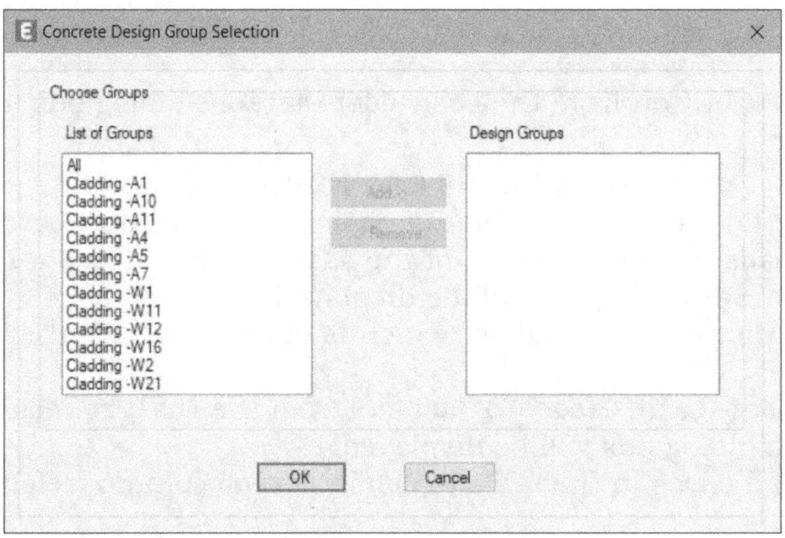

Figure-46. Concrete Design Group Selection dialog box

- Select all the groups from which design members will be selected while holding the **CTRL** key and click on the **Add ->** button. Based on analysis, the frame members will be selected automatically. Click on the **OK** button to apply settings.

- Click on the **Select Design Combinations** tool from the **Concrete Frame Design** cascading menu of the **Design** menu to select load combinations for design study. The **Design Load Combinations Selection** dialog box will be displayed; refer to Figure-47.

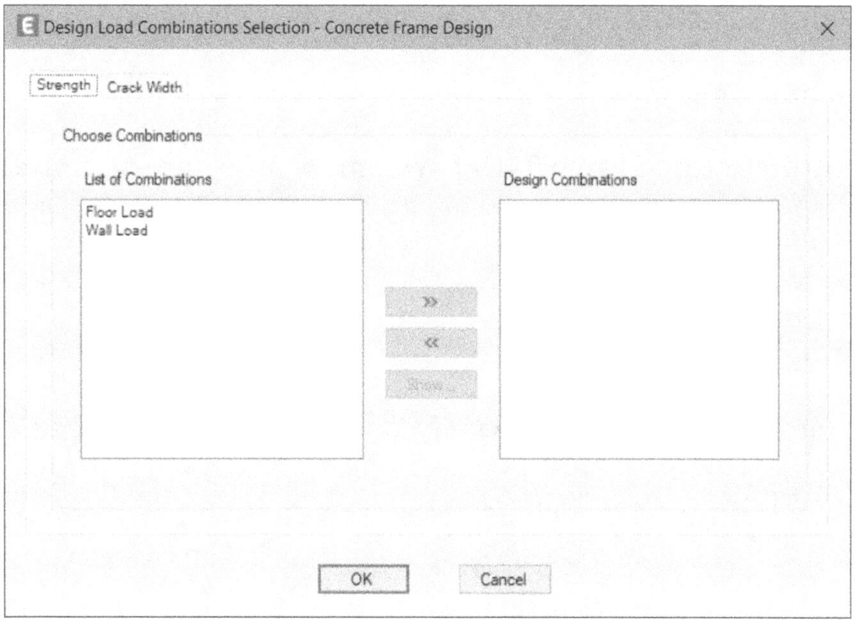

Figure-47. Design Load Combinations Selection dialog box for concrete frame

- Select all the load combinations to used for designing from the **List of Combinations** list box and click on the **>>** button.
- Click on the **OK** button from the dialog box to apply design load combinations.
- Click on the **Start Design/Check** tool from the **Concrete Frame Design** cascading menu of the **Design** menu to run design study.
- Click on the **Verify Analysis vs Design Section** tool from the **Concrete Frame Design** cascading menu of the **Design** menu to check whether design sections qualify for analysis and sustain the load. A message box will be displayed telling you that all the members match the analysis parameters.
- Click on the **Verify All Members Passed** tool from the **Concrete Frame Design** cascading menu of the **Design** menu. A message box will be displayed telling you that some concrete members failed the design check.
- Click on the **Yes** button from the message box to select the frame members which did not pass the check. Click on the **Change Design Section** tool from the **Concrete Frame Design** cascading menu of the **Design** menu to change the sections for members which have failed the design check.

Note that you can unlock the model and create new section properties as discussed earlier if you do not have desired section properties for frame members.

SELF-ASSESSMENT

Q1. What is the key feature of a tall building from a structural engineer's point of view?
A. Foundation depth
B. Lateral stability structural system
C. Interior decoration
D. Plumbing arrangement

Q2. In a Moment Resisting Frame (MRF) system, beams and columns are connected to provide:
A. Thermal insulation
B. Vertical drainage
C. Lateral resistance
D. Soundproofing

Q3. Which structural elements are widely used as lateral resisting systems in tall buildings?
A. Staircases and lifts
B. Bracings and shear walls
C. Doors and windows
D. Footings and piles

Q4. Why are outriggers used in tall buildings?
A. To increase roof height
B. To connect the core to exterior columns
C. To reduce dead load
D. To improve aesthetics only

Q5. Tube structures in tall buildings are designed to act like:
A. A simply supported beam
B. A hollow cylinder cantilevered perpendicular to the ground
C. A suspension bridge
D. A truss roof

Q6. Which of the following is NOT a type of tube structure?
A. Framed tube
B. Braced tube
C. Bundle tube
D. Arch tube

Q7. Which structural system uses triangular geometry effectively?
A. Tube system
B. Diagrid system
C. Flat slab system
D. Arch frame system

Q8. The diagrid system is considered an adapted version of:
A. Shell structures
B. Bracing systems
C. Dome structures
D. Cable systems

Q9. Super frame structures mainly use:
A. Timber beams and slabs
B. Mega columns and mega girders
C. Cable stays and pylons
D. Light gauge steel members

Q10. Floor diaphragms in tall buildings are designed to:
A. Reduce snow accumulation
B. Transmit lateral load to resisting systems
C. Carry only dead loads
D. Improve exterior appearance

Q11. Horizontal trusses or deep beams are used when:
A. The roof is sloped
B. The building is underground
C. Floor diaphragm is interrupted by openings
D. Wind loads are absent

Q12. Dead loads in buildings include:
A. Furniture and occupants
B. Wind pressure
C. Self-weight of structural members
D. Earthquake forces

Q13. Live loads are characterized by variation in:
A. Material density only
B. Magnitude and location
C. Temperature only
D. Foundation depth

Q14. Which of the following is an example of live load?
A. Beam self-weight
B. Column self-weight
C. Furniture
D. Core wall

Q15. Snow load is generally considered as:
A. Dead load
B. Dynamic load
C. Live load
D. Thermal load

Q16. Snow loads in ASCE/SEI 7–10 are determined using:
A. Soil maps
B. Wind tunnel tests
C. Zone maps with 50-year recurrence interval
D. Traffic analysis

Q17. Wind load effects depend on:
A. Density and flow of air
B. Angle of incidence
C. Shape and stiffness of structure
D. All of the above

Q18. For buildings taller than 200 m, wind analysis generally requires:
A. Soil testing
B. Wind tunnel testing
C. Earthquake simulation only
D. Thermal analysis

Q19. Earthquake load is classified as:
A. Static load
B. Thermal load
C. Dynamic load
D. Impact load only

Q20. A very flexible building with large masses will experience:
A. Large accelerations and small displacement
B. Small accelerations and large displacement
C. No displacement
D. Only vertical movement

Q21. Which of the following is a linear dynamic analysis method?
A. Pushover analysis
B. Response Spectrum Analysis
C. Plastic analysis
D. Collapse analysis

Q22. Response Spectrum Analysis defines response spectrum by:
A. Soil stabilization
B. Enveloping and smoothing earthquake spectra
C. Wind pressure reduction
D. Reinforcement detailing

Q23. Pushover analysis is a type of:
A. Linear analysis
B. Nonlinear analysis
C. Thermal analysis
D. Fatigue analysis

Q24. In pushover analysis, horizontal loads are applied until:
A. Foundation failure
B. Roof collapse
C. Target displacement is obtained
D. Beam cracking occurs

Q25. Time History Analysis may be:
A. Only linear
B. Only nonlinear
C. Linear or nonlinear
D. Neither linear nor nonlinear

Q26. Time History Analysis uses:
A. Constant acceleration values
B. Ground acceleration time history
C. Dead load combinations only
D. Snow load maps

Q27. One way to make a building earthquake resistant is by using:
A. Brittle structures
B. Ductile structures
C. Floating foundations only
D. Glass walls

Q28. A ductile structure is designed to dissipate energy through:
A. Plastic deformation only
B. Elastic deformation
C. Thermal expansion
D. Foundation settlement

Q29. A building is classified as a soft story building if stiffness is less than:
A. 50% of average stiffness
B. 60% of average stiffness
C. 70% of average stiffness
D. 80% of average stiffness

Q30. Soft story buildings are vulnerable to:
A. Fire spread
B. Snow accumulation
C. Collapse during earthquakes
D. Water leakage

Q31. To avoid soft-story collapse, structures should have:
A. Abrupt stiffness changes
B. Vertical and horizontal regularity
C. Excessive openings
D. Weak columns

Q32. Which of the following is an active or passive device for reducing earthquake excitation?
A. Base isolation
B. Shock absorber
C. Damping system
D. All of the above

Q33. Nearby tall buildings can influence wind flow and create:
A. Greenhouse effect
B. Venturi effect
C. Doppler effect
D. Magnus effect

Q34. Progressive collapse analysis studies the structure when:
A. Floors are painted
B. Columns are removed
C. Roof load decreases
D. Snow load increases

Q35. Which type of analysis can be used for progressive collapse studies?
A. Linear static
B. Nonlinear dynamic
C. Linear dynamic
D. All of the above

Q36. Fire safety in tall buildings is important to ensure:
A. Decorative appearance
B. Minimum property and life damage
C. Lower construction cost only
D. Faster construction

Q37. Cladding can help in:
A. Preventing sound transmission
B. Providing thermal insulation
C. Preventing spread of fire
D. All of the above

Q38. Which code is applicable for buildings with height less than 200 m for wind loading?
A. EN1991-1-4
B. IS 456
C. ACI 318
D. Eurocode 2

Q39. Which analysis method can be nonlinear dynamic?
A. Time History Analysis
B. Dead Load Analysis
C. Snow Load Analysis
D. Static Equilibrium Analysis

Q40. In tall buildings, more than one type of lateral stability system is generally employed to provide:
A. Better aesthetics only
B. Lateral strength
C. Increased roof area
D. Reduced live load

FOR STUDENT NOTES

For Student Notes

Chapter 9

Editing and Advanced Structure Design

Topics Covered

The major topics covered in this chapter are:

- ***Introduction***
- ***Editing Tools***
- ***Practical***
- ***Practice***

INTRODUCTION

After creating model, there are various tools that can help in modifying structure and properties of model. You can use these tools to create copies of objects and modify them. The editing tools are available in the **Edit** menu; refer to Figure-1. These tools are discussed next.

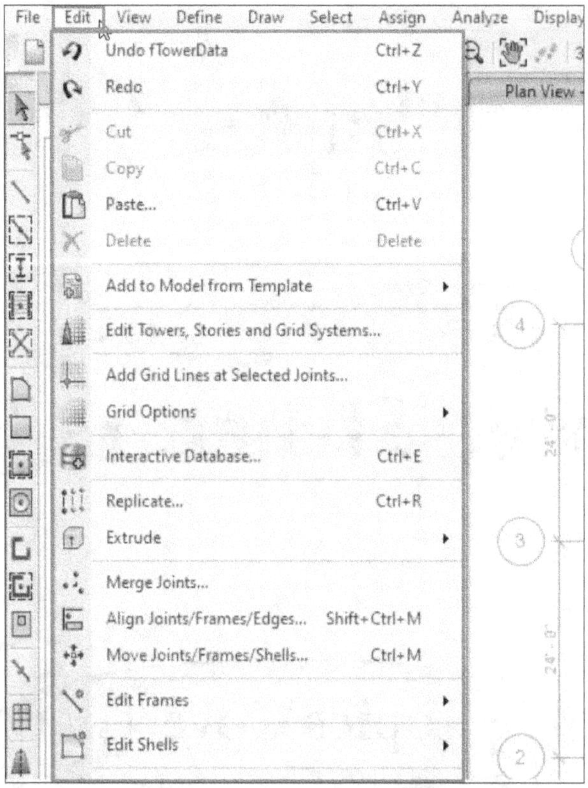

Figure-1. Edit menu

UNDO-REDO TOOLS

The **Undo** tool in **Edit** menu is used to reverse the most recent operations. For example, if you have moved any object accidently in the model and want it to be back to its original position then you can use **Undo** tool. You can also use the shortcut key **CTRL+Z**.

The **Redo** tool in **Edit** menu is used to reverse the undo operation. If you have used the **Undo** tool accidently then you can use **Redo** tool to reverse the operation. You can also use **CTRL+Y** shortcut key to perform redo operation.

CUT-COPY-PASTE

The **Cut** tool in **Edit** menu is used to delete selected object and save it into clipboard for performing paste later. You can also use **CTRL+X** shortcut key to perform the same operation.

The **Copy** tool in **Edit** menu is used to copy selected object and save it into clipboard for performing paste later. You can also use **CTRL+C** shortcut key to perform the same operation.

The **Paste** tool in **Edit** menu is used to place objects saved in clipboard into the model at desired location. You can also use **CTRL+V** shortcut key to perform this operation. On clicking this tool, the **Paste Coordinates** dialog box will be displayed; refer to Figure-2.

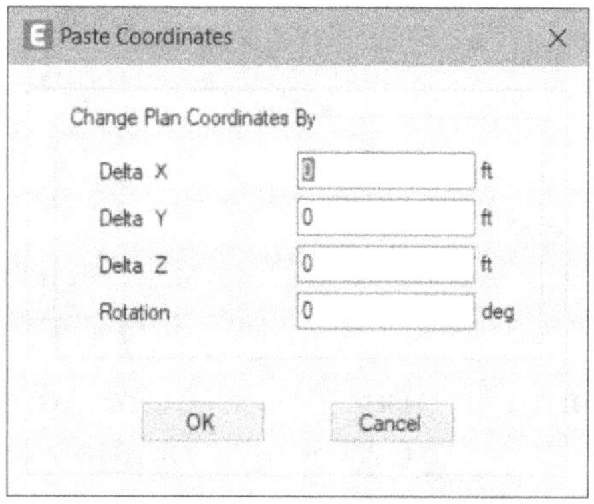

Figure-2. Paste Coordinates dialog box

Specify desired parameters in the **Delta X** and **Delta Y** edit boxes to specify distance from center of selected objects and click on the **OK** button to create the copy.

DELETING OBJECTS

The **Delete** tool in **Edit** menu is used to delete selected objects from the model. You can also use **DELETE** key from the keyboard.

ADDING STRUCTURES TO MODEL

The tools in the **Add to Model from Template** cascading menu are used to add 2D and 3D structures to the model; refer to Figure-3. The tools in this menu are discussed next.

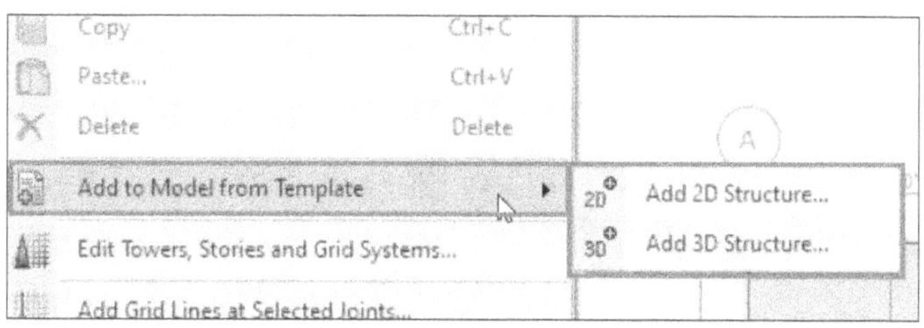

Figure-3. Add to Model from Template cascading menu

Adding 2D Structure

The **Add 2D Structure** tool is used to add 2D structure to model at specified location. The procedure to use this tool is given next.

- Click on the **Add 2D Structure** tool from the **Add to Model from Template** cascading menu in the **Edit** menu. The **Add to Model - 2D Structure** dialog box will be displayed; refer to Figure-4.

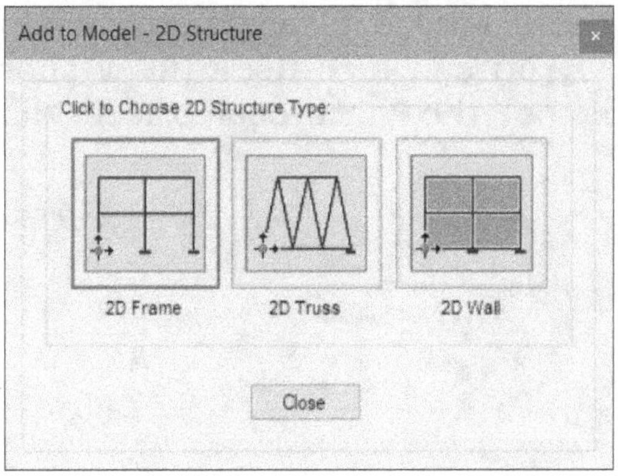

Figure-4. Add to Model-2D Structure dialog box

- Select desired button from the dialog box to create respective type of structure. The options for selected structure type will be displayed; refer to Figure-5.

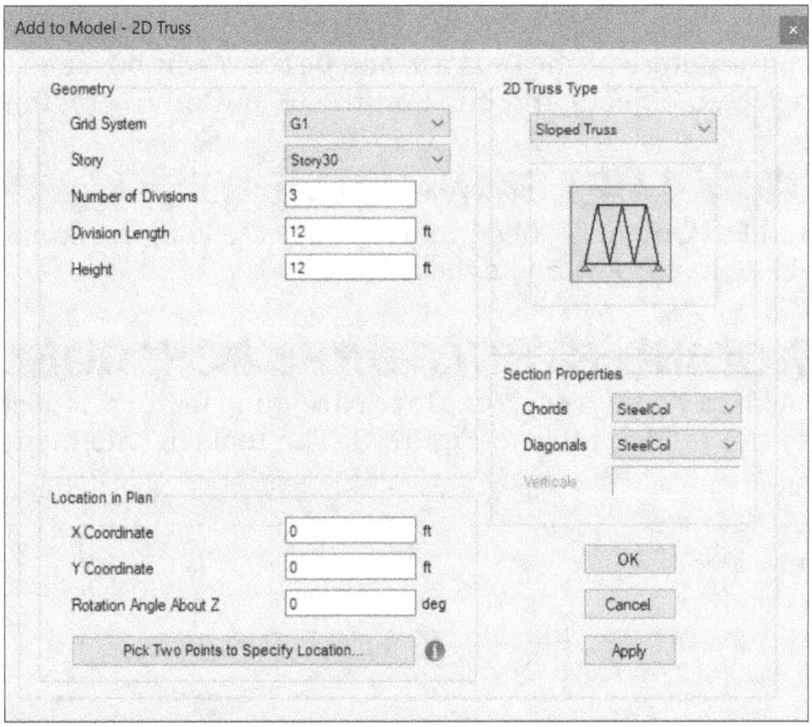

Figure-5. Add to Model-2D Truss dialog box

- Select the type of structure member from the drop-down in **2D Truss Type/Wall Type/Frame Type** area depending on the button selected in **Add to Model - 2D Structure** dialog box earlier.
- Set the parameters as discussed earlier to place the structure member at desired location in the **Location in Plane** area.
- Specify the number of divisions, length of each division, height of structure, story, grid system, and tower for creating structure member.
- Set the other parameters and click on the **OK** button from the dialog box. The 2D structure of selected type will be added to the model; refer to Figure-6 and the **Add to Model - 2D Structure** dialog box will be displayed again. Press **ESC** to exit the dialog box.

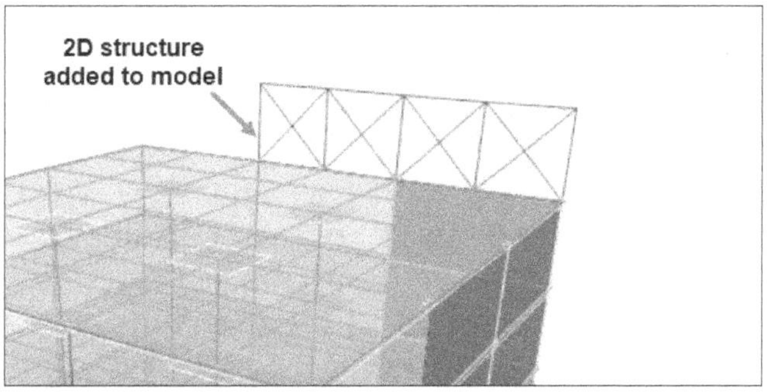

Figure-6. Adding 2D structure

Adding 3D Structure

The **Add 3D Structure** tool is used to used to add predefined type 3 dimensioning structures to the model. The template structures available are same as displayed when starting a new project. The procedure to use this tool is given next.

- Click on the **Add 3D Structure** tool from the **Add to Model from Template** cascading menu in the **Edit** menu. The **Add to Model - 3D Structure** dialog box will be displayed; refer to Figure-7.

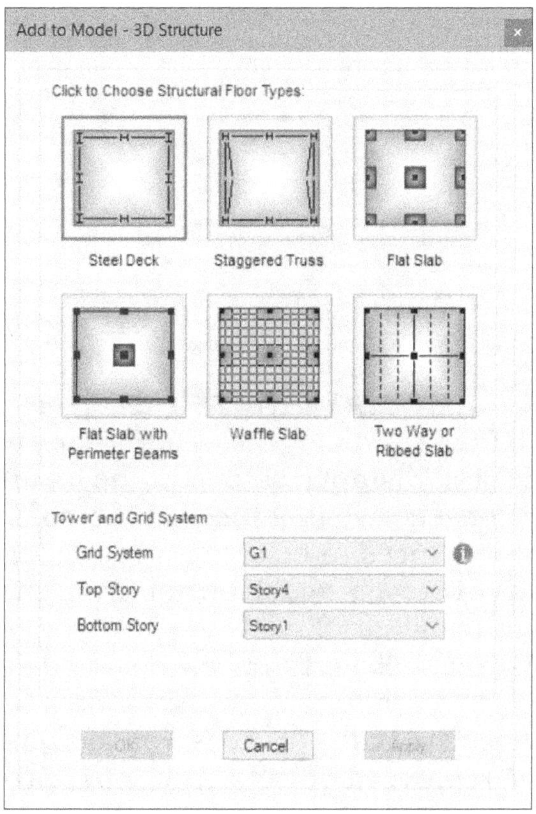

Figure-7. Add to Model-3D Structure dialog box

- Set desired parameters like tower, grid system, top story, and bottom story to define the location of structure members.
- Select desired button from the top area in dialog box to define the type of structure to be added. The options displayed on selecting a button in this dialog box have been discussed earlier.
- Click on the **OK** button from the dialog box to create the structure.

REPLICATING OBJECTS

The **Replicate** tool in **Edit** menu is used to create multiple copies of selected objects at specified distance or angle. You can also use this tool to create mirror copy or copy objects to selected story level. The procedure to use this tool is given next.

- Click on the **Replicate** tool from **Edit** menu. The **Replicate** dialog box will be displayed; refer to Figure-8. There are four tabs in this dialog box to perform linear pattern, radial pattern, mirror copy, and copy objects to a story. The methods to perform each action are discussed next.

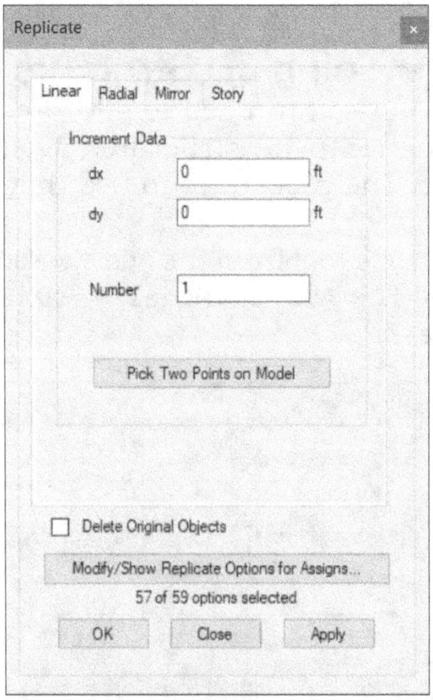

Figure-8. Replicate dialog box

Creating Linear Pattern

- Click on the **Linear** tab from the dialog box. The options will be displayed as shown in Figure-8. Specify desired distance values in the **dx** and **dy** edit boxes to define distance between two consecutive instances of the copied objects.
- Specify the number of instances to be created in the **Number** edit box.
- Click on the **Pick Two Points on Model** button from the dialog box to specify dx and dy values using an imaginary line.
- Select the **Delete Original Objects** check box to delete selected objects after creating copies.
- Click on the **Modify/Show Replicate Options for Assigns** button to define assigns to be included with the replicated objects. The **Replicate Assigns** dialog box will be displayed; refer to Figure-9.

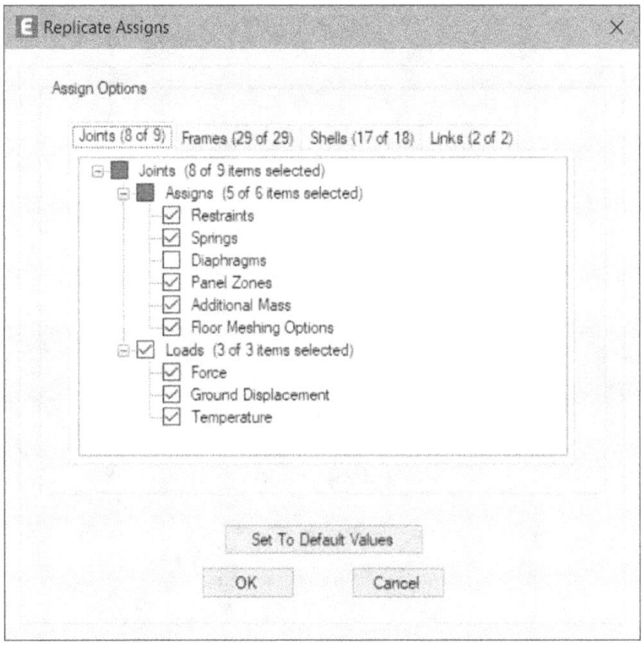

Figure-9. Replicate Assigns dialog box

- Select check boxes for the assigns to be included from various tabs of the dialog box and click on the **OK** button.
- After setting the parameters in dialog box, click on the **Apply** button to create pattern; refer to Figure-10.

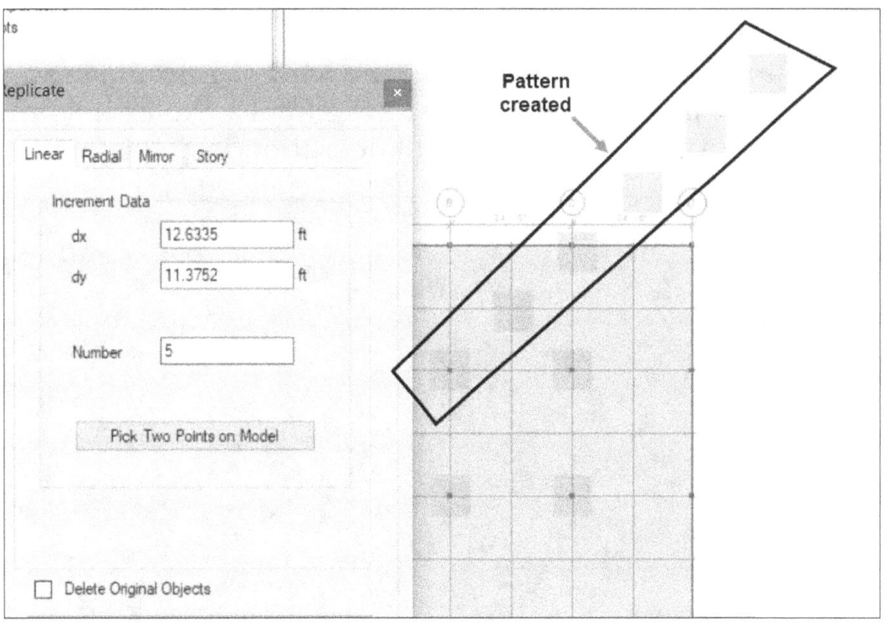

Figure-10. Linear pattern of slab

Creating Radial Pattern

- Click on the **Radial** tab in the dialog box. The options will be displayed as shown in Figure-11.

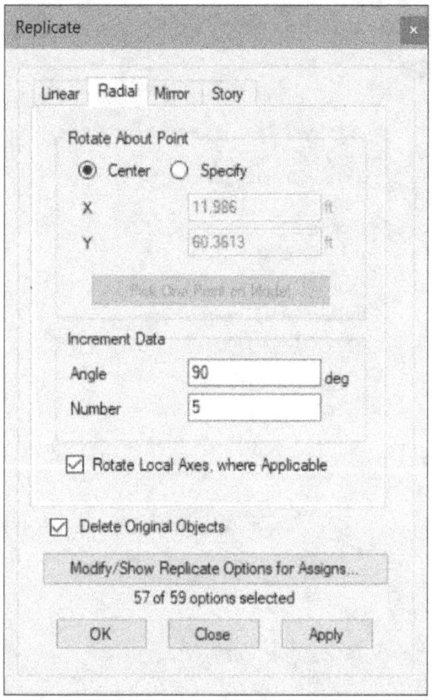

Figure-11. Radial tab in Replicate dialog box

- Select the **Center** radio button to use center of model as center for radial pattern. Select the **Specify** radio button to define the center of model at desired location.
- Specify the angle between two consecutive instances and total number of instances in the **Angle** and **Number** edit boxes, respectively.
- After setting desired parameters, click on the **Apply** button to create the pattern. The radial pattern will be created; refer to Figure-12. Note that when you hover cursor over apply button, then preview of operation is displayed.

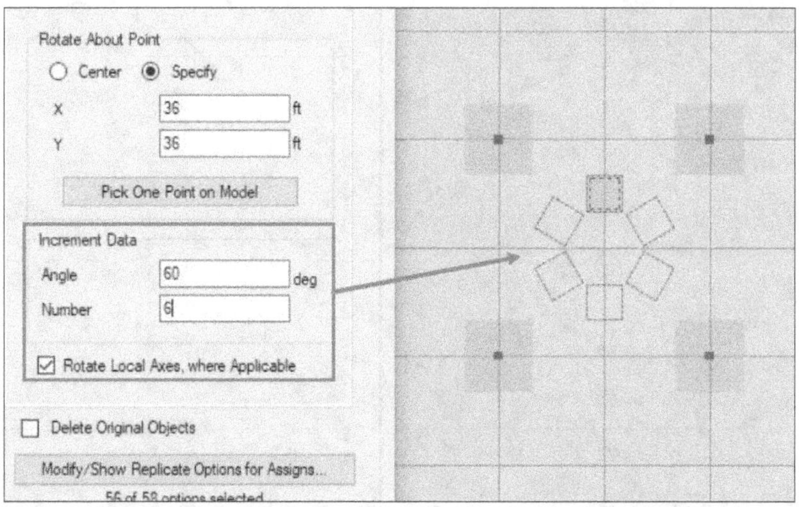

Figure-12. Preview of radial pattern

Creating Mirror Copy

- Click on the **Mirror** tab in the dialog box. The options will be displayed as shown in Figure-13.

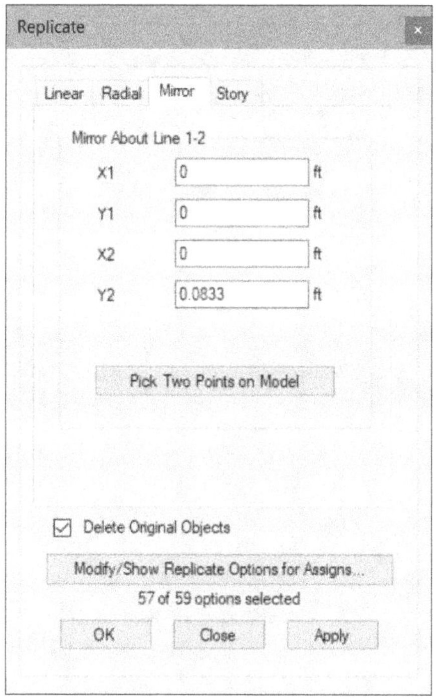

Figure-13. Mirror tab

- Set desired coordinates for mirror line in the edit boxes or click on the **Pick Two Points on Model** button to create mirror line.
- Set the parameters as discussed earlier, click on the **Apply** button. The mirror copy will be created; refer to Figure-14.

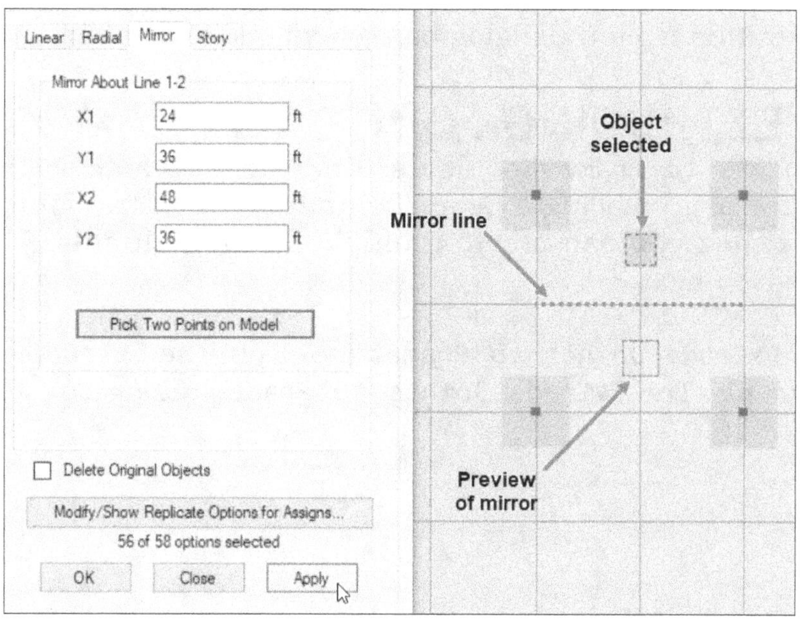

Figure-14. Preview of mirror

Replicating Objects to a Story

- Click on the **Story** tab in the dialog box to copy object on a story. The options will be displayed as shown in Figure-15.

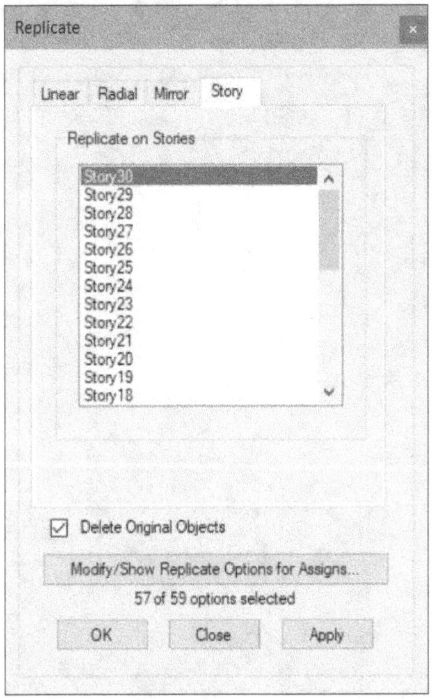

Figure-15. Story tab

- Select the objects that you want to replicate from the graphics area.
- Select the story to which you want to copy selected objects and click on the **Apply** button.

Click on the **OK** button from the dialog box to exit the dialog box.

EXTRUDING JOINTS TO FRAMES

The **Extrude Joints to Frames** tool is used to extrude frame section between two joints. This is similar to placing a beam/column between two joints but when you perform extrude then you can create multiple beams/columns. The procedure to use this tool is given next.

- Click on the **Extrude Joints to Frames** tool from the **Extrude** cascading menu of the **Edit** menu. The **Extrude Joints to Frames** dialog box will be displayed; refer to Figure-16.

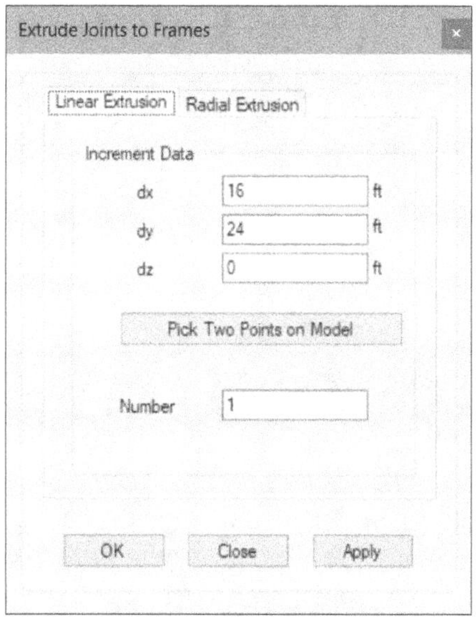

Figure-16. Extrude Joints to Frames dialog box

Linear Extrusion

- Select the joints to be extruded from the graphics area.
- Specify the distance values along X, Y, and Z axes in **dx**, **dy**, and **dz** edit boxes, respectively. You can also use **Pick Two Points on Model** button to specify distance by creating an imaginary line.
- After setting desired parameters, click on the **Apply** button. Beams/columns will be created at the joints; refer to Figure-17.

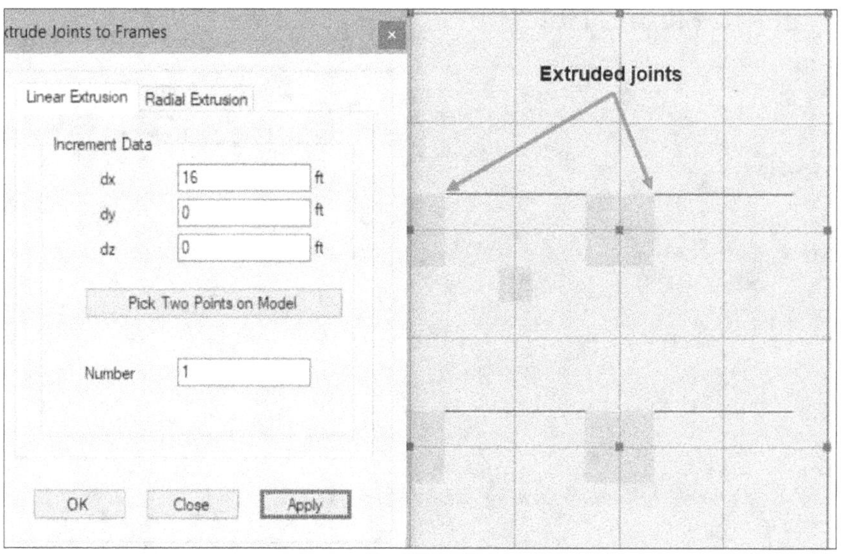

Figure-17. Extruding joints to frames

Radial Extrusion

- Click on the **Radial Extrusion** tab in the dialog box. The options will be displayed as shown in Figure-18.

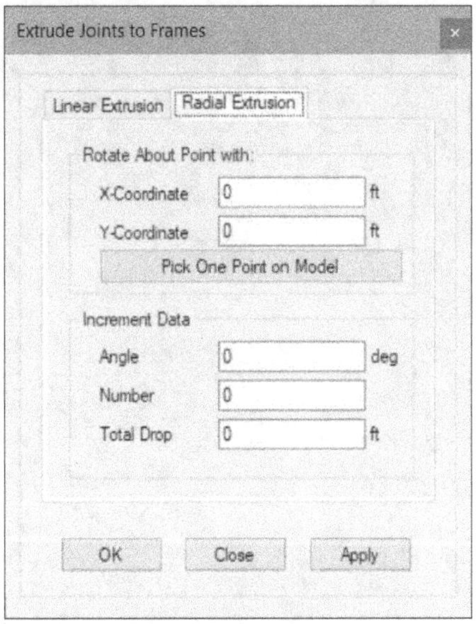

Figure-18. Radial Extrusion tab

- Specify the coordinates for center about which you want to rotate the extruded joints in the edit boxes of **Rotate About Point with** area. You can also use the **Pick One Point on Model** button to specifying center point by clicking in the model.
- Specify desired value in the **Angle** and **Number** edit boxes to define angle between two instances and total number of instances in radial extrusion, respectively.
- Specify the slope value of radially extruded frame members in the **Total Drop** edit box. This total drop is used to define the distance between start point and end point of extruded frame members along Z axis.
- After setting desired parameters, click on the **OK** button. The radial extrusion will be created; refer to Figure-19.

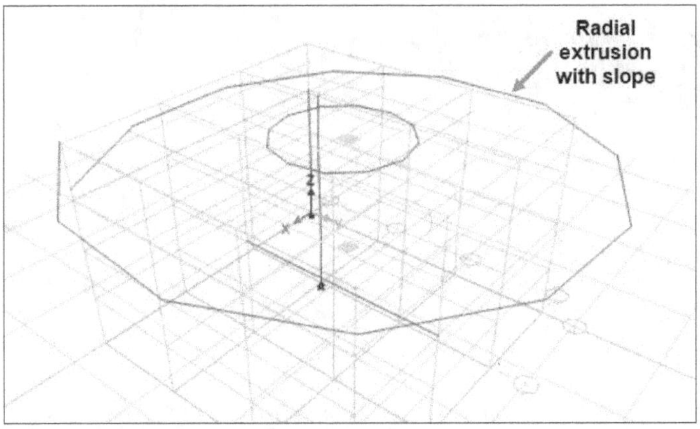

Figure-19. Radial extrusion with slope

EXTRUDING FRAMES TO SHELLS

The **Extrude Frames to Shells** tool works similar to **Extrude Joints to Frames** tool discussed earlier but in this case, you need to select frame members for extruding them to shells. The procedure to use this tool is given next.

- Click on the **Extrude Frames to Shells** tool from the **Extrude** cascading menu of the **Edit** menu. The **Extrude Frames to Shells** dialog box will be displayed; refer to Figure-20.

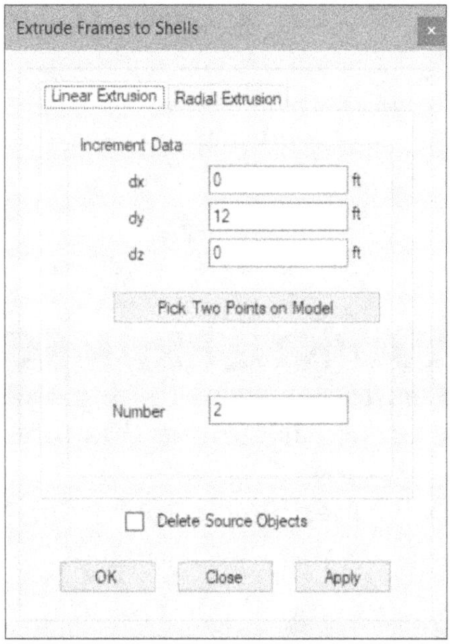

Figure-20. Extrude Frames to Shells dialog box

- Select the frame member to be extruded to shell member (wall, slab, and so on) and specify the parameters as discussed earlier in the dialog box.
- Click on the **OK** button from the dialog box to create extrusion; refer to Figure-21.

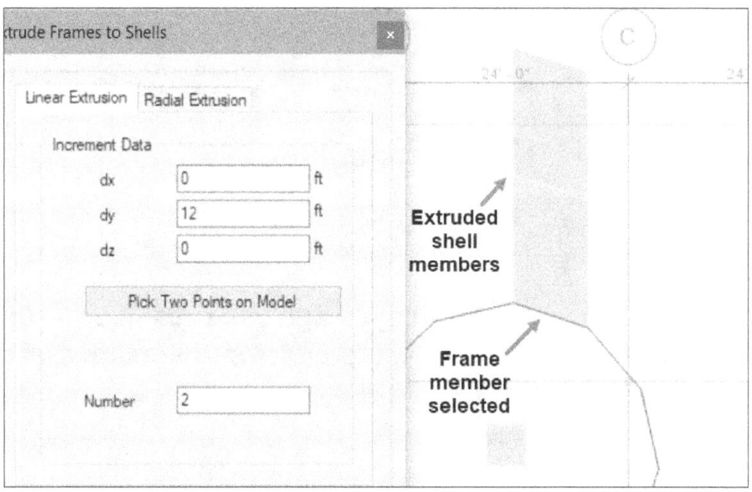

Figure-21. Extruding frames to shells

MERGING NEARBY JOINTS

The **Merge Joints** tool is used to merge all the joints within the vicinity of specified tolerance value. This tool is useful for cleaning up the model. On clicking **Merge Joints** tool from **Edit** menu, the **Merge Selected Joints** dialog box is displayed; refer to Figure-22. Select the joints to be merged, specify the tolerance value within which joints will be searched for merging, and click on the **OK** button from the dialog box to merge joints.

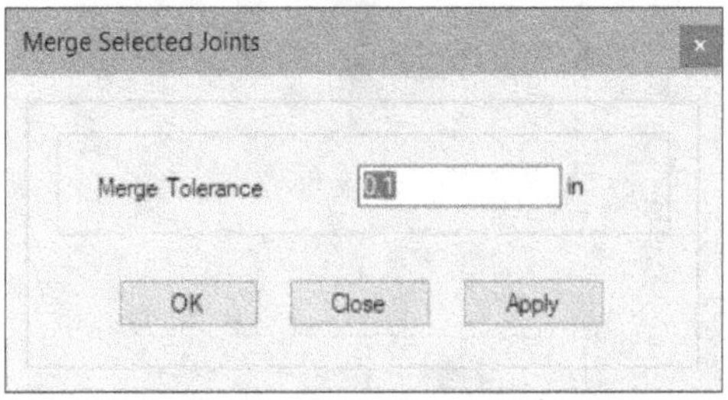

Figure-22. Merge Selected Joints dialog box

ALIGNING JOINTS/FRAMES/EDGES

The **Align Joints/Frames/Edges** tool is used to align selected joints/frames/edges to a predefined reference. The procedure to use this tool is given next.

- Click on the **Align Joints/Frames/Edges** tool from the **Edit** menu. The **Align Joints/Frames/Edges** dialog box will be displayed; refer to Figure-23.

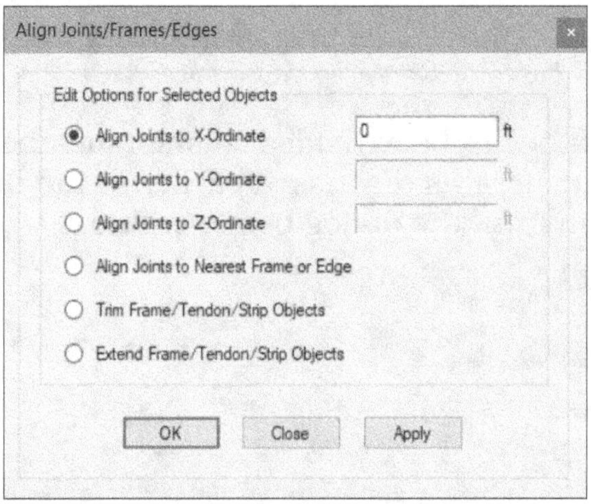

Figure-23. Align Joints-Frames-Edges dialog box

- Select **Align Joints to X-Ordinate**, **Align Joints to Y-Ordinate**, or **Align Joints to Z-Ordinate** for aligning selected joints to specified ordinate value along respective axis; refer to Figure-24.

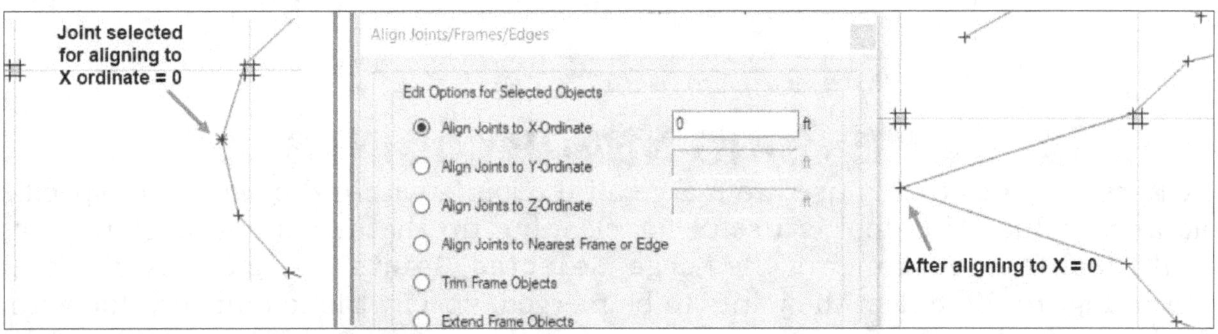

Figure-24. Aligning joint to X ordinate 0

- Select the **Align Joints to Nearest Frame or Edge** radio button to align selected joints to nearest selected frame or edge. After selecting this radio button, select the joint to be aligned and frame/edge to which the joint will be aligned, and then click on the **Apply** button; refer to Figure-25.

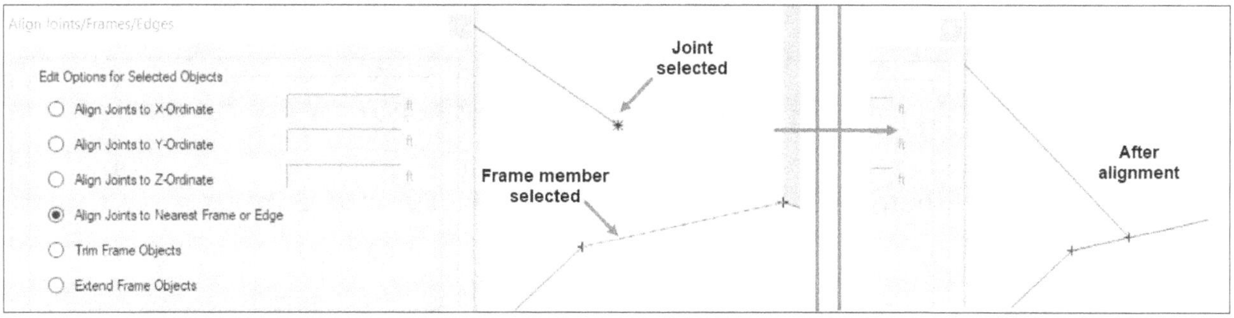

Figure-25. Aligning joint to nearest frame

- Select the **`Trim Frame Objects`** radio button to remove extra segment of a frame member intersecting with other frame member. After selecting this radio button, select the frame member to be trimmed and its joint to be moved on trimming, then select the frame member to be used as trimming tool, and then click on the **`Apply`** button; refer to Figure-26.

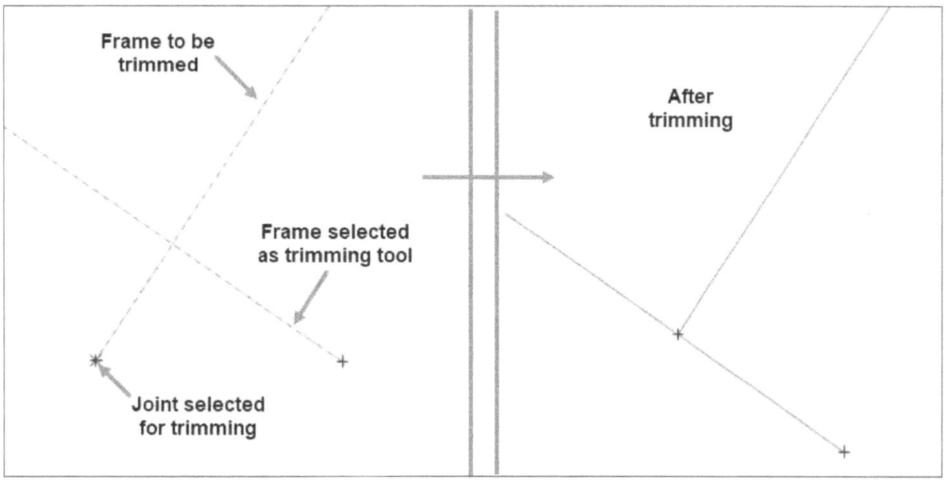

Figure-26. Trimming using joints

- Select the **`Extend Frame Objects`** radio button to extend selected frame object by selected joint. After selecting this radio button, select the frame object to be extended and its joint, then select the frame object up to which extension should be performed, and then click on the **`Apply`** button; refer to Figure-27.

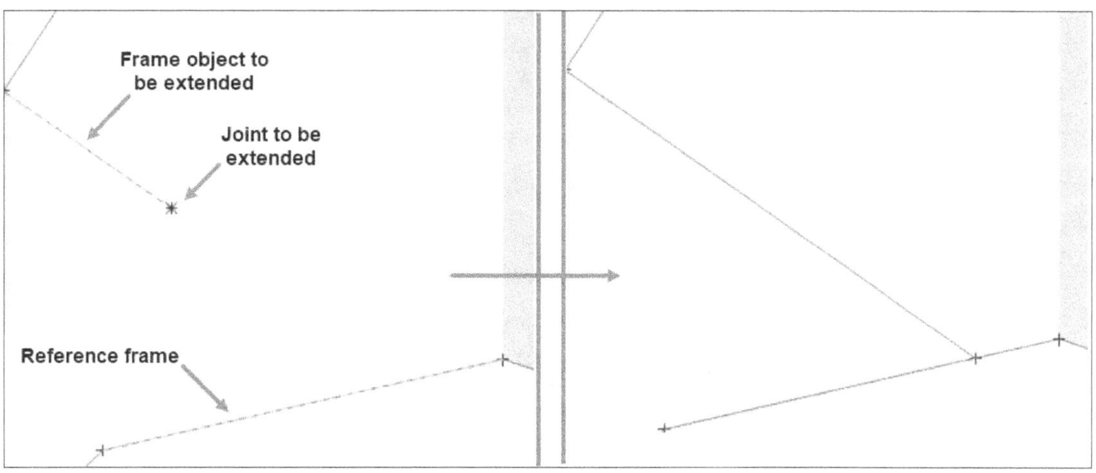

Figure-27. Extending frame object

- After performing desired operations, click on the **OK** button from the dialog box to exit.

MOVING JOINTS/FRAMES/SHELLS

The **Move Joints/Frames/Shells** tool is used to move selected joint/frame/shell objects to specified distance along X, Y, and Z axes. The procedure to use this tool is given next.

- Click on the **Move Joints/Frames/Shells** tool from the **Edit** menu. The **Move Joints/Frames/Shells** dialog box will be displayed; refer to Figure-28.

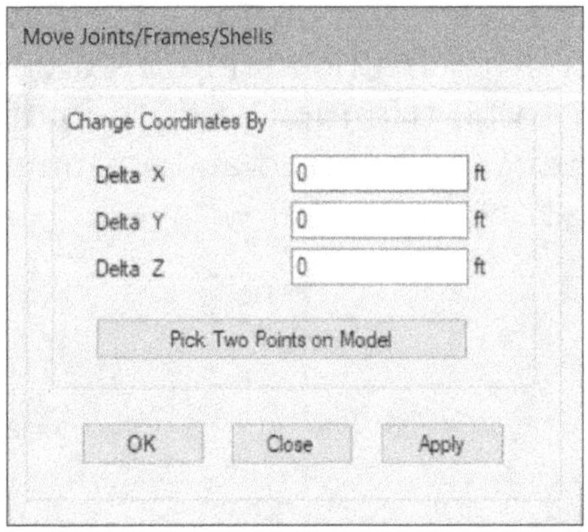

Figure-28. Move Joints/Frames/Shells dialog box

- Specify desired value in the **Delta X**, **Delta Y**, and **Delta Z** edit boxes to move the object in respective directions.
- Select the joint/frame/shell object(s) to be moved and click on the **Apply** button. The selected object(s) will move by specified distance.
- Click on the **OK** button from the dialog box to exit.

FRAME EDITING TOOLS

There are various tools available in **Edit Frame** cascading menu of the **Edit** menu to perform operations like divide frame, join frames, reverse frame connectivity, and so on; refer to Figure-29. These tools are useful when frame objects are available in segment at construction site. Various tools in this cascading menu are discussed next.

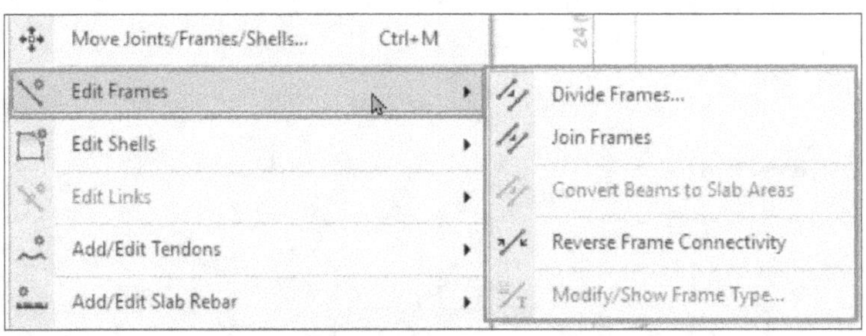

Figure-29. Edit Frames cascading menu

Dividing Frames

The **Divide Frames** tool is used to divide selected frame objects into specified number of segments or at specified points. The procedure to use this tool is given next.

- Click on the **Divide Frames** tool from the **Edit Frames** cascading menu of the **Edit** menu. The **Divide Selected Frames** dialog box will be displayed; refer to Figure-30.

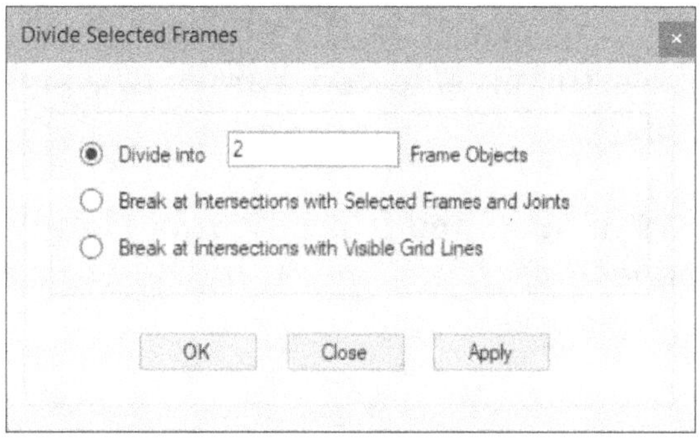

Figure-30. Divide Selected Frames dialog box

- Select the **Divide into** radio button and specify number of equal segments to be created in the edit box.
- Select the **Break at Intersections with Selected Frames and Joints** radio button to break selected frame objects at their intersection point. After selecting this radio button, select the intersecting frame objects and click on the **Apply** button. The objects will be divided at intersection; refer to Figure-31.

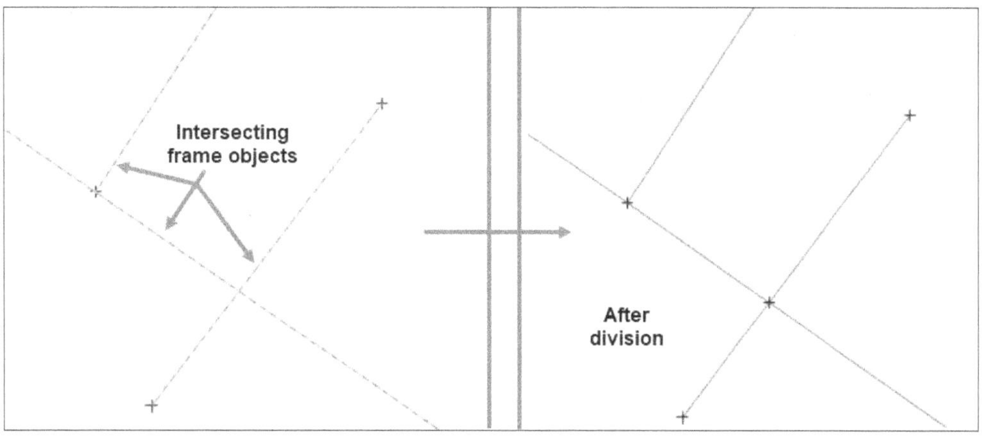

Figure-31. Dividing frame objects

- Select the **Break at Intersections with Visible Grid Lines** radio button to divide the frame objects at intersection points with visible grid lines. After selecting this radio button, select the frame objects intersecting with grid lines and click on the **Apply** button; refer to Figure-32.

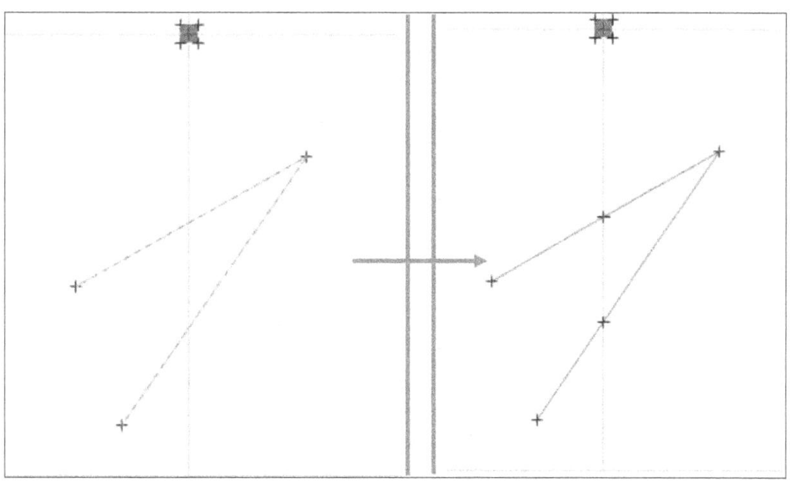

Figure-32. Dividing frame object at grid lines

- Click on the **OK** button from the dialog box to exit.

Joining Frames

The **Join Frames** tool is used to join selected frames to a single frame. The procedure to use this tool is given next.

- Select the frame objects to be joined from the model and click on the **Join Frames** tool from the **Edit Frames** cascading menu in the **Edit** menu. The frame objects will be created; refer to Figure-33.

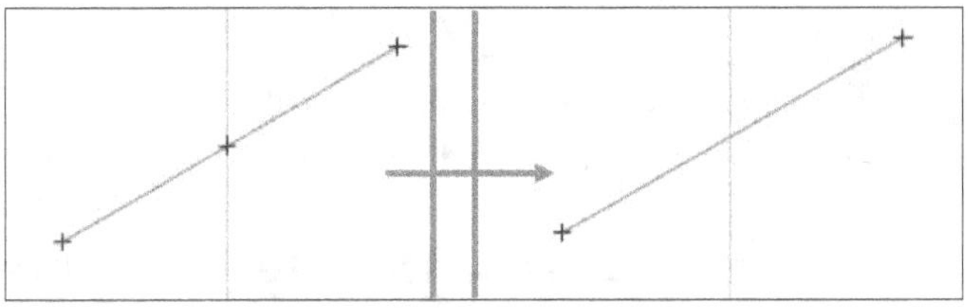

Figure-33. Joining frame objects

Converting Beams to Slab Areas

The **Convert Beams to Slab Areas** tool is used to convert selected concrete beams to slab areas. Note that during analysis, this affects load distribution and documentation details. The procedure to use this tool is given next.

- Select concrete beams from the model and click on the **Convert Beams to Slab Areas** tool from the **Edit Frames** cascading menu of **Edit** menu. The beams will be converted to slab areas.

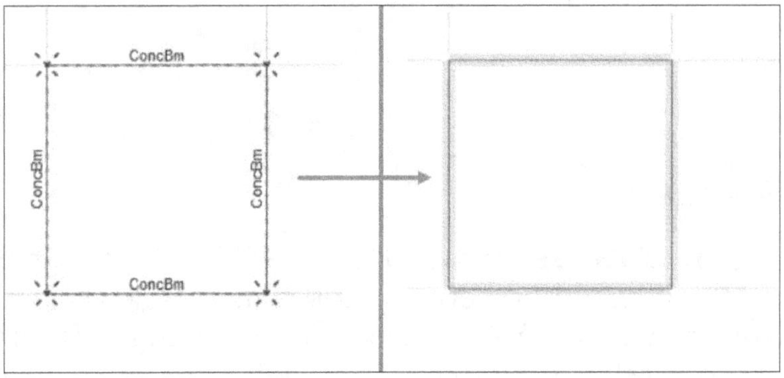

Figure-34. Beams to slab area conversion

Reversing Frame Connectivity

The **Reverse Frame Connectivity** tool is used to reverse the local coordinate system and side of selected frame object. To do so, select the frame object to be reversed and click on the **Reverse Frame Connectivity** tool from the **Edit Frames** cascading menu of the **Edit** menu. The frame object will be reversed; refer to Figure-35.

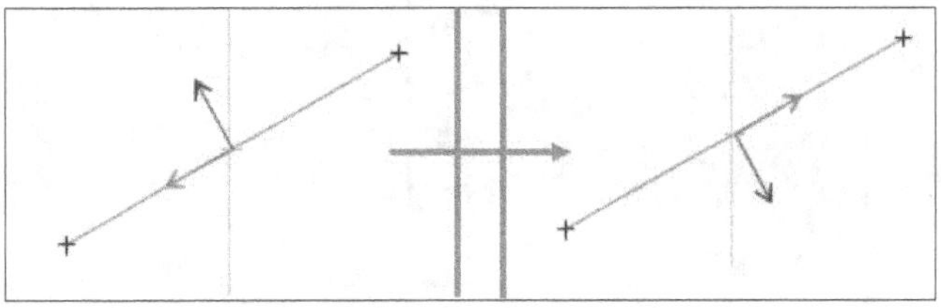

Figure-35. Reversing frame object connectivity

Modifying/Showing Frame Type

The **Modify/Show Frame Type** tool is used to change the frame type and check the properties of frame object. After selecting frame object from model, click on the **Modify/Show Frame Type** tool from the **Edit Frames** cascading menu of the **Edit** menu. The **Frame Object Type Options** dialog box will be displayed; refer to Figure-36.

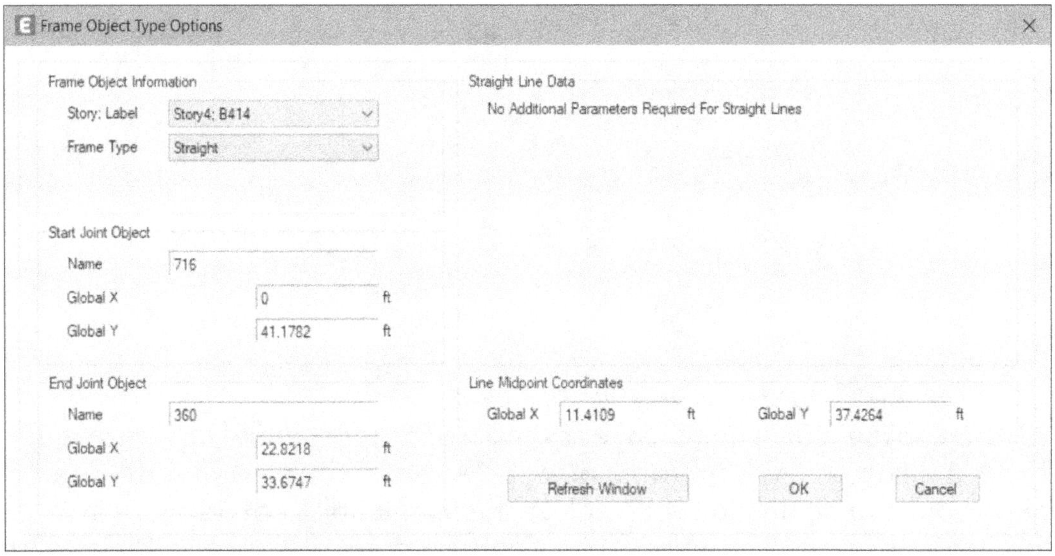

Figure-36. Frame Object Type Options dialog box

Select desired option from the **Frame Type** drop-down to specify type of frame object. Based on selected type, specify the parameters in dialog box and click on the **OK** button to exit the dialog box.

EDITING SHELL OBJECTS

The tools in **Edit Shell** cascading menu are used to modify shell objects (like slabs, planks, and so on); refer to Figure-37. Various tools in this menu are discussed next.

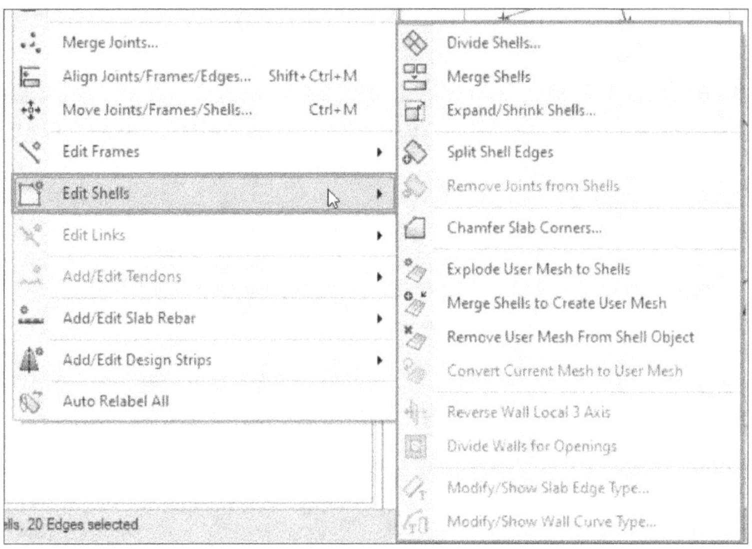

Figure-37. Edit Shells cascading menu

Dividing Shells

The **Divide Shells** tool is used to divide selected shell objects into specified number of segments. The procedure to use this tool is given next.

- Select the shell objects to be divided and click on the **Divide Shells** tool from the **Edit Shells** cascading menu of the **Edit** menu. The **Divide Selected Shells** dialog box will be displayed; refer to Figure-38.

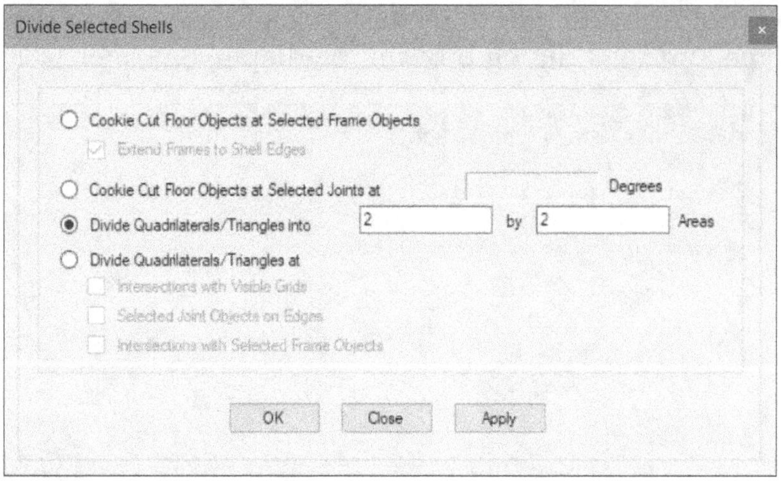

Figure-38. Divide Selected Shells dialog box

- Select the **Cookie Cut Floor Objects at Selected Frame Objects** radio button to divide shell objects using frame objects. After selecting this radio button, select the shell object to be divided then select the frame object dividing it, and then click on the **Apply** button. The shell object will be divided; refer to Figure-39. Select the **Extend Frames to Shell Edges** check box to extend the frame object if it is not completely dividing the shell object.

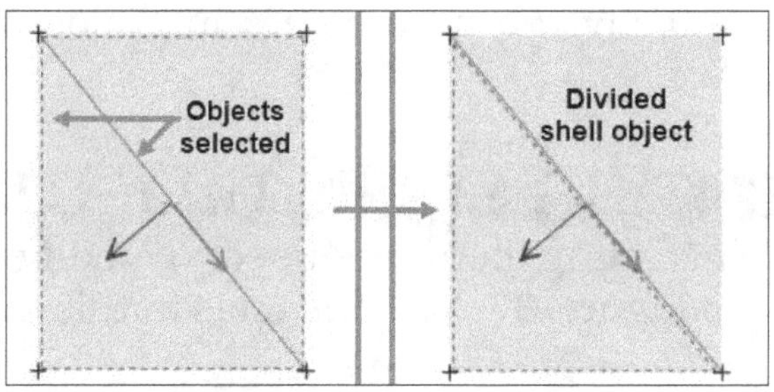

Figure-39. Divided shell object

- Select the **Cookie Cut Floor Objects at Selected Joints** radio button to divide selected shell objects at specified angle from horizontal axis references to selected joint; refer to Figure-40.

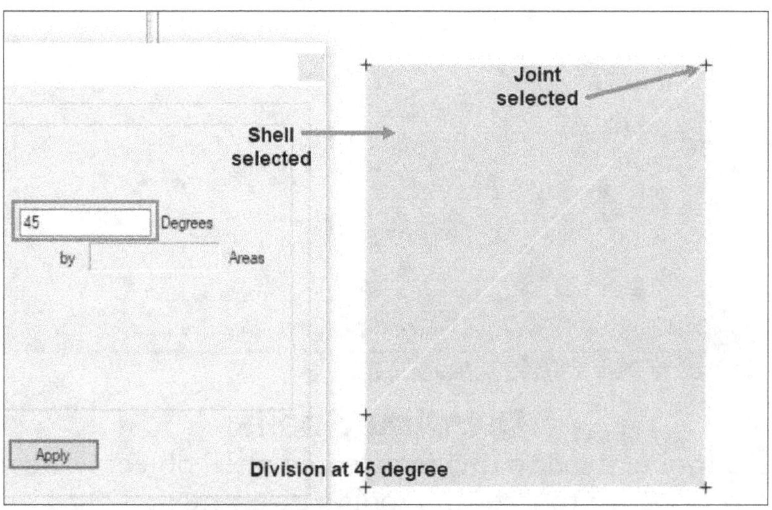

Figure-40. Dividing shell object at angle

- Select the **Divide Quadrilaterals/Triangles into** radio button to divide shell objects to specified number of n x m areas. After selecting this radio button, select the shell object and click on the **Apply** button. Similarly, you can use **Divide Quadrilaterals/Triangles at** radio button.
- Click on the **OK** button from the dialog box to exit the dialog box.

Merging Shells

The **Merge Shells** tool is used to combine two or more shell objects. The procedure to use this tool is given next.

- After selecting the shell objects in contact with each other, click on the **Merge Shells** tool from the **Edit Shells** cascading menu of the **Edit** menu. The **Merge Shells by Removing Common Edges** dialog box will be displayed; refer to Figure-41.

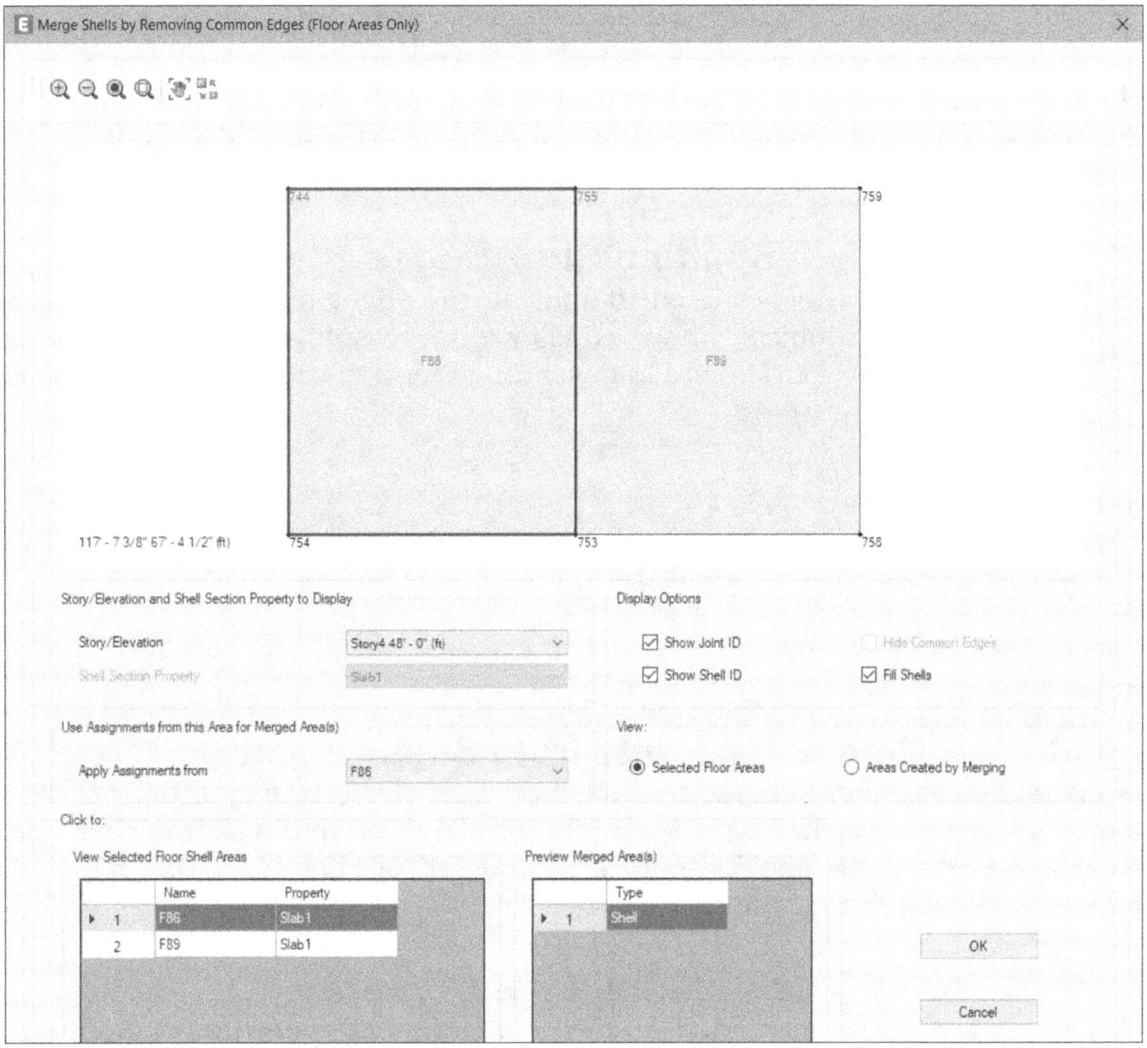

Figure-41. Merge Shells by Removing Common Edges dialog box

- Select desired option from the **Apply Assignments from** drop-down to specify the shell object whose properties will be assigned to merged shell object.
- Click on the **OK** button from the dialog box to create the merged shell object.

Expanding/Shrinking Shell Objects

The **Expand/Shrink Shells** tool is used to increase/decrease the size of selected shell object. The procedure to use this tool is given next.

- Click on the **Expand/Shrink Shells** tool from the **Edit Shells** cascading menu of the **Edit** menu after selecting the shell object. The **Expand/Shrink Shells** dialog box will be displayed; refer to Figure-42.

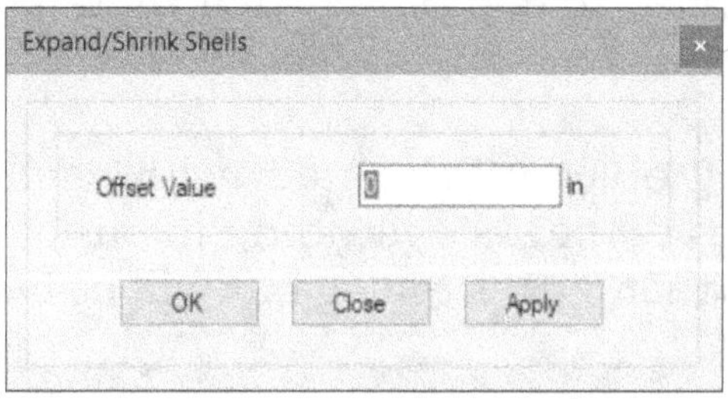

Figure-42. Expand Shrink Shells dialog box

- Specify desired offset value in the edit box to specify increase/decrease the size uniformly. Note that a negative value will decrease the size of shell object.
- After setting parameters, click on the **OK** button.

Splitting Shell Edges

The **Split Shell Edges** tool is used to split all the edges of shell objects into two. After selecting the shell object whose edges are to be split and click on the **Split Shell Edges** tool from the **Edit Shells** cascading menu of the **Edit** menu. The edges will get split; refer to Figure-43.

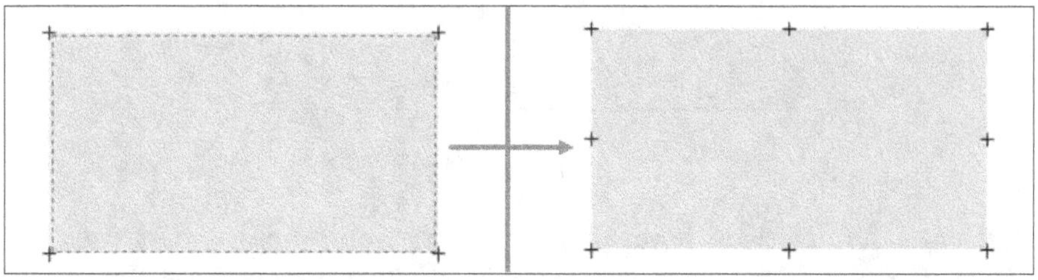

Figure-43. Splitting shell edges

Removing Joints from Shell Objects

The **Remove Joints from Shells** tool is used to remove selected joints from selected shell objects. Select the shell object then its joint(s) to be removed and then click on the **Remove Joints from Shells** tool from the **Edit Shells** cascading menu of the **Edit** menu. Selected joints will be removed.

Chamfering Slab Corners

The **Chamfer Slab Corners** tool is used to apply chamfers at corners of selected shell objects. The procedure to use this tool is given next.

- Select the shell objects whose corners are to be chamfered and click on the **Chamfer Slab Corners** tool from the **Edit Shells** cascading menu in the **Edit** menu. The **Chamfer Slab Corners** dialog box will be displayed; refer to Figure-44.

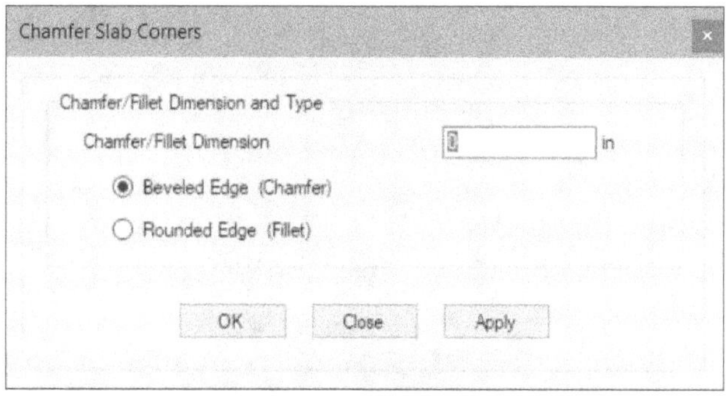

Figure-44. Chamfer Slab Corners dialog box

- Select the **Beveled Edge (Chamfer)** radio button to apply chamfer and specify desired length of chamfer in **Chamfer/Fillet Dimension** edit box. Select the **Rounded Edge (Fillet)** radio button and specify desired radius value in the **Chamfer/Fillet Dimension** edit box to create rounded corners.
- After setting desired parameters, select the corners of selected shell objects to which chamfer/round will be applied and click on the **OK** button. The chamfer/fillet will be applied.

Similarly, you can use the other tools of **Edit Shells** cascading menu and **Edit** menu.

ADVANCED STRUCTURE DESIGN

Earlier tall buildings were box like structures created using vertical columns and straight line beams but due to availability of advanced structure analysis software, the things have changed a lot. Now, you can find tilted buildings, taper buildings, twisted buildings, and many irregular shaped buildings (scan the QR codes below).

Twisted Tilted Irregular

ETABS is capable to model most of these new designs for structural analysis. Here, we will discuss the methods to create some of the modern structures.

Creating Twisting building structures with Cylindrical Grid

In ETABS, twisted building structures are created using the combinations of angle and **Replicate** tool. Here, we will create the model of twisted building as shown in Figure-45.

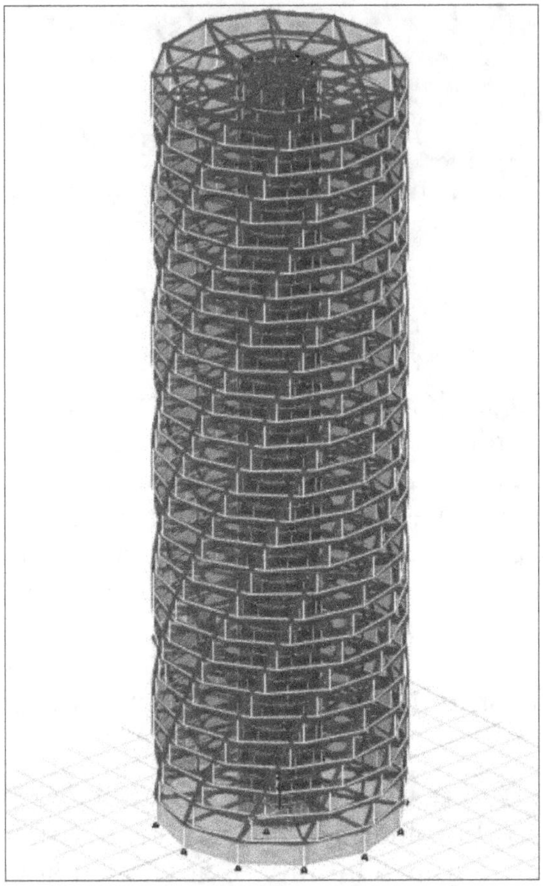

Figure-45. Twisted building model

The steps are given next.

- Start a new project with high number of stories like 30 stories; refer to Figure-46. Select the **Blank** button from **Add Structural Objects** area in the dialog box as we will create cylindrical grid later.

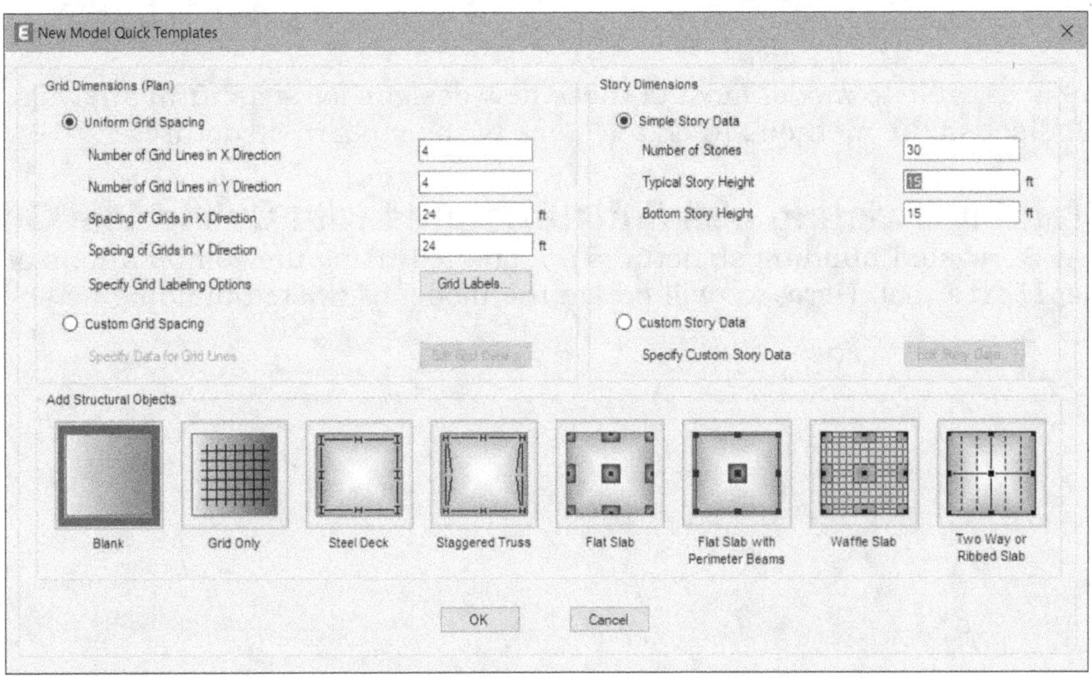

Figure-46. High rise building using template

- After setting parameters as shown in Figure-46 and click on the **OK** button. The model will be displayed as blank in viewports.
- Click on the **Edit Stories and Grid Systems** tool from the **Edit** menu. The **Edit Story and Grid System Data** dialog box will be displayed.
- Click on the **Add New Grid System** tool from the dialog box. The **Add New Grid System Quick** dialog box will be displayed.
- Select the **Cylindrical** radio button and specify parameters as shown in Figure-47 in the dialog box. Note that the value specified in the **Number of Grid Lines in T Direction** edit box define the total span of cylindrical grid. If you want to create a cylindrical grid of 180 degree then specify 7 in the **Number of Grid Lines in T Direction** edit box which will give (7-1) x 30 = 180 degree. The value 30 degree is specified in **Spacing of Grids in T Direction** edit box. If the spacing between two consecutive grid lines is 10 degree then you will need 18+1=19 grid lines for 180 degree span of cylindrical grid. In this case, we want to create a full 360 degree cylindrical grid system so we have specified 13 for number of grid line in T direction.

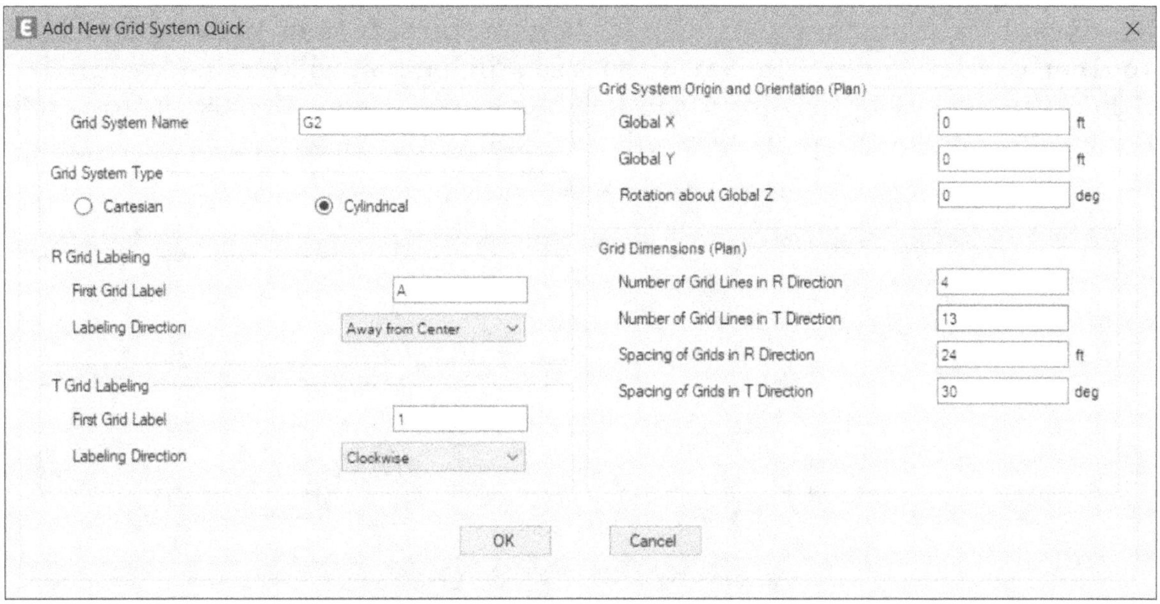

Figure-47. Parameters for cylindrical grid

- After specifying the parameters, click on the **OK** button from **Add New Grid System Quick** dialog box. The **Edit Story and Grid System Data** dialog box will be displayed again. Click on the **OK** button from the dialog box. The grid system will be displayed in graphics area; refer to Figure-48.

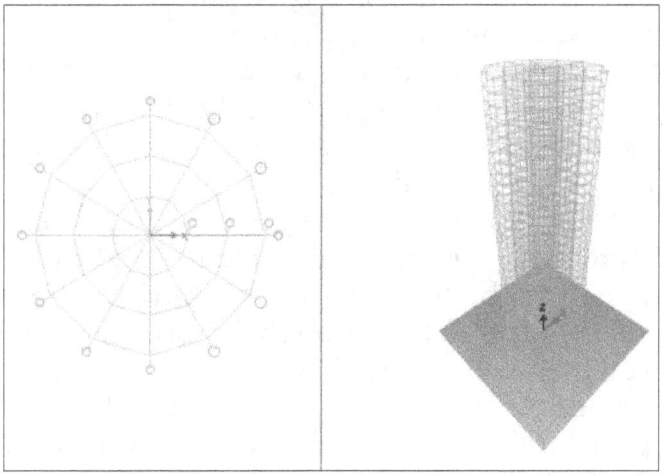

Figure-48. Grid system created

- In this grid system, we want to create shear wall in the inner section and beams for rest of the building and then we will apply twisting angle to each floor individually based on height and angle required. Click on the **Set Plan View** button from the **Toolbar** or **View** menu. The **Set Plan View** dialog box will be displayed.
- Select the **Story1** option from the list box and click on the **OK** button; refer to Figure-49. The story1 plan view will be displayed in viewport.

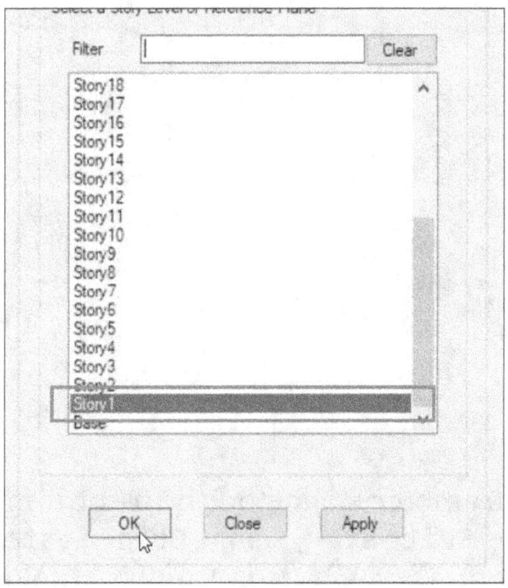

Figure-49. Story1 to be selected

- Click on the **Draw Walls** tool from the **Draw Floor/Wall Objects** cascading menu in the **Draw** menu. You will be asked to select grids for creating walls.
- Make sure **Pier** is selected in the **Type of Area** drop-down of **Properties of Object** box and select the **Arc (Center & 2 Points)** option from the **Line Drawing Type** drop-down.
- Select the center point of grid and specify the 2 points as shown in Figure-50. Press **ESC** to exit the tool, the wall will be created; refer to Figure-51.

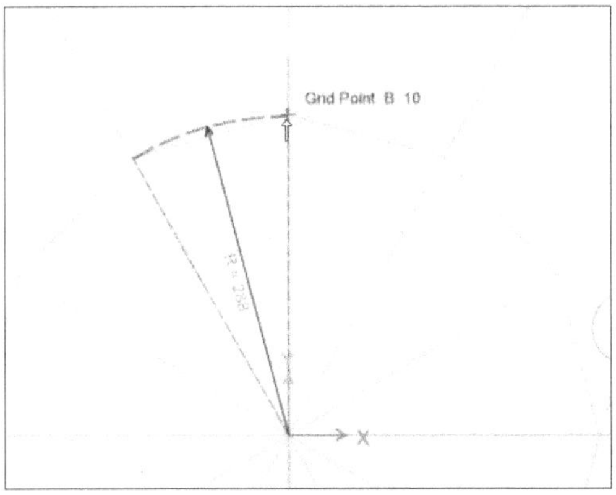

Figure-50. Creating pier walls

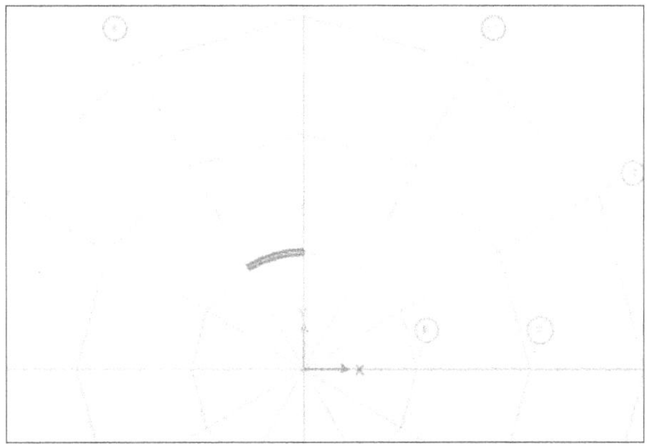

Figure-51. Segment of circular wall created

- Select the newly created wall and click on the **Replicate** tool from the **Edit** menu. The **Replicate** dialog box will be displayed.
- Select the **Radial** tab from the dialog box and specify the parameters as shown in Figure-52.

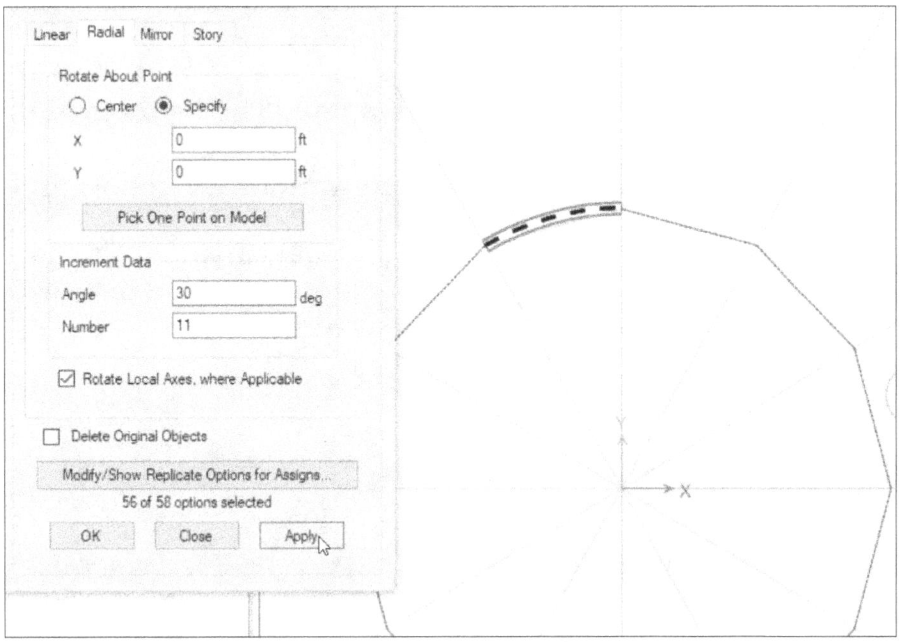

Figure-52. Applying radial replication

- Click on the **OK** button from the dialog box to create the walls.

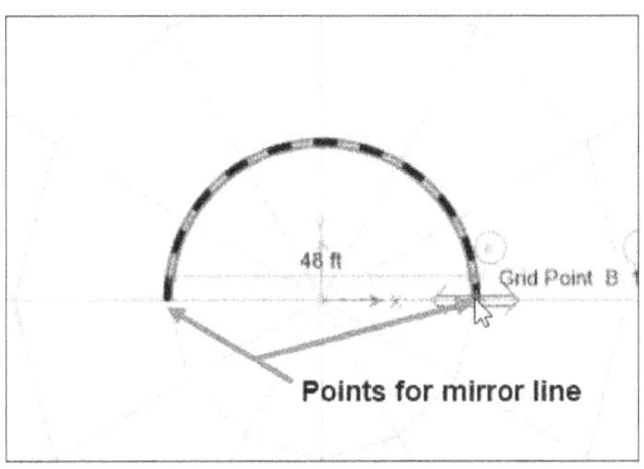

Figure-53. Creating mirror line

- Click on the **Quick Draw Beams/Columns** tool from the **Draw Beam/Column/Brace Objects** cascading menu of the **Draw** menu. You will be asked to select grids for creating beams.
- Select the **SteelBm** option from the **Property** drop-down of **Properties of Object** box because we are creating a steel structure in this case.
- Draw a selection window covering all the grids to create beams; refer to Figure-54.

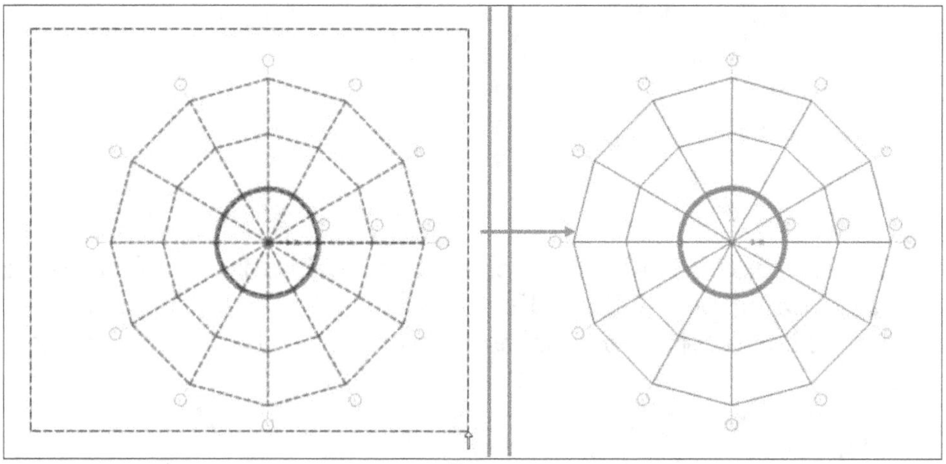

Figure-54. Creating beams

- Click on the **Replicate** tool from the **Edit** menu. The **Replicate** dialog box will be displayed.
- Select all the objects in graphics area using window selection and click on the **Story** tab.
- Select all the stories except Story 1 and Base by dragging in the list box; refer to Figure-55 and click on the **Apply** button.

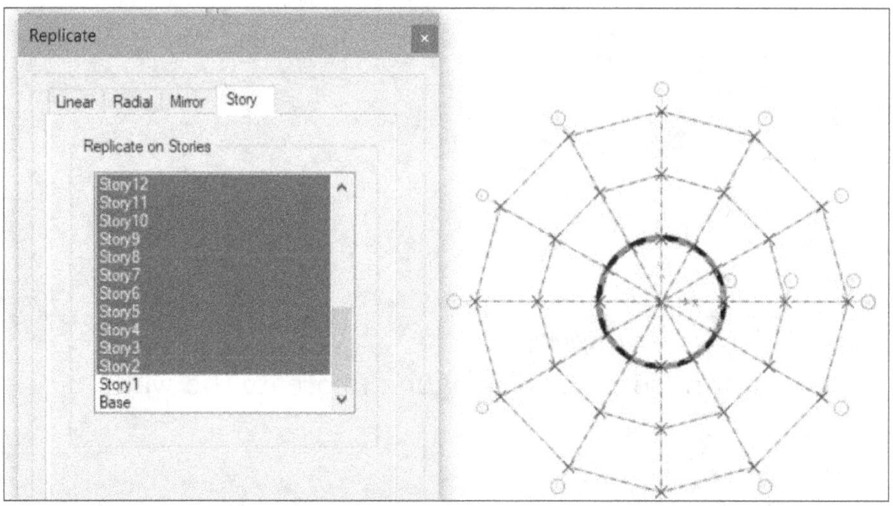

Figure-55. Selecting stories for replication

- Now, we are ready to perform twisting of model. Click on the **Radial** tab in the dialog box and specify the parameters as shown in Figure-56. Note that we are going to create a total twist of 180 degree between 30 stories so we have specified increment angle as 180/30 = 6 degree.

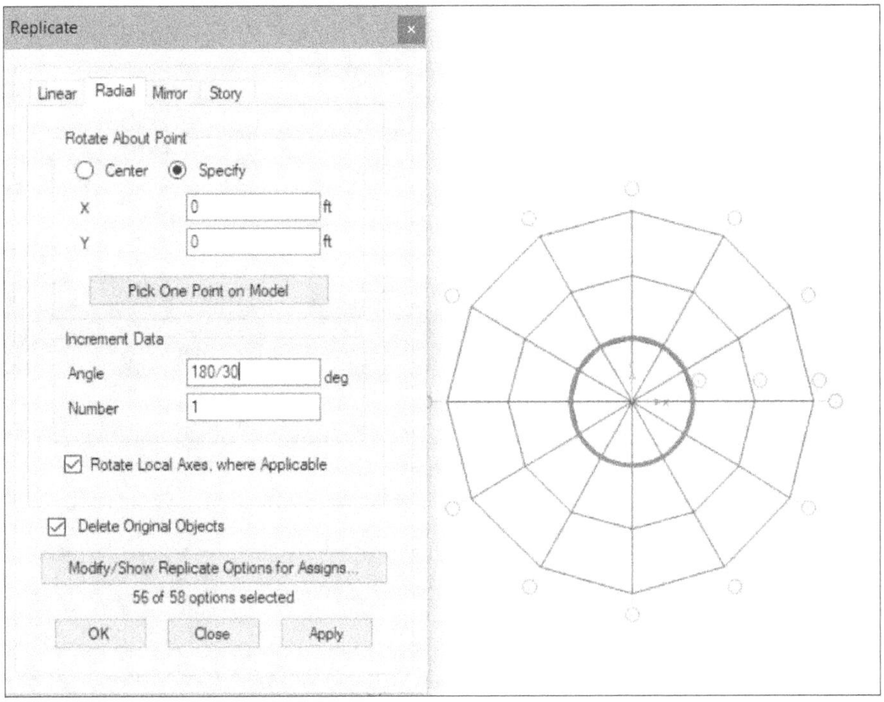

Figure-56. Specifying twisting angle

- Click on the **Select Plan View** tool from the **Toolbar** or **View** menu. The **Select Plan View** dialog box will be displayed.
- Select **Story2** from the list box in **Select Plan View** dialog box and click on the **Apply** button.
- Select the model using window selection in plan view viewport, specify the angle as 6 in the **Angle** edit box and click on the **Apply** button.
- Select the **Story3** option from list box in **Select Plan View** dialog box and click on the **Apply** button. Story3 plan view will be displayed.
- Specify the angle as **12** in **Angle** edit box of **Replicate** dialog box, select all the objects using window selection and click on the **Apply** button.
- Similarly, select the Story4 plan view and rotate it by 12+6 = 18 degree and so on up to story 30 which will give you twisted building frame; refer to Figure-57.

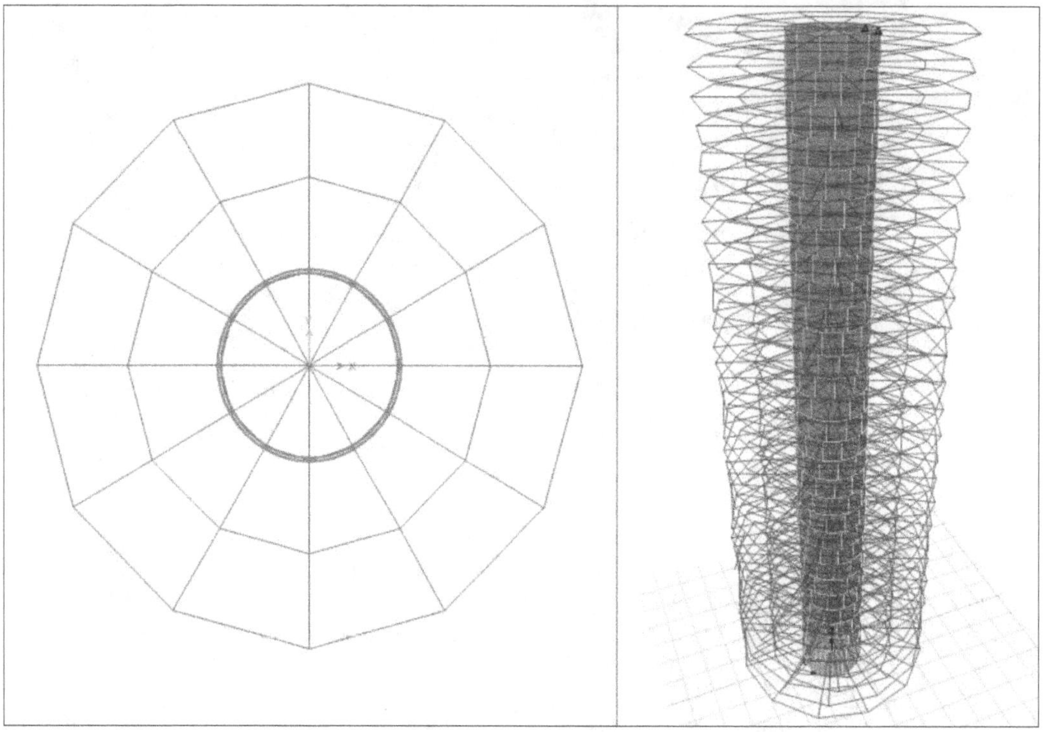

Figure-57. Twisted building frame

- Add steel columns connecting consecutive outer joints of beams using **Draw Beam/ Column/Brace** tool in **Draw Beam/Column/Beam Objects** cascading menu of **Draw** menu. Make sure you are working in 3D view to connect them; refer to Figure-58.

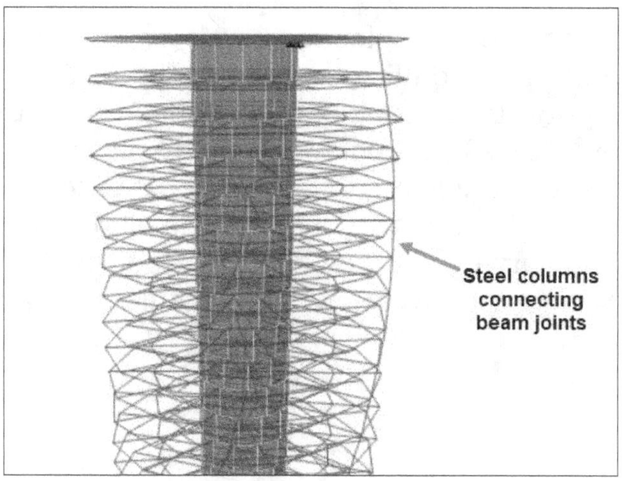

Figure-58. Creating columns in building

- Now, we will create a circular pattern of all the columns recently created. Click on the **Frame Labels** option from the **Select > Select > Labels** menu; refer to Figure-59. The **Select by Frame Label** dialog box will be displayed.

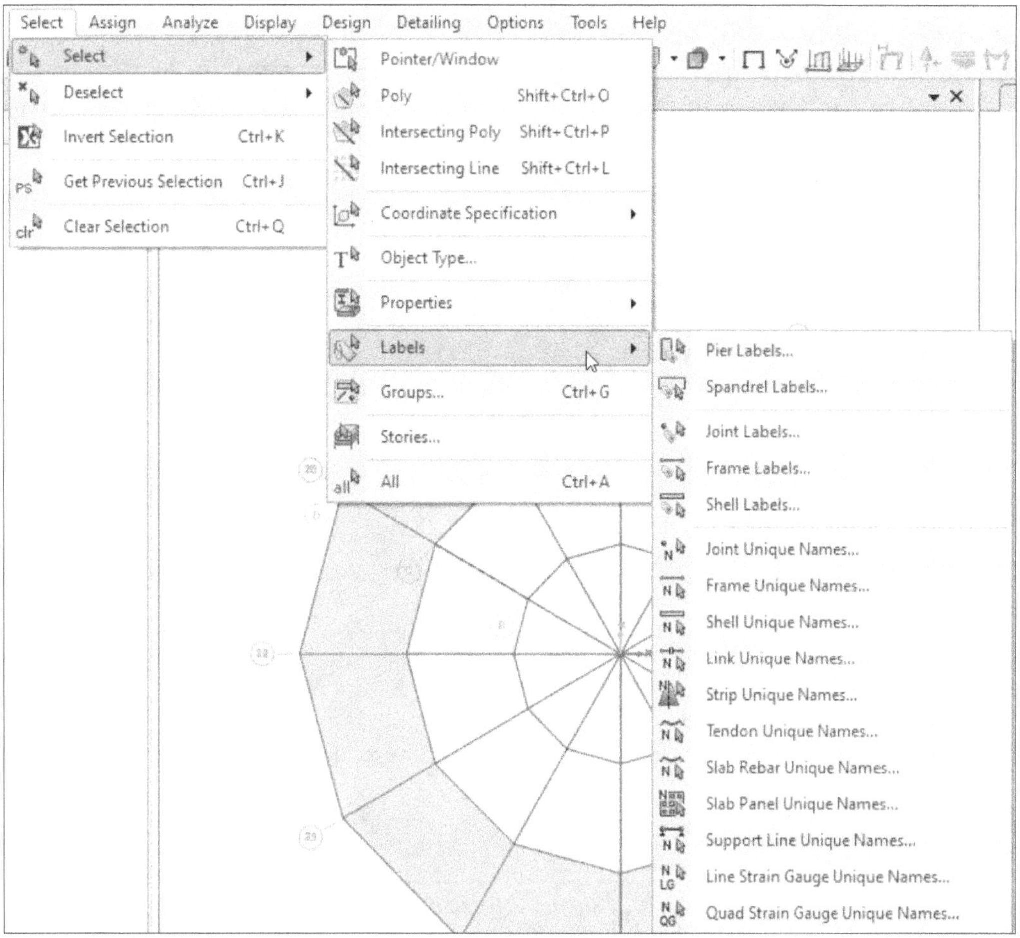

Figure-59. Frame Labels option

- Select all the Frame Labels starting with D using drag selection from the list box; refer to Figure-60 and click on the **Select** button. The recently created columns will get selected. Click on the **Close** button to exit the dialog box.

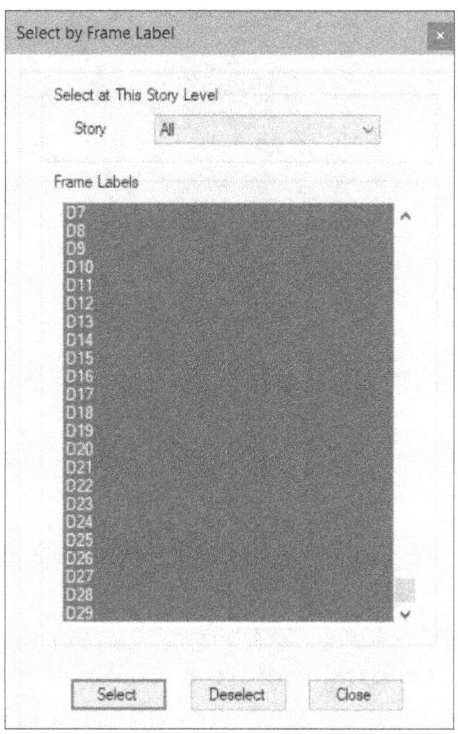

Figure-60. Frame labels to be selected

- To create circular pattern, click on the **Replicate** tool from **Edit** menu. The **Replicate** dialog box will be displayed.
- Select the **Radial** tab and specify parameters as shown in Figure-61. This will create 11 more instances of the columns each separated by 30 degree angle which matches with the number of joints in outer beams.

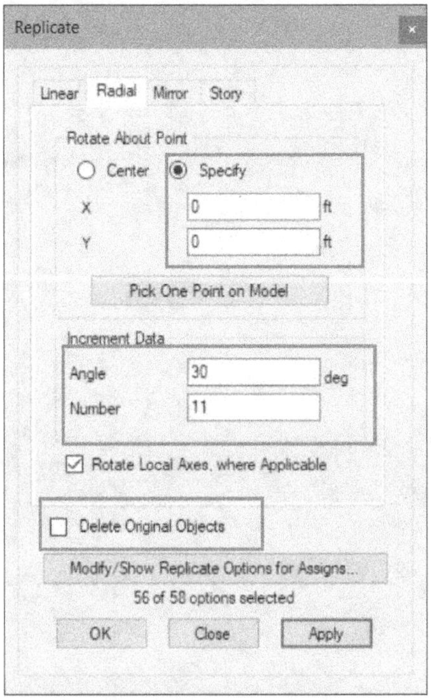

Figure-61. Column replication parameters

- Click on the **OK** button from the dialog box. The model will be displayed as shown in Figure-62.

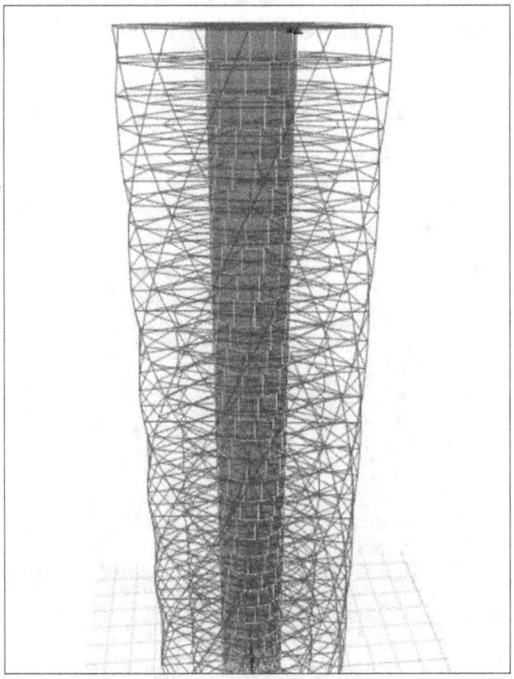

Figure-62. Model after replication

- Click on the **Auto Draw Cladding** tool from the **Draw** menu. The **Cladding Options** dialog box will be displayed.

- Select the **Use Beams** radio button and click on the **OK** button to automatically create cladding between floors; refer to Figure-63.

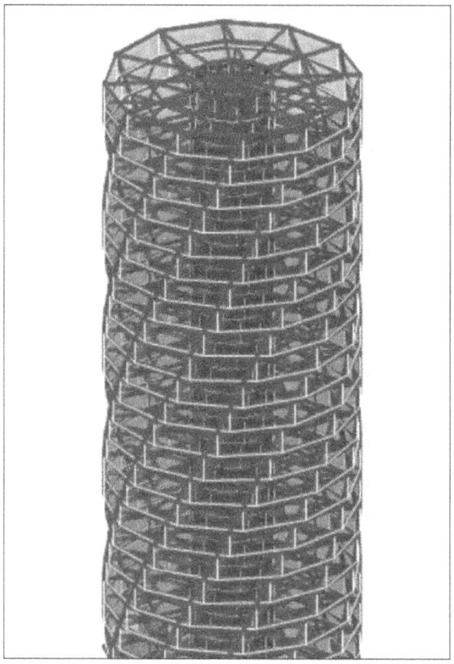

Figure-63. Model after creating cladding

Creating Tapered Building

For creating tapered building, we reduce the size of building while going upward. In ETABS, this can be performed by using the custom grid system. The only factor to calculate for creating tapered buildings is the amount of reduction to be applied to each floor. Microsoft Excel is an excellent tool for performing this task. Assume that we want to create a building of 3 degree taper starting from 1st floor to 20th floor of building. By trigonometry rule,

Perpendicular/Base = Tan(x), here x is angle in degree

Here, base size is 12 feet for second floor and we want to calculate the reduction in floor size for taper which is perpendicular for this formula; refer to Figure-64.

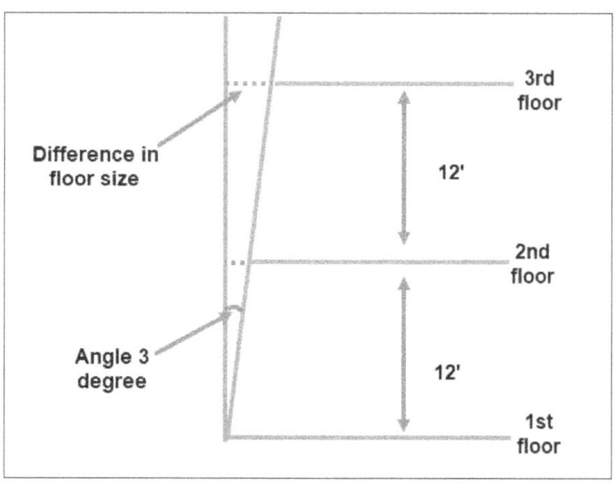

Figure-64. Trigonometric representation of floors

Lets assume reduction in 2nd floor size is R_2 , for 3rd floor is R_3, and so on. So, based on the trigonometric formula,

$R_2/12' = \tan(3)$ ==> $R_2 = 12' \times \tan(3)$ ==> $R_2 = 0.63$

Similarly,

$R_3 = 24' \times \tan(3)$ ==> $R_3 = 2 \times R_2 = 1.26$

$R_4 = 3 \times R_2 = 1.89$ and so on till $R_{20} = 11.97$. You do not need to calculate all the values as we will require only R_2 value for setting spacing in grid lines. Now, we will create a grid using these parameters and then create a structure on it.

The method to create one such structure is given next.

- Start a new project with settings as shown in Figure-65 and click on the **OK** button from **Model Initialization** dialog box. The **New Model Quick Templates** dialog box will be displayed.

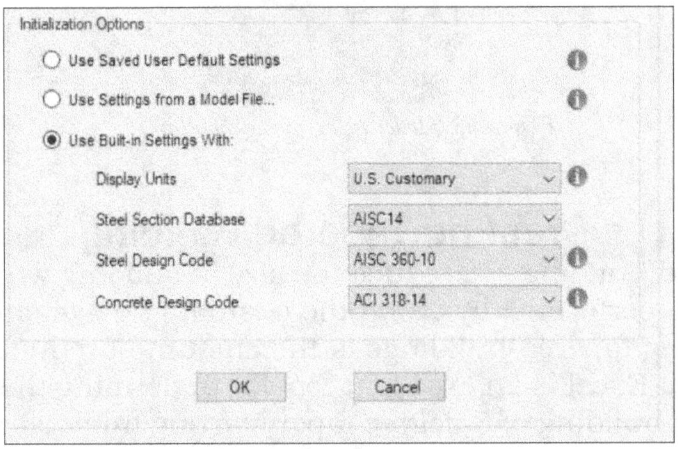

Figure-65. Starting new project

- Set the number of stories as **20** in the **Number of Stories** edit box and **Grid Only** button in the **Add Structural Objects** area of the dialog box.
- Select the **Custom Grid Spacing** radio button and click on the **Edit Grid Data** button from the dialog box. The **Grid System Data** dialog box will be displayed.
- Specify the parameters as shown in Figure-66.

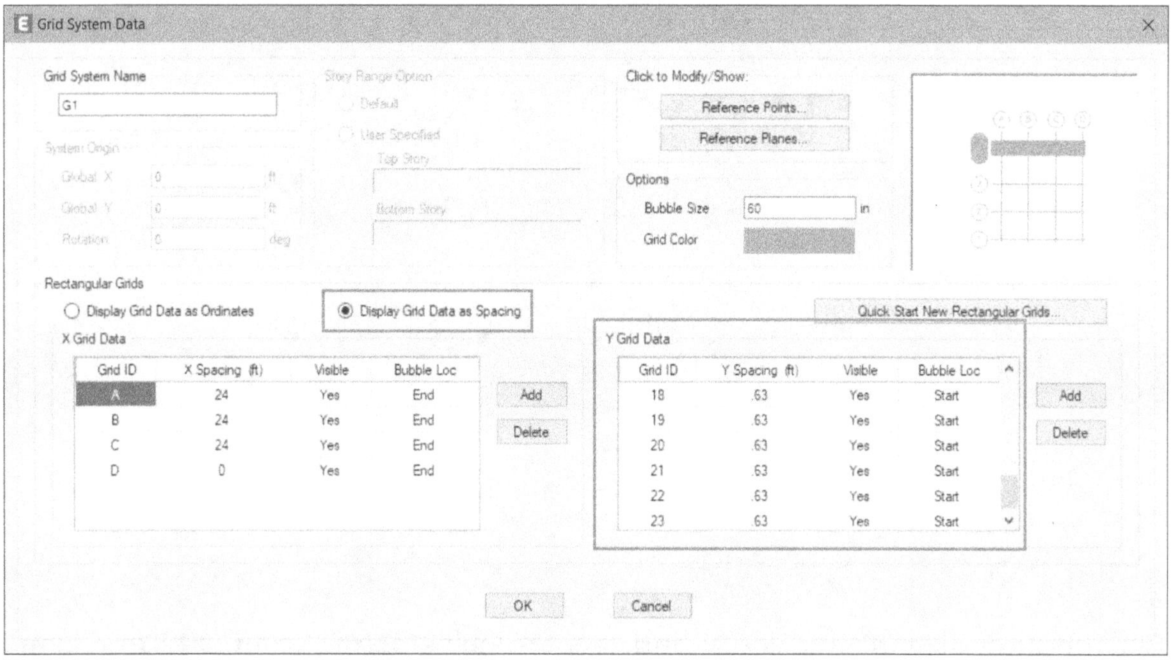

Figure-66. Setting Y grid data

- Note that we have created extra grid lines along Y axis with spacing of 0.63 feet up to 23 grid IDs. So, we will start creating tapered floors starting from grid ID 4 to grid ID 23 making total 20 floors for drawing floor slabs. After setting parameters, click on the **OK** button. The **New Model Quick Templates** dialog box will be displayed again.
- Click on the **OK** button from the dialog box. The grid system will be displayed in graphics area; refer to Figure-67.

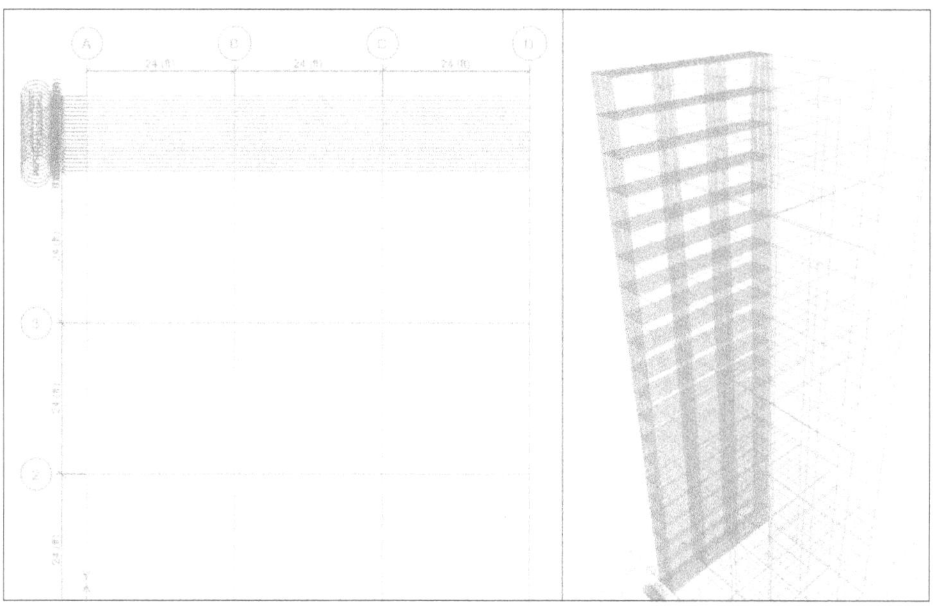

Figure-67. Grid data created

- Click in the Plan view viewport and set the view to plan view of 1st story using **Set Plan View** tool from the **View** menu.
- Click on the **Draw Rectangular Floor/Wall** tool from the **Draw Floor/Wall Objects** cascading menu of the **Draw** menu and create rectangular floor; refer to Figure-68.

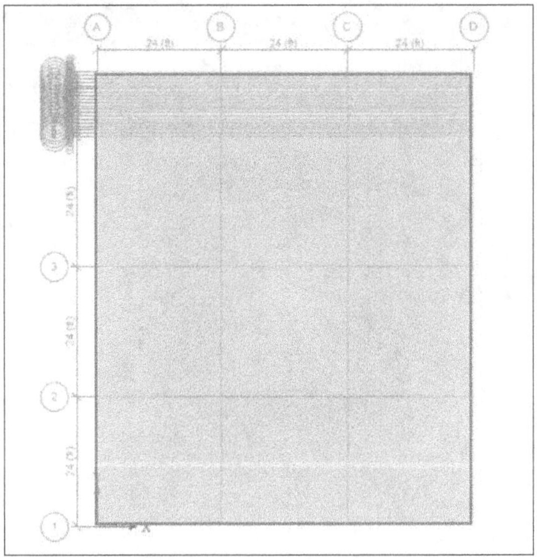

Figure-68. Creating 1st floor

- Set the plan view to 2nd story and create floor starting from D22 grid point to A1; refer to Figure-69.

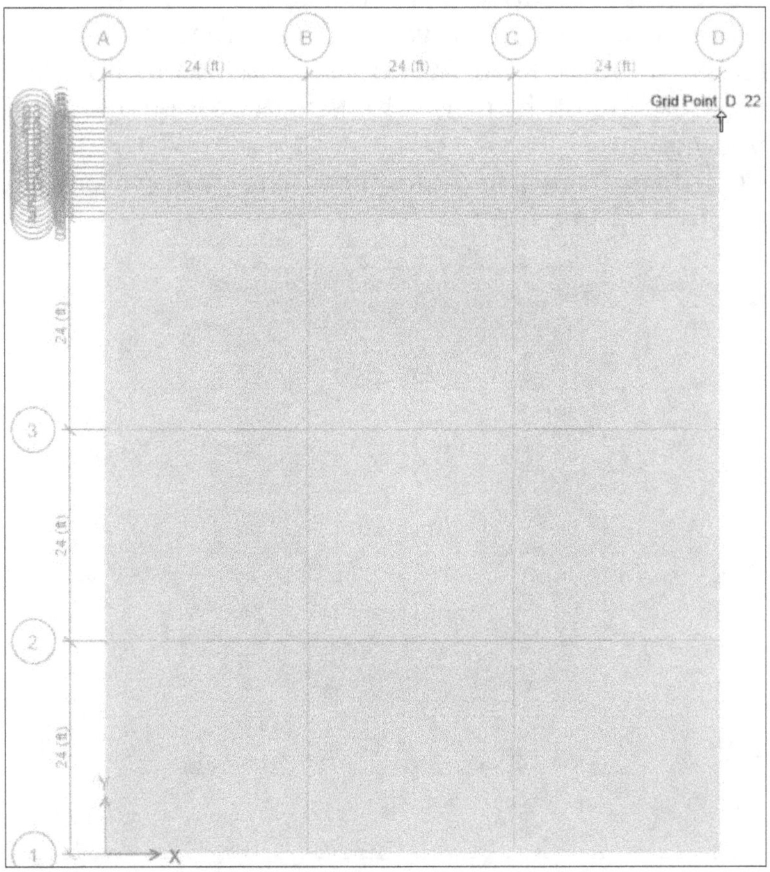

Figure-69. Creating floor for 2nd story

- Similarly, create the floors with size lower by one grid point consecutively for each story; refer to Figure-70.

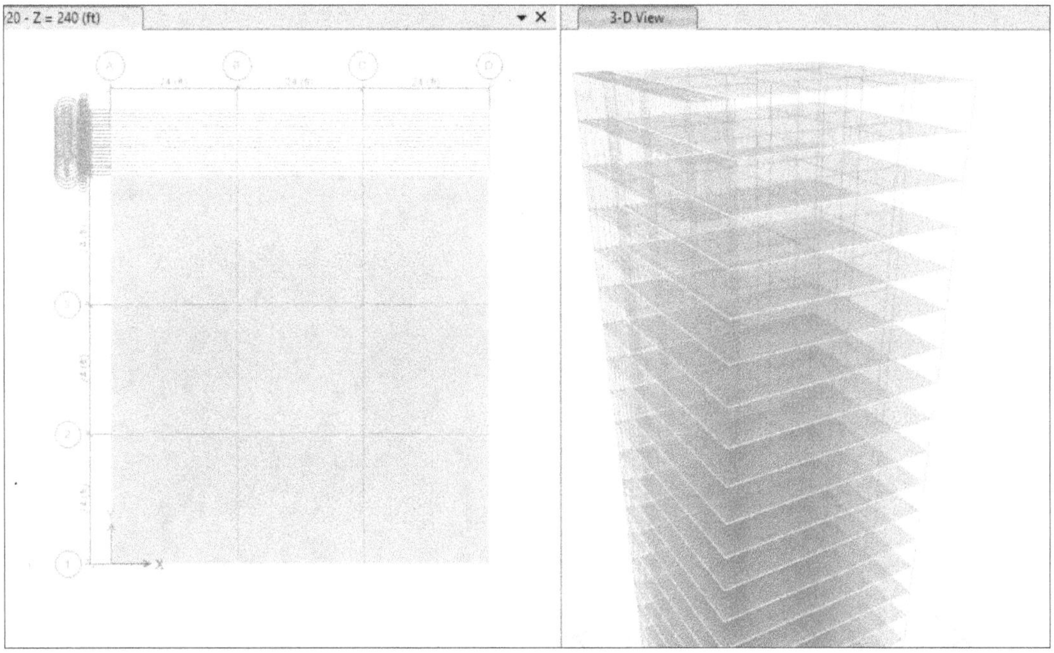

Figure-70. After creating floors

- Click in the 3-D View viewport and press **CTRL+D** to hide grid system.
- Set the 3-D view to yz fast view with 0 Aperture using the **Set 3D View** tool from **View** menu; refer to Figure-71.

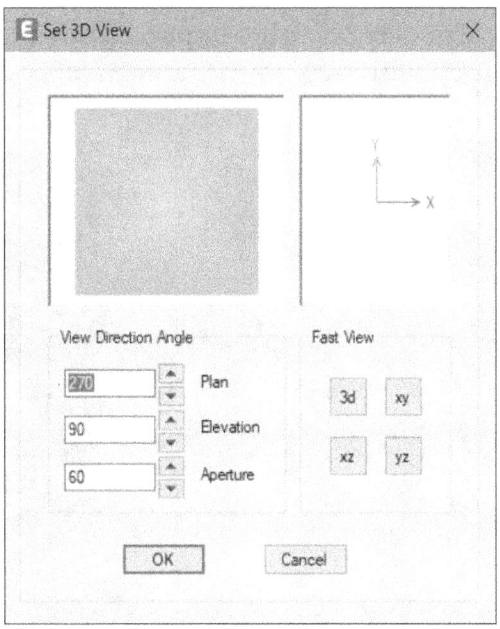

Figure-71. Setting 3D view

- Click on the **Set Display Options** tool from the **Toolbar** or **View** menu and select the **Extrude Frames** and **Extrude Shells** check boxes. Click on the **OK** button from the dialog box. The model will be displayed as shown in Figure-72 which confirms that model is properly tapered between floors.
- Rotate the model by dragging using **SHIFT+MMB** (Middle Mouse Button) so that joints for created tapered wall are visible; refer to Figure-73.

Figure-72. Model in yz 3D view with extruded shells

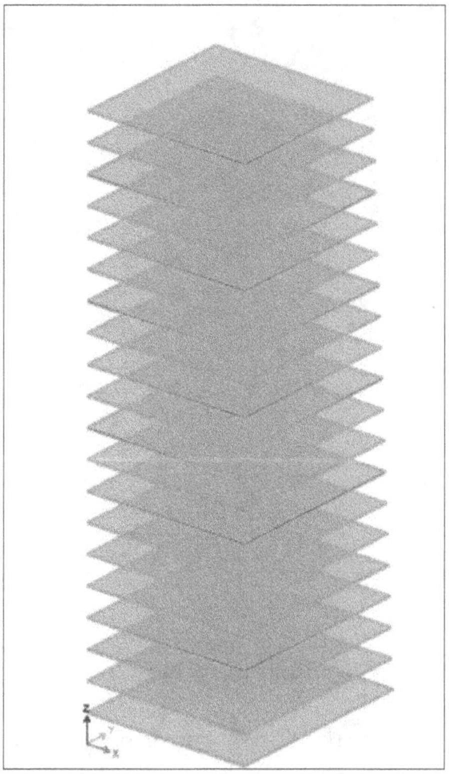

Figure-73. 3D view of tapered model

- Click on the **Draw Floor/Wall** tool from the **Draw Floor/Wall Objects** cascading menu of the **Draw** menu and create wall by specifying points as shown in Figure-74. Creating walls on all sides will give you model as shown in Figure-75.

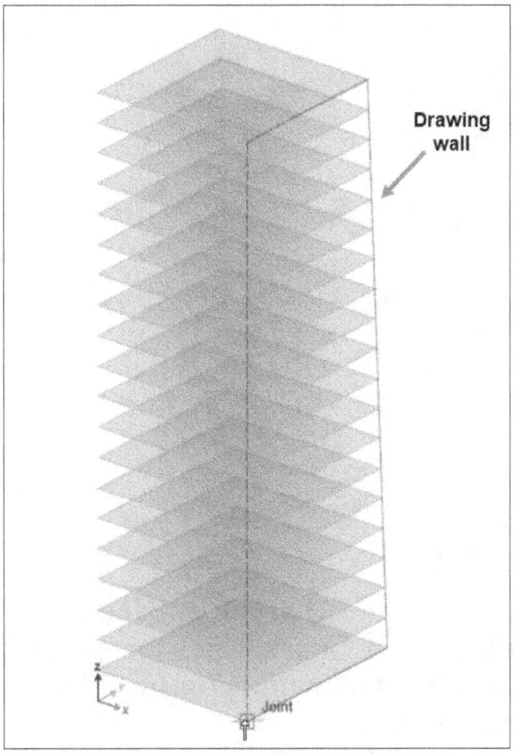

Figure-74. Drawing wall

Figure-75. Taper building with walls

- In place of creating walls, you can also use **Auto Draw Cladding** tool of **Draw** menu to create model with cladding; refer to Figure-76.

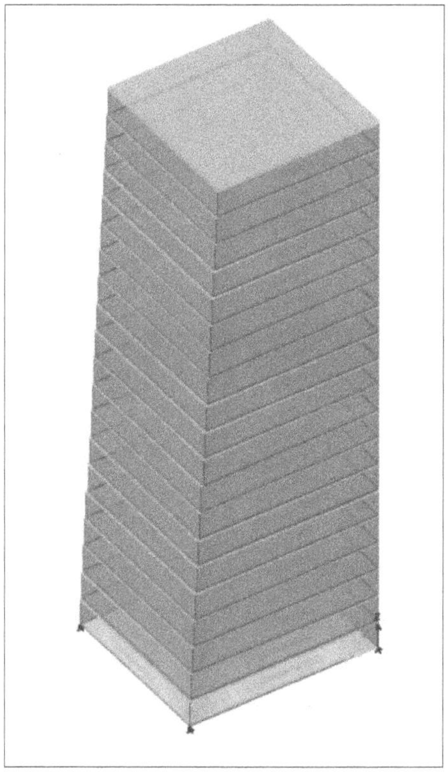

Figure-76. Final model of building

- You can add the beams, columns, and other structural members to complete the design.

One of the greatest benefit of creating this structure is availability of more than 1/2 feet space for aesthetics of building like, small plants can be placed in this space to make it look beautiful. This chapter opens doors for creative structures, so create your own unique designs and assess their structural stability under various loads.

SELF-ASSESSMENT

Q1. Which tool is used to reverse the most recent operation in ETABS?
A. Redo
B. Undo
C. Paste
D. Delete

Q2. What is the shortcut key for Undo operation?
A. CTRL+Y
B. CTRL+X
C. CTRL+Z
D. CTRL+V

Q3. Which tool is used to reverse the Undo operation?
A. Redo
B. Copy
C. Delete
D. Move

Q4. What is the shortcut key for Redo operation?
A. CTRL+C
B. CTRL+V
C. CTRL+Y
D. CTRL+Z

Q5. Which tool deletes selected objects and stores them in clipboard?
A. Copy
B. Paste
C. Cut
D. Replicate

Q6. What is the shortcut key for Cut operation?
A. CTRL+X
B. CTRL+Z
C. CTRL+C
D. CTRL+Y

Q7. Which tool is used to copy selected objects into clipboard?
A. Undo
B. Copy
C. Delete
D. Align

Q8. What is the shortcut key for Copy operation?
A. CTRL+V
B. CTRL+X
C. CTRL+Y
D. CTRL+C

Q9. Which tool is used to place clipboard objects into the model?
A. Paste
B. Copy
C. Cut
D. Undo

Q10. What is the shortcut key for Paste operation?
A. CTRL+Z
B. CTRL+V
C. CTRL+Y
D. CTRL+C

Q11. Which dialog box appears when Paste tool is used?
A. Replicate dialog box
B. Move dialog box
C. Paste coordinates dialog box
D. Align dialog box

Q12. Which edit boxes are used to define distance in Paste Coordinates dialog box?
A. Dx and dy
B. Delta x and delta y
C. X and y
D. Width and height

Q13. Which tool is used to delete selected objects from the model?
A. Delete
B. Replicate
C. Merge
D. Align

Q14. Which keyboard key can also be used for deleting objects?
A. ENTER
B. SHIFT
C. DELETE
D. TAB

Q15. Which cascading menu is used to add structures to the model?
A. Edit shells
B. Extrude
C. Add to model from template
D. Drawing

Q16. Which tool is used to add a 2D structure to the model?
A. Add 3D structure
B. Divide shells
C. Add 2D structure
D. Join frames

Q17. Which dialog box appears after clicking Add 2D Structure tool?
A. Add to model - 2D structure dialog box
B. Frame object type options dialog box
C. Replicate dialog box
D. Merge joints dialog box

Q18. Which parameters can be specified while creating 2D structure?
A. Number of divisions
B. Height of structure
C. Grid system
D. All of these

Q19. Which tool is used to add predefined 3D structures?
A. Add 3D structure
B. Divide frames
C. Expand shells
D. Move joints/frames/shells

Q20. Which dialog box appears when Add 3D Structure tool is selected?
A. Add to model - 3D structure dialog box
B. Merge shells dialog box
C. Align dialog box
D. Paste coordinates dialog box

Q21. Which tool is used to create multiple copies of selected objects?
A. Trim
B. Replicate
C. Divide
D. Move

Q22. Which dialog box appears after clicking Replicate tool?
A. Divide frames dialog box
B. Replicate dialog box
C. Move joints dialog box
D. Merge shells dialog box

Q23. Which tab in Replicate dialog box is used for linear pattern creation?
A. Radial
B. Mirror
C. Story
D. Linear

Q24. Which edit boxes define distance between consecutive copies in linear pattern?
A. Delta x and delta y
B. Dx and dy
C. Width and height
D. Angle and number

Q25. Which option is used to delete original objects after replication?
A. Remove original objects
B. Delete copied objects
C. Delete original objects
D. Erase source objects

Q26. Which button is used to define assigns included with replicated objects?
A. Pick two points on model
B. Modify/show replicate options for assigns
C. Apply assigns
D. Replicate assigns

Q27. Which tab in Replicate dialog box is used to create radial pattern?
A. Story
B. Radial
C. Mirror
D. Linear

Q28. Which radio button uses center of model for radial pattern?
A. Specify
B. Center
C. Radial
D. Midpoint

Q29. Which edit boxes define angle and number of instances in radial pattern?
A. Dx and dy
B. Width and height
C. Angle and number
D. X and y

Q30. Which tab in Replicate dialog box is used for mirror copy?
A. Story
B. Radial
C. Mirror
D. Linear

Q31. Which button can define mirror line using model points?
A. Pick one point on model
B. Pick two points on model
C. Apply mirror
D. Extend frame

Q32. Which tab in Replicate dialog box copies objects to another story?
A. Radial
B. Story
C. Linear
D. Mirror

Q33. Which tool is used to extrude frame section between joints?
A. Extrude joints to frames
B. Divide frames
C. Join frames
D. Align joints

Q34. Which dialog box appears for Extrude Joints to Frames tool?
A. Extrude joints to frames dialog box
B. Divide selected shells dialog box
C. Frame type options dialog box
D. Merge shells dialog box

Q35. Which edit boxes define extrusion distances in linear extrusion?
A. X y z
B. Width height depth
C. Dx dy dz
D. Angle number radius

Q36. Which tab in Extrude Joints to Frames dialog box is used for radial extrusion?
A. Radial extrusion
B. Linear extrusion
C. Shell extrusion
D. Frame extrusion

Q37. Which edit box defines slope value in radial extrusion?
A. Total rise
B. Total drop
C. Delta z
D. Height

Q38. Which tool works similar to Extrude Joints to Frames but uses frame members?
A. Divide shells
B. Extrude frames to shells
C. Join frames
D. Merge joints

Q39. Which tool is used to merge nearby joints within a tolerance value?
A. Merge joints
B. Align joints
C. Divide joints
D. Trim joints

Q40. Which tool is useful for cleaning up the model?
A. Divide shells
B. Merge joints
C. Reverse connectivity
D. Move shells

Q41. Which tool is used to align joints, frames, or edges?
A. Align joints/frames/edges
B. Merge joints
C. Replicate
D. Extrude frames

Q42. Which radio button aligns joints to nearest frame or edge?
A. Align joints to x-ordinate
B. Align joints to nearest frame or edge
C. Extend frame objects
D. Trim frame objects

Q43. Which radio button removes extra segment of intersecting frame member?
A. Extend frame objects
B. Align joints to z-ordinate
C. Trim frame objects
D. Reverse connectivity

Q44. Which radio button extends selected frame object?
A. Extend frame objects
B. Trim frame objects
C. Align joints
D. Divide frames

Q45. Which tool moves selected objects along X, Y, and Z axes?
A. Move joints/frames/shells
B. Align joints
C. Merge shells
D. Divide shells

Q46. Which dialog box appears after clicking Move Joints/Frames/Shells tool?
A. Move joints/frames/shells dialog box
B. Divide selected frames dialog box
C. Merge selected joints dialog box
D. Paste coordinates dialog box

Q47. Which edit boxes are used to define movement distances?
A. Dx dy dz
B. Delta x delta y delta z
C. Angle number radius
D. X y ordinate

Q48. Which cascading menu contains divide frame and join frame tools?
A. Edit shells
B. Extrude
C. Edit frames
D. Draw objects

Q49. Which tool divides selected frame objects into segments?
A. Join frames
B. Reverse frame connectivity
C. Divide frames
D. Convert beams

Q50. Which dialog box appears after clicking Divide Frames tool?
A. Divide selected frames dialog box
B. Divide selected shells dialog box
C. Merge shells dialog box
D. Frame object type options dialog box

FOR STUDENT NOTES

Index

Symbols

A

B

C

D

E

F

G

H

I

T

U

V

W

Z

OTHER BOOKS BY CADCAMCAE WORKS

Autodesk Revit 2027 Black Book
Autodesk Revit 2026 Black Book
Autodesk Revit 2025 Black Book

Autodesk Inventor 2027 Black Book
Autodesk Inventor 2026 Black Book
Autodesk Inventor 2025 Black Book

Autodesk Fusion Black Book (V 2701.1.18)
Autodesk Fusion PCB Black Book (V 2701.1.18)

AutoCAD Electrical 2027 Black Book
AutoCAD Electrical 2026 Black Book
AutoCAD Electrical 2025 Black Book

Basics of Autodesk Inventor Nastran 2027
Basics of Autodesk Inventor Nastran 2026

Autodesk CFD 2027 Black Book
Autodesk CFD 2026 Black Book

Tinkercad Black Book: (2nd Edition)

SolidWorks 2026 Black Book
SolidWorks 2025 Black Book
SolidWorks 2024 Black Book
SolidWorks 2023 Black Book

SolidWorks Simulation 2026 Black Book
SolidWorks Simulation 2025 Black Book
SolidWorks Simulation 2024 Black Book
SolidWorks Simulation 2023 Black Book

SolidWorks Flow Simulation 2026 Black Book
SolidWorks Flow Simulation 2025 Black Book
SolidWorks Flow Simulation 2024 Black Book

SolidWorks CAM 2026 Black Book
SolidWorks CAM 2025 Black Book
SolidWorks CAM 2024 Black Book
SolidWorks CAM 2023 Black Book

SolidWorks Electrical 2026 Black Book
SolidWorks Electrical 2025 Black Book
SolidWorks Electrical 2024 Black Book

SolidWorks Workbook 2022

Mastercam 2023 for SolidWorks Black Book
Mastercam 2022 for SolidWorks Black Book
Mastercam 2017 for SolidWorks Black Book

Mastercam 2026 Black Book
Mastercam 2025 Black Book
Mastercam 2024 Black Book
Mastercam 2023 Black Book

Creo Parametric 12.0 Black Book
Creo Parametric 11.0 Black Book
Creo Parametric 10.0 Black Book
Creo Parametric 9.0 Black Book
Creo Parametric 8.0 Black Book

Creo Manufacturing 12.0 Black Book
Creo Manufacturing 11.0 Black Book
Creo Manufacturing 10.0 Black Book
Creo Manufacturing 9.0 Black Book

ETABS V23 Black Book
ETABS V22 Black Book
ETABS V21 Black Book

FreeCAD 1.0 Black Book
FreeCAD 0.21 Black Book
FreeCAD 0.20 Black Book
FreeCAD 0.19 Black Book

LibreCAD 2.2 Black Book

Ansys Discovery 2025 Black Book

NX 2506 Black Book

www.ingramcontent.com/pod-product-compliance
Lightning Source LLC
LaVergne TN
LVHW080008210926
848481LV00011B/2968